Lecture Notes in Computer Scie

Stefano Berardi Ferruccio Damiani
Ugo de'Liguoro (Eds.)

Types for Proofs and Programs

International Conference, TYPES 2008
Torino, Italy, March 26-29, 2008
Revised Selected Papers

Springer

Volume Editors

Stefano Berardi
Ferruccio Damiani
Ugo de'Liguoro
Università di Torino, Dipartimento di Informatica
Corso Svizzera 185, 10149 Torino, Italy
E-mail: {stefano, damiani, deligu}@di.unito.it

Library of Congress Control Number: Applied for

CR Subject Classification (1998): F.3.1, F.4.1, D.3.3, I.2.3

LNCS Sublibrary: SL 1 – Theoretical Computer Science and General Issues

ISSN 0302-9743
ISBN-10 3-642-02443-2 Springer Berlin Heidelberg New York
ISBN-13 978-3-642-02443-6 Springer Berlin Heidelberg New York

springer.com

Printed in Germany

Typesetting: Camera-ready by author, data conversion by Scientific Publishing Services, Chennai, India
Printed on acid-free paper SPIN: 12697175 06/3180 5 4 3 2 1 0

Preface

These proceedings contain a selection of refereed papers presented at or related to the Annual Workshop of the TYPES project (EU coordination action 510996), which was held during March 26–29, 2008 in Turin, Italy. The topic of this workshop, and of all previous workshops of the same project, was formal reasoning and computer programming based on type theory: languages and computerized tools for reasoning, and applications in several domains such as analysis of programming languages, certified software, mobile code, formalization of mathematics, mathematics education. The workshop was attended by more than 100 researchers and included more than 40 presentations. We also had three invited lectures, from A. Asperti (University of Bologna), G. Dowek (LIX, Ecole polytechnique, France) and J. W. Klop (Vrije Universiteit, Amsterdam, The Netherlands). From 27 submitted papers, 19 were selected after a reviewing process. Each submitted paper was reviewed by three referees; the final decisions were made by the editors. This workshop is the last of a series of meetings of the TYPES working group funded by the European Union (IST project 29001, ESPRIT Working Group 21900, ESPRIT BRA 6435). The proceedings of these workshops were published in the *Lecture Notes in Computer Science* series:

TYPES 1993 Nijmegen, The Netherlands, LNCS 806,
TYPES 1994 Båstad, Sweden, LNCS 996,
TYPES 1995 Turin, Italy, LNCS 1158,
TYPES 1996 Aussois, France, LNCS 1512,
TYPES 1998 Kloster Irsee, Germany, LNCS 1657,
TYPES 1999 Lökeborg, Sweden, LNCS 1956,
TYPES 2000 Durham, UK, LNCS 2277,
TYPES 2002 Berg en Dal, The Netherlands, LNCS 2646,
TYPES 2003 Turin, Italy, LNCS 3085,
TYPES 2004 Jouy-en-Josas, France, LNCS 3839,
TYPES 2006 Nottingham, UK, LNCS 4502,
TYPES 2007 Cividale del Friuli, Italy, LNCS 4941.

ESPRIT BRA 6453 was a continuation of ESPRIT Action 3245, *Logical Frameworks: Design, Implementation and Experiments.* TYPES 2008 was made possible by the contribution of many people. We thank all the participants of the workshops, and all the authors who submitted papers for consideration for these proceedings. We would like to also thank the referees for their effort in preparing careful reviews.

March 2009

Stefano Berardi
Ferruccio Damiani
Ugo de'Liguoro

Organization

Referees

A. Abel
J. Adamek
T. Altenkirch
D. Ancona
F. Aschieri
A. Asperti
J. Avigad
F. Barbanera
E. Beffara
S. Berghofer
Y. Bertot
F. Besson
L. Bettini
S. B. Ana
B. E. Brady
J. Caldwell
F. Cardone
P. Casteran
J. Charles
A. Chlipala
A. Ciaffaglione
M. Comini
T. Coquand
P. Corbineau
U. Dal Lago
D. de Carvalho
G. Dowek
C. Dubois
A. Felty
J.-C. Filliatre
C. Fuhs
M. Gebser
M. Gelfond
H. Geuvers
G. Gupta
J. Harrison
T. Hoang
M. Hofmann
B. Jacobs
D. Kesner
E. Komendantskaya
J. Laird
S. Lengrand
M. Lenisa
P. B. Levy
R. Matthes
D. Mazza
C. Mcbride
J. McKinna
M. Miculan
D. Miller
A. Miquel
J.-F. Monin
A. Murawski
K. Nakazawa
R. O'Connor
N. Oury
M. Piccolo
A. Poetzsch-Heffter
L. Pottier
C. Raffalli
F. Ricca
W. Ricciotti
E. Ritter
L. Robaldo
P. Rudnicki
A. Saurin
A. Setzer
E. Tassi
K. Terui
L. Thery
A. Trybulec
T. Uustalu
F.-J. de Vries
R. de Vrijer
H. Zantema

Table of Contents

Type Inference by Coinductive Logic Programming*

Davide Ancona, Giovanni Lagorio, and Elena Zucca

DISI, Univ. of Genova, v. Dodecaneso 35, 16146 Genova, Italy
{davide,lagorio,zucca}@disi.unige.it

Abstract. We propose a novel approach to constraint-based type inference based on coinductive logic. Constraint generation corresponds to translation into a conjunction of Horn clauses P, and constraint satisfaction is defined in terms of the coinductive Herbrand model of P. We illustrate the approach by formally defining this translation for a small object-oriented language similar to Featherweight Java, where type annotations in field and method declarations can be omitted.

In this way, we obtain a very precise type inference and provide new insights into the challenging problem of type inference for object-oriented programs. Since the approach is deliberately declarative, we define in fact a formal specification for a general class of algorithms, which can be a useful road map to researchers.

Furthermore, despite we consider here a particular language, the methodology could be used in general for providing abstract specifications of type inference for different kinds of programming languages.

Keywords: Type inference, coinduction, nominal and structural typing, object-oriented languages.

1 Introduction

Type inference is a valuable method to ensure static guarantees on the execution of programs (like the absence of some type errors) and to allow sophisticated compiler optimizations. In the context of object-oriented programming, many solutions have been proposed to perform type analysis (we refer to the recent article of Wang and Smith [20] for a comprehensive overview), but the increasing interest in dynamic object-oriented languages is asking for even more precise and efficient type inference algorithms [3,14].

Two important features which have to be supported by type inference are *parametric* and *data polymorphism* [1]; the former allows invocation of a method on arguments of unrelated types, the latter allows assignment of values of unrelated types to a field. While most solutions proposed in literature support well parametric polymorphism, only few inference algorithms are able to deal properly with data polymorphism; such algorithms, however, turn out to be quite complex and cannot be easily described.

* This work has been partially supported by MIUR EOS DUE - Extensible Object Systems for Dynamic and Unpredictable Environments.

S. Berardi, F. Damiani, and U. de'Liguoro (Eds.): TYPES 2008, LNCS 5497, pp. 1–18, 2009.

In this paper we propose a novel approach to type inference, by exploiting coinductive logic programming. Our approach is deliberately declarative, that is, we do not define any algorithm, but rather try to capture a space of possible solutions to the challenging problem of precise type inference of object-oriented programs.

The basic idea is that the program to be analyzed can be translated into an approximating logic program and a goal; then, type inference corresponds to find an instantiation of the goal which belongs to the coinductive model of the logic program. Coinduction allows to deal in a natural way with both recursive types [11,12] and mutually recursive methods.

The approach is fully formalized for a purely functional object-oriented language similar to Featherweight Java [16], where type annotations can be omitted, and are used by the programmer only as subtyping constraints. The resulting type inference is very powerful and allows, for instance, very precise analysis of heterogeneous container objects (as linked lists).

The paper is structured as follows: Section 2 defines the language and gives an informal presentation of the type system, based on standard recursive and union types. In Section 3 the type system is reconsidered in the light of coinductive logic programming, and the translation is fully formalized. Type soundness w.r.t. the operational semantics is claimed (proofs are sketched in Appendix B). Finally, Section 4 draws some conclusions and discusses future developments.

2 Language Definition and Types

In this section we present a simple object-oriented (shortly OO) language together with the definition of types. Constraint generation and satisfaction are only informally illustrated; they will be formally defined in the next section, on top of coinductive logic programming.

2.1 Syntax and Operational Semantics

The syntax is given in Figure 1. Syntactic assumptions listed in the figure are verified before performing type inference. We use bars for denoting sequences: for instance, $\overline{e}^m$ denotes $e_1, \ldots, e_m$, $\overline{T\ x}^n$ denotes[1] $T_1\ x_1, \ldots, T_n\ x_n$, and so on.

The language is basically Featherweight Java (FJ) [16], a small Java subset which has become a standard example to illustrate extensions and new technologies for Java-like languages. Since we are interested in type inference, type annotations for parameters, fields, and returned values can be omitted; furthermore, to make the type inference problem more interesting, we have introduced the conditional expression `if` (e) e_1 `else` e_2, and a more expressive form of constructor declaration.

We assume countably infinite sets of *class names* c, *method names* m, *field names* f, and *parameter names* x. A program is a sequence of class declarations

[1] If not explicitly stated, the bar "distributes over" all meta-variables below it.

$$
\begin{array}{rl}
prog ::= & \overline{cd}^n\ e \\
cd ::= & \texttt{class}\ c_1\ \texttt{extends}\ c_2\ \{\ \overline{fd}^n\ cn\ \overline{md}^m\ \}\quad (c_1 \neq \texttt{Object}) \\
fd ::= & T\ f; \\
cn ::= & c(\overline{T\ x}^n)\ \{\texttt{super}(\overline{e}^m);\overline{f = e';}^k\} \\
md ::= & T_0\ m(\overline{T\ x}^n)\ \{e\} \\
e ::= & \texttt{new}\ c(\overline{e}^n) \mid x \mid e.f \mid e_0.m(\overline{e}^n) \mid \texttt{if}\ (e)\ e_1\ \texttt{else}\ e_2 \mid \texttt{false} \mid \texttt{true} \\
T ::= & N \mid \epsilon \\
N ::= & c \mid \texttt{bool} \\
v ::= & \texttt{new}\ c(\overline{v}^n) \mid \texttt{false} \mid \texttt{true}
\end{array}
$$

Assumptions: $n, m, k \geq 0$, inheritance is not cyclic, names of declared classes in a program, methods and fields in a class, and parameters in a method are distinct.

Fig. 1. Syntax of OO programs

together with a main expression from which the computation starts. A class declaration consists of the name of the declared class and of its direct superclass (hence, only single inheritance is supported), a sequence of field declarations, a constructor declaration, and a sequence of method declarations. We assume a predefined class `Object`, which is the root of the inheritance tree and contains no fields, no methods and a constructor with no parameters. A field declaration consists of a type annotation and a field name. A constructor declaration consists of the name of the class where the constructor is declared, a sequence of parameters with their type annotations, and the body, which consists of an invocation of the superclass constructor and a sequence of field initializations, one for each field declared in the class.[2] A method declaration consists of a return type annotation, a method name, a sequence of parameters with their type annotations, and an expression (the method body).

Expressions are standard; boolean values and conditional expressions have been introduced just to show how the type system allows precise typing in case of branches. Integer values and the related standard primitives will be used in the examples, but are omitted in the formalization, since their introduction would only imply a straightforward extension of the type system. As in FJ, *this* is considered as a special implicit parameter.

A type annotation T can be either a nominal type N (the primitive type `bool` or a class name c) or empty.

Finally, the definition of values v is instrumental to the (standard) small steps operational semantics of the language, indexed over the class declarations defined by the program, shown in Figure 2.

For reasons of space, side conditions have been placed together with premises, and standard contextual closure have been omitted. To be as general as possible, no evaluation strategy has been fixed. Auxiliary functions *cbody* and *mbody* are defined in Appendix A.

[2] This is a generalization of constructors of FJ, whose arguments exactly match in number and type the fields of the class, and are used as initialization expressions.

$$\text{(field-1)}\ \frac{cbody(cds,c) = (\overline{x}^n, \{\texttt{super}(\ldots);\overline{f = e';}^k\}) \qquad f = f_i \qquad 1 \le i \le k}{\texttt{new}\ c(\overline{e}^n).f \longrightarrow_{cds} e'_i[\overline{e}^n/\overline{x}^n]}$$

$$\text{(field-2)}\ \frac{\begin{array}{l} cbody(cds,c) = (\overline{x}^n, \{\texttt{super}(\overline{e'}^m);\overline{f = \ldots;}^k\}) \\ \forall i \in 1..k \quad f \neq f_i \qquad \texttt{class}\ c\ \texttt{extends}\ c'\ \{\ldots\} \in cds \\ \texttt{new}\ c'(e'_1[\overline{e}^n/\overline{x}^n],\ldots,e'_m[\overline{e}^n/\overline{x}^n]).f \longrightarrow_{cds} e \end{array}}{\texttt{new}\ c(\overline{e}^n).f \longrightarrow_{cds} e}$$

$$\text{(invk)}\ \frac{mbody(cds,c,m) = (\overline{x}^n, e) \qquad e_{this} = \texttt{new}\ c(\overline{e}^k)}{\texttt{new}\ c(\overline{e}^k).m(\overline{e'}^n) \longrightarrow_{cds} e[\overline{e'}^n/\overline{x}^n][e_{this}/this]}$$

$$\text{(if-1)}\ \frac{}{\texttt{if (true)}\ e_1\ \texttt{else}\ e_2 \longrightarrow_{cds} e_1} \qquad \text{(if-2)}\ \frac{}{\texttt{if (false)}\ e_1\ \texttt{else}\ e_2 \longrightarrow_{cds} e_2}$$

Fig. 2. Reduction rules for OO programs

Rule (field-1) corresponds to the case where the field f is declared in the same class of the constructor, whereas rule (field-2) covers the disjoint case where the field has been declared in some superclass. The notation $e[\overline{e}^n/\overline{x}^n]$ denotes parallel substitution of x_i by e_i (for $i = 1..n$) in expression e.

In rule (invk), the parameters and the body of the method to be invoked are retrieved by the auxiliary function *mbody*, which performs the standard method look-up. If the method is found, then the invocation reduces to the body of the method where the parameters are substituted by the corresponding arguments, and *this* by the receiver object (the object on which the method is invoked).

The remaining rules are trivial.

The one step reduction relation on programs is defined by: $(cds\ e) \rightarrow (cds\ e')$ iff $e \rightarrow_{cds} e'$. Finally, $\rightarrow^*$ and $\rightarrow^*_{cds}$ denote the reflexive and transitive closures of $\rightarrow$ and $\rightarrow_{cds}$, respectively.

2.2 Types

Types, class environments and constraints are defined in Figure 3.

Value types (meta-variable τ) must not be confused with nominal types (meta-variable N) in the OO syntax. Nominal types are used as type annotations by

$$\begin{array}{rl}
\tau ::= & X \mid bool \mid obj(c,\rho) \mid \tau_1 \vee \tau_2 \mid \mu X.\tau \quad (\mu X.\tau \text{ contractive}) \\
\rho ::= & [\overline{f{:}\tau}^n] \\
\Delta ::= & \overline{c{:}(c', fts, ct, mts)}^n \\
fts ::= & [\overline{f{:}T}^n] \\
ct ::= & \forall \overline{X}^n.C \Rightarrow ((\prod_{i=1..k} X'_i) \rightarrow obj(c,\rho)) \quad (\{\overline{X'}^k\} \subseteq \{\overline{X}^n\}) \\
mts ::= & [\overline{m{:}mt}^n] \\
mt ::= & \forall \overline{X}^n.C \Rightarrow ((\prod_{i=1..k} X'_i) \rightarrow \tau) \quad (\{\overline{X'}^k\} \subseteq \{\overline{X}^n\}, n \ge k \ge 1) \\
C ::= & \{\overline{\gamma}^n\} \\
\gamma ::= & inst_of(\tau, N) \mid new(c, [\overline{\tau}^n], \tau) \mid fld_acc(\tau_1, f, \tau_2) \\
 & \mid invk(\tau_0, m, [\overline{\tau}^n], \tau) \mid cond(\tau_1, \tau_2, \tau_3, \tau)
\end{array}$$

Fig. 3. Definition of types, class environments and constraints

programmers, whereas value types are used in the type system and are transparent to programmers. Nominal types are approximations[3] of the much more precise value types. This is formally captured by the constraint $inst\_of(\tau, N)$ (see in the following).

A value type can be a type variable X, the primitive type *bool*, an object type $obj(c, \rho)$, a union type $\tau_1 \vee \tau_2$, or a recursive type $\mu X.\tau$.

An object type $obj(c, \rho)$ consists of the class c of the object and of a record type $\rho = [\overline{f{:}\tau}^n]$ specifying the types of the fields. Field types need to be associated with each object, to support data polymorphism; the types of methods can be retrieved from the class c of the object (see the notion of class environment below).

Union types [10,15] have the conventional meaning: an expression of type $\tau_1 \vee \tau_2$ is expected to assume values of type τ_1 or τ_2.

Recursive types are standard [2]: intuitively, $\mu X.\tau$ denotes the recursive type defined by the equation $X = \tau$, thus fulfilling the equivalences $\mu X.\tau \equiv \tau[\mu X.\tau/X]$ and $\mu X.\tau \equiv \mu X'.\tau[X'/X]$, where substitutions are capture avoiding. As usual, to rule out recursive types whose equation has no unique solution[4], we consider only *contractive* types [2]: $\mu X.\tau$ is contractive iff (1) all free occurrences of X in τ appear inside an object type $obj(c, \rho)$, (2) all recursive types in τ are contractive.

A class environment Δ is a finite map associating with each defined class name c all its relevant type information: the direct superclass; the type annotations associated with each declared field (*fts*); the type of the constructor (*ct*); the type of each declared method (*mts*).

Constructor types can be seen as particular method types. The method type $\forall \overline{X}^n.C \Rightarrow ((\prod_{i=1..k} X'_i) \rightarrow \tau)$ is read as follows: for all type variables $\overline{X}^n$, if the finite set of constraints C is satisfied, then the type of the method is a function from $\prod_{i=1..k} X'_i$ to τ. Without any loss of generality, we assume distinct type variables for the parameters; furthermore, the first type variable corresponds to the special implicit parameter *this*, therefore the type $\forall \overline{X}^n.C \Rightarrow ((\prod_{i=1..k} X'_i) \rightarrow \tau)$ corresponds to a method with $k-1$ parameters. Finally, note that C and τ may contain other universally quantified type variables (hence, $\{\overline{X'}^k\}$ is a subset of $\{\overline{X}^n\}$).

Constructor types correspond to functions which always return an object type and do not have the implicit parameter *this* (hence, k corresponds to the number of parameters).

Constraints are based on our long-term experience on compositional type-checking and type inference of Java-like languages [6,9,5,17,7]. Each kind of compound expression comes with a specific constraint:

- $new(c, [\overline{\tau}^n], \tau)$ corresponds to object creation, c is the class of the invoked constructor, $\overline{\tau}^n$ the types of the arguments, and τ the type of the newly created object;
- $fld\_acc(\tau_1, f, \tau_2)$ corresponds to field access, τ_1 is the type of the receiver, f the field name, and τ_2 the resulting type of the whole expression;

[3] Except for the type *bool*.

[4] For instance, $\mu X.X$ or $\mu X.X \vee X$.

- $invk(\tau_0, m, [\overline{\tau}^n], \tau)$ corresponds to method invocation, τ_0 is the type of the receiver, m the method name, $\overline{\tau}^n$ the types of the arguments, and τ the type of the returned value;
- $cond(\tau_1, \tau_2, \tau_3, \tau)$ corresponds to conditional expression[5], τ_1 is the type of the condition, τ_2 and τ_3 the types of the "then" and "else" branches, respectively, and τ the resulting type of the whole expression.

The constraint $inst\_of(\tau, N)$ does not correspond to any kind of expression, but is needed for checking that value type τ is approximated by nominal type N.

As it is customary, in the constraint-based approach type inference is performed in two distinct steps: constraint generation, and constraint satisfaction.

Constraint Generation. Constraint generation is the easiest part of type inference. A program *cds e* is translated into a pair (Δ, C), where Δ is obtained from *cds*, and C from e. As we will formally define in the next section, Δ can be represented by a set of Horn clauses, and C by a goal. To give an intuition, consider the following method declaration:

```
class List extends Object {
   altList(i,x){
      if(i<=0) new EList()
      else new NEList(x,this.altList(i-1,x.succ()))
   }
}
```

The method type of `altList` is inferred by collecting all constraints generated from its body:

$$\forall This, I, X, R_1, R_2, R_3, R_4, R_5.$$
$$\left\{\begin{array}{l} inst\_of(This, List), inst\_of(I, int), new(EList, [\,], R_1), invk(X, succ, [\,], R_2), \\ invk(This, altList, [int, R_2], R_3), new(NEList, [X, R_3], R_4), cond(bool, R_1, R_4, R_5) \end{array}\right\}$$
$$\Rightarrow ((This \times I \times X) \rightarrow R_5)$$

For simplicity we have simplified the set of constraints, omitting the constraints of `i<=0` and `i-1`. The constraint $inst\_of(This, List)$ forces the receiver object to be an instance of (a subclass of) `List`, since the method is declared in class `List`. The other constraints derive from each compound subexpression in the body of the method.

Constraint Satisfaction. After generating the pair (Δ, C) from the program *cds e*, to ensure that the execution of *cds e* is type-safe, one needs to prove that the set of constraints C is satisfiable in the class environment Δ. Typically, in constraint-based type inference of object-oriented programs, constraint satisfaction is defined operationally: most approaches directly provide an algorithm, or, at their best, a framework which can be instantiated by various algorithms

[5] This constraint could be easily avoided in practice, but has been introduced to show how a general methodology can be adopted, by associating with each kind of compound expression a specific constraint.

[20], but a declarative definition of constraint satisfaction is often missing. Even though this operational approach guarantees that type inference is decidable, providing a declarative definition of satisfiability based on a logical model allows one to abstract away from any possible implementation, and to give a simpler specification of the underlying type system. In this paper we take the opposite approach, by defining constraint satisfaction in terms of coinductive logic. In this way, we obtain a very powerful type system which, in fact, is not decidable, but can be approximated by precise type inference algorithms [8,4].

In the last part of this section we provide just an example to show how coinductive logic supports very precise typing. Let us add to the class List above the following class declarations:

```
class EList extends List {
   EList(){super();}
}

class NEList extends List {
   el; next;
   NEList(e,n){super();
                        el=e;next=n;}}

class A extends Object {
   A(){super();}
   succ(){new B()}
}

class B extends Object {
   B(){super();}
   succ(){new A()}
}
```

In such a program, the main expression **new** List().altlist(i,**new** A()) returns an empty list if $i \leq 0$; otherwise, a non empty list is returned whose length is i and whose elements are alternating instances of class A and B (starting from an A instance). Similarly, **new** List().altlist(i,**new** B()) returns an alternating list starting with a B instance.

The results of these two expressions can be specified by the following two precise types, respectively:

$$\begin{aligned}\tau_A = \mu X.\, & obj(EList, [\,])\vee \\ & obj(NEList, [el{:}obj(A, [\,]), next{:}\, obj(EList, [\,])\vee \\ & \qquad obj(NEList, [el{:}obj(B, [\,]), next{:}X])])\\ \tau_B = \mu X.\, & obj(EList, [\,])\vee \\ & obj(NEList, [el{:}obj(B, [\,]), next{:}\, obj(EList, [\,])\vee \\ & \qquad obj(NEList, [el{:}obj(A, [\,]), next{:}X])])\end{aligned}$$

By unfolding and coinduction, the following two type equivalences hold:

$$\tau_A \equiv obj(EList, [\,]) \vee obj(NEList, [el{:}obj(A, [\,]), next{:}\tau_B])$$
$$\tau_B \equiv obj(EList, [\,]) \vee obj(NEList, [el{:}obj(B, [\,]), next{:}\tau_A])$$

We show now that in the class environment corresponding to the example program, the constraints

$$invk(obj(List, [\,]), altList, [int, obj(A, [\,])], X_A)$$
$$invk(obj(List, [\,]), altList, [int, obj(B, [\,])], X_B)$$

generated from the two expressions are satisfiable for $X_A = \tau_A$ and $X_B = \tau_B$. For the first constraint we have to prove that the constraints of the method type of `altList` are satisfiable for $This = obj(List, [\,])$, $I = int$, and $X = obj(A, [\,])$. That is, the following set is satisfiable.

$$\left\{\begin{array}{l} inst\_of(obj(List, [\,]), List), inst\_of(int, int), new(EList, [\,], R_1), \\ invk(obj(A, [\,]), succ, [\,], R_2), invk(obj(List, [\,]), altList, [int, R_2], R_3), \\ new(NEList, [obj(A, [\,]), R_3], R_4), cond(bool, R_1, R_4, R_5) \end{array}\right\}$$

The two *inst_of* constraints are trivially satisfied, whereas $new(EList, [\,], R_1)$ and $invk(obj(A, [\,]), succ, [\,], R_2)$ are satisfiable for $R_1 = obj(EList, [\,])$ and $R_2 = obj(B, [\,])$. Then, by coinduction, $invk(obj(List, [\,]), altList, [int, R_2], R_3)$ is satisfiable for $R_3 = \tau_B$. Consequently, $new(NEList, [obj(A, [\,]), R_3], R_4)$ is satisfiable for $R_4 = obj(NEList, [el{:}obj(A, [\,]), next{:}\tau_B])$, and $cond(bool, R_1, R_4, R_5)$ for $R_5 = obj(EList, [\,]) \vee obj(NEList, [el{:}obj(A, [\,]), next{:}\tau_B]) \equiv \tau_A$. This last equivalence can be proved by unfolding and coinduction. The proof for the other constraint is symmetric.

3 Reconsidered Type Inference System

In this section we reconsider the type inference system described in the previous section in the light of coinductive logic.

The first basic idea consists in representing a class environment as a conjunction of Horn clauses (that is, a logic program), a set of type constraints as a conjunction of atoms (predicates applied to terms), and value types as terms. In this way, constraint generation corresponds to a translation from an OO program *cds* e to a pair (P, B), where P is a logic program corresponding to the class environment generated from *cds*, and B is a conjunction of atoms corresponding to the constraints generated from e.

We assume two countably infinite sets of *predicate* p and *function* f symbols, respectively, each one with an associated *arity* $n \geq 0$, and a countably infinite set of *logical variables* X. Functions with arity 0 are called *constants*. We write p/n, f/n to mean that predicate p, function f have arity n, respectively. For symbols we follow the usual convention: function and predicate symbols always begin with a lowercase letter, whereas variables always begin with an uppercase letter.

A logic program is a finite conjunction of *clauses* of the form $A \leftarrow B$, where A is the *head* and B is the *body*. The head is an *atom*, while the body is a finite and possibly empty conjunction of atoms; the empty conjunction is denoted by *true*. A clause with an empty body (denoted by $A \leftarrow true$) is called a *fact*. An atom has the form[6] $p(\overline{t}^n)$ where the predicate p has arity n and $\overline{t}^n$ are *terms*.

For list terms we use the standard notation $[\,]$ for the empty list and $[\_|\_]$ for the list constructor, and adopt the syntax abbreviation $[\overline{t}^n]$ for $[t_1|[\ldots[t_n|[\,]]]]$.

In coinductive Herbrand models, terms are possibly infinite trees. The definition of tree which follows is quite standard [13,2]. A path p is a finite and

[6] Parentheses are omitted for predicate symbols of arity 0; the same convention applies for function applications, see below.

possibly empty sequence of natural numbers. The empty path is denoted by ϵ, $p_1 \cdot p_2$ denotes the concatenation of p_1 and p_2, and $|p|$ denotes the length of p. A tree t is a partial function from paths to logical variables and function symbols, satisfying the following conditions:

1. the domain of t (denoted by $dom(t)$) is prefix-closed and not empty;
2. for all paths p in $dom(t)$ and for all natural numbers n,

$$p \cdot n \in dom(t) \text{ iff } t(p) = f/m \text{ and } n < m.$$

If $p \in dom(t)$, then the subtree t' of t rooted at p is defined by $dom(t') = \{p' \mid p \cdot p' \in dom(t)\}$, $t'(p') = t(p \cdot p')$; t' is said a *proper* subset of t iff $p \neq \emptyset$.

Note that recursive types defined with μ correspond to *regular* trees (see below), while here we are considering also types corresponding to non regular trees, therefore the set of types is much more expressive than that defined in the previous section, and, in fact, allows much more precise typings [4]. This is perfectly reasonable for a declarative definition of type inference; implementations of the system can only be sound approximations restricted to regular trees. A tree is regular (a.k.a. rational) if and only if it has a finite number of distinct subtrees. Regular terms can be finitely represented by means of term unification problems [19], that is, finite sets of equations [13,2] of the form $X = t$ (where t is a finite term which is not a variable). Note that logic programs are built over finite terms; infinite terms are only needed for defining coinductive Herbrand models [19] (co-Herbrand models for short, see Section 3.4).

3.1 Restricted Co-herbrand Universe

Given an OO program *prog*, the co-Herbrand universe [19] of its logic counterpart is the set of all terms built on [ ], *bool*, all constant symbols corresponding to class, field, and method names declared in *prog*, and the symbols of arity 2 $[\_|\_]$, $\_ : \_$, *obj*, and $\_ \vee \_$.

The co-Herbrand universe contains also terms which are non contractive types, as that defined by $X = X \vee X$. The definition of contractive type given in Section 2 can be generalized in a natural way to non regular terms as follows. A term t is contractive iff there exists no countable infinite sequence of natural numbers s s.t. there exists n s.t. for all paths p which are prefixes[7] of s, if $|p| \geq n$, then $p \in dom(t)$, and $t(p) = \vee/2$.

3.2 Restricted Co-herbrand Base

Given an OO program *prog*, the restricted co-Herbrand base of its logical encoding is the set of all ground atoms built on the contractive terms of the restricted co-Herbrand universe and on the following predicate symbols:

- all symbols of the type constraints defined in Figure 3 with the corresponding arity: *inst_of*/2, *new*/3, *fld_acc*/3, *invk*/4, *cond*/4;

[7] Recall that paths are finite sequences.

- *class*/1, where *class*(c) means that c is a defined class;
- *ext*/2, where *ext*(c_1, c_2) means that c_1 extends c_2;
- *subclass*/2, where *subclass*(c_1, c_2) means that c_1 is equal to or is a subclass of c_2;
- *has_fld*/3, where *has_fld*(c, f, T) means that class c has field f with type annotation T;
- *fld*/3, where *fld*(ρ, f, τ) means that the record type ρ has field f of type τ;
- *dec_fld*/3, where *dec_fld*(c, f, T) means that class c contains the declaration of field f with type annotation T;
- *dec_meth*/2 where *dec_meth*(c, m) means that c contains the declaration of method m;
- *meth*/4 where *meth*($c, m, [\tau_0, \overline{\tau}^n], \tau$) means that class c has a method m which returns a value of type τ when invoked on receiver of type τ_0 and with arguments of types $\overline{\tau}^n$.

These predicates are needed for translating class environments in logic programs (see Figure 4).

3.3 Constraint Generation

Constraint generation is defined in Figure 4. For the translation we assume bijections from the three sets of class, field and method names declared in the program to three corresponding sets containing constants of the co-Herbrand universe, and bijections from the two sets of parameter names and type variables to two corresponding sets containing logical variables. Given a class name c, a field name f, a method name m, a parameter name x, and a type variable X, we denote with $\widehat{c}, \widehat{f}, \widehat{m}$ the corresponding constants in the co-Herbrand universe, and with $\widehat{x}$ and $\widehat{X}$ the corresponding logical variables. For simplicity, we assume that the implicit parameter *this* is mapped to the logical variable *This* ($\widehat{\mathit{this}} = \mathit{This}$).

The rules define a judgment for each syntactic category of the OO language:

- *prog* $\leadsto$ (P, B): a program is translated in a pair where the first component is a logic program, and the second is a conjunction of atoms which is satisfiable in P iff *prog* is well-typed (see Section 3.4);
- *fds* **in** $c \leadsto$ *Cl*, *mds* **in** $c \leadsto P$: a field declaration is translated in a clause, whereas a method declaration is translated in a logic program (consisting of two clauses); both kinds of translation depend on the name of the class where the declaration is contained;
- *cn* **in** *fds* $\leadsto$ *Cl*: a constructor declaration is translated in a clause and is defined only if all fields in *fds* are initialized by the constructor in the same order[8] as they are declared in *fds*;
- e **in** $V \leadsto (t \,|\, B)$: an expression is translated in a pair where the first component is a term corresponding to the value type of the expression, and the second is a conjunction of atoms corresponding to the generated constraints. Constraint generation succeeds only if all free variables of e are contained in the set of variables V.

[8] This last restriction is just for simplicity.

$$\text{(prog)}\ \frac{\overline{cd \leadsto P}^n \quad e \textbf{ in } \emptyset \leadsto (t \mid B)}{\overline{cd}^n\ e \leadsto (P_{default} \cup (\cup_{i=1..n} P_i), B)} \qquad \text{(field)}\ \frac{}{T\ f;\ \textbf{in}\ c \leadsto \mathit{dec\_fld}(\widehat{c}, \widehat{f}, \widehat{T}) \leftarrow \mathit{true}.}$$

$$\text{(class)}\ \frac{\overline{fd \textbf{ in } c_1 \leadsto P^F}^n \quad cn \textbf{ in } \overline{fd}^n \leadsto Cl \quad \overline{md \textbf{ in } c_1 \leadsto P^M}^m \quad \begin{array}{l} P^F = \cup_{i=1..n} P^F_i \\ P^M = \cup_{i=1..m} P^M_i \end{array}}{\texttt{class}\ c_1\ \texttt{extends}\ c_2\ \{\ \overline{fd}^n\ cn\ \overline{md}^m\ \} \leadsto \left\{ \begin{array}{l} \mathit{class}(\widehat{c_1}) \leftarrow \mathit{true}. \\ \mathit{ext}(\widehat{c_1}, \widehat{c_2}) \leftarrow \mathit{true}. \end{array} \right\} \cup P^F \cup \{Cl\} \cup P^M}$$

$$\text{(constr-dec)}\ \frac{\overline{e \textbf{ in } \{\overline{x}^n\} \leadsto (t \mid B)}^m \quad \overline{e' \textbf{ in } \{\overline{x}^n\} \leadsto (t' \mid B')}^k}{\begin{array}{l} c(\overline{T\ x}^n)\ \{\texttt{super}(\overline{e}^m); \overline{f = e;}^k\} \textbf{ in } \overline{T'\ f;}^k \leadsto \\ \quad \mathit{new}(\widehat{c}, [\overline{\widehat{x}}^n], \mathit{obj}(\widehat{c}, [\overline{\widehat{f}{:}t'}^k \mid R])) \leftarrow \overline{\mathit{inst\_of}(\widehat{x}, \widehat{T})}^n, \overline{B}^m, \mathit{ext}(\widehat{c}, C), \\ \qquad\qquad \mathit{new}(C, [\overline{t}^m], \mathit{obj}(C, R)), \overline{B', \mathit{inst\_of}(t', \widehat{T'})}^k. \end{array}}$$

$$\text{(meth-dec)}\ \frac{e \textbf{ in } \{\mathit{This}, \overline{\widehat{x}}^n\} \leadsto (t \mid B)}{\begin{array}{l} T_0\ m(\overline{T\ x}^n)\{e\} \textbf{ in } c \leadsto \\ \quad \mathit{dec\_meth}(\widehat{c}, \widehat{m}) \leftarrow \mathit{true}. \\ \quad \mathit{meth}(\widehat{c}, \widehat{m}, [\mathit{This}, \overline{\widehat{x}}^n], t) \leftarrow \mathit{inst\_of}(\mathit{This}, \widehat{c}), \overline{\mathit{inst\_of}(\widehat{x}, \widehat{T})}^n, B, \mathit{inst\_of}(t, \widehat{T_0}). \end{array}}$$

$$\text{(new)}\ \frac{\overline{e \textbf{ in } V \leadsto (t \mid B)}^n}{\texttt{new}\ c(\overline{e}^n) \textbf{ in } V \leadsto (R \mid \overline{B}^n, \mathit{new}(\widehat{c}, [\overline{t}^n], R))}\ R \text{ fresh}$$

$$\text{(var)}\ \frac{}{x \textbf{ in } V \leadsto (\widehat{x} \mid \mathit{true})}\ x \in V \qquad \text{(field-acc)}\ \frac{e \textbf{ in } V \leadsto (t \mid B)}{e.f \textbf{ in } V \leadsto (R \mid B, \mathit{fld\_acc}(t, \widehat{f}, R))}\ R \text{ fresh}$$

$$\text{(invk)}\ \frac{e_0 \textbf{ in } V \leadsto (t_0 \mid B_0) \quad \overline{e \textbf{ in } V \leadsto (t \mid B)}^n}{e_0.m(\overline{e}^n) \textbf{ in } V \leadsto (R \mid B_0, \overline{B}^n, \mathit{invk}(t_0, \widehat{m}, [\overline{t}^n], R))}\ R \text{ fresh}$$

$$\text{(if)}\ \frac{e \textbf{ in } V \leadsto (t \mid B) \quad e_1 \textbf{ in } V \leadsto (t_1 \mid B_1) \quad e_2 \textbf{ in } V \leadsto (t_2 \mid B_2)}{\texttt{if}\ (e)\ e_1\ \texttt{else}\ e_2 \textbf{ in } V \leadsto (R \mid B, B_1, B_2, \mathit{cond}(t, t_1, t_2, R))}\ R \text{ fresh}$$

$$\text{(true)}\ \frac{}{\texttt{true} \textbf{ in } V \leadsto (\mathit{bool} \mid \mathit{true})} \qquad \text{(false)}\ \frac{}{\texttt{false} \textbf{ in } V \leadsto (\mathit{bool} \mid \mathit{true})}$$

Fig. 4. Constraint generation

In rule (class), $\overline{fd \textbf{ in } c_1 \leadsto P^F}^n$ abbreviates $fd_1 \textbf{ in } c_1 \leadsto P^F_1, \ldots, fd_n \textbf{ in } c_1 \leadsto P^F_n$ (the abbreviation $\overline{md \textbf{ in } c_1 \leadsto P^M}^m$ has a similar meaning).

In rule (constr-dec), $e_1 \textbf{ in } \{\overline{x}^n\} \leadsto (t_1 \mid B_1), \ldots, e_m \textbf{ in } \{\overline{x}^n\} \leadsto (t_m \mid B_m)$ is abbreviated by $\overline{e \textbf{ in } \{\overline{x}^n\} \leadsto (t \mid B)}^m$ (the same comment applies to the other premises of the rule).

Most of the rules are self-explanatory; we comment only rules for programs and for constructor and method declarations.

In rule (prog) $P_{default}$ (see Figure 5) contains those clauses shared by any program, whereas $\cup_{i=1..n} P_i$ are the clauses obtained by translating the class declarations of the program. Note that the type t of the main expression e is

$$\begin{array}{l}
class(object) \leftarrow true. \\
subclass(X, X) \leftarrow class(X). \\
subclass(X, object) \leftarrow class(X). \\
subclass(X, Y) \leftarrow ext(X, Z), subclass(Z, Y). \\
inst_of(bool, bool) \leftarrow true. \\
inst_of(obj(C1, X), C2) \leftarrow subclass(C1, C2). \\
inst_of(T1 \vee T2, C) \leftarrow inst_of(T1, C), inst_of(T2, C). \\
fld_acc(obj(C, R), F, T) \leftarrow has_fld(C, F, TA), fld(R, F, T), inst_of(T, TA). \\
fld_acc(T1 \vee T2, F, FT1 \vee FT2) \leftarrow fld_acc(T1, F, FT1), fld_acc(T1, F, FT1). \\
fld([F{:}T|R], F, T) \leftarrow true. \\
fld([F1{:}T1|R], F2, T) \leftarrow fld(R, F2, T), F1 \neq F2. \\
invk(obj(C, S), M, A, R) \leftarrow meth(C, M, [obj(C, S)|A], R). \\
invk(T1 \vee T2, M, A, R1 \vee R2) \leftarrow invk(T1, M, A, R1), invk(T2, M, A, R2). \\
new(object, [\,], obj(object, [\,])) \leftarrow true. \\
has_fld(C, F, T) \leftarrow dec_fld(C, F, T). \\
has_fld(C, F, T1) \leftarrow ext(C, P), has_fld(P, F, T1), \neg dec_fld(C, F, T2). \\
meth(C, M, [This|A], R) \leftarrow \\
\qquad inst_of(This, C), ext(C, P), meth(P, M, [This|A], R), \neg dec_meth(C, M). \\
cond(T1, T2, T3, T2 \vee T3) \leftarrow inst_of(T1, bool).
\end{array}$$

Fig. 5. Clauses in $P_{default}$ shared by all programs

discarded in the consequence of the rule, since only the constraints generated from e are needed to check the type safety of the program.

Note that not all formulas in Figure 5 are Horn clauses; indeed, for brevity we have used the negation of predicates *dec_fld* and *dec_meth*, and the inequality for field names. However, since the set of all field and method names declared in a program is finite, the predicates *not_dec_fld*, *not_dec_meth* and $\neq$ could be trivially defined by conjunctions of facts, therefore all formulas could be turned into Horn clauses.

A constructor declaration generates a single clause whose head has the form $new(\widehat{c}, [\overline{\widehat{x}}^n], obj(\widehat{c}, [\overline{\widehat{f}{:}t'}^k|R]))$, where c is the class of the constructor, $\overline{x}^n$ are its parameters, and $obj(\widehat{c}, [\overline{\widehat{f}{:}t'}^k|R])$ is the type of the object created by the constructor. This is obviously an object type corresponding to an instance of c, where the types associated with the fields $\overline{f}^k$ declared in c are determined by the initialization expressions $\overline{e'}^k$ (see the second premise), whereas the types associated with the inherited fields are determined by the invocation of the constructor of the direct superclass. Such invocation corresponds to the atom $new(C, [\overline{t}^m], obj(C, R))$; indeed, the atom $ext(\widehat{c}, C)$ is satisfied only if C is instantiated with the direct superclass of c, and the value types $\overline{t}^m$ of the arguments passed to the constructor of C are determined by the expressions $\overline{e}^m$ (see the first premise). Hence, R is the record type associating types with all fields inherited from C. The remaining atoms of the body of the clause are generated either from the expressions $\overline{e}^m$ and $\overline{e'}^k$ (conjunctions of atoms $\overline{B}^m, \overline{B'}^k$), or from the type annotations of the parameters $\overline{x}^n$ and of the fields $\overline{f}^k$ declared in c;

for convenience, we define the translation $\widehat{\epsilon}$ of the empty annotation to always return a fresh variable so that in this case no constraint is actually imposed.

Finally, notice that the clause is correctly generated only if: (1) the free variables of the expressions contained in the constructor body are contained in the set $\{\overline{x}^n\}$ of the parameters (therefore, *this* cannot be accessed); (2) all fields declared in the class are initialized exactly once and in the same order as they are declared.

Rule (meth-dec) is quite similar to (constr-dec) except for: (1) two clauses are generated, one for the predicate *dec_meth* and the other for the predicate *meth*. Notice that *dec_meth* specifies just the names of all methods declared in c, whereas *meth* specifies the names and the types of all methods (either declared or inherited) of c; (2) the variable *this* can be accessed in the body of the method; for this reason, *This* appears as the first parameter in the head of the clause for the predicate *meth*, and *this* is in the set of free variables which can appear in the body e of the method. Obviously, the variable *this* will always contain an instance of (a subclass of) c (see the atom $inst\_of(This, \widehat{c})$).

3.4 Constraint Satisfaction

A substitution θ is a total map from the set of logical variables into the set of contractive terms s.t. $\{X \mid \theta(X) \neq X\}$ is finite. The application of a substitution θ to a term t returns the term $t\theta$ defined as follows:

- $dom(t\theta) =$
  $\{p \mid p \in dom(t) \text{ or } p = p' \cdot p'' \text{ with } p' \in dom(t),\ t(p') = X, \text{ and } p'' \in dom(\theta(X))\}$
- $(t\theta)(p) = \begin{cases} t(p) & \text{if } p \in dom(t),\ t(p) \text{ is not a variable} \\ \theta(X)(p'') & \text{if } p = p' \cdot p'',\ p' \in dom(t),\ t(p') = X,\ p'' \in dom(\theta(X)) \end{cases}$

A ground instance of a clause $A \leftarrow \overline{A}^n$ is a ground clause Cl (that is, Cl does not contain logical variables) s.t. $Cl = A\theta \leftarrow \overline{A\theta}^n$ for a substitution[9] θ.

Constraint satisfaction is defined in terms of restricted co-Herbrand models. A restricted co-Herbrand model of a logic program P is a subset of the restricted co-Herbrand base of P which is a fixed-point of the immediate consequence operator T_P from the restricted co-Herbrand base into itself, defined by

$$T_P(S) = \{A \mid A \leftarrow \overline{A}^n \text{ is a ground instance of a clause of } P, \overline{A}^n \in S\}.$$

We have to show that for any program P, T_P is well-defined, that is, is closed w.r.t. contractive terms. This comes from the following proposition.

Proposition 1. *If t is contractive, then $t\theta$ is contractive.*

Proof. See Appendix B. □

Since T_P is obviously monotonic w.r.t. set inclusion, by the Knaster-Tarski theorem there always exists the greatest fixed-point of T_P, which is the greatest restricted co-Herbrand model $M^{co}(P)$ [19] of P.

We say that B is satisfiable in P iff there exists a substitution θ s.t. $B\theta \subseteq M^{co}(P)$.

---

[9] $\overline{A\theta}^n$ denotes $A_1\theta, \ldots, A_n\theta$.

3.5 Soundness of the System

Soundness follows by progress and subject reduction theorems below; the former states that a well-typed program cannot get stuck, the latter states that if a well-typed program reduces, then it reduces to a well-typed program. The proofs of these two theorems come directly from the main lemmas in Appendix B, whose proofs are a generalization of those which can be found in a companion paper [8].

Theorem 1 (Progress). *If $cds\ e \leadsto (P, B)$ and B is satisfiable in P, then either e is a value or $e \rightarrow_{cds} e'$ for some e'.*

Theorem 2 (Subject reduction). *If $cds\ e \leadsto (P, B)$, B is satisfiable in P, and $e \rightarrow_{cds} e'$, then $cds\ e' \leadsto (P, B')$, and B' is satisfiable in P.*

We say that $cds\ e$ is a normal form iff there exists no e' s.t. $(cds\ e) \rightarrow (cds\ e')$. Soundness ensures that reduction of well-typed programs never gets stuck.

Theorem 3 (Soundness). *If $cds\ e \leadsto (P, B)$, B is satisfiable in P, $(cds\ e) \rightarrow^* (cds\ e')$, and $cds\ e'$ is a normal form, then e' is a value.*

Proof. By induction on the number n of reduction steps. The claim for $n = 0$ holds by progress. If $n > 0$, then there exists e'' s.t. $(cds\ e) \rightarrow (cds\ e'')$, and $(cds\ e'') \rightarrow^* (cds\ e')$ in $n - 1$ steps. By subject reduction we have that $cds\ e'' \leadsto (P, B')$ and B' is satisfiable in P, therefore we can conclude by inductive hypothesis. □

4 Conclusion and Further Developments

We have defined a constraint-based type system for an object-oriented language similar to Featherweight Java, where type annotations in class declarations can be omitted. The type system is specified in a declarative way, by translating programs in sets of Horn clauses and considering their coinductive Herbrand models. This was made possible by our notion of constraints which has been introduced in previous works on principal typing of Java-like languages [9,5].

To our knowledge, this is the first attempt to exploit coinductive logic programming for type inference of object-oriented languages. The resulting type system is very precise and supports well data polymorphism, by allowing precise type inference of heterogeneous container objects (for instance, linked lists containing instances of unrelated classes).

We believe that this approach deserves further developments in several directions.

One of the most interesting and challenging issue concerns the implementation of the type inference defined here in a declarative way. Since the type system is defined on infinite and non regular types, clearly it is not decidable. Nevertheless, devising algorithms restricted to regular types which are sound w.r.t. the type system would represent an important advance in the topic. A possible implementation can be based on the recent results on the operational semantics of coinductive logic programming [19,18]. We have followed this approach to implement a prototype[10] in Java and Prolog, which is an approximation of the type system able to type the examples presented in this paper. We refer to the companion paper [8] for more details on the implementation.

Scalability and applicability are two other important issues. For the former, it would be interesting to study more complex translations able to deal with flow sensitive analysis and imperative features. To prove that our approach is applicable to other kinds of languages, a first step would consist in defining type inference based on coinductive logic programming for a simple functional language.

References

1. Agesen, O.: The cartesian product algorithm. In: Olthoff, W. (ed.) ECOOP 1995. LNCS, vol. 952, pp. 2–26. Springer, Heidelberg (1995)
2. Amadio, R., Cardelli, L.: Subtyping recursive types. ACM Transactions on Programming Languages and Systems 15(4), 575–631 (1993)
3. Ancona, D., Ancona, M., Cuni, A., Matsakis, N.: RPython: a Step Towards Reconciling Dynamically and Statically Typed OO Languages. In: OOPSLA 2007 Proceedings and Companion, DLS 2007: Proceedings of the 2007 Symposium on Dynamic Languages, pp. 53–64. ACM, New York (2007)
4. Ancona, D., Lagorio, G.: Type systems for object-oriented languages based on coinductive logic. Technical report, DISI - Univ. of Genova (2008) (submitted for publication)
5. Ancona, D., Damiani, F., Drossopoulou, S., Zucca, E.: Polymorphic bytecode: Compositional compilation for Java-like languages. In: ACM Symp. on Principles of Programming Languages 2005, January 2005. ACM Press, New York (2005)
6. Ancona, D., Lagorio, G., Zucca, E.: True separate compilation of Java classes. In: PPDP 2002 - Principles and Practice of Declarative Programming, pp. 189–200. ACM Press, New York (2002)
7. Ancona, D., Lagorio, G., Zucca, E.: Type inference for polymorphic methods in Java-like languages. In: Italiano, G.F., Moggi, E., Laura, L. (eds.) ICTCS 2007 - 10th Italian Conf. on Theoretical Computer Science 2003, eProceedings. World Scientific, Singapore (2007)
8. Ancona, D., Lagorio, G., Zucca, E.: Type inference for Java-like programs by coinductive logic programming. Technical report, Dipartimento di Informatica e Scienze dell'Informazione, Università di Genova (2008)
9. Ancona, D., Zucca, E.: Principal typings for Java-like languages. In: ACM Symp. on Principles of Programming Languages 2004, pp. 306–317. ACM Press, New York (2004)
10. Barbanera, F., Dezani-Cincaglini, M., de'Liguoro, U.: Intersection and union types: Syntax and semantics. Information and Computation 119(2), 202–230 (1995)
11. Brandt, M., Henglein, F.: Coinductive axiomatization of recursive type equality and subtyping. In: de Groote, P., Hindley, J.R. (eds.) TLCA 1997. LNCS, vol. 1210, pp. 63–81. Springer, Heidelberg (1997)

[10] Available at `http://www.disi.unige.it/person/LagorioG/J2P`

12. Brandt, M., Henglein, F.: Coinductive axiomatization of recursive type equality and subtyping. Fundam. Inform. 33(4), 309–338 (1998)
13. Courcelle, B.: Fundamental properties of infinite trees. Theoretical Computer Science 25, 95–169 (1983)
14. Furr, M., An, J., Foster, J.S., Hicks, M.: Static type inference for Ruby. In: SAC 2009: Proceedings of the 2009 ACM symposium on Applied computing. ACM Press, New York (to appear, 2009)
15. Igarashi, A., Nagira, H.: Union types for object-oriented programming. Journ. of Object Technology 6(2), 47–68 (2007)
16. Igarashi, A., Pierce, B.C., Wadler, P.: Featherweight Java: a minimal core calculus for Java and GJ. ACM Transactions on Programming Languages and Systems 23(3), 396–450 (2001)
17. Lagorio, G., Zucca, E.: Just: safe unknown types in java-like languages. Journ. of Object Technology 6(2), 69–98 (2007); special issue: OOPS track at SAC 2006
18. Simon, L., Bansal, A., Mallya, A., Gupta, G.: Co-logic programming: Extending logic programming with coinduction. In: Arge, L., Cachin, C., Jurdziński, T., Tarlecki, A. (eds.) ICALP 2007. LNCS, vol. 4596, pp. 472–483. Springer, Heidelberg (2007)
19. Simon, L., Mallya, A., Bansal, A., Gupta, G.: Coinductive logic programming. In: Etalle, S., Truszczyński, M. (eds.) ICLP 2006. LNCS, vol. 4079, pp. 330–345. Springer, Heidelberg (2006)
20. Wang, T., Smith, S.: Polymorphic constraint-based type inference for objects. Technical report, The Johns Hopkins University (2008) (submitted for publication)

A Auxiliary Functions

$$\text{(mbody-1)}\ \frac{\texttt{class}\ c\ \texttt{extends}\ c'\ \{\ \ldots T_0\ m(\overline{T\ x}^n)\{e\}\ldots\} \in cds}{mbody(cds, c, m) = (\overline{x}^n, e)}$$

$$\text{(mbody-2)}\ \frac{\texttt{class}\ c\ \texttt{extends}\ c'\ \{\ \ldots mds\ \} \in cds \qquad m \notin mds \qquad mbody(cds, c', m) = (\overline{x}^n, e)}{mbody(cds, c, m) = (\overline{x}^n, e)}$$

$$\text{(cbody)}\ \frac{\texttt{class}\ c\ \texttt{extends}\ c'\ \{\ \ldots c(\overline{T\ x}^n)\ \{\texttt{super}(\overline{e}^m); \overline{f = e';}^k\}\ldots\} \in cds}{cbody(cds, c) = (\overline{x}^n, \{\texttt{super}(\overline{e}^m); \overline{f = e';}^k\})}$$

Fig. 6. Auxiliary functions

B Proofs of Claims of Section 3

Proposition 1. If t is contractive, then $t\theta$ is contractive.

Proof. By contradiction, let us assume that $t\theta$ is not contractive, hence, there exists a countable and infinite sequence s of natural numbers and a natural number n s.t. for all paths p which are prefixes of s if $|p| \geq n$, then $p \in dom(t')$, and $t'(p) = \_ \vee \_/2$, for $t' = t\theta$. Let us consider the two following exhaustive and disjoint cases:

- If $p \in dom(t)$ for all paths which are prefixes of s, then t does not contain any variable along p for all finite prefixes p of s, therefore, by definition of $t\theta$, we have $t(p) = t'(p)$ for all finite prefixes p of s, but this contradicts the hypothesis that t is contractive.
- Otherwise, let us consider the longest path p' among all finite prefixes of s s.t. $p' \in dom(t)$, and let $l = |p'|$ (p' exists since we are assuming that there exists a finite prefix of s which does not belong to $dom(t)$, and, by definition of tree, $dom(t)$ is not empty and prefix-closed). Then, by definition of $t\theta$, $p' \in dom(t)$ and $t(p') = X$ for a certain logic variable X, and for all finite prefixes p of s, if $|p| \geq l$, then there exists p'' s.t. $p = p' \cdot p''$, $p'' \in dom(t'')$, and $t''(p'') = t'(p)$, where $t'' = \theta(X)$. Therefore, for all finite prefixes p of s, if $|p| \geq \max(0, n - l)$, then $p \in dom(t'')$, and $t''(p) = \_ \vee \_/2$, which contradicts the hypothesis that $\theta(X)$ is contractive. □

Progress. To prove progress we need the following lemmas.

Lemma 1. *If* $\mathcal{C}[e]$ ***in*** $V \rightsquigarrow (t \mid B)$*, then* e ***in*** $V \rightsquigarrow (t' \mid B')$*, with* $B' \subseteq B$*.*

Proof. By case analysis on the contexts and by induction on their structure. □

Lemma 2. *If* $cds \rightsquigarrow P$*, and* $invk(\widehat{c}, \widehat{m}, [t_1, \ldots, t_n], t)$ *is satisfiable in* P*, then* $mbody(cds, c, m) = (\overline{x}^n, e)$ *for some variables* $\overline{x}^n$ *and expression* e*.*

Proof. By induction on the height of the inheritance tree. □

Theorem 1 [Progress]. If $cds\ e \rightsquigarrow (P, B)$ and B is satisfiable in P, then either e is a value or $e \rightarrow_{cds} e'$ for some e'.

Proof. A generalization of the proof which can be found in a companion paper [8]. □

Subject Reduction. To prove subject reduction we need to introduce a subtyping relation $\leq$ between value types, since after a reduction step the inferred type of the reduced expression may become more specific.

Consider for instance the following expression $e =$ **if true** 1 **else false**. We have $e \rightarrow_{cds} 1$, e **in** $V \rightsquigarrow (X \mid cond(bool, int, bool, X))$ and 1 **in** $V \rightsquigarrow (int \mid true)$. Now $cond(bool, int, bool, X)$ is satisfiable for $X = int \vee bool$, but 1 **in** $V \rightsquigarrow (int \vee bool \mid true)$ does not hold. However, the subtyping relation $int \leq int \vee bool$ holds.

Since subtyping has to be defined on infinite terms, we adopt a coinductive definition [11,12]. We define $\leq$ as the greatest binary relation defined on the restricted co-Herbrand universe satisfying the following rules:

$$(\text{bool})\frac{}{bool \leq bool} \qquad (\text{obj})\frac{\overline{t_1 \leq t_2}^{m}}{obj(c, [\overline{f{:}t_1}^{n}]) \leq obj(c, [\overline{f{:}t_2}^{m}])}\ n \geq m$$

$$(\vee\text{R1})\frac{t \leq t_1}{t \leq t_1 \vee t_2} \qquad (\vee\text{R2})\frac{t \leq t_2}{t \leq t_1 \vee t_2} \qquad (\vee\text{L})\frac{t_1 \leq t \quad t_2 \leq t}{t_1 \vee t_2 \leq t}$$

$$(\text{distr})\frac{obj(c, [\overline{f{:}t}^{n}, f{:}t_f, \overline{f'{:}t'}^{m}]) \leq t \quad obj(c, [\overline{f{:}t}^{n}, f{:}t'_f, \overline{f'{:}t'}^{m}]) \leq t}{obj(c, [\overline{f{:}t}^{n}, f{:}t_f \vee t'_f, \overline{f'{:}t'}^{m}]) \leq t}$$

Lemma 3. *If* $cds \rightsquigarrow P$, e ***in*** $V \rightsquigarrow (t \mid B)$, $B\theta \subseteq M^{co}(P)$, *and* $e \rightarrow_{cds} e'$, *then there exist* t', B' *and* θ' *s.t.* e' ***in*** $V \rightsquigarrow (t' \mid B')$, $B'\theta' \subseteq M^{co}(P)$, *and* $t'\theta' \leq t\theta$.

Theorem 2 [Subject reduction]. If $cds\ e \rightsquigarrow (P, B)$, B is satisfiable in P, and $e \rightarrow_{cds} e'$, then $cds\ e' \rightsquigarrow (P, B')$, and B' is satisfiable in P.

Proof. A corollary of lemma 3. □

About the Formalization of Some Results by Chebyshev in Number Theory

Andrea Asperti* and Wilmer Ricciotti

Dipartimento di Scienze dell'Informazione
Mura Anteo Zamboni 7, Bologna
{asperti,ricciott}@cs.unibo.it

Abstract. We discuss the formalization, in the Matita Interactive Theorem Prover, of a famous result by Chebyshev concerning the distribution of prime numbers, essentially subsuming, as a corollary, Bertrand's postulate. Even if Chebyshev's result has been later superseded by the stronger prime number theorem, his machinery, and in particular the two functions ψ and θ still play a central role in the modern development of number theory. Differently from other recent formalizations of other results in number theory, our proof is entirely arithmetical. It makes use of most part of the machinery of elementary arithmetics, and in particular of properties of prime numbers, factorization, products and summations, providing a natural benchmark for assessing the actual development of the arithmetical knowledge base.

1 Introduction

Let $\pi(n)$ denote the number of primes not exceeding n. The prime number theorem, proved by Hadamard and la Vallé Poussin in 1896 states that $\pi(n)$ is asymptotically equal to $n/\log(n)$, that is the ratio between the two functions tends to 1 when n tends to infinity. In this paper we address a weaker result, due to Chebyshev around 1850, stating that the *order of magnitude* of $\pi(n)$ is $n/\log n$, meaning that we can find two constants c_1 and c_2 such that, for any n

$$c_1 \frac{n}{\log(n)} \leq \pi(n) \leq c_2 \frac{n}{\log n}$$

Even if Chebyshev's theorem is sensibly simpler than the prime number theorem, already formalized by Avigad et al. in Isabelle [3] and by Harrison in HOL Light [5], it is far form trivial (in Hardy and Wright's famous textbook [7], it takes pages 340-344 of chapter 22). In particular, our point was to give a fully arithmetical (and constructive) proof of this theorem. Even if Selberg's proof of the prime number theorem is "elementary", meaning that it requires no sophisticated tools of analysis except for the properties of logarithms, a fully arithmetical proof of this results looks problematics, considering that the statement involves *in an essential way* the Naperian logarithm. On the other side, the logarithm

* On leave at INRIA-Microsoft Research Center, Orsay, France.

S. Berardi, F. Damiani, and U. de'Liguoro (Eds.): TYPES 2008, LNCS 5497, pp. 19–31, 2009.

in Chebyshev's theorem can be in any base, and can be also essentially avoided (at least from the statement), asserting the existence of two constants c_1 and c_2 such that, for any n

$$2^{c_1 n} \leq n^{\pi(n)} \leq 2^{c_2 n}$$

that is what we actually proved.

As an important byproduct, we also give the first *purely arithmetical* formal proof of Bertrand's postulate, stating that for any n, there exists a prime number between n and $2n$[1].

The paper aims at providing a discussion of the subject in a form suitable to its formalization, without actually entering in implementation details (hence avoiding a direct discussion of the Matita system, but for a few descriptive examples).

2 Primes and the Factorial Function

In the rest of the paper, all functions are defined on natural numbers. In particular, n/m denotes the integer part of the division between n and m, and $\log_a n$ denotes the maximum i such $a^i \leq n$.

Chebyshev's approach to the study of the distribution of prime numbers consists in exploiting the decomposition of the number $n!$ as a product of prime numbers. The idea is that the numbers $1, 2, \ldots, n$ include just $\frac{n}{p}$ multiples of p, $\frac{n}{p^2}$ multiples of p^2, an so on. Hence (the variable bound by the product is written in bold)

$$n! = \prod_{\mathbf{p} \leq n} \prod_{\mathbf{i} < \log_p n} p^{n/p^{i+1}} \tag{1}$$

The previous one is a good example of a typical mathematical argumentation (see e.g. [7], p. 342). Looking more carefully, you see that it provides you (almost) no information, since it is essentially a mere rephrasing of the statement: it is a gentle invitation to work it out by yourself, just a bit less unsympathetic than a brutal "trivial".

The formal proof requires a bit more work. The starting point is that every integer n may be uniquely decomposed as the product of all its prime factors. Le us write $ord_p(n)$ for the multiplicity of p in n; then

$$n = \prod_{\mathbf{p} \leq n} p^{ord_p(n)} = \prod_{\mathbf{p} \leq n} \prod_{\substack{\mathbf{i} < \log_p n \\ p^{i+1} | n}} p \tag{2}$$

for p prime. At the time we started this work, the mathematical library of Matita already contained the proof of the Fundamental Theorem of Arithmetic, namely the existence and uniqueness of the decomposition in prime factors. This was

[1] Providing a good upper bound to the search for the next prime, in systems based on logics like the Calculus of Inductive Constructions, is essential to define a reasonably efficient enumeration function for all primes.

proved by giving a factorization function returning for each natural number n a list of multiplicities of its prime factors (for a given factorization strategy), a function computing the products of the elements in the list, and proving that they are inverse of each other. However, passing from this result to the formulation of equation 2 is not so evident. Since, on the other hand, all the needed machinery was already in the library, we opted for a direct proof. The idea is to work by induction on the upper bound of the product. However, we cannot directly work on n, since this must be the *constant* argument of $ord_p(n)$. So have to rephrase the statement in the form

$$\forall m > c(n), n = \prod_{\mathbf{p} \leq m} p^{ord_p(n)}$$

Where $c(n)$ is a suitable function of n. The naive idea to take $c(n) = n$ does not work: in fact, in order to ensure that the induction works properly, we must take a *minimum* bound, that in this case is the largest prime factor of n. This is the actual statement we proved:

```
theorem lt_max_to_pi_p_primeb:
\forall q,m.
 O < m \to
  max m (\lambda i.primeb i \land divides_b i m) < q \to
   m = pi_p q (\lambda i.primeb i \land divides_b i m)
        (\lambda p.exp p (ord m p)).
```

From the previous result we obtain equation 2 as a simple corollary. So,

$$\begin{aligned} n! &= \prod_{1 \leq \mathbf{m} \leq n} m \\ &= \prod_{1 \leq \mathbf{m} \leq n} \prod_{\mathbf{p} \leq m} \prod_{\substack{\mathbf{i} < \log_p m \\ p^{i+1} \mid m}} p \\ &= \prod_{\mathbf{p} \leq n} \prod_{p \leq \mathbf{m} \leq n} \prod_{\substack{\mathbf{i} < \log_p m \\ p^{i+1} \mid m}} p \\ &= \prod_{\mathbf{p} \leq n} \prod_{\mathbf{i} < \log_p n} \prod_{\substack{\mathbf{m} \leq n \\ p^{i+1} \mid m}} p \\ &= \prod_{\mathbf{p} \leq n} \prod_{\mathbf{i} < \log_p n} p^{n/p^{i+1}} \end{aligned}$$

In, particular, for $2n$ we have:

$$(2n)! = \prod_{\mathbf{p} \leq 2n} \prod_{\mathbf{i} < \log_p 2n} p^{2n/p^{i+1}} \tag{3}$$

But

$$\frac{2n}{p^{i+1}} = 2\frac{n}{p^{i+1}} + \left(\frac{2n}{p^{i+1}} \mod 2\right)$$

Moreover, if $n \leq p$ or $\log_p n \leq i$ we have

$$\frac{n}{p^{i+1}} = 0$$

Hence, if we define

$$B(n) = \prod_{\mathbf{p} \leq n} \prod_{\mathbf{i} < \log_p n} p^{(n/p^{i+1} \mod 2)}$$

equation (3) becomes

$$(2n)! = n!^2 B(2n) \tag{4}$$

$B(2n)$ is thus the binomial coefficient $\binom{2n}{n}$.

2.1 Upper and Lower Bounds for B

For all n, $(2n)! \leq 2^{2n-1}n!^2$. For technical reasons, we need however a slightly stronger result, namely,

$$(2n)! \leq 2^{2n-2}n!^2$$

that holds for any n larger than 4. The proof is by induction.

The base case amounts to check that $10! \leq 2^8 5!^2$, which can be proved by a mere computation (after some simplification).

In the inductive case

$$\begin{aligned}(2 \cdot (n+1))! &= (2n+2)(2n+1)(2n)! \\ &\leq (2n+2)(2n+1)2^{2n-2}n!^2 \\ &\leq (2n+2)(2n+2)2^{2n-2}n!^2 \\ &= 2^{2n}(n+1)!^2\end{aligned}$$

So, by equation (4), we conclude that, for any n

$$B(2n) \leq 2^{2n-1} \tag{5}$$

and when n is larger than 4,

$$B(2n) \leq 2^{2n-2} \tag{6}$$

Similarly, we prove that, for any $n > 0$,

$$2^{2n}n!^2 \leq 2n(2n)!$$

The proof is by induction on n. For $n = 1$ both sides reduce to 4. For $n > 1$,

$$\begin{aligned}
2^{2n+2}(n+1)!^2 &= 4(n+1)^2 2^{2n} n! \\
&= 4(n+1)^2 2n(2n)! \\
&= 4(n+1)(n+1)2n(2n)! \\
&\leq 4(n+1)(n+1)(2n+1)(2n)! \\
&= 2(n+1)(2n+2)(2n+1)(2n)! \\
&= 2(n+1)(2n+2)!
\end{aligned}$$

By equation (4) we conclude that

$$2^{2n} \leq 2nB(2n) \tag{7}$$

and since for any n, $2n \leq 2^n$,

$$2^n \leq B(2n) \tag{8}$$

3 Chebyshev's Ψ Function

Let us now consider the following function

$$\Psi(n) = \prod_{\mathrm{p} \leq n} p^{\log_p n}$$

where the product is over all *primes* less or equal to n. Chebyshev ψ function is the naperian logarithm of Ψ, but as we mentioned in the introduction, we try to avoid the use of logarithms as far as possible. The relation between Ψ and π should be clear:

$$\Psi(n) = \prod_{\mathrm{p} \leq n} p^{\log_p n} \leq \prod_{\mathrm{p} \leq n} n = n^{\pi(n)} \tag{9}$$

Since moreover, $n < a^{\log_a n+1}$ we also have $n < a^{2\log_a n}$, so that, easily,

$$n^{\pi(n)} \leq \Psi(n)^2 \tag{10}$$

Let us now rewrite $\Psi(n)$ in the following equivalent form:

$$\Psi'(n) = \prod_{\mathrm{p} \leq n} \prod_{\mathrm{i} < \log_p n} p$$

It is then clear that, for any n,

$$B(n) \leq \Psi'(n) = \Psi(n)$$

Hence, the lower bound for B immediately gives a lower bound for Ψ, namely

$$2^n \leq 2^{2n}/2n \leq \Psi(2n) \tag{11}$$

For the upper bound, let us first observe that

$$\Psi(2n) = \Psi(n) \prod_{\mathbf{p} \leq 2n} \prod_{\mathbf{i} < \log_p 2n} p^{j(n,p,i)} \tag{12}$$

where $j(n, p, i)$ is 1 if $n < p^{i+1}$ and 0 otherwise. Indeed

$$\begin{aligned}
\Psi(2n) &= \prod_{\mathbf{p} \leq 2n} \prod_{\mathbf{i} < \log_p 2n} p \\
&= \left(\prod_{\mathbf{p} \leq 2n} \prod_{\mathbf{i} < \log_p 2n} p^{j(n,p,i)} \right) \left(\prod_{\mathbf{p} \leq 2n} \prod_{\mathbf{i} < \log_p 2n} p^{1-j(n,p,i)} \right) \\
&= \Psi(n) \prod_{\mathbf{p} \leq 2n} \prod_{\mathbf{i} < \log_p 2n} p^{j(n,p,i)}
\end{aligned}$$

Then observe that

$$\prod_{\mathbf{p} \leq 2n} \prod_{\mathbf{i} < \log_p 2n} p^{j(n,p,i)} \leq B(2n) = \prod_{\mathbf{p} \leq 2n} \prod_{\mathbf{i} < \log_p 2n} p^{(2n/p^{i+1} \mod 2)} \tag{13}$$

since if $n < p^{i+1}$ then $2n/p^{i+1} \mod 2 = 1$. So we may conclude that

$$\Psi(2n) \leq B(2n)\Psi(n) \tag{14}$$

and in particular, for any n

$$\Psi(2n) \leq 2^{2n-1}\Psi(n) \tag{15}$$

and for $4 < n$

$$\Psi(2n) \leq 2^{2n-2}\Psi(n) \tag{16}$$

We may now use inductively these estimates to prove

$$\Psi(n) \leq 2^{2n-3} \tag{17}$$

For the proof, we need the monotonicity of Ψ, that is easily proved:

$$\Psi(n) = \prod_{\mathbf{p} \leq n} p^{\log_p n} \leq \prod_{\mathbf{p} \leq n} p^{\log_p (n+1)} \leq \prod_{\mathbf{p} \leq n+1} p^{\log_p (n+1)} = \Psi(n+1) \tag{18}$$

Then we check that the property holds for any $n \leq 8$, which can be done by direct computation. If n is larger than 8 we distinguish two cases, according to n is even or odd. We only consider the case $n = 2m+1$ that is the most interesting one. Observe first that $8 < 2m + 1$ implies $4 < m$. Then we have:

$$\begin{aligned}
\Psi(n) &= \Psi(2m+1) \\
&\leq \Psi(2m+2) \\
&\leq 2^{2m}\Psi(m+1) \\
&\leq 2^{2m}2^{2(m+1)-3} \\
&\leq 2^{2(2m+1)-3}
\end{aligned}$$

In conclusion, we have

$$2^{n/2} \leq \Psi(n) \leq n^{\pi(n)} \leq \Psi(n)^2 \leq 2^{4n-6} \leq 2^{4n} \tag{19}$$

4 Bertrand's Postulate

Our approach to Chebyshev's theorem, as most modern presentations of the subject, essentially follows Chebyshev's original idea, but in a rudimentary form which provides a result that is numerically less precise, though of a similar nature. In particular, Chebyshev was able to prove the asymptotic estimates

$$(c_1 + o(1))\frac{n}{\log n} \leq \pi(n) \leq (c_2 + o(1))\frac{n}{\log n} \qquad (n \to \infty)$$

with

$$\begin{aligned} c_1 &= \log(2^{1/2}3^{1/3}5^{1/5}30^{-1/30}) \approx 0.92129 \\ c_2 &= 6/5c_1 \approx 1.10555 \end{aligned}$$

In particular, since $c_2 < 2c_1$, this implies that

$$\pi(2n) > \pi(n)$$

for all large n. Actually, by direct computation, Chebyshev proved that the inequality remains true for all n, confirming a famous conjecture known as *Bertrand's postulate.*

With our rough estimates, we could only prove the existence of a prime number between n and $5n$, for n sufficiently large. There exists however an alternative approach to the proof of Bertrand's postulate due to Erdös [4] (see also [7], p. 344) that is well suited to a formal encoding in arithmetics[2].

Let

$$k(n,p) = \sum_{i<\log_p n} (n/p^{i+1} \mod 2)$$

Then, B can also be written as

$$B(n) = \prod_{p \leq n} p^{k(n,p)}$$

We now split this product in two parts B_1 and B_2, according to $k(n,p) = 1$ or $k(n,p) > 1$. Suppose that Bertrand postulate is $false$, hence there is no prime between n and $2n$. Moreover, if $\frac{2n}{3} < p \leq n$, then $2n/p = 2$ and for $i > 1$ and $n \geq 6$ $2n/p^i = 0$ since

$$2n \leq \left(\frac{2n}{3}\right)^2 \leq p^i$$

[2] Erdös' argument was already exploited by Théry in his proof of Bertrand postulate [11]; however he failed to provide a fully arithmetical proof, being forced to make use of the (classical, axiomatic) library of Coq reals to solve the remaining inequalities. Similarly, Riccardi's formalization of Bertrand's postulate in Mizar [8] makes an essential use of real numbers.

so $k(2n,p) = 0$. Summing up, under the assumption that Bertrand postulate is *false*,

$$\begin{aligned} B_1(2n) &= \prod_{\substack{p \leq 2n \\ k(2n,p)=1}} p \\ &= \prod_{p \leq 2n/3} p \\ &\leq \Psi(2n/3) \\ &\leq 2^{2(2n/3)} \end{aligned}$$

On the other side, note that $k(n,p) \leq \log_p n$, so if $k(2n,p) \geq 2$ we also have $\log_p 2n \geq 2$ that implies $p \leq \sqrt{2n}$. So

$$\begin{aligned} B_2(2n) &= \prod_{\substack{p \leq 2n \\ 2 \leq k(2n,p)}} p^{k(2n,p)} \\ &\leq \prod_{p \leq \sqrt{2n}} 2n \\ &= (2n)^{\pi(\sqrt{2n})} \end{aligned}$$

For $n \geq 15$, $\pi(n) \leq n/2 - 1$. Hence, for any $n \geq 2^7 > 15^2$, we have

$$B_2(2n) \leq (2n)^{\sqrt{2n}/2-1}$$

Putting everything together, supposing Bertrand's postulate is false, we would have, for any $n \geq 2^7$

$$\begin{aligned} 2^{2n} &\leq 2nB(2n) \\ &= 2nB_1(2n)B_2(2n) \\ &\leq 2^{2(2n/3)}(2n)^{\sqrt{2n}/2} \end{aligned}$$

Observe that

$$2^{2n} = 2^{2(2n/3)}2^{2n/3}$$

so, by cancellation,

$$2^{2n/3} \leq (2n)^{\sqrt{2n}/2}$$

and taking logarithms

$$\frac{2n}{3} \leq \frac{\sqrt{2n}}{2}(\log(2n)+1)$$

We want to find, *by arithmetical means*, an integer m such that for all values larger than m the previous equation is false; moreover, the integer m must be sufficiently small to allow to check the remaining cases automatically in a feasible time.

We must prove

$$\frac{\sqrt{2n}}{2}(\log(2n)+1) < \frac{2n}{3}$$

The strict inequality is the first source of trouble, so we prove instead

$$\frac{\sqrt{2n}}{2}(\log(2n)+1) \leq \frac{2n}{4}$$

using the fact that

$$\frac{n}{m+1} < \frac{n}{m}$$

for any $n \geq m^2$ (in our case, $n \geq 8$). By means of simple manipulations, it is easy to transform the last equation in the following simpler form

$$2(\log(2n)+1) \leq \sqrt{2n}$$

or equivalently

$$2(\log n + 2)^2 \leq n$$

We now use the fact that for any $a > 0$ and any $n \geq 4a$

$$2^a n^2 \leq 2^n$$

to get, for any $n \geq 2^8$

$$2(\log n + 2)^2 \leq 4(\log n)^2 = 2^2(\log n)^2 \leq 2^{\log n} \leq n$$

4.1 Automatic Check

To complete the proof, we have still to check that Bertrand's postulate remains true for all integers less then 2^8. This is very simple in principle: it is sufficient to

1. Generate the list of all primes up to the first prime larger than 2^8 (in reverse order).
2. Check that for any pair p_i, p_{i+1} of consecutive primes in such list, $p_i < 2p_{i+1}$.

Both the generation of the list and its check can be performed automatically. All we have to do is to prove that our algorithm for generating primes is correct and complete, and that the previous check is equivalent to Bertrand's postulate, on the given interval.

Since before this formalization, Matita has contained in its library the machinery necessary to perform this check – particularly a function `primeb` capable of deciding whether its argument is a prime number or not. `primeb` is implemented in the trivial way: it computes the smallest factor of its argument n by repeatedly dividing it by any $m \leq n$, and finally checks whether it equals n or not. The proof of correctness is, of course, straightforward; however, this comes at the cost of an inefficient algorithm, whose use is practical only for small values of n.

As it is often the case, to get better performance we must resort to a different algorithm, whose proof of correctness is less trivial. The sieve of Eratosthenes came as a good candidate, since it directly computes the list of the first primes up to a given number, which is precisely what we need. Furthermore, it has a simple implementation and an elementary, though a bit involved, proof of correctness, which is also interesting in itself as a small case of software verification. This is the actual code of the sieve, written in the Matita language:

```
let rec sieve_aux l1 l2 t on t \def
  match t with
  [ O => l1 (* this case is vacuous *)
  | S t1 => match l2 with
    [ nil => l1
    | cons n tl => sieve_aux (n::l1)
        (filter nat tl (\lambda x.notb (divides_b n x))) t1]].

definition sieve : nat \to list nat \def
  \lambda m.sieve_aux [] (list_n m) m.
```

The function sieve_aux takes in input a list of primes (initially empty), a list of integers yet to sieve (initially comprising all natural numbers between 2 and a given number m), and an integer that is supposed to be larger than the length of the second list (initially m). This last parameter is used as recursive parameter to ensure termination. The algorithm simply takes the first element of the second list, adds it to the first list, and removes from the second list all its multiples.

Here is the function checking that each element of the list is less than twice its successor (we also check that the last element is 2):

```
let rec check_list l \def
  match l with
  [ nil \Rightarrow true
  | cons (hd:nat) tl \Rightarrow
    match tl with
     [ nil \Rightarrow eqb hd 2
     | cons hd1 tl1 \Rightarrow
      (leb (S hd1) hd \land leb hd (2*hd1) \land check_list tl)
    ]
  ].
```

In order for these procedures to be useful, some properties must hold. First we need to prove correctness and completeness of sieve, which in turn requires us to understand and prove the recursion invariant of sieve_aux. Informally:

Given a natural number m and two lists $l1$ and $l2$, such that
- for any natural number p, p is contained in $l1$ if and only if it is prime and less than any number contained in $l2$
- for any natural number x, x is contained in $l2$ if and only if $2 \leq x \leq m$ and x isn't multiple of any number contained in $l1$

then, assuming $l1$ and $l2$ are respectively sorted decreasingly and increasingly, and t is less than the length of $l2$, `sieve_aux l1 l2 t` is a sorted list of decreasing numbers and p is contained in `sieve_aux l1 l2 t` if and only if p is prime and less than m.

The invariant is relatively complex, due to the mutual dependency of the properties of the two lists $l1$ and $l2$. A proof may be obtained by induction on t and then by cases on $l2$. In the interesting part, for $t = t' + 1$ and $l2 = h :: l$, the statement is obtained by means of the induction hypothesis. The following lemmata are also needed:

1. p is contained in $h :: l1$ if and only if it is prime, less or equal than m, and less than any number contained in l'
2. x is in l' if and only if it is greater or equal than 2, less or equal than m, and it is not divisible by any number contained in $h :: l1$
3. $length\, l' \leq t'$
4. $h :: l1$ is sorted decreasingly
5. l' is sorted increasingly

where l' is l from which any number divisible by h has been removed, preserving the order, that is `filter nat l (\lambda x.notb (divides_b h x))`.

The tricky lemmata are 1 and 2. For the first one, we proceed by cases:

- if $p = h$, p is contained in $h :: l$ (that is $l2$), therefore it is less than m and it isn't divisible by any number in $l1$; since $h :: l$ is sorted, h is also less than any number contained in l (and, in particular, less than any number in l'); this implies p is also a prime number. The opposite direction of the logical equivalence is trivial.
- if $p \neq h$, the implication from left to right is trivial since, under this hypothesis, if p is contained in $h :: l1$, it must be contained in $l1$: by the hypothesis on $l1$, this implies the thesis. In the opposite direction, we must prove that if p is prime, less than m and less than any number contained in l', then p is contained in $l1$. First, $p < h$, otherwise by the hypothesis on l and the definition of l' we could prove p is contained in l', thus obtaining $p < p$, which is absurd. Furthermore, for any x contained in $h :: l$, $h \leq x$, because $h :: l$ is sorted increasingly by hypothesis. Thus we get, for all x in $h :: l$, $p < x$, which implies by the hypothesis on $l1$ that p is contained in $l1$.

The second lemma is less complicated. In the left-to-right implication, the nontrivial part is to see that, if x is contained in l', then it isn't a multiple of any p contained in $h :: l1$. By cases, if $p = h$, the thesis follows by definition of l'; if p is contained in $l1$, it is sufficient to apply the hypothesis on $l1$. The opposite direction of the implication is obtained combining the hypotheses to show that x must be in $h :: l$. Then, x must be different from h (otherwise, we could prove that x doesn't divide itself). Since x must be in l and h doesn't divide x, x must also be in l'.

Last, we prove that if `checklist l = true`, then for any number p contained in l and greater than 2, there exists some number q contained in l, such that $q < p \leq 2q$. The proof is easy by induction.

Combining the correctness and completeness of the sieve and this last property, we finally get that Bertrand's postulate holds for all integers less than 2^8, just by checking that `check_list (sieve (S (exp 2 8))) = true`, a test which only takes some seconds.

5 Conclusions

In this paper we presented the formalization, in the Matita interactive theorem prover, of some results by Chebyshev about the distribution of prime numbers. Even if Chebyshev's main result has been later superseded by the stronger prime number theorem, his machinery, and in particular the two functions ψ and θ still play a central role in the modern development of number theory.

As also testified by our own development, Matita is a mature system that already permits the formalization of proofs of not trivial complexity (see . for another recent formalization effort). Although the Matita arithmetical library was already well developed at the time we started the work (see [2]), several integrations were required, concerning the following subjects:

- logarithms, square root (632 lines)
- inequalities involving integer division (339 lines)
- magnitude of functions (255 lines)
- decomposition of a number n as a product of its primes (250 lines)
- binomial coefficients (260 lines)
- properties of the factorial function (303 lines)
- integrations to the library for $\sum$ and $\prod$ (148 lines)
- operations over lists (224 lines)

Apart from these prerequisites, the proofs of Chebyshev's theorem and Bertrand's conjecture take respectively 2073 and 2389 lines (of which 1863 just devoted to the validity check of the conjecture for integers less then 2^8). A good amount of work was also spent in the investigation of related fields (Abel summations, properties of the Θ function, upper and lower bounds for Euler's e constant) that at the end have not been used in the main proof, but still have an interest in themselves. The following table summarizes the dimension of the development, and the total effort in time:

	prereq.	**chebys.**	**Bertrand**	**check**	**other**	**total**
lines	2411	2073	743	526	1863	7616
hours	54	51	21	16	48	190

In Hardy's book [7], the proof of Bertrand's postulate takes 42 lines, while Chebyshev's theorem takes precisely three pages (90 lines): this gives a de Bruijn factor of 20-25, that is in line with other developments in related subjects (see [3,2]). The most interesting datum is however the average time required to formalize a line of mathematical text, that in our case is about 1.5 hours (in [2], on a different arithmetical subject, we gave an estimation of 2 hours per line). The impressive cost of the formalization is the main obstacle towards a larger diffusion of automatic provers in the mathematical community, and all the research effort in the area of formalized reasoning is finally aimed to reduce this cost. Computing this value on large formalizations is an important an effective way to measure the state of the art and to testify its advancement.

References

1. Apostol, T.M.: Introduction to Analytic Number Theory. Springer, Heidelberg (1976)
2. Asperti, A., Armentano, C.: A Page In Number Theory. Journal of Formalized Reasoning 1 (2008) (to appear)
3. Avigad, J., Donnelly, K., Gray, D., Raff, P.: A formally verified proof of the prime number theorem. ACM Transactions on Computational Logic 9(1) (2007) (to appear in the ACM Transactions on Computational Logic)
4. Erdös, P.: Beweis eines Satzes von Tschebyschef. Acta Scientifica Mathematica 5, 194–198 (1932)
5. Harrison, J.: Formalizing an analytic proof of the Prime Number Theorem (extended abstract). In: Participant's proceedings of TTVSI Festschrift in honour of Mike Gordon's 60th birthday (2008)
6. Jameson, G.J.O.: The Prime Number Theorem. London Mathematical Society Student Texts, vol. 53. Cambridge University Press, Cambridge (2003)
7. Hardy, G.H., Wright, E.M.: An introduction to the theory of numbers. Oxford University Press, Oxford (1938) (Fourth edition 1975)
8. Riccardi, M.: Pocklington's Theorem and Bertrand's Postulate. Formalized Mathematics 14(2), 47–52 (2006)
9. Sacerdoti Coen, C., Tassi, E.: A constructive and formal proof of Lebesgues Dominated Convergence Theorem in the interactive theorem prover Matita. Journal of Formalized Reasoning 1(1), 51–89 (2008)
10. Tenenbaum, G., Mendès France, M.: The Prime Numbers and Their Distribution. Student Mathematical Library. American Mathematical Society (2000)
11. Théry, L.: Proving Pearl: Knuth's Algorithm for Prime Numbers. In: Basin, D., Wolff, B. (eds.) TPHOLs 2003. LNCS, vol. 2758, pp. 304–318. Springer, Heidelberg (2003)

A New Elimination Rule for the Calculus of Inductive Constructions

Bruno Barras[1], Pierre Corbineau[2], Benjamin Grégoire[3], Hugo Herbelin[1], and Jorge Luis Sacchini[3]

[1] INRIA Saclay – Île-de-France
{Bruno.Barras,Hugo.Herbelin}@inria.fr
[2] Université Joseph Fourier, INPG, CNRS
Pierre.Corbineau@imag.fr
[3] INRIA Sophia Antipolis – Méditerranée
{Benjamin.Gregoire,Jorge-Luis.Sacchini}@inria.fr

Abstract. In Type Theory, definition by dependently-typed case analysis can be expressed by means of a set of equations — the semantic approach — or by an explicit pattern-matching construction — the syntactic approach. We aim at putting together the best of both approaches by extending the pattern-matching construction found in the Coq proof assistant in order to obtain the expressivity and flexibility of equation-based case analysis while remaining in a syntax-based setting, thus making dependently-typed programming more tractable in the Coq system. We provide a new rule that permits the omission of impossible cases, handles the propagation of inversion constraints, and allows to derive Streicher's K axiom. We show that subject reduction holds, and sketch a proof of relative consistency.

1 Introduction

The Calculus of Inductive Constructions (CIC) [13,10] is an extension of the Calculus of Constructions with inductive types and universes. Inductive types can be added to the system by specifying their constructors (introduction rules). To reason about inductive types, CIC includes a mechanism for performing pattern matching. It allows to define a function on an inductive type by giving computation rules for the constructors, in a similar way as in functional programming languages, such as Haskell or ML.

It is well known that dependent types add a new dimension to the pattern matching mechanism. This was first observed by Coquand [2], and later studied by other authors [5,7,3,8]. A simple example is provided by the definition of lists indexed with their length, which we call here vectors. In CIC, given a type X, vectors are introduced by a constant vector of type $\mathsf{nat} \to \mathsf{Type}$, where $\mathsf{vector}\, n$ represents lists of n elements of type X. The constructors are $\mathsf{nil} : \mathsf{vector}\, 0$ for the empty vector, and $\mathsf{cons} : \Pi(n : \mathsf{nat}).X \to \mathsf{vector}\, n \to \mathsf{vector}\, (\mathsf{S}\, n)$ for adding an element to a vector. One of the slogans of using inductive families and dependently typed languages is the fact that functions can be given a more

S. Berardi, F. Damiani, and U. de'Liguoro (Eds.): TYPES 2008, LNCS 5497, pp. 32–48, 2009.

precise typing. The usual tail function, that removes the first element of a non-empty vector can be given the type $\Pi(n : \mathsf{nat}).\mathsf{vector}\,(\mathsf{S}\,n) \to \mathsf{vector}\,n$, thus ensuring that it cannot be applied to an empty vector. In Coquand's setting, we could write the tail function as

$$\mathsf{tail}\,n\,(\mathsf{cons}\,k\,x\,t) = t\ .$$

Note the missing case for nil. This definition is accepted because the type system can ensure that the vector argument, being a term of type $\mathsf{vector}\,(\mathsf{S}\,n)$, cannot reduce to nil.

In CIC, the direct translation of the above definition is rejected, because of the missing case. Instead, we are forced to make an explicit proof that the nil case is not necessary. This makes the function more difficult to write by hand, and the reasoning necessary to rule out impossible cases hinders the intended computational rules. As a consequence, CIC is not well suited to be the basis for a programming language with dependent types.

Our objective is to adapt the work that has been done in dependent pattern matching to the CIC framework, thus reducing the gap between current implementations of CIC, such as Coq [1], and programming languages such as Epigram [6,7] and Agda [8] — at least, in terms of programming facilities. In particular, we propose a new rule for pattern matching that automatically handles the reasoning steps mentioned above (Sect. 4). The new rule, which allows the user to write more direct and more efficient functions, combines explicit restriction of pattern-matching to inductive subfamilies, (as independently investigated by the second author for deriving axiom K and by the third and fifth authors for simulating Epigram in Coq without computational penalty) and translation of unification constraints into local definitions of the typing context (as investigated by the first and fourth authors). At the end, we prove that the type system satisfies subject reduction and outline a proof of relative consistency (Sect. 5).

2 A Primer on Pattern Matching in CIC

In this section, we study in detail how to write functions by pattern matching in CIC. The presentation is intentionally informal because we want to give some intuition on the problem at hand, and our proposed solution.

Let us consider the definition of tail. The naive solution is to write $\mathsf{tail}\,n\,v$ as

$$\texttt{match}\ v\ \texttt{with}\ |\ \mathsf{nil} \Rightarrow\ ?\ |\ \mathsf{cons}\ k\ x\ t \Rightarrow t\ .$$

There are two problems with this definition. The first is that we need to complete the nil branch with a term explicitly ruling out this case. The second is that the body of the cons branch is not well-typed, since we are supposed to return a term of type $\mathsf{vector}\,n$, while t has type $\mathsf{vector}\,k$. Let us see how to solve them.

For the first problem, it should be possible to reason by absurdity: if v is a non-empty vector (as evidenced by its type), it cannot be nil. More specifically, we reason on the indices of the inductive families, and the fact that the indices can determine which constructors were used to build the term (the inversion principle). In this case, v has type $\mathsf{vector}\,(\mathsf{S}\,n)$, while nil has type $\mathsf{vector}\,0$. Since distinct constructors build distinct objects (the "no confusion" property), we can prove that $0 \neq \mathsf{S}\,n$, and, as a consequence, v cannot reduce to nil. This is translated to the definition of tail by generalizing the type of each branch to include a proof of equality between the indices. The definition of tail looks something like

$$\begin{aligned}\texttt{match}\ v\ \texttt{with}\ &|\ \mathsf{nil} \Rightarrow \lambda(H : 0 = \mathsf{S}\,n).\ \textit{here a proof of contradiction from } H\\ &|\ \mathsf{cons}\,k\,x\,t \Rightarrow \lambda(H : \mathsf{S}\,k = \mathsf{S}\,n).t,\end{aligned}$$

where, in the nil branch, we reason by absurdity from the hypothesis H.

We have solved the first problem, but we still suffer the second. Luckily, the same generalization argument used for the nil branch provides a way out. Note that, in the cons branch, we now have a new hypothesis H of type $\mathsf{S}\,k = \mathsf{S}\,n$. From it, we can prove that $k = n$, since the constructor S is injective (again, the no-confusion property). Then, we can use the obtained equality to build, from t, a term of type $\mathsf{vector}\,n$. In the end, the body of this branch is a term built from H and t that *changes* the type of t from $\mathsf{vector}\,k$ to $\mathsf{vector}\,n$.

This solves both problems, but the type of the function obtained is $\mathsf{S}\,n = \mathsf{S}\,n \to \mathsf{vector}\,n$, which is not the desired one yet. So, all we need to do is just to apply the function to a trivial proof of equality for $\mathsf{S}\,n = \mathsf{S}\,n$.

It is important to notice that this function, as defined above, still has the desired computational behavior: given a term $v = \mathsf{cons}\,n\,h\,t$, we have $\mathsf{tail}\,n\,v \to^+ t$. In particular, in the body of the cons branch, the extra equational burden necessary to change the type of t collapses to the identity. However, the definition is clouded with equational reasoning expressions that do not relate to the computational behavior of the function, but are necessary to convince the typechecker that the function does not compromise the type correctness of the system.

Our proposition is a new rule for pattern matching that allows to write dependent pattern matching in a direct way, avoiding pollution of the underlying program with proofs of equality statements and confining the justifications of the correctness of the dependencies to the typing rules. We would then be able to write the tail function as

$$\mathsf{tail} := \lambda(n : \mathsf{nat})(v : \mathsf{vector}\,(\mathsf{S}\,n)).\texttt{match}\ v\ \texttt{with}\ |\ \mathsf{cons}\,k\,x\,t \Rightarrow t\ \texttt{where}\ k := n,$$

where some constructors are omitted (like nil above), and for the present constructors, some additional information is given (like $k := n$ above). The typing rules justify that the nil case is not necessary, and that the definition $k := n$ is valid to use in the typing of the cons branch.

In the general case, checking whether a pattern-matching branch is useless is undecidable [2,11,7,9]. To remain in a decidable framework, we propose to only address the detection of clauses whose inaccessibility is provable using a simple evidence based on first-order unification of the inductive structure of the indices. The idea is to generate, for each constructor in a pattern matching definition, a set of equations between the indices of the inductive type in question, in the same way as shown at the beginning of this section. The goal is then to find a unification substitution for these equations. In the case of tail, the unification for nil fails, while the unification for cons succeeds. This approach is based on work by McBride and McKinna [7] and Norell [8], and is described in detail in Sect. 4.

3 The Calculus of Inductive Constructions

In this section, we give a (necessarily) short description of CIC, specially focusing on inductive types and pattern matching.

The sorts of CIC are $\mathtt{Set}$, $\mathtt{Prop}$ and $\mathtt{Type}_i$, for $i \in \mathbb{N}$. The terms are variables, λ-abstractions $\lambda x : T.M$, applications $M\,N$, products $\Pi x : T.U$ (we write $T \to U$ if x is not used in U), local definitions[1] $[x := N : T]\,M$, and constructions related to inductive types that are described below. We use $\mathrm{FV}(M)$ to denote the set of free variables of M, and $M[x := N]$ to denote the term obtained by substituting every free occurrence of x in M with N.

A context is a sequence of *declarations*, i.e., *assumptions* of the form $(x : T)$ or *definitions* of the form $(x := M : T)$; the empty context is denoted by $[]$. We use $\mathcal{D}om\,(\Gamma)$ to denote the ordered sequence of variables declared in Γ.

We use $m, n, k, p, q, t, u, v, A, B, M, N, P, T, U, \ldots$ to denote terms, $x, y, z, \ldots$ to denote variables, $\Gamma, \Delta, \Theta, \ldots$ to denote contexts and the letter s and its variants to denote sorts. We use $\boldsymbol{X}$ to denote a sequence of X, ε to denote the empty sequence, and $\#\,(\boldsymbol{X})$ to denote the length of the sequence $\boldsymbol{X}$. We use de Bruijn telescopes: the notation $\Pi\Delta.T$ (resp. $\lambda\Delta.T$) abbreviates the iterated expansion of the declarations in Δ into products (resp. abstractions) or local definitions.

CIC comes equipped with a notion of convertibility between terms, written $\Gamma \vdash T \approx U$ and a notion of subtyping, written $\Gamma \vdash T \leq U$.

We consider two typing judgments:

- $\Gamma \vdash t : T$ means that, under context Γ, the term t has type T;
- $\Gamma \vdash \boldsymbol{t} : \Delta$ means that, under context Γ, the terms $\boldsymbol{t}$ form an instance of Δ.[2]

Inductive Types. Terms of CIC also include names of inductive types, names of constructors, fixpoint declarations $\mathtt{fix}_n\, f : T := M$, and pattern matching $\mathtt{match}\ M\ \mathtt{as}\ x\ \mathtt{in}\ I\,\boldsymbol{p}\,\boldsymbol{y}\ \mathtt{return}\ P\ \mathtt{with}\ \{C_i\,\boldsymbol{z_i} \Rightarrow t_i\}_i$. We use the letter I and its variants to denote inductive types, and C to denote constructors.

Inductive definitions are declared in a *signature*. A signature Σ is a sequence of declarations of the form

$$\mathsf{Ind}(I[\Delta_p] : \Pi\Delta_a.s := \{C_i : \Pi\Delta_i.I\,\mathcal{D}om\,(\Delta_p)\ \boldsymbol{u_i}\}_i)$$

where I is the name of the inductive type, Δ_p is the context of its parameters, Δ_a is the context of its indices, s is a sort denoting the universe where the type is

[1] Local definitions are not part of the usual definition of CIC. We have included them here because they play an important part in the typing of pattern matching.

[2] In particular, if $f : \Pi\Delta.T$ then $f\ \boldsymbol{t}$ is well-typed.

defined. In the general case, due to the dependency over parameters and indices, I is an inductive family. The type of I is then $\Pi\Delta_p\Delta_a.s$. To the right of the $:=$ symbol are names and types of the constructor. We assume in the sequel that all typing rules are parameterized by a fixed signature Σ.

We describe in detail the pattern matching mechanism. In a term of the form $(\mathtt{match}\, M\, \mathtt{as}\, x\, \mathtt{in}\, I\, \boldsymbol{p}\, \boldsymbol{y}\, \mathtt{return}\, P\, \mathtt{with}\, \{C_i\, \boldsymbol{x_i} \Rightarrow t_i\}_i)$, M is the term to destruct, P is the return type (which depends on $\boldsymbol{y}$ and x), and t_i represent the body of the i-th branch, with $\boldsymbol{x_i}$ the arguments of the i-th constructor, bound in t_i. The reduction rule associated, denoted ι, is

$$(\mathtt{match}\, C_j\, \boldsymbol{t}\, \boldsymbol{u}\, \mathtt{as}\, x\, \mathtt{in}\, I\, \boldsymbol{p}\, \boldsymbol{y}\, \mathtt{return}\, P\, \mathtt{with}\, \{C_i\, \boldsymbol{x_i} \Rightarrow t_i\}_i) \rightarrow_\iota t_j[\boldsymbol{x_j} := \boldsymbol{u}],$$

where, by typing invariants, $\boldsymbol{t} \approx \boldsymbol{p}$ and $\#(\boldsymbol{x_j}) = \#(\boldsymbol{u})$.

Figure 1 shows the typing rules of pattern matching in CIC, where $\Delta_i^* = \Delta_i[\mathcal{D}om\,(\Delta_p) := \boldsymbol{p}]$, $\boldsymbol{u_i^*} = \boldsymbol{u_i}[\mathcal{D}om\,(\Delta_p) := \boldsymbol{p}]$, and $\Delta_a^* = \Delta_a[\mathcal{D}om\,(\Delta_p) := \boldsymbol{p}]$. To be accepted, a `match` construction has to satisfy the predicate $\textsc{Elim}(I, s)$ that restricts the class of objects that can be constructed using pattern matching; the exact definition of $\textsc{Elim}$ is not important in our context (see, e.g., [10]).

The typing rule for pattern matching is complicated by the fact that the return type can depend on the term being destructed. When P does not depend on x nor $\boldsymbol{y}$, i.e. they are not used in P, then the typing reduces to something close to non-dependent languages like Haskell or ML. But P can depend on x (of type an instance of I), and, therefore, it should also depend on the indices of the type of x (i.e. $\boldsymbol{y}$). In each branch, we instantiate both x with the corresponding constructor applied to the arguments of the branch, and $\boldsymbol{y}$ with the indices of the inductive type corresponding to that constructor. Finally, the type of the whole `match` is obtained by replacing x with M in P, and $\boldsymbol{y}$ accordingly.

$$\frac{\begin{array}{c}\mathsf{Ind}(I[\Delta_p] : \Pi\Delta_a.s := \{C_i : \Pi\Delta_i.I\,\mathcal{D}om\,(\Delta_p)\,\boldsymbol{u_i}\}_i) \in \Sigma \\ \Gamma \vdash M : I\,\boldsymbol{p}\,\boldsymbol{u} \qquad \Gamma(\boldsymbol{y} : \Delta_a^*)(x : I\,\boldsymbol{p}\,\boldsymbol{y}) \vdash P : s \\ \textsc{Elim}(I, s) \qquad \Gamma(\boldsymbol{x_i} : \Delta_i^*) \vdash t_i : P[\boldsymbol{y} := \boldsymbol{u_i^*}][x := C_i\,\boldsymbol{p}\,\boldsymbol{x_i}]\end{array}}{\Gamma \vdash \mathtt{match}\; M\; \mathtt{as}\; x\; \mathtt{in}\; I\,\boldsymbol{p}\,\boldsymbol{y}\; \mathtt{return}\; P\; \mathtt{with}\; \{C_i\,\boldsymbol{x_i} \Rightarrow t_i\}_i : P[\boldsymbol{y} := \boldsymbol{u}][x := M]}$$

Fig. 1. Typing rules for pattern matching in CIC

4 A New Elimination Rule

In this section we present the new rule for pattern matching. We modify the syntax of terms with the construction

$$\mathtt{match}\; M\; \mathtt{as}\; x\; \mathtt{in}\; [\Delta]\, I\,\boldsymbol{p}\,\boldsymbol{t}\; \mathtt{where}\; \Delta := \boldsymbol{q}\; \mathtt{return}\; P\; \mathtt{with}\; \{C_i\,\boldsymbol{x_i} \Rightarrow b_i\}_i$$

where the body of a branch (b_i above) can be either the symbol $\bot$ or a term of the form N `where` $\boldsymbol{d}$, with $\boldsymbol{d}$ a sequence composed of variables and variable definitions (e.g. $x := N$). We write $\mathcal{D}om\,(\boldsymbol{d})$ for the set of variables that are declared in $\boldsymbol{d}$, and refer to $\boldsymbol{d}$ as the *definitions* of the branch.

The rôle of $[\Delta]\, I\, \boldsymbol{p}\, \boldsymbol{t}$ is to characterize the subfamily of I, with parameters in Δ, over which the pattern matching is done. Some constructors may not belong to that subfamily, so the body of the corresponding branches is simply $\bot$. On the other hand, some constructors may (partially) belong to the subfamily, so the bodies of the corresponding branches are of the form N `where` $\boldsymbol{d}$, where N is the body proper and $\boldsymbol{d}$ defines some restrictions on the arguments of the constructor that need to be satisfied in order to belong to the subfamily.

Before showing the typing rule for this new construction, that is more complex than the one presented in the previous section, we show how to write our running example, the tail function:

$$\begin{array}{l} \mathsf{tail} := \lambda(n : \mathsf{nat})(v : \mathsf{vector}\,(\mathsf{S}\, n)). \\ \quad \mathtt{match}\ v\ \mathtt{as}\ x\ \mathtt{in}\ [(n_0 : \mathsf{nat})]\, \mathsf{vector}\,(\mathsf{S}\, n_0)\ \mathtt{where}\ n_0 := n\ \mathtt{return}\ \mathsf{vector}\, n_0 \\ \quad \mathtt{with}\ \mid \mathsf{nil} \Rightarrow \bot \mid \mathsf{cons}\, k\, x\, v' \Rightarrow v'\ \mathtt{where}\ n_0 := k \end{array}$$

Comparing with the generic term above, M is v, Δ is $(n_0 : \mathsf{nat})$, and $\boldsymbol{q}$ is n. We explain how to check that this definition is accepted. Note that we are targeting a particular subfamily of the inductive type, $[(n_0 : \mathsf{nat})]\, \mathsf{vector}\,(\mathsf{S}\, n_0)$, parametrized by n_0, which is the subfamily of non-empty vectors. We need to make sure that v belongs to that family. In the general case, this means instantiating Δ with $\boldsymbol{q}$ and checking that M has type $I\, \boldsymbol{p}\, (\boldsymbol{t}[\mathcal{D}om\,(\Delta) := \boldsymbol{q}])$. In the particular case of tail, we check that v has type $\mathsf{vector}\,(\mathsf{S}\, n_0)[n_0 := n]$.

Let us look at the return type. In the general case, the return type P depends on x and Δ. Hence, we need to check that

$$\Gamma \Delta (x : I\, \boldsymbol{p}\, \boldsymbol{t}) \vdash P : s,$$

where Γ is the context where we are typing the whole `match`. In the particular case of tail, the return type depends on n_0 but not on x.

Finally, we look at the branches. In the nil case, it is clear that this constructor does not belong to the subfamily $[(n_0 : \mathsf{nat})]\, \mathsf{vector}\,(\mathsf{S}\, n_0)$, since its type is $\mathsf{vector}\, 0$. Hence, the branch is simply $\bot$. We call this type of branch *impossible*. How do we check that a branch is impossible? We try to unify the indices of the subfamily under consideration ($\mathsf{S}\, n_0$ in this case) with the indices of the constructor (0 in this case), for the variables in $\mathcal{D}om\,(\Delta)$. As we said, this problem is undecidable in general, so we proceed by first-order unification with constructor theory. In this case, $\mathsf{S}\, n_0$ and 0 are not unifiable (constructors are disjoint), and therefore, the branch is effectively impossible.

In the cons case, we have the body proper v' and the definition $n_0 := k$. The return type of this branch should be $\mathsf{vector}\, n_0$, where x is replaced by $\mathsf{cons}\, k\, x\, v'$, and n_0 is replaced by k as dictated by the definition. To check this kind of branch, we need to check that the given definition is correct. This is done, as for impossible branches, by unification. In this case, unifying $\mathsf{S}\, n_0$ with $\mathsf{S}\, k$ (this last value corresponds to the index of $\mathsf{cons}\, k\, x\, v'$), for n_0. Since S is injective, the result is the substitution $\{n_0 \mapsto k\}$. If the unification succeeds, we apply the substitution obtained in the return type. In the case of the cons branch, its type should be $\mathsf{vector}\,(\mathsf{S}\,n_0)[n_0 := k]$. As we will see below, parts of the `where` clauses can be omitted. In this particular case, the entire clause can be omitted.

Note that the procedure to check branches is similar for both kinds. First, we unify the indices of the subfamily with the indices of the constructor. If they are not unifiable, the branch is impossible. If the unification succeeds, we obtain a substitution σ that is applied to the return type in order to check the body proper of the branch. There is one third possibility, though. Since the unification problem is undecidable, it is possible that the procedure gets stuck. In that case, we simply give up, and the typechecking fails.

We now proceed to explain formally the typing rule for this `match`. The presentation is divided into three parts: first, we describe substitutions, then the unification judgment, and finally, we show the typechecking of branches and put everything together. Then, we discuss the associated reduction rule.

Substitutions. A *pre-substitution* is a function σ, from variables to terms, such that $\sigma(x) \neq x$ for a finite number of variables x. The set of variables for which $\sigma(x) \neq x$ is the *domain* of σ and is denoted $\mathcal{D}om\,(\sigma)$. Given a term t and a substitution σ, we write $t\sigma$ to mean the term obtained by substituting every free variable x of t with $\sigma(x)$. The set of free variables of a pre-substitution σ is defined as $\mathrm{FV}(\sigma) = \cup_{x \in \mathcal{D}om(\sigma)} \mathrm{FV}(x\sigma)$.

A *substitution* σ from Γ to Δ, denoted $\sigma : \Gamma \to \Delta$ is a pre-substitution with $\mathcal{D}om\,(\sigma) \subseteq \Gamma$, idempotent (i.e., $\mathrm{FV}(\sigma) \cap \mathcal{D}om\,(\sigma) = \emptyset$), such that for every $(x : T) \in \Gamma$, $\Delta \vdash x\sigma : T\sigma$, and for every $(x := t : T) \in \Gamma$, $\Delta \vdash x\sigma \approx t\sigma : T\sigma$.[3] We use $\sigma, \rho, \ldots$ to denote (pre-)substitutions. We sometimes write $\Gamma \vdash \sigma : \Delta \to \Theta$ to denote a substitution $\sigma : \Gamma\Delta \to \Gamma\Theta$, with $\mathcal{D}om\,(\sigma) \subseteq \mathcal{D}om\,(\Delta)$.

Substitutions can be composed: if $\sigma : \Gamma \to \Delta$ and $\rho : \Delta \to \Theta$, then $\sigma\rho : \Gamma \to \Theta$, where $(\sigma\rho)(x) = (x\sigma)\rho$. Two substitutions $\sigma, \rho : \Gamma \to \Delta$ are convertible, written $\sigma \approx \rho$, if for every $x \in \mathcal{D}om\,(\Gamma)$, $\Delta \vdash x\sigma \approx x\rho$.

Unification. We now describe in detail the unification judgment. A unification problem is written

$$\Gamma; \Delta, \zeta \vdash [\boldsymbol{u} = \boldsymbol{u}' : \Theta],$$

meaning that $\boldsymbol{u}$ and $\boldsymbol{u}'$ have type Θ under context $\Gamma\Delta$, and $\zeta \subseteq \mathcal{D}om\,(\Delta)$ is the set of variables that are open to unification. Context Γ is intended to be the "outer context", i.e. the context where we want to type a `match` construction, while context Δ is defined inside the `match`. We only allow to unify variables in Δ, so that the unification is invariant under substitutions and reductions that happen outside the `match`. This is important in the proofs of the Substitution Lemma and Subject Reduction.

The unification judgment is defined by the rules of Fig. 2. These rules are based on the unification given in [7,8], with a notation close to that in [8]. Trying to unify $\boldsymbol{u}$ and $\boldsymbol{u}'$ may have one of three possible outcomes:

positive success. A derivation $\Gamma; \Delta, \zeta \vdash [\boldsymbol{u} = \boldsymbol{u}' : \Theta] \mapsto \Delta', \zeta' \vdash \sigma$ is obtained, meaning that σ is a substitution ($\Gamma \vdash \sigma : \Delta \to \Delta'$) that unifies $\boldsymbol{u}$ and $\boldsymbol{u}'$ with domain $\zeta \setminus \zeta'$ and ζ' is the set of variables that are still open to unification;

[3] The judgment $\Gamma \vdash t \approx u : T$ is shorthand for $\Gamma \vdash t : T$, $\Gamma \vdash u : T$ and $\Gamma \vdash t \approx u$.

negative success. A derivation $\Gamma; \Delta, \zeta \vdash [\boldsymbol{u} = \boldsymbol{u'} : \Theta] \mapsto \bot$ is obtained, meaning that $\boldsymbol{u}$ and $\boldsymbol{u'}$ are not unifiable;

failure. No rule is applicable, hence no derivation is obtained (the unification problem is too difficult).

In the rules (U-VarL) and (U-VarR), a reordering of the context Δ may be required in order to obtain a (well-typed) substitution. This is achieved by the (partial) operation $\Delta_{\Gamma|x:=t}$ defined as

$$\begin{aligned}(\Delta_0(x:T)\Delta_1)_{\Gamma|x:=t} = \Delta_0\Delta^t(x := t : T)\Delta_t& \\ \text{where } (\Delta^t, \Delta_t) = \textsc{Strengthen}(\Delta_1, t)& \\ \Gamma\Delta_0\Delta^t \vdash t : T& \\ \Gamma\Delta_0\Delta^t(x:T) \vdash \Delta_t&\end{aligned}$$

The Strengthen operation [7] is defined as

$$\textsc{Strengthen}([], t) = ([], [])$$

$$\textsc{Strengthen}((x:U)\Delta, t) = \begin{cases} ((x:U)\Delta_0, \Delta_1) & \text{if } x \in \mathrm{FV}(\Delta_0) \cup \mathrm{FV}(t) \\ (\Delta_0, (x:U)\Delta_1) & \text{if } x \notin \mathrm{FV}(\Delta_0) \cup \mathrm{FV}(t) \end{cases}$$
$$\text{where } (\Delta_0, \Delta_1) = \textsc{Strengthen}(\Delta, t)$$

We give some informal explanations of the unification rules. Rules (U-VarL) and (U-VarR) are the basic rules, concerning the unification of a variable with a term. As a precondition, the variable must be a variable open to unification (i.e., it must belong to ζ) and the equation must not be circular (i.e., x does not belong to the set $\mathrm{FV}(v)$), although this last condition is also ensured by the operation $\Delta_{\Gamma|x:=v}$.

Rules (U-Discr) and (U-Inj) codify the no-confusion property of inductive types: rule (U-Discr) states that constructors are disjoint (negative success), while rule (U-Inj) states that constructors are injective.

If the first four rules are not applicable, then the unification can succeed only if the terms are convertible. This is shown in rule (U-Conv). Finally, rules (U-Empty) and (U-Tel) concern the unification of sequence of terms. Missing from Fig. 2 are the corresponding rules to (U-Inj) and (U-Tel) that propagate a negative unification (i.e., $\bot$).

The typing rule. In Fig. 3 we show the typing rule for the new elimination rule, and introduce a new judgment for typechecking branches. This new judgment has the form

$$\Gamma; \Delta_i; \Delta; [\boldsymbol{u} = \boldsymbol{v} : \Theta] \vdash b : T$$

The intuition is that we take the unification problem $\Gamma; \Delta_i\Delta, \zeta \vdash [\boldsymbol{u} = \boldsymbol{v} : \Theta]$ (where ζ depends on the kind of branch considered), and take the result of the unification into account while checking the body of the branch. This judgment is defined by the rules (B-$\bot$) and (B-Sub) in Fig. 3. In rule (B-$\bot$),

$$\text{(U-VarL)}\quad \frac{x \in \zeta \qquad x \notin \mathrm{FV}(v)}{\Gamma; \Delta, \zeta \vdash [x = v : T] \mapsto \Delta_{\Gamma|x:=v}, \zeta \setminus \{x\} \vdash \{x \mapsto v\}}$$

$$\text{(U-VarR)}\quad \frac{x \in \zeta \qquad x \notin \mathrm{FV}(v)}{\Gamma; \Delta, \zeta \vdash [v = x : T] \mapsto \Delta_{\Gamma|x:=v}, \zeta \setminus \{x\} \vdash \{x \mapsto v\}}$$

$$\text{(U-Discr)}\quad \frac{C_1 \neq C_2}{\Gamma; \Delta, \zeta \vdash [C_1\, \boldsymbol{u} = C_2\, \boldsymbol{v} : T] \mapsto \bot}$$

$$\text{(U-Inj)}\quad \frac{\Gamma; \Delta, \zeta \vdash [\boldsymbol{u} = \boldsymbol{v} : \Theta] \mapsto \Delta', \zeta' \vdash \sigma \qquad \mathsf{Type}(C) = \Pi\Theta.I\, \boldsymbol{p}\, \boldsymbol{t}}{\Gamma; \Delta, \zeta \vdash [C\, \boldsymbol{u} = C\, \boldsymbol{v} : T] \mapsto \Delta', \zeta' \vdash \sigma}$$

$$\text{(U-Conv)}\quad \frac{\Gamma\Delta \vdash u \approx v}{\Gamma; \Delta, \zeta \vdash [u = v : T] \mapsto \Delta, \zeta \vdash id}$$

$$\text{(U-Empty)}\quad \frac{}{\Gamma; \Delta, \zeta \vdash [\varepsilon = \varepsilon : []] \mapsto \Delta, \zeta \vdash id}$$

$$\text{(U-Tel)}\quad \frac{\begin{array}{c}\Gamma; \Delta, \zeta \vdash [u = v : T] \mapsto \Delta_1, \zeta_1 \vdash \sigma_1 \\ \Gamma; \Delta_1, \zeta_1 \vdash [\boldsymbol{u}\sigma_1 = \boldsymbol{v}\sigma_1 : \Theta[x := u]\sigma_1] \mapsto \Delta_2, \zeta_2 \vdash \sigma_2\end{array}}{\Gamma; \Delta, \zeta \vdash [u\,\boldsymbol{u} = v\,\boldsymbol{v} : (x : T)\Theta] \mapsto \Delta_2, \zeta_2 \vdash \sigma_1\sigma_2}$$

Fig. 2. Unification rules

that corresponds to impossible branches, we take ζ to be $\mathcal{D}om\,(\Delta_i) \cup \mathcal{D}om\,(\Delta)$, and we check that the unification succeeds negatively.

In rule (B-Sub), that corresponds to possible branches, we take ζ to be $\mathcal{D}om\,(\Delta)$ together with the domain of the definitions of the branch ($\boldsymbol{d}$ in this case). Context Δ corresponds to the variables that define the subfamily under analysis. We check that the unification succeeds positively, leaving no variables open. We also check that the definitions $\boldsymbol{d}$ are valid using the judgment $\Gamma\Delta' \vdash \boldsymbol{d} \leq \sigma$; this means that, for every variable definition $(x := N)$ of $\boldsymbol{d}$, we have $\Gamma\Delta' \vdash N \approx x\sigma$. Then, we typecheck the body proper of the branch using the context given by the unification (Δ').

Finally, in the rule (T-Match) we put everything together. We have $\Delta_i^* = \Delta_i[\mathcal{D}om\,(\Delta_p) := \boldsymbol{p}]$, $\boldsymbol{u}_i^* = \boldsymbol{u}_i[\mathcal{D}om\,(\Delta_p) := \boldsymbol{p}]$, and $\Delta_a^* = \Delta_a[\mathcal{D}om\,(\Delta_p) := \boldsymbol{p}]$.

The subfamily under analysis is defined by $[\Delta]\, I\, \boldsymbol{p}\, \boldsymbol{t}$, hence, we check that M belongs to it by checking that $\Gamma \vdash \boldsymbol{u} \approx \boldsymbol{t}[\Delta := \boldsymbol{q}]$. We also check that $\boldsymbol{q}$ has the correct type; and also that P is a type. The return type P depends on x and $\mathcal{D}om\,(\Delta)$, similarly to the old rule, where P depended on x and $\boldsymbol{y}$ (indices of the inductive type). In the branches, as in the old rule, x is replaced by the corresponding constructor applied to the arguments. Here, in contrast with the old rule where it was clear how to replace $\boldsymbol{y}$, there are no obvious values we can give to the variables in $\mathcal{D}om\,(\Delta)$. Therefore, we try the unification between $\boldsymbol{t}$ (that defines the subfamily under analysis) and $\boldsymbol{u_i}$ (the indices in the type of x). Since, for possible branches, the unification does not leave open variables, we effectively find a value for each variable in $\mathcal{D}om\,(\Delta)$.

Reduction. The reduction rule is the same as the original elimination rule of CIC, except that it is only applicable to possible constructors.

$$(\text{B-}\bot)\ \frac{\Gamma; \Delta_i\Delta, \mathcal{D}om\,(\Delta_i) \cup \mathcal{D}om\,(\Delta) \vdash [\boldsymbol{u} = \boldsymbol{v} : \Theta] \mapsto \bot}{\Gamma; \Delta_i; \Delta; [\boldsymbol{u} = \boldsymbol{v} : \Theta] \vdash \bot : P}$$

$$(\text{B-Sub})\ \frac{\begin{array}{c}\Gamma; \Delta_i\Delta, \mathcal{D}om\,(\boldsymbol{d}) \cup \mathcal{D}om\,(\Delta) \vdash [\boldsymbol{u} = \boldsymbol{v} : \Theta] \mapsto \Delta', \emptyset \vdash \sigma \\ \Gamma\Delta' \vdash t : P \qquad \Gamma\Delta' \vdash \boldsymbol{d} \leq \sigma\end{array}}{\Gamma; \Delta_i; \Delta; [\boldsymbol{u} = \boldsymbol{v} : \Theta] \vdash t \texttt{ where } \boldsymbol{d} : P}$$

$$(\text{T-Match})\ \frac{\begin{array}{c}\mathsf{Ind}(I[\Delta_p] : \Pi\Delta_a.s := \{C_i : \Pi\Delta_i.I\,\mathcal{D}om\,(\Delta_p)\ \boldsymbol{u_i}\}_i) \in \Sigma \\ \Gamma \vdash M : I\,\boldsymbol{p}\,\boldsymbol{u} \qquad \Gamma \vdash \boldsymbol{u} \approx \boldsymbol{t}[\Delta := \boldsymbol{q}] \qquad \Gamma\Delta(x : I\,\boldsymbol{p}\,\boldsymbol{t}) \vdash P : s \\ \Gamma \vdash \boldsymbol{q} : \Delta \qquad \Gamma; (\boldsymbol{z_i} : \Delta_i^*); \Delta; [\boldsymbol{u_i^*} = \boldsymbol{t} : \Delta_a^*] \vdash b_i : P[x := C_i\,\boldsymbol{p}\,\boldsymbol{z_i}]\end{array}}{\Gamma \vdash \begin{pmatrix}\texttt{match } M \texttt{ as } x \texttt{ in } [\Delta]\,I\,\boldsymbol{p}\,\boldsymbol{t} \texttt{ where } \Delta := \boldsymbol{q} \\ \texttt{return } P \texttt{ with } \{C_i\ \boldsymbol{z_i} \Rightarrow b_i\}_i\end{pmatrix} : P[\Delta := \boldsymbol{q}][x := M]}$$

Fig. 3. Typing rules for the new elimination rule

$$(\texttt{match } C_j\,\boldsymbol{t}\,\boldsymbol{u} \texttt{ as } x \texttt{ in } I\,\boldsymbol{p}\,\boldsymbol{y} \ldots \texttt{with } \{C_i\,\boldsymbol{x_i} \Rightarrow b_i\}_i) \ \to_\iota\ t_j[\boldsymbol{x_j} := \boldsymbol{u}]$$

where $b_j = (t_j \texttt{ where } \sigma_j)$, $\#\,(\boldsymbol{t}) = \#\,(\boldsymbol{p})$ and $\#\,(\boldsymbol{x_i}) = \#\,(\boldsymbol{u})$.

In the compatible closure of the reduction, we do not use the definitions of each branch. Hence, we have the rule

$$\frac{\Gamma(\boldsymbol{z_i} : \Delta_i) \vdash t_j \to t_j'}{\Gamma \vdash \texttt{match} \ldots C\,\boldsymbol{z_i} \Rightarrow t_j \texttt{ where } \boldsymbol{d_j} \to \texttt{match} \ldots C\,\boldsymbol{z_i} \Rightarrow t_j' \texttt{ where } \boldsymbol{d_j}}$$

Allowing the definitions as part of the context when reducing the body of the branch t_j would mean to break confluence on pseudoterms (although, in that case, confluence remains valid for well-typed terms).

Remark 1. In a branch of the form N **where** $\boldsymbol{d}$, only $\mathcal{D}om\,(\boldsymbol{d})$ is needed to compute the unification. The defined values of $\boldsymbol{d}$, if there are any, are checked to be valid with respect to the substitution given by the unification with the judgment $\boldsymbol{d} \leq \sigma$. This is similar to the situation of *inaccessible patterns* in [3,8].

In the examples below, for the sake of readability, we sometimes omit definitions that are inferred by the unification; in some other cases, the definitions are "inlined" in the arguments of a constructor.

Remark 2. Note that the usual rule for pattern matching in CIC is a special case of the new rule: we just set Δ to be the context of indices of the inductive type, i.e. Δ_a, and $\boldsymbol{t}$ to be $\mathcal{D}om\,(\Delta)$. It is not difficult to see that the unification succeeds positively for each branch.

4.1 Examples

We illustrate the new elimination rule with some examples. We have already seen how to type the tail function. We show two sets of examples, one about Streicher's K axiom and heterogeneous equality, and the other about the less-or-equal relation on natural numbers. To simplify the syntax, we assume that missing constructors are impossible.

Streicher's K axiom and heterogeneous equality. Axiom K, also known as *uniqueness of reflexivity proofs*, has type

$$\Pi(A : Set)(x : A)(P : \mathsf{eq}\, A\, x\, x \to \mathtt{Prop}).P(\mathsf{refl}\, A\, x) \to \Pi(p : \mathsf{eq}\, A\, x\, x).P\, p \ .$$

It is not derivable in CIC, as shown by Hofmann and Streicher [4]. However, it is no surprise that we can derive it using the new rule:

$$\begin{aligned} K := \lambda(A : \mathtt{Set})(x : A)(P : \mathsf{eq}\, A\, x\, x \to \mathtt{Prop})(H : P(\mathsf{refl}\, A\, x))(p : \mathsf{eq}\, A\, x\, x). \\ \mathtt{match}\ p\ \mathtt{as}\ p_0\ \mathtt{in}\ []\, \mathsf{eq}\, A\, x\, x\ \mathtt{return}\ P\, p_0\ \mathtt{with}\ \mathsf{refl} \Rightarrow H \ . \end{aligned}$$

Note the pattern $[]\,\mathsf{eq}\, A\, x\, x$ in the elimination. We fix the index of eq to be x, therefore p_0 has type $\mathsf{eq}\, A\, x\, x$, i.e. a reflexivity proof, and $P\, p_0$ is well typed. The new rule allows us to restrict the analysis to reflexivity proofs.

In [5], McBride introduced the heterogeneous equality, defined by

$$\mathsf{Ind}(\mathsf{Heq}(A : \mathtt{Set})(x : A) : \Pi(B : \mathtt{Set}).B \to \mathtt{Prop} := \mathsf{Hrefl} : \mathsf{Heq}\, A\, x\, A\, x) \ .$$

Note that the derived induction principle for this equality is not very useful. Therefore, McBride proposed a more conservative elimination rule: only homogeneous equations can be eliminated. The elimination rule used for Heq, denoted by Subst, has type

$$\Pi(A : \mathtt{Set})(x\, y : A)(P : A \to \mathtt{Set}).P\, x \to \mathsf{Heq}\, A\, x\, A\, y \to P\, y \ .$$

In [5] it is shown that this elimination rule is equivalent to axiom K; therefore, it is not derivable in CIC. Using the new rule, we can derive it as:

$$\begin{aligned} \mathsf{Subst} := \lambda(A : \mathtt{Set})(x\, y : A)(P : A \to \mathtt{Set})(M : P\, x)(H : \mathsf{Heq}\, A\, x\, A\, y). \\ \mathtt{match}\ H\ \mathtt{as}\ h_0\ \mathtt{in}\ [(y_0 : A)]\, \mathsf{Heq}\, A\, x\, A\, y_0\ \mathtt{where}\ y_0 := y\ \mathtt{return}\ P\, y_0 \\ \mathtt{with}\ \mathsf{Hrefl} \Rightarrow M \ . \end{aligned}$$

Similarly to axiom K, we use the new rule to restrict the subfamily under analysis; in this case, we restrict to homogeneous equalities, expressed by the pattern $[(y_0 : A)]\, \mathsf{Heq}\, A\, x\, A\, y_0$. Note that the first index of Heq is fixed to be A.

Less-or-equal relation on natural numbers. We show two examples concerning the relation less-or-equal for natural numbers defined inductively by

$$\begin{aligned} \mathsf{Ind}(\mathsf{leq} : \mathsf{nat} \to \mathsf{nat} \to \mathtt{Prop} := \mathsf{leq0} : \Pi(n : \mathsf{nat}).\mathsf{leq}\, 0\, n, \\ \mathsf{leqS} : \Pi(m\, n : \mathsf{nat}).\mathsf{leq}\, m\, n \to \mathsf{leq}\, (\mathsf{S}\, m)\, (\mathsf{S}\, n)) \ . \end{aligned}$$

First, we show that the successor of a number is not less-or-equal than the number itself. That is, we want to find a term of type

$$\Pi(n : \mathsf{nat}).\mathsf{leq}\, (\mathsf{S}\, n)\, n \to \mathsf{False} \ .$$

One possible solution is to take

$$\begin{array}{l}
\texttt{fix}\ f : \Pi(n:\mathsf{nat}).\mathsf{leq}(\mathsf{S}\,n)\,n \to \mathsf{False} := \\
\quad \lambda(n:\mathsf{nat})(H:\mathsf{leq}(\mathsf{S}\,n)\,n). \\
\quad\ \texttt{match}\ H\ \texttt{in}\ [(n_0:\mathsf{nat})]\ \mathsf{leq}(\mathsf{S}\,n_0)\,n_0\ \texttt{where}\ n_0 := n\ \texttt{return}\ \mathsf{False}\ \texttt{with} \\
\quad\ \mid \mathsf{leqS}\,x\,y\,H \Rightarrow f\,y\,H\ \texttt{where}\ (x := \mathsf{S}\,y)(n_0 := \mathsf{S}\,y)
\end{array}$$

In the $\mathsf{leq0}$ branch, the unification problem considered is

$$\{x, n_0\} \vdash [0, x = \mathsf{S}\,n_0, n_0],$$

where x is a fresh variable that stands for the argument of $\mathsf{leq0}$. Clearly, unification succeeds negatively because of the first equation. On the leqS branch, the unification problem is

$$\{x, n_0\} \vdash [\mathsf{S}\,x, \mathsf{S}\,y = \mathsf{S}\,n_0, n_0],$$

which succeeds positively with the substitution $\{x \mapsto \mathsf{S}\,y, n_0 \mapsto \mathsf{S}\,y\}$. Note that the unification gives us the value for n_0 that is necessary for the branch to have the required type, but also finds a relation between the arguments of the constructor x and y. Therefore, the body of the branch is typed in a context containing the declarations

$$(y:\mathsf{nat})(x := \mathsf{S}\,y : \mathsf{nat})(H : \mathsf{leq}\,x\,y)\ .$$

Note the reordering of x and y. In this context, the recursive call to f is well typed.

The second example shows that the relation leq is transitive. That is, we want to find a term of type $\Pi(x\,y\,z:\mathsf{nat}).\mathsf{leq}\,x\,y \to \mathsf{leq}\,y\,z \to \mathsf{leq}\,x\,z$. One possible solution is to take

$$\begin{array}{l}
\texttt{fix}\ \mathit{trans} : \Pi(m\,n\,k:\mathsf{nat}).\mathsf{leq}\,m\,n \to \mathsf{leq}\,n\,k \to \mathsf{leq}\,m\,k := \\
\quad \lambda(m\,n\,k:\mathsf{nat})(H_1:\mathsf{leq}\,m\,n)(H_2:\mathsf{leq}\,n\,k). \\
\quad (\texttt{match}\ H_1\ \texttt{in}\ [(m_1\,n_1:\mathsf{nat})]\ \mathsf{leq}\,m_1\,n_1\ \texttt{return}\ \mathsf{leq}\,n_1\,k \to \mathsf{leq}\,m_1\,k\ \texttt{with} \\
\quad\ \mid \boxed{\mathsf{leq0}\,x} \Rightarrow \lambda(h_2:\mathsf{leq}\,x\,k).\ \mathsf{leq0}\,k \\
\quad\ \mid \boxed{\mathsf{leqS}\,x\,y\,H} \Rightarrow \lambda(h_2:\mathsf{leq}\,(\mathsf{S}\,y)\,k). \\
\qquad\quad \texttt{match}\ h_2\ \texttt{in}\ [(k_2:\mathsf{nat})]\ \mathsf{leq}\,(\mathsf{S}\,y)\,k_2\ \texttt{return}\ \mathsf{leq}\,(\mathsf{S}\,x)\,k_2\ \texttt{with} \\
\qquad\quad \mid \boxed{\mathsf{leqS}\,(x' := y)\,y'\,H'} \Rightarrow\ \mathsf{leqS}\,x\,y'\,(\mathit{trans}\,H\,H')\ )\,H_2
\end{array}$$

For the sake of readability, we have used implicit arguments (e.g., in the recursive call to *trans*), and omitted definitions that can be inferred by unification.

Nevertheless, this definition looks complicated. It consists of a nested case analysis on $\langle H_1, H_2\rangle$. However, to make the definition go through, we need to generalize the type of the hypothesis H_2 in the return type of the case analysis of H_1, so that we can match the common value n in the types of H_1 and H_2.

The case $\langle \mathsf{leq0}, \_ \rangle$ is simple; the case $\langle \mathsf{leqS}, \mathsf{leq0} \rangle$ is impossible; finally, the case $\langle \mathsf{leqS}, \mathsf{leqS} \rangle$ is the most complicated. Note however, that things are simplified by stating that variable x' should be unified. The unification then finds a value for x' and checks that is convertible with y. The body of the branch is typed in a context containing the declarations

$$(x\,y : \mathsf{nat})(H : \mathsf{leq}\,x\,y)(x' := y : \mathsf{nat})(y' : \mathsf{nat})(H' : \mathsf{leq}\,x'\,y') \ .$$

It is easy to see that, in this context, the recursive call to *trans* is well typed.

Let us compare this definition with its counterpart in Agda. Transitivity of leq can be defined in Agda as

$$\mathit{trans} : (m\,n\,k : \mathsf{nat}) \to \mathsf{leq}\,m\,n \to \mathsf{leq}\,n\,k \to \mathsf{leq}\,m\,k$$

$$\mathit{trans}\ \lfloor 0 \rfloor\ \lfloor x \rfloor\ k\ \boxed{(\mathsf{leq0}\,x)}\ \_ = \mathsf{leq0}\,k$$

$$\mathit{trans}\ \lfloor \mathsf{S}\,x \rfloor\ \lfloor \mathsf{S}\,y \rfloor\ \lfloor \mathsf{S}\,y' \rfloor\ \boxed{(\mathsf{leqS}\,x\,y\,H)}\ \boxed{(\mathsf{leqS}\ \lfloor y \rfloor\ y'\,H')} = \mathsf{leqS}\,x\,y'\,(\mathit{trans}\,H\,H')$$

Besides writing the return types in both cases, and the fact that we generalize the type of the second argument, our definition looks very much like a direct translation of the Agda version to a nested case definition (compare the highlighted parts).

5 Metatheory

In this section we state some metatheoretical properties about the system. We prove Subject Reduction (Lemma 5), and sketch a translation into a simpler theory (Lemma 6), from which consistency follows as a corollary.

The substitution lemma is still valid with the new rule.

Lemma 1. *If $\Gamma \vdash t : T$ and $\sigma : \Gamma \to \Delta$, then $\Delta \vdash t\sigma : T\sigma$.*

The following lemmas formally state the intuitive meaning of the unification judgment. If a unification succeeds positively, then the result is a unifier (Lemma 2); moreover, in a sense, it is a most general unifier (Lemma 3). If a unification succeeds negatively, then there is no unifier (Lemma 4).

Lemma 2. *Let $\Gamma; \Delta, \zeta \vdash [\boldsymbol{u} = \boldsymbol{v} : \Theta] \mapsto \Delta', \zeta' \vdash \sigma$ be a unification judgment, with $\Gamma\Delta \vdash \boldsymbol{u} : \Theta$, and $\Gamma\Delta \vdash \boldsymbol{v} : \Theta$. Then $\Gamma \vdash \sigma : \Delta \to \Delta'$, and $\Gamma\Delta' \vdash \boldsymbol{u}\sigma \approx \boldsymbol{v}\sigma$.*

Lemma 3. *Let $\Gamma; \Delta, \zeta \vdash [\boldsymbol{u} = \boldsymbol{v} : \Theta] \mapsto \Delta', \zeta' \vdash \sigma$ be a unification judgment, and $\rho : \Gamma\Delta \to \Gamma$ a substitution such that $\Gamma \vdash \boldsymbol{u}\rho \approx \boldsymbol{v}\rho$. Then, there exists $\rho' : \Gamma\Delta' \to \Gamma$ such that $\rho \approx \sigma\rho'$.*

Lemma 4. *Let $\Gamma; \Delta, \zeta \vdash [\boldsymbol{u} = \boldsymbol{v} : \Theta] \mapsto \bot$ be a unification judgment. Then, there exists no $\rho : \Gamma\Delta \to \Gamma$ such that $\Gamma \vdash \boldsymbol{u}\rho \approx \boldsymbol{v}\rho$.*

The proof of Subject Reduction proceeds by induction on the typing derivation. The only difficult case is, of course, the new elimination rule.

Lemma 5 (Subject Reduction). *If $\Gamma \vdash M : T$, and $\Gamma \vdash M \to M'$, then $\Gamma \vdash M' : T$.*

To prove consistency, we define a type-preserving translation of our system, to the system CIC+Heq, which is CIC together with the elimination rule of heterogeneous equality, i.e., the term Subst defined in Sect. 4.1.[4] Therefore, our consistency result is relative to the consistency of CIC+Heq. The translation is similar to the translation described in [3].

The translation function is written $[\![\ ]\!]$ and defined by structural induction on the terms. The only interesting case is that of the new elimination rule. The intuitive idea is to generate the term we usually build in CIC by generating equalities between the indices of the inductive type, as described in Sect. 2. Given a term

$$\texttt{match}\ M\ \texttt{as}\ x\ \texttt{in}\ [\Delta]\, I\, \boldsymbol{p}\, \boldsymbol{t}\ \texttt{where}\ \Delta := \boldsymbol{q}\ \texttt{return}\ P\ \texttt{with}\ \{C_i\, \boldsymbol{x_i} \Rightarrow b_i\}_i$$

its translation along $[\![\ ]\!]$ is (we use $=$ to denote Heq omitting types for readability)

$$\begin{array}{l}(\texttt{match}\ [\![M]\!]\ \texttt{as}\ z\ \texttt{in}\ I\ [\![\boldsymbol{p}]\!]\ \boldsymbol{y} \\ \texttt{return}\ \Pi[\![\Delta]\!](x : I\ [\![\boldsymbol{p}]\!]\ [\![\boldsymbol{t}]\!]).\boldsymbol{y} = [\![\boldsymbol{t}]\!] \to x = z \to [\![P]\!]\ \texttt{with} \\ \{C_i\ \boldsymbol{z_i} \Rightarrow B_i\}_i)\ [\![\boldsymbol{q}]\!]\ [\![M]\!]\ (\mathsf{Hrefl}\ [\![\boldsymbol{t}[\mathcal{D}om\,(\Delta) := q]]\!])\ (\mathsf{Hrefl}\ [\![M]\!])\end{array}$$

Note that we generalize equalities between the indices, in the same way as shown in Sect. 2. This is where we need heterogeneous equality (for instance, observe that x and z have different types in $x = z$). We also generalize over Δ and x, and then apply the resulting term to $[\![\boldsymbol{q}]\!]$ (for Δ), $[\![M]\!]$ (for x), and trivial proofs of equality (for the equalities between indices). Each branch B_i takes the form

$$\lambda[\![\Delta]\!](x : I\ [\![\boldsymbol{p}]\!]\ [\![\boldsymbol{t}]\!])(\boldsymbol{H} : [\![\boldsymbol{u_i}]\!] = [\![\boldsymbol{t}]\!])(H : x = (C_i\ [\![\boldsymbol{p}]\!]\ \boldsymbol{z_i}))\ldots.$$

We also define a translation of the unification judgment, that takes as input the sequence of equalities $\boldsymbol{H}$, and returns a sequence of terms whose type corresponds to the substitution of the branch (if the unification succeeds positively); or a proof of contradiction (if the unification succeeds negatively). In the latter case, we are done, while in the former, we use the returned sequence of terms to rewrite in the translation of the body proper, and obtain a term of the right return type. We can prove that typing is preserved by this translation:

Lemma 6. *If $\Gamma \vdash M : T$, then $[\![\Gamma]\!] \vdash_{CIC+Heq} [\![M]\!] : [\![T]\!]$.*

Since $[\![\forall P : \texttt{Prop}.P]\!] = \forall P : \texttt{Prop}.P$, consistency of our system with respect to consistency of CIC+Heq follows immediately.

Corollary 1. *If CIC+Heq is consistent, then the new rule is consistent. That is, there is no term M such that $[] \vdash M : (\forall P : Prop.P)$.*

[4] Recall that this rule is not derivable in CIC, while it is in our system.

6 Related Work

Pattern matching and axiom K. Coquand [2] was the first to consider the problem of pattern matching with dependent types. He already observed that the axiom K is derivable in his setting. Hofmann and Streicher [4] later proved that pattern matching is not a conservative extension of Type Theory, by showing that K is not derivable in Type Theory. Finally, Goguen *et al.* [3] proved that pattern matching can be translated into a Type Theory with K as an axiom, showing that K is sufficient to support pattern matching — this result was already discovered by McBride [5]. Given this series of results, it is not surprising that axiom K is derivable with the rule we propose.

Epigram and Agda. Two modern presentations of Coquand's work, which are important inspirations for this work, are the programming languages Epigram [6] and Agda [8].

The pattern matching mechanism of Epigram, described by McBride and McKinna in [7], provides a way to reason by case analysis, not only on constructors, but using more general elimination principles. In that sense, it is more general than our approach. It also defines a mechanism to perform case analysis on intermediate expressions. This is not necessary in our case, where we have a more primitive notion of pattern matching (we can simply do a case analysis on any expression). Finally, it also defines a simplification method based on first-order unification, that we have reformulated here.

Agda's pattern matching mechanism, described in [8], allows definitions by a sequence of (possibly overlapping) equations, and uses the *with* construct to analyze intermediate expressions, in a similar way to [7]. The first-order unification algorithm used in Agda served as basis of our own presentation. Internally, pattern matching definitions are translated in Agda to nested case definitions, which is what we directly write in our approach.

The *with* construct developed in Epigram and Agda does not increase the expressive power of those systems — internally, it is translated into more primitive expressions. However, it does provide a concise and elegant way of writing functions. In comparison, definitions written using our proposed rule are more verbose and difficult to write by hand (cf. the example on transitivity of less-or-equal in Sect. 4.1). On the other hand, since our rule handles much of the work necessary to typecheck an Agda-style definition (e.g., unification of inversion constraints, elimination of impossible cases), it should not be difficult to translate from an Agda-style definition to a nested case definition using the new rule.

Coq. The current implementation of Coq [1] provides mechanisms to define functions by pattern matching. The basic pattern-matching algorithm, initially written by Cristina Cornes and extended by Hugo Herbelin, supports omission of impossible cases by encoding the proofs of negative success of the first-order unification process within the return predicate of the `match` expression (see Coq version 8.2). Another algorithm of Coq, provided by the `Program` construction of Matthieu Sozeau [12], allows to exploit inversion constraints using heterogeneous equality for typing dependent pattern-matching in a way similar to what is done in Epigram. Because Coq lacks the reduction rule of axiom K, not all definitions built by this algorithm are computable. Our rule not only simplifies the underlying computational structure of programs typed using explicit insertion of heterogeneous equalities but also removes the limitations in code execution that the absence of reduction rule for K induces.

Other approaches. Oury [9] proposed a different approach to remove impossible cases based on set approximations. His approach allows the removal of cases in situations where unification is not sufficient. As mentioned in [9], it remains to be seen if the combination of both techniques can be used to remove more cases.

7 Conclusions and Future Work

We have presented a new rule for performing pattern matching in CIC. Functions on inductive families are simpler to write and more efficient using the new rule. Also, the underlying theory is slightly increased by providing axiom K and its reduction rule, which means that the new system is more amenable to use as the basis for a programming language with dependent types.

For future work, the obvious first step is implementation. Since the new rule is not much different from the current elimination rule, adapting it to existent implementations, e.g. Coq, should not be difficult. However, taking full advantage of the new possibilities would mean to redesign many tactics. Also, it could be of interest to implement Agda-style definitions, on top of the new rule.

In another direction, there is lots of room for improving the unification. We could add the treatment of circular equations, such as $n = \mathsf{S}\, n$, that are provably false in CIC. Also, it could be of interest to have a more general notion of injective and discriminative constants, so that we are able to write functions by pattern matching when the indices are not necessarily inductive objects.

References

1. The Coq Development Team. The Coq Reference Manual, version 8.1. Distributed electronically (February 2007), `http://coq.inria.fr/doc`
2. Coquand, T.: Pattern matching with dependent types. In: Nordström, B., Petersson, K., Plotkin, G. (eds.) Informal Proceedings Workshop on Types for Proofs and Programs, Båstad, Sweden (1992)
3. Goguen, H., McBride, C., McKinna, J.: Eliminating dependent pattern matching. In: Futatsugi, K., Jouannaud, J.-P., Meseguer, J. (eds.) Algebra, Meaning, and Computation. LNCS, vol. 4060, pp. 521–540. Springer, Heidelberg (2006)
4. Hofmann, M., Streicher, T.: The groupoid model refutes uniqueness of identity proofs. In: LICS, pp. 208–212. IEEE Computer Society, Los Alamitos (1994)
5. McBride, C.: Dependently Typed Functional Programs and their Proofs. PhD thesis, University of Edinburgh (1999)
6. McBride, C.: Epigram: Practical programming with dependent types. In: Vene, V., Uustalu, T. (eds.) AFP 2004. LNCS, vol. 3622, pp. 130–170. Springer, Heidelberg (2005)

7. McBride, C., McKinna, J.: The view from the left. J. Funct. Program. 14(1), 69–111 (2004)
8. Norell, U.: Towards a practical programming language based on dependent type theory. PhD thesis, Chalmers University of Technology (2007)
9. Oury, N.: Pattern matching coverage checking with dependent types using set approximations. In: Stump, A., Xi, H. (eds.) PLPV, pp. 47–56. ACM, New York (2007)
10. Paulin-Mohring, C.: Inductive definitions in the system Coq - rules and properties. In: Bezem, M., Groote, J.F. (eds.) TLCA 1993. LNCS, vol. 664, pp. 328–345. Springer, Heidelberg (1993)
11. Schürmann, C., Pfenning, F.: A coverage checking algorithm for LF. In: Basin, D.A., Wolff, B. (eds.) TPHOLs 2003. LNCS, vol. 2758, pp. 120–135. Springer, Heidelberg (2003)
12. Sozeau, M.: Subset coercions in coq. In: Altenkirch, T., McBride, C. (eds.) TYPES 2006. LNCS, vol. 4502, pp. 237–252. Springer, Heidelberg (2007)
13. Werner, B.: Une Théorie des Constructions Inductives. PhD thesis, Université Paris 7 (1994)

A Framework for the Analysis of Access Control Models for Interactive Mobile Devices*

Juan Manuel Crespo[1,2], Gustavo Betarte[3], and Carlos Luna[3]

[1] FCEIA, Universidad Nacional de Rosario, Argentina
[2] IMDEA Software, Madrid, Spain
juanmanuel.crespo@imdea.org
[3] Instituto de Computación, Universidad de la República, Uruguay
{gustun,cluna}@fing.edu.uy

Abstract. The Java Micro Edition platform (JME), a Java enabled technology, provides the Mobile Information Device Profile (MIDP) standard that facilitates applications development and specifies a security model for the controlled access to sensitive resources of the device. The model builds upon the notion of protection domain, which in turn can be grasped as a set of permissions. An alternative model has been proposed that extends MIDP's by introducing permissions with multiplicities and adding flexibility to the way in which permissions are granted by the user of the device and used by the applications running on it. This paper presents a framework, formalized using the proof-assistant Coq, suitable for defining and comparing the access control policies that can be enforced by (variants of) those security models and to prove desirable properties they should satisfy. The proofs of some of those properties are also stated and discussed in this work.

Keywords: Access control models, mobile devices, formal proofs.

1 Introduction

Devices such as cell phones or personal digital assistants often have access to sensitive personal information and are subscribed to paid services in order to communicate with other entities. In addition to this, users are able to download and install applications from unreliable sources at their will. Java Micro Edition (JME) [10] is a version of the Java platform targeted at resource-constrained devices which comprises two kinds of components: configurations and profiles. The Mobile Information Device Profile (MIDP) [7, 6] defines an application life cycle, a security model and APIs that offer the functionality required by mobile applications, including networking, user interface, push activation and persistent local storage. Many mobile device manufacturers have adopted MIDP since the specification was made available. A formal specification of the JME-MIDP 2.0 security model developed using the proof-assistant Coq is presented and described in detail in [12].

* This work was partially funded by the Project PDT 63/118 STEVE, DINACYT, Uruguay.

S. Berardi, F. Damiani, and U. de'Liguoro (Eds.): TYPES 2008, LNCS 5497, pp. 49–63, 2009.

In [2], a security model for interactive mobile devices is put forward which can be grasped as an extension of the JME-MIDP model. The work presented in that paper has focused in developing a formal model for studying, in particular, interactive user querying mechanisms for permission granting for application execution on mobile devices. Like in the MIDP case, the notion of permission is central to this model and MIDP is extended by introducing permissions with multiplicities and by adding flexibility to the way in which permissions are granted by the user and used by the applications.

One of the main objectives of the work reported here has been to build a framework which would provide a formal setting to define and analyse the permission models defined by MIDP and the one presented in [2]. This framework, which is formally defined using the Calculus of Inductive Constructions [4, 5], adopts, with variations, most of the security and programming constructions defined in [2]. The principal difference is that most of those constructions are now parameterized by a permission grant policy. In this paper it is shown how the framework can be used to define a type of permission grant policies and to represent the four user permission modes of MIDP and the policies defined in [2] as objects of that type. The paper also presents the definition of an order relation, based on a notion of safe programs, which can be used to perform a comparative analysis of grant policies. In particular, it is described the proof, which has been constructed using the proof-assistant Coq [11], of the theorem that establishes how the grant policies mentioned above are related according to the defined order. The complete definition of the framework as well as the statement and proof of the properties are available in `www.fing.edu.uy/inco/grupos/mf/projects/PermModel/ACM-Coq.zip`

The structure of the rest of the paper is organized as follows. Section 2 provides a brief account of the permission models that are the object of the analysis presented in this work. Section 3 describes the formal setting and the security concepts that constitute the basis of the access control mechanisms used to define those models. In section 4 the grant policies and the order relation are formally defined. A theorem that establishes the conditions that suffice to prove that two grant policies are in the order relation is also discussed. In section 5 it is presented the proof of the theorem that establishes how the concrete permission grant policies studied in this work are related. Section 6 concludes and describes further work.

2 Security Models for Interactive Mobile Devices

This section provides a brief account of the permission models that are the object of the analysis presented in this work.

2.1 The JME–MIDP Security Model

In MIDP, applications (MIDlets) are packaged and distributed as suites. A MIDlet suite can contain one or more MIDlets and is distributed as two files, an application descriptor file and an archive file that contains the actual classes and resources. A suite that needs access to protected APIs or functions must declaratively request the corresponding permissions in its descriptor. MIDlet suites may request permissions either as required or as optional. In the first version of MIDP [7], any application not installed by the device manufacturer or a service provider runs in a sandbox that prohibits access to security sensitive APIs or functions of the device. Although this sandbox security model effectively prevents any rogue application from jeopardising the security of the device, it is excessively restrictive and does not allow many useful applications to be deployed after issuance of the device.

Version 2.0 of MIDP [6] introduces a new security model based on the concept of protection domain. A protection domain can be grasped as an abstraction of the execution context of an application, and it determines the access rights to the protected functions of the device. Each sensitive API or function on the device may define permissions in order to prevent it from being used without authorisation. A protection domain consists of both a set of permissions which are granted unconditionally, without intervention of the device's user (called **allowed** permissions), and a set of permissions which require authorisation from the user (called **user**). Permissions may be granted by the user to an active MIDlet suite in either of the following three modes:

- **blanket:** the permission is granted for as long as the application remains installed in the device
- **session:** the permission is granted for as long as the application is running
- **one-shot:** the permission is granted for only one use of the function

An installed MIDlet suite is bound to a unique protection domain. Untrusted MIDlet suites are bound to a protection domain with permissions equivalent to those in a MIDP 1.0 sandbox. Trusted MIDlet suites may be identified by means of cryptographic signatures and bound to more permissive protection domains. This security model enables applications developed by trusted third parties to be downloaded and installed after issuance of the device without compromising its security.

The set of permissions effectively granted to a suite is determined from its protection domain, the permissions the suite request in its descriptor and the authorisations granted by the user.

For a more detailed description of the mechanisms defined by the security model the reader is referred to [7, 6]. A formal specification of the MIDP 2.0 security model is presented in [12] and a certified access controller for the enforcement of policies admitted by that model is described in details in [9].

2.2 An Alternative Model

In [2], a security model for interactive mobile devices is put forward which can be grasped as an extension of that of MIDP. The work presented in that paper has focused in developing a formal model for studying, in particular, interactive user querying mechanisms for permission granting for application execution on mobile devices. Like in the MIDP case, the notion of permission is central to this model. A generalisation of the one-shot permission described above is proposed that consists in associating to a permission a multiplicity which states how many times that permission can be used.

The proposed model has two basic constructs for manipulating permissions: **grant** and **consume**. The grant construct models the interactive querying of the user, asking whether he grants a particular permission with a certain multiplicity. The consume construct models the access to a sensitive function which is protected by the security police, and therefore requires (consumes) permissions.

A semantics of the model constructs is proposed as well as a logic for reasoning on properties of the execution flow of programs using those constructs. The basic security property the logic allows to prove is that a program will never attempt to access a resource for which it does not have a permission. The authors also provide a static analysis that makes it possible to verify that a particular combination of the grant-consume constructs does not violate that security property. For developing that kind of analysis the constructs are integrated into a program model based on control-flow graphs. This model has also been used in previous work on modelling access control for Java, see for instance [8, 3].

One of the main objectives of the work that is being reported here, has been to build a framework which would provide a formal setting to define the permission models defined by MIDP and the one presented in [2] (and variants of it) in an uniform way and to perform a formal analysis and comparison of those models. This framework, which is formally defined using the Calculus of Inductive Constructions [4, 5], adopts, with variations, most of the security and programming constructions defined in [2]. In particular it has been modified so as to be parameterized by permission granting policies, while in the original work this relation is fixed.

3 A Framework for Access Control Modeling

This section introduces the formal setting used to define the security concepts that constitute the basis of certain access control mechanisms, to proceed then to described how those mechanisms are used to define the permission granting models which are object of analysis of this work.

3.1 The Formal Language Used

Standard notation is used for equality and logical connectives ($\wedge, \vee, \neg, \rightarrow, \forall, \exists$). Anonymous functions and predicates use standard lambda notation (e.g. $\lambda\ (x : T)\ .\ x, \lambda\ (x : nat)\ .\ x > 10$). In case there is more than one binder, the standard abbreviation $\lambda\ (x : nat)\ (y : nat)\ .\ x + y$ is used.

An inductive relation I is defined by giving introduction rules of the form:

$$\frac{P_1 \ldots P_m}{I\ x_1 \ldots x_n}$$

where the variables occurring free are implicitly universally quantified. Similarly, inductive types are defined by giving constructors in the following form:

$$T \stackrel{def}{=} \mid C_1 : A_{1,1} \rightarrow \ldots A_{1,n_1} \rightarrow T$$
$$\vdots$$
$$\mid C_m : A_{m,1} \rightarrow \ldots A_{m,n_m} \rightarrow T$$

where $C_1 \ldots C_n$ are the constructors of T.

A (dependent) record type R is defined as follows:

$$R \stackrel{def}{=} \{field_1 : A_1, \ldots, field_n : A_n\}$$

This definition generates a non-recursive inductive type with a single constructor $mkR : A_1 \rightarrow \ldots A_n \rightarrow R$ and projection functions $field_i : R \rightarrow A_i$. Application of projection functions is abbreviated using dot notation: $field_i\, r = r.field_i$. When the type is clear for the context $\langle x_1, \ldots, x_n \rangle$ is written instead of $mkR\ x_1, \ldots, x_n$.

In the fomalization developed it has been used inductive types that have *valid* and *invalid* cases. In the rest of this paper it is adopted the convention that a type with the same name but prefixed with valid is the type consisting only of the valid cases. Which are the valid constructors is usually clear from the context, otherwise it is specified.

The following parametric inductive types are assumed to be predefined:

- *option* T with constructors *None* : *option* T and *Some* : $T \rightarrow$ *option* T,
- finite lists over T, *list* T. The empty list is denoted by [] and the (infix) constructor that inserts an element a at the front of a list s is denoted by $a \triangleright s$. Finite snoc lists over T, *snocList* T, that is, lists that are constructed by inserting elements at the back, are also used. [] denotes the empty snoc list and $s \triangleleft\ a$ denotes the insertion of an element a at the back of the snoc list s.

3.2 Permissions

Every (controlled) resource of the device is given a type. Let *ResType* be the set of types of resources. If *rt* is a resource type, *Resources rt* and *Actions rt* define the set of resources of type *rt* available on the device and the actions that can performed over them, respectively. The permissions of a resource type are defined as follows:

$$PermRes\ \ (rt : ResType) \stackrel{def}{=}$$
$$\mid valid : list\ (Resources\ rt) \rightarrow list\ (Actions\ rt) \rightarrow PermRes\ rt$$
$$\mid invalid : PermRes\ rt$$

That is, given a resource type *rt*, an object of type *PermRes rt* is a set (represented by a list) of actions and resources over *rt*, or the constant *invalid*. A relation $\sqsubseteq_{PermRes}$ is defined by applying set inclusion component-wise. This relation defines a lattice structure where *invalid* is the bottom element $\bot_{PermRes}$ and $\sqcup_{PermRes}$ a lub operator which is obtained applying set union component-wise.

As already mentioned, a notion of multiplicity of granted permission is introduced in [2]. A multiplicity is defined to be either a natural number, a special value ∞ that denotes an irrestricted permission, or an error value $\bot$. A type Mul is defined:

$$Mul \stackrel{def}{=} \begin{array}{l} \mid \bot \ : Mul \\ \mid val : nat \rightarrow Mul \\ \mid \infty \ : Mul \end{array}$$

It is straightforward to see that a lattice can be constructed over Mul with $\bot$ and ∞ as the bottom and top elements, respectively. The obvious extensions of functions and predicates defined over naturals to functions and predicates over Mul, such as $\sqsubseteq_{Mul}$, $+_{Mul}$, $-_{Mul}$, $pred_{Mul}$, are also defined.

An accumulated permission for a resource type is comprised of two components: the set of resources and actions allowed and a multiplicity. One such permission (of resource type rt) is then grasped as an object of the following record type:

$$PermMul\ (rt : ResType) \stackrel{def}{=} \{permRes : PermRes\ rt; mul : Mul\}$$

The lattice of permissions of a resource type can be obtained by defining the order $\sqsubseteq_{PermMul}$:

$$\begin{array}{rl} pm_1 \sqsubseteq_{PermMul} pm_2 \stackrel{def}{=} & pm_1.permRes \sqsubseteq_{PermRes} pm_2.permRes \\ \wedge & pm_1.mul \sqsubseteq_{Mul} pm_2.mul \end{array}$$

where pm_1 and pm_2 are objects of type $PermMul\ rt$. Now, the permission state of the device is defined. One such state is ultimately a mapping that associates a permission to each resource type. Therefore, it is defined as the following dependent function type:

$$Perm \stackrel{def}{=} \forall (rt : ResType), PermMul\ rt$$

It is said that two permissions p_1 and p_2 are (extensionally) equal if for every resource type rt it holds that $p_1\ rt\ = p_2\ rt$.

An order $\sqsubseteq_{Perm}$ can be defined as the product-wise extension of $\sqsubseteq_{PermMul}$ as follows:

$$p_1 \sqsubseteq_{Perm} p_2 \stackrel{def}{=} \forall (rt : ResType), (p_1\ rt) \sqsubseteq_{PermMul} (p_2\ rt)$$

In order to model the operations that affect the state of the permissions an *update* function is introduced:

$$update\ (p : Perm)(rt : ResType)(pres : PermRes\ rt)(m : Mul) : Perm$$

The intended (and formalized) behaviour of this function is that of an usual store updating operator: the permission state remains unchanged for every resource type different from rt, and for rt yields $\langle pres, m\rangle$.

If rt is a resource type and p a permission state, then the following inductive relation *Error* is defined

$$\frac{(p\ rt).permRes = invalid\ rt}{Error\ p} \qquad \frac{(p\ rt).mul = \bot}{Error\ p}$$

The intuition is that an error situation may occur when either there is an attempt to perform an action over a resource of type rt and no valid permission is associated to it (first rule) or when there are no granted permissions for that resource (second rule).

3.3 Programs

A program in, among others, [2, 1] is represented by a control-flow graph that captures the manipulations of permissions and the handling of method calls and returns as well as exceptions.

A control-flow graph is a tuple $G = (NO, EX, KD, TG, CG, EG, n_0)$ where:

- NO is the set of nodes of the graph (one for each instruction),
- EX is the set of exceptions,
- KD is a function of type $KD : NO \rightarrow Instr$ that associates each node to an instruction,
- $TG : NO \rightarrow NO \rightarrow Prop$ is the propositional function that characterizes the set of intra-procedural edges (i.e. $n_1\ TG\ n_2$ if control can be transferred from instruction at node n_1 to instruction at node n_2 within the currect procedure),
- CG is the set of inter-procedural edges (which can be used to capture dynamic method calls),
- $EG : EX \rightarrow NO \rightarrow NO \rightarrow Prop$ are the intra-procedural exception edges,
- $n_0 : NO$ is the graph entry node.

The instructions are formally defined in the framework by means of the following inductive type:

$$
\begin{aligned}
Instr \stackrel{def}{=}\ & \\
& \mid Grant : \forall (rt : ResType), validPermRes\ rt \rightarrow MulValid \rightarrow Instr \\
& \mid Consume : \forall (rt : ResType), validPermRes\ rt \rightarrow Instr \\
& \mid Call : Instr \\
& \mid Return : Instr \\
& \mid Throw : EX \rightarrow Instr
\end{aligned}
$$

where $MulValid$ is the type of valid multiplicities, that is, different from the multiplicity $\bot$. The definition of the operational semantics of programs strongly depends on those of the permission granting and consumption mechanisms. They are briefly discussed and described in what follows.

In [2] two variants are discussed concerning the effect of the update operation after a permission has been granted: either the permissions before the update instruction are discarded or they are accumulated. At a first sight these *permission granting policies* have advantages and drawbacks. Furthermore, independently of this particular discussion, it is at this point that the permission model proposed by the authors introduces a generalization with respect to that of MIDP: the multiplicity of a permission. One of the main objectives of the work presented here has been to design a framework that would make it possible to provide a uniform setting where those different permissions models could be formally defined and compared. To that end, the constructions defined to provide semantics to the computational behaviour of the programs as well as to reason over that behaviour have been parameterized by permission granting policies. One such parameter shall be formally represented by an object of the following type:

$$\begin{aligned} grantPolicy \stackrel{def}{=} \forall (rt : ResType),& \\ & validPermRes\ rt \rightarrow NZMulValid \rightarrow Perm \rightarrow Perm \end{aligned}$$

where an object of type $NZMulValid$ is a valid multiplicity constructed with a non-zero natural.

As to the consumption of permissions, the following is the definition of the consume operation:

$$\begin{aligned} consume\ \ & (rt : ResType)(pr : validPermRes\ rt)(p : Perm) : Perm \stackrel{def}{=} \\ & if\ (pr\ \sqsubseteq_{PermRes}\ (p\ rt).permRes) \\ & then\ update\ p\ rt\ (p\ rt).permRes\ (pred_{Mul}\ (p\ rt).mul) \\ & else\ update\ p\ rt\ (invalid\ rt)\ (pred_{Mul}\ (p\ rt).mul) \end{aligned}$$

The consume operation is monotonic on permissions. This is stated (and proved) in the following lemma:

Lemma 1

$$\begin{aligned} & Lemma\ consumeMon : \\ & \quad \forall (rt : ResType)(pr : validPermmRes\ rt)(p\ p' : Perm), \\ & \quad p \sqsubseteq_{Perm} p' \rightarrow (consume\ rt\ pr\ p) \sqsubseteq_{Perm} (consume\ rt\ pr\ p') \end{aligned}$$

Following [2] the small-step operational semantics of a control-flow graph has been defined basically as a relation that defines transitions between states consisting of a standard control-flow stack of nodes enriched with the permissions held at that point in the execution. This definition has been extended by making it depend on a permission granting policy g. Formally, it has been defined as an inductive propositional function $\rightsquigarrow_g$ whose rules are depicted in Fig. 1. An important property of this semantics is that it is non-intrusive, that is to say, the permission state does not interfere with execution. In other words, a transition will not be blocked by the absence of permissions. This is formally stated, and proved, in the following lemma:

$$\frac{KD\ n = Grant\ rt\ pr\ m \quad TG\ n\ n'}{(n \triangleright s)\ None\ p \rightsquigarrow_g (n' \triangleright s)\ None\ (g\ rt\ pr\ m\ p)}$$

$$\frac{KD\ n = Consume\ rt\ pr \quad TG\ n\ n'}{(n \triangleright s)\ None\ p \rightsquigarrow_g (n' \triangleright s)\ None\ (consume\ rt\ pr\ p)}$$

$$\frac{KD\ n = Call \quad CG\ n\ n'}{(n \triangleright s)\ None\ p \rightsquigarrow_g (n' \triangleright n \triangleright s)\ None\ p} \qquad \frac{KD\ r = Return \quad TG\ n\ n'}{(r \triangleright n \triangleright s)\ None\ p \rightsquigarrow_g (n' \triangleright s)\ None\ p}$$

$$\frac{KD\ n = Throw\ ex \quad EG\ ex\ n\ h}{(n \triangleright s)\ None\ p \rightsquigarrow_g (h \triangleright s)\ None\ p} \qquad \frac{KD\ n = Throw\ ex \quad \forall (h : NO), \neg EG\ ex\ n\ h}{(n \triangleright s)\ None\ p \rightsquigarrow_g (n \triangleright s)\ (Some\ ex)\ p}$$

$$\frac{\forall (h : NO), \neg EG\ ex\ n\ h}{(t \triangleright n \triangleright s)\ (Some\ ex)\ p \rightsquigarrow_g (n \triangleright s)\ (Some\ ex)\ p} \qquad \frac{EG\ ex\ n\ h}{(t \triangleright n \triangleright s)\ (Some\ ex)\ p \rightsquigarrow_g (h \triangleright s)\ None\ p}$$

Fig. 1. Semantics of instructions

Lemma 2

Lemma nonIntrusive :

$$\forall (g : grantPolicy)(s\ s' : list\ NO)(ex\ ex' : option\ EX)(p\ p' : Perm),$$
$$s\ ex\ p \rightsquigarrow_g s' ex' p' \rightarrow \forall (p : Perm), (\exists (p' : Perm), s\ ex\ p \rightsquigarrow_g s'\ ex'\ p')$$

3.4 Traces

In [2] global results on the execution of programs are expressed on traces, which in turn are defined in terms of the operational semantics described above (instantiated for a particular grant policy) as follows: *a partial trace of a control-flow graph is a sequence (of type snocList (NO, option EX)) of nodes* $[] \triangleleft \langle n_0, None \rangle \triangleleft \langle n_1, e_1 \rangle \triangleleft \cdots \triangleleft \langle n_k, e_k \rangle$ *such that for all* $0 \leq\ i < k$ *there exists* $\rho, \rho' \epsilon\ Perm$, $s, s' \epsilon\ (list\ NO)$ *and verifying* $n_i \triangleright s, e_i, \rho \rightsquigarrow n_{i+1} \triangleright s', e_{i+1}, \rho'$.

The stacks s and s' in the above definition are existentially quantified because they are not defined to be components of the elements of a trace. This quantification however induces a loss of information w.r.t. the operational semantics. An example[1] should clarify this situation. Consider the control-flow graph:

$$NO = \{A, B, C, D\},\ TG = \{(B, C), (C, D)\},\ EX = CG = EG = \{\},\ n_0 = A$$
$$KD = \{(A, Return), (B, x), (C, Consume\ rt\ y), (D, Return)\}$$

where x : *Instr*, rt : *ResType*, y : *validPermRes rt*, and with initial permission $p_{init} = \lambda\ (rt : ResType)\ .\ \langle (valid\ rt\ []\ []), (val\ 0) \rangle$. Fig. 2 depicts the control-flow graph in question. From this definition it can be noticed that $[] \triangleleft \langle A, None \rangle$ is the only admissible trace yielding a valid permission state. According to the definition of partial trace stated above, the object $([] \triangleleft \langle A, None \rangle \triangleleft \langle C, None \rangle \triangleleft \langle D, None \rangle)$ is admitted as a partial trace of the defined control-flow graph. This trace can be built using the transition rules for the *Consume* and *Return* instructions (see Fig. 1). However, this latter trace yields an error situation, because the transition from node C to node D attempts to consume a not available permission.

[1] This example is due to Santiago Zanella.

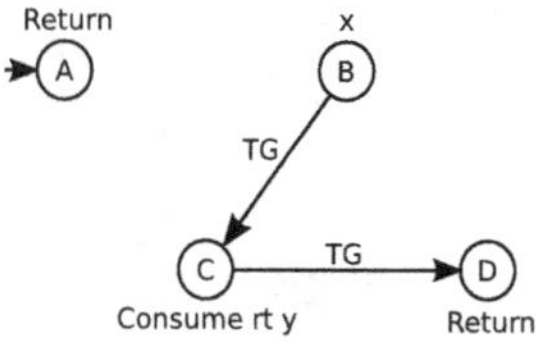

Fig. 2. Control-flow graph example

The definition of program execution traces that are proposed in the framework presented here remedies the situation described above by including the node stack as a component of the elements of the trace. This is formally represented by the following type: $Trace \stackrel{def}{=} snocList\ \{noT : NO, stT : list\ NO, exT : option\ EX\}$.

The notion of parameterized partial trace is then inductively defined over elements of type $Trace$ as follows:

$$\frac{}{PTrace_g\ []} \qquad \frac{}{PTrace_g([] \triangleleft \langle n_0, [], None\rangle)}$$

$$\frac{PTrace_g\ (tr \triangleleft \langle n, s, ex\rangle) \quad \exists (p\ p' : Perm), n \triangleright s\ ex\ p \rightsquigarrow_g n' \triangleright s'\ ex'\ p'}{PTrace_g\ (tr \triangleleft \langle n, s, ex\rangle \triangleleft \langle n', s', ex'\rangle)}$$

Let tr be a trace and g be a grant policy, if $PTrace_g\ tr$ holds then it shall be said that tr is a valid trace according to g.

Given a trace tr and a grant policy g, the function $PermsOf_g : Perm \rightarrow Trace \rightarrow Perm$ computes the permission state resulting from the execution of the program that tr represents:

$$\begin{array}{l} PermsOf_g(p_{init} : Perm)(tr : Trace) : Perm \stackrel{def}{=} \\ match\ tr\ with \\ \quad |[] \Rightarrow p_{init} \\ \quad |tr' \triangleleft e \Rightarrow match\ KD\ e.noT\ with \\ \qquad |Consume\ rt\ pr \Rightarrow consume\ rt\ pr\ (PermsOf_g\ p_{init}\ tr') \\ \qquad |Grant\ rt\ pr\ m \Rightarrow g\ rt\ pr\ m\ (PermsOf_g\ p_{init}\ tr') \\ \qquad |\_ \ \Rightarrow PermsOf_g\ p_{init}\ tr' \\ \quad end \\ end \end{array}$$

Finally, given a grant policy g, a trace is said to be safe if none of its prefixes yields a faulty permission state:

$$\begin{array}{l} Safe_g(tr : Trace)(p_{init} : Perm) \stackrel{def}{=} \\ \qquad \forall tr' : Trace, (prefix\ tr'\ tr) \rightarrow \neg Error(PermsOf_g\ p_{init}\ tr') \end{array}$$

4 Permission Grant Policies

Two kinds of grant policies are analysed in [2]: given a resource type rt, one of the policies establishes that when a new permission is granted to resources of

rt, all previous granted permissions are overwritten. This policy is called here $grant_{ow}$. The another policy, called here $grant_{ac}$, establishes that new granted permissions for rt are accumulated with the ones previously obtained for that resource type. These policies are formally defined as follows:

$$
\begin{aligned}
grant_{ow} : grantPolicy \stackrel{def}{=} \\
&\lambda\,(p : Perm)\,(rt : ResType)\,(pr : PermRes\;rt)\,(m : Mul)\;.\\
&update\;p\;rt\;pr\;m\\
grant_{ac} : grantPolicy \stackrel{def}{=} \\
&\lambda\,(p : Perm)\,(rt : ResType)\,(pr : PermRes\;rt)\,(m : Mul)\;.\\
&update\;p\;rt\;(pr \sqcup_{PermRes} (p\;rt).permRes)\;(m +_{mul} (p\;rt).mul)
\end{aligned}
$$

The permission modes defined by MIDP are also defined below as grant policies. The $grant_{bk}$ term represents the blanket permission mode, which specifies unrestricted access to a given resource type. The one-shot permission mode, which specifies a single access to a given resource type, is represented by the term $grant_{os}$.

$$
\begin{aligned}
grant_{bk} : grantPolicy \stackrel{def}{=} \\
&\lambda\,(p : Perm)\,(rt : ResType)\,(pr : PermRes\;rt)\,(m : Mul)\;.\\
&update\;p\;rt\;(pr \sqcup_{PermRes} (p\;rt).permRes)\;\infty\\
grant_{os} : grantPolicy \stackrel{def}{=} \\
&\lambda\,(p : Perm)\,(rt : ResType)\,(pr : PermRes\;rt)\,(m : Mul)\;.\\
&update\;p\;rt\;pr\;1
\end{aligned}
$$

It should be noticed that both the allowed mode and the session permission mode specified by MIDP 2.0 can be modeled as a blanket grant policy. In the first case, the granted permission would hold for the rest of the life cycle of the application to which is granted the permission and, in the second case, the scope would be that of a session during which that application is active.

In order to perform a comparative analysis of grant policies of the kind of the ones just defined, the following relation is defined:

$$
\begin{aligned}
g_1 \sqsubseteq_g g_2 \stackrel{def}{=} &\forall (tr : Trace)(p : Perm),\\
&Ptrace_{g_1}\;tr \rightarrow Safe_{g_1}\;p\;tr \rightarrow Safe_{g_2}\;p\;tr
\end{aligned}
$$

This order establishes that given a control-flow graph, for every valid trace of the graph according to g_1 and every initial set of permissions it holds that if the trace is safe by granting the permissions using g_1 as policy, then it must also be safe if the permissions are granted using the policy g_2. Intuitevely, g_1 yields a more restrictive permission model.

The following lemma states that the order relation between permission states preserves error situations. It can also be proved that $\sqsubseteq_g$ is a partial order (reflexive, transitive and antisymmetric). These results shall be of help when relating the grant policies described so far.

Lemma 3

Lemma lePermError : $\forall (p_1\ p_2 : Perm), Error\ p_1 \rightarrow p_1 \sqsubseteq_{Perm} p_2 \rightarrow Error\ p_2$

The relation $\sqsubseteq_g$ defines a lattice structure with the policies $grant_{os}$ and $grant_{bk}$ as the bottom and top elements respectively.

The following theorem states a sufficient condition (a criterion) to prove that two permission granting policies, g_1 and g_2 say, are in the order relation ($g_1 \sqsubseteq_g g_2$):

1. the error situations that arise using g_2 as a policy are also error situations if g_1 is used, and
2. if a grant policy g_1 is applied, then every permission available at the end of a trace is also available if g_2 is used instead of g_1.

This theorem is important in order to compare different security policies.

Theorem 1

Theorem lePolicyCrit :

$$\begin{aligned} &\forall (g_1\ g_2 : grantPolicy) \\ &(H_{errors} : \forall (rt : ResType)\ (pr : validPermRes\ rt)\ (m : NZMulValid) \\ &\qquad (p : Perm), Error(g_2\ rt\ pr\ m\ p) \rightarrow Error(g_1\ rt\ pr\ m\ p)) \\ &(H_{perms} : \forall (p : Perm)\ (tr : Trace), \\ &\qquad (PermsOf_{g_1}\ p\ tr) \sqsubseteq_{Perm} (PermsOf_{g_2}\ p\ tr)), \\ &g_1 \sqsubseteq_g g_2 \end{aligned}$$

Proof. The proof proceeds by induction over ($PTrace_{g_1}\ tr$), which is obtained after unfolding $g_1 \sqsubseteq_g g_2$. If the trace *tr* is empty, then the theorem holds trivially. In the case the trace is a singleton node, the proof uses hypothesis H_{errors} and proceeds by doing case analysis on the instruction type associated with that node; the interesting case corresponds to the *Grant* instruction, since *consume* is monotonic w.r.t. $\sqsubseteq_{Perm}$ and the rest of the instructions do not affect the permission state.

The inductive step follows basically from the lemma *lePermError*, the hypothesis H_{perms}, and the induction hypothesis. □

5 Relating Permission Grant Policies

Using the formal setting defined so far it is now possible to state and prove a theorem that establishes how the four policies described in the previous section are related according to the order relation $\sqsubseteq_g$.

Theorem 2.

Theorem grantPolicyRel : $grant_{os} \sqsubseteq_g grant_{ow} \sqsubseteq_g grant_{ac} \sqsubseteq_g grant_{bk}$

Proof. The proof of this theorem proceeds first by proving the three inequalities $grant_{os} \sqsubseteq_g grant_{ow}$, $grant_{ow} \sqsubseteq_g grant_{ac}$ and $grant_{ac} \sqsubseteq_g grant_{bk}$, and then applying the transitivy of the order $\sqsubseteq_g$. Each inequality is proved applying the theorem that establishes the sufficient conditions to prove that two grant policies are in the order relation (theorem *lePolicyCrit*), and following a similar strategy. Here it shall be presented in detail the proof of the first inequality, indications on how to proceed for the remainding two cases shall also be provided.

The application of the lemma *lePolicyCrit* to prove $grant_{os} \sqsubseteq_g grant_{ow}$ generates in turn the following proof obligations:

1. $\forall (rt : ResType)\ (pr : validPermRes\ rt)\ (m : NZMulValid)\ (p : Perm),$ $Error(g_2\ rt\ pr\ m\ p) \rightarrow Error(g_1\ rt\ pr\ m\ p))$
2. $\forall (p : Perm)\ (tr : Trace), (PermsOf_{g_1}\ p\ tr) \sqsubseteq_{Perm} (PermsOf_{g_2}\ p\ tr))$

The proof of (1) proceeds by first applying the lemma *lePermError*. This leads to have to prove that $(grant_{os}\ rt\ pr\ m\ p) \sqsubseteq_{Perm} (grant_{ow}\ rt\ pr\ m\ p)$. Unfolding the definition of $grant_{ow}$ and $grant_{os}$, and applying the lemmas that characterize the function *update*, we have to prove $\langle pr, 1\rangle \sqsubseteq_{PermMul} \langle pr, m\rangle$ and $(p\ rt) \sqsubseteq_{PermMul} (p\ rt)$. The latter follows directly because $\sqsubseteq_{PermMul}$ is reflexive. As to the former, as $m : NZMulValid$ so the least number it can be is 1, in which case, since $\sqsubseteq_{PermMul}$ is reflexive, the obligation is discharged.

For (2), the proof poceeds by induction on *tr*:

- $tr = []$, the inequality simplifies to $p \sqsubseteq_{Perm} p$ and since $\sqsubseteq_{Perm}$ is reflexive, this obligation is discharged.
- $tr = tr' \lhd \langle n, st, ex\rangle$, the proof proceeds by case analysis on $KD\ n$. The relevant cases are *Grant* and *Consume*, since the rest of the instructions do not affect the permission state. The *Consume* case is straightforward since the function *consume* is monotonic, and by induction hypothesis it is known that $(PermsOf_{grant_{os}}\ p\ tr') \sqsubseteq_{Perm} (PermsOf_{grant_{ow}}\ p\ tr')$. The *Grant* case is proved using transitivity of $\sqsubseteq_{Perm}$, the induction hypothesis and the following two lemmas:
  - $\forall (rt : ResType)(pr : validPermRes\ rt)(m : NZMulValid)(p : Perm),$ $(grant_{os}\ rt\ pr\ m\ p) \sqsubseteq_{Perm} (grant_{ow}\ rt\ pr\ m\ p)$
  - $\forall (rt : ResType)(pr : validPermRes\ rt)(m : NZMulValid)(p\ p' : Perm), p \sqsubseteq_{Perm} p' \rightarrow (grant_{ow}\ rt\ pr\ m\ p) \sqsubseteq_{Perm} (grant_{ow}\ rt\ pr\ m\ p')$.

  The proofs of these lemmas are omitted due to space restrictions.

The structure of the proof of the two remaining inequalities are quite similar to the one just described above. In both cases the bulk of the proof reduces to prove auxiliary lemmas similar to the ones of the proof obligation (2) for the involved grant policies. □

This theorem and its proof provide a formal evidence that, in the first place, of the four policies, MIDP's one-shot is the most restrictive policy and MIDP's blanket is the most permissive one. In addition to that, these two policies have been formally related with the permission grant policies defined in [2]. Furthermore, the theorem also formally relates these two latter granting policies, showing that the accumulative one is more permissive than the overwriting one.

The difference between accumulating permissions and overwriting permissions is subtle. The problem with accumulating permissions is that at any program point to approximate the permissions available for a given resource type it has to be considered all the consumptions and all the permissions granted for that resource type. Whereas in the overwriting grant policy it is enough to consider the last grant operation and the subsequent consume operations. This suggest that a static permission analysis might be simpler using the overwriting grant policy.

6 Conclusion and Further Work

This paper reports work concerning the formal specification and analysis of access control models for interactive mobile devices.

Here it has been presented an unprecedented framework, formalized using the proof-assistant Coq, that provides a uniform setting to define and analyse access control models which incorporate interactive permission requesting/granting mechanisms. In particular, the work presented here has focused on two distinguished permission models: the one defined by version 2.0 of MIDP and the one defined by Besson et al. in [2]. A drawback of MIDP permission model is that the user is forced to decide between tedious continuous interruption in interactive programs in order to grant a (one-shot) permission or otherwise to trust applications and concede almost irrestricted permission for it to access sensible resources. The model proposed in [2] is more flexible than MIDP's, allowing additional possibilities in the way permissions are granted. A characterization of both models in terms of a formal definition of grant policy has also been provided.

Another kind of permission policies can also be expressed in the framework. In particular, it can be adapted to introduce a notion of permission revocation, a permission mode not considered in MIDP. A revoke can be modeled in the permission overwriting approach, for instance, by assigning a zero multiplicity to a resource type. In the accumulative approach, revocation might be modeled using negative multiplicities. To introduce revocations, in turn, enables, without further changes to the framework, to model a notion of permission scope. One such scope would be grasped as the session interval delimited by an activation and a revocation of that permission.

An order relation $\sqsubseteq_g$ on grant policies has also been presented in this work. Two theorems have been established and their proofs discussed: one that states a sufficient condition to prove that two permission granting policies are related by that order, and another one that establishes a precise comparison of permission granting policies defined by the models. In particular it is formally proved that the accumulative grant policy is more permissive than the overwriting one. Furthermore, it has been shown that $\sqsubseteq_g$ defines a lattice structure with the policies $grant_{os}$ and $grant_{bk}$ as the bottom and top elements respectively, providing then a formal algebraic setting in which grant policies can be precisely related and compared.

Further work is the study and specification, using the formal setting provided by the framework, of algorithms for enforcing the security policies derived from different sort of permission models to control the access to sensitive resources of the devices. Moreover, one main objective is to extend the framework so as to be able to construct certified prototypes from the formal definitions of those algorithms.

References

[1] Bartoletti, M., Degano, P., Ferrari, G.-L.: Static analysis for stack inspection. Design and Implementation of Programming Languages 54 (2001)

[2] Besson, F., Dufay, G., Jensen, T.: A formal model of access control for mobile interactive devices. In: Gollmann, D., Meier, J., Sabelfeld, A. (eds.) ESORICS 2006. LNCS, vol. 4189, pp. 110–126. Springer, Heidelberg (2006)

[3] Besson, F., Jensen, T., Le Métayer, D., Thorn, T.: Model ckecking security properties of control flow graphs. Journal of Computer Security 9, 217–250 (2001)

[4] Coquand, T., Huet, G.: The Calculus of Constructions. In: Information and Computation, vol. 76, pp. 95–120. Academic Press, London (1988)

[5] Coquand, T., Paulin-Mohring, C.: Inductively defined types. In: Martin-Löf, P., Mints, G. (eds.) COLOG 1988. LNCS, vol. 417, pp. 50–66. Springer, Heidelberg (1990)

[6] JSR 118 Expert Group. Mobile information device profile for java 2 micro edition. version 2.0. Technical report, Sun Microsystems, Inc. and Motorola, Inc. (2002)

[7] JSR 37 Expert Group. Mobile information device profile for java 2 micro edition. version 1.0. Technical report, Sun Microsystems, Inc. (2000)

[8] Jensen, T., Le Métayer, D., Thorn, T.: Verification of control flow based security properties. In: Proc. of the 20th IEEE Symp. on Security and Privacy, pp. 89–103. IEEE Computer Society, New York (1999)

[9] Roushani, R., Betarte, G., Luna, C.: A Certified Access Controller for JME-MIDP 2.0 enabled Mobile Devices. In: I Chilean Workshop on Formal Methods, Punta Arenas, Chile. IEEE Computer Society, Los Alamitos (2008) (to be published)

[10] Sun Microsystems, Inc. Java Platform Micro Edition (last accessed October 2008), `http://java.sun.com/javame/index.jsp`

[11] The Coq Development Team. The Coq Proof Assistant Reference Manual – Version V8.1 (2006)

[12] Zanella Béguelin, S., Betarte, G., Luna, C.: A formal specification of the MIDP 2.0 security model. In: Dimitrakos, T., Martinelli, F., Ryan, P.Y.A., Schneider, S. (eds.) FAST 2006. LNCS, vol. 4691, pp. 220–234. Springer, Heidelberg (2007)

Proving Infinitary Normalization

Jörg Endrullis[1], Clemens Grabmayer[2], Dimitri Hendriks[1], Jan Willem Klop[1], and Roel de Vrijer[1]

[1] Vrije Universiteit Amsterdam, Department of Computer Science
De Boelelaan 1081a, 1081 HV Amsterdam, The Netherlands
joerg@few.vu.nl, diem@cs.vu.nl, jwk@cs.vu.nl, rdv@cs.vu.nl
[2] Universiteit Utrecht, Department of Philosophy
Heidelberglaan 8, 3584 CS Utrecht, The Netherlands
clemens@phil.uu.nl

Abstract. We investigate the notion of 'infinitary strong normalization' (SN^∞), introduced in [6], the analogue of termination when rewriting infinite terms. A (possibly infinite) term is SN^∞ if along every rewrite sequence each fixed position is rewritten only finitely often. In [9], SN^∞ has been investigated as a system-wide property, i.e. SN^∞ for all terms of a given rewrite system. This global property frequently fails for trivial reasons. For example, in the presence of the collapsing rule $\mathsf{tail}(x{:}\sigma) \to \sigma$, the infinite term $t = \mathsf{tail}(0{:}t)$ rewrites to itself only. Moreover, in practice one usually is interested in SN^∞ of a certain set of initial terms. We give a complete characterization of this (more general) 'local version' of SN^∞ using interpretations into weakly monotone algebras (as employed in [9]). Actually, we strengthen this notion to *continuous* weakly monotone algebras (somewhat akin to [5]). We show that tree automata can be used as an automatable instance of our framework; an actual implementation is made available along with this paper.

1 Introduction

In first-order term rewriting a major concern is how to prove termination, or in another terminology, originating in the tradition of the λ-calculus, how to prove strong normalization (SN), i.e. the property that all rewrite sequences must end eventually in a normal form. Numerous advanced techniques and tools have been developed to prove SN, including interpretations of terms in monotone algebras [7,8] and in weakly monotone algebras [4].

Another development in term rewriting, in line with the increased attention for coalgebraic and coinductive notions and techniques, was concerned with the generalization of finitary to infinitary rewriting, where normal forms are infinite objects such as streams or infinite trees. Such trees need not be well-founded. At first sight, termination is then no longer an issue. But a notion analogous to strong normalization emerges, bearing in mind the same goal of reaching normal forms. This is infinitary normalization, SN^∞, stating that eventually always a normal form will be reached, although, depending on the chosen rewriting strategy, this may take an infinite or even a transfinitely infinite number of steps.

S. Berardi, F. Damiani, and U. de'Liguoro (Eds.): TYPES 2008, LNCS 5497, pp. 64–82, 2009.

The property SN^∞ has been investigated in Klop and de Vrijer [6], where it is shown that it can be rephrased as: all transfinite rewrite sequences converge, or, equivalently, along every transfinite rewrite sequence each fixed term position is rewritten only finitely often.

Zantema [9] initiated the development of proof methods for infinitary normalization by adapting the weakly monotone algebras to the infinitary setting. As a matter of fact, Zantema also studies a weaker notion than SN^∞, which he calls SN^ω, and which states that all rewrite sequences of length ω are convergent, in the sense that throughout the infinite reduction any position is rewritten at most finitely often.[1]

The properties SN^∞ and SN^ω can be viewed *locally*, as properties of individual terms or of sets of terms in a TRS, or *globally*: the entire TRS is SN^∞ (or SN^ω) if all its terms are. In [9] only the global versions are investigated, obtaining characterization theorems for the global properties SN^ω and SN^∞.

The first objective of this paper is to adapt the method of weakly monotone algebras for proving *local* versions of SN^∞ and SN^ω, which means that we can parametrize these properties to arbitrary sets S of finite or infinite terms. The gain is that the global system-wide version may fail, whereas the local version for a set S of intended terms may still succeed. Thus we are able to fine-tune the infinitary termination result for just the terms we want, removing the spoiling effect of unintended terms. Note that the global properties are special cases of the local ones. In that sense our results generalize those of [9].

The characterization theorems in [9] impose a certain continuity requirement on the algebras. However, we found that for the characterization of the stronger property SN^∞ that requirement does not suffice. In order to obtain a full characterization of SN^∞ we will strengthen the requirement to what we call below *continuous* weakly monotone algebras. They appear to be connected to an early study of continuous algebraic semantics by Goguen et al. [5].

The second contribution of this paper is the employment of tree automata to actually prove SN^∞ for a set S of infinite terms. Here the tree automaton $\mathcal{T}$ plays a double role: first, it specifies the set S of intended terms, namely as those infinite terms generated by $\mathcal{T}$, and second, it provides a 'termination certificate' for S. Moreover, and here is the bridge between this second part and the first part described above, the tree automaton $\mathcal{T}$ gives rise to a continuous weakly monotone algebra that guarantees the property SN^∞ for S. Thus the tree automata method is an 'instance' of the general set-up using continuous weakly monotone algebras.

An explicit goal of our study is finding automatable methods to establish infinitary normalization properties. Indeed, finding such a tree automaton can be automated, and we provide and discuss the actual implementation of the search process using SAT solvers. The implementation is available via the web page: `http://infinity.few.vu.nl/sni/`

[1] This property SN^ω does not imply that in ω many steps a normal form will always be reached (see Remark 2.5). Therefore "ω-convergence" would seem a more appropriate name. To keep consistency we stick here to the terminology used in [9].

2 Infinitary Rewriting

We will consider a finite or infinite term as a function on a prefix-closed subset of $\mathbb{N}^*$ taking values in a first-order signature. A *signature* Σ is a finite set of symbols each having a fixed *arity* $\sharp(f) \in \mathbb{N}$. We use $\Sigma_n := \{f \in \Sigma \mid \sharp(f) = n\}$ for the set of *n-ary function symbols.*

Let $\mathcal{X}$ be a set of symbols, called *variables*, such that $\mathcal{X} \cap \Sigma = \varnothing$. Then, a *term over* Σ is a partial map $t : \mathbb{N}^* \to \Sigma \cup \mathcal{X}$ such that the *root* is defined, $t(\epsilon) \in \Sigma \cup \mathcal{X}$, and for all $p \in \mathbb{N}^*$ and all $i \in \mathbb{N}$ we have $t(pi) \in \Sigma \cup \mathcal{X}$ if and only if $t(p) \in \Sigma_n$ for some $n \in \mathbb{N}$ and $1 \leq i \leq n$. The set of (not necessarily well-founded) terms over Σ and $\mathcal{X}$ is denoted by $Ter^\infty(\Sigma, \mathcal{X})$. Usually we will write $Ter^\infty(\Sigma)$ for the set of terms over Σ and a countably infinite set of variables, which is assumed to be fixed as underlying the definition of terms.

The set of *positions* $\mathcal{P}os(t)$ *of a term* $t \in Ter^\infty(\Sigma)$ is the domain of t, that is, the set of values $p \in \mathbb{N}^*$ such that $t(p)$ is defined: $\mathcal{P}os(t) := \{p \in \mathbb{N}^* \mid t(p) \in \Sigma \cup \mathcal{X}\}$. Note that, by the definition of terms, the set $\mathcal{P}os(t)$ is prefix closed. A term t is called *finite* if the set $\mathcal{P}os(t)$ is finite. We write $Ter(\Sigma)$ for the set of finite terms. For positions $p \in \mathcal{P}os(t)$ we use $t|_p$ to denote the *subterm of* t *at position* p, defined by $t|_p(q) := t(pq)$ for all $q \in \mathbb{N}^*$.

For $f \in \Sigma_n$ and terms $t_1, \ldots, t_n \in Ter^\infty(\Sigma)$ we write $f(t_1, \ldots, t_n)$ to denote the term t defined by $t(\epsilon) = f$, and $t(ip) = t_i(p)$ for all $1 \leq i \leq n$ and $p \in \mathbb{N}^*$. For *constants* $c \in \Sigma_0$ we simply write c instead of $c()$. We use $x, y, z, \ldots$ to range over variables. We write $s \equiv t$ for *syntactic equivalence* of terms s and t, that is, if $\forall p \in \mathbb{N}^*.\ s(p) = t(p)$ and $s \equiv_{\leq n} t$ for *syntactic equivalence up to depth* n, that is, if for all positions p with length $|p| \leq n$ we have $s(p) = t(p)$.

A *substitution* is a map $\sigma : \mathcal{X} \to Ter^\infty(\Sigma, \mathcal{X})$. For terms $t \in Ter^\infty(\Sigma, \mathcal{X})$ and substitutions σ we define $t\sigma$ as the result of replacing each $x \in \mathcal{X}$ in t by $\sigma(x)$. Formally, $t\sigma$ is defined, for all $p \in \mathbb{N}^*$, by: $t\sigma(p) = \sigma(t(p_0))(p_1)$ if there exist $p_0, p_1 \in \mathbb{N}^*$ such that $p = p_0p_1$ and $t(p_0) \in \mathcal{X}$, and $t\sigma(p) = t(p)$, otherwise. Let $\Box$ be a fresh symbol, $\Box \notin \Sigma \cup \mathcal{X}$. A *context* C is a term from $Ter^\infty(\Sigma, \mathcal{X} \cup \{\Box\})$ containing precisely one occurrence of $\Box$. By $C[s]$ we denote the term $C\sigma$ where $\sigma(\Box) = s$ and $\sigma(x) = x$ for all $x \in \mathcal{X}$.

Dropping in the definition of terms the requirement that the number of subterms coincides with the arity of the symbols, we obtain the general notion of *labelled trees*. For trees we reuse the notation introduced above for terms.

Definition 2.1. An *infinitary term rewrite system (TRS)* is a set R of rewrite rules over a first-order signature Σ (and a set of variables $\mathcal{X}$): a *rewrite rule* is a pair $\langle \ell, r \rangle$ of terms $\ell, r \in Ter^\infty(\Sigma)$, usually written as $\ell \to r$, such that for *left-hand side* ℓ and *right-hand side* r we have $\ell(\epsilon) \notin \mathcal{X}$ and $Var(r) \subseteq Var(\ell)$.

Restriction. In this paper we restrict attention to TRSs R in which for all rules $\ell \to r \in R$ both ℓ and r are finite terms.

Definition 2.2. On the set of terms $Ter^\infty(\Sigma)$ we define a *metric* d by $d(s, t) = 0$ whenever $s \equiv t$, and $d(s, t) = 2^{-k}$ otherwise, where $k \in \mathbb{N}$ is the least length of all positions $p \in \mathbb{N}^*$ such that $s(p) \neq t(p)$.

Definition 2.3. Let R be a TRS over Σ. For terms $s, t \in Ter^{\infty}(\Sigma)$ and $p \in \mathbb{N}^*$ we write $s \to_{R,p} t$ if there exist $\ell \to r \in R$, a substitution σ and a context C with $C(p) = \Box$ such that $s \equiv C[\ell\sigma]$ and $t \equiv C[r\sigma]$. A step $s \to_{R,\epsilon} t$ is called a *root step*. We write $s \to_R t$ if there exists a position p such that $s \to_{R,p} t$.

A *transfinite rewrite sequence* (of length α) is a sequence of rewrite steps $(t_\beta \to_{R,p_\beta} t_{\beta+1})_{\beta<\alpha}$ such that for every limit ordinal $\lambda < \alpha$ we have that if β approaches λ from below (i) the distance $d(t_\beta, t_\lambda)$ tends to 0 and, moreover, (ii) the depth of the rewrite action, i.e. the length of the position p_β, tends to infinity. The sequence is called *strongly convergent* if the conditions (i) and (ii) are fulfilled for every limit ordinal $\lambda \leq \alpha$. In this case we write $t_0 \twoheadrightarrow\!\!\!\!\to_R t_\alpha$, or $t_0 \to^\alpha t_\alpha$ to explicitly indicate the length α of the sequence. Note that this ordinal will always be countable (see [6,7]). In the sequel we will use the familiar fact that countable limit ordinals have cofinality ω.

A transfinite rewrite sequence that is not strongly convergent will be called *divergent*. Note that all proper initial segments of a divergent reduction are yet strongly convergent.

Definition 2.4. A TRS R is *infinitary strongly normalizing on* $S \subseteq Ter^{\infty}(\Sigma)$, denoted $\mathsf{SN}^{\infty}_R(S)$, if every rewrite sequence starting from a term $t \in S$ is strongly convergent. We write $\mathsf{SN}^{\omega}_R(S)$ if all rewrite sequences of length $\leq \omega$ starting from a term $t \in S$ are strongly convergent. We write SN^{∞}_R shortly for $\mathsf{SN}^{\infty}_R(Ter^{\infty}(\Sigma))$, that is, infinitary normalization on all terms. Likewise SN^{ω}_R. Furthermore, the subscript R may be suppressed if it is clear from the context.

Remark 2.5. The notion SN^{ω} was introduced in [9]. Note that it does not imply that every reduction of length ω converges to a normal form, as examplified by a reduction $f(a, b) \to^{\omega} f(g^{\omega}, g^{\omega})$ in the TRS $\{a \to g(a),\ b \to g(b),\ f(x, x) \to c\}$. For the TRS R obtained by adding the extra rewrite rule $c \to c$ we will even have SN^{ω}_R without SN^{∞}_R. For this reason the terminology SN^{ω} seems a bit deceptive. We suggest to call it *ω-convergence*. For rewrite systems with rules that are left-linear and have finite left-hand sides the notions SN^{ω} and SN^{∞} coincide.

Infinitary strong normalization is related to root termination, as follows.

Definition 2.6. Let R be a TRS over Σ and $S \subseteq Ter^{\infty}(\Sigma)$. The *$\omega$-family $\mathcal{F}^{\omega}_R(S)$ of S* is the set of all subterms of $\twoheadrightarrow_R$-reducts of terms $t \in S$. Likewise the *∞-family $\mathcal{F}^{\infty}_R(S)$ of S* is the set of all subterms of $\twoheadrightarrow\!\!\!\!\to_R$-reducts of terms $t \in S$. We suppress the subscript R whenver R is clear from the context.

Definition 2.7. We call a term $t \in Ter^{\infty}(\Sigma)$ *root terminating* if t admits no rewrite sequence of length $\leq \omega$ which contains infinitely many root steps. Likewise, t is called *∞-root terminating* if t does not admit a transfinite reduction containing infinitely many root steps.

We obtain the following lemma, a refinement of Theorem 2 in [6].

Lemma 2.8. *A set of terms $S \subseteq Ter^{\infty}(\Sigma)$ is $\mathsf{SN}^{\infty}_R(S)$ if and only if all ∞-family members $t \in \mathcal{F}^{\infty}(S)$ are ∞-root terminating. Likewise we have $\mathsf{SN}^{\omega}_R(S)$ if and only if all ω-family members $t \in \mathcal{F}^{\omega}(S)$ are root terminating.*

Proof. For the 'only if'-direction, assume there exists a term $t \in \mathcal{F}^\infty(S)$ which admits a rewrite sequence $t \twoheadrightarrow$ containing infinitely many root steps. Then there exists a divergent rewrite sequence $s \twoheadrightarrow C[t] \twoheadrightarrow$ for some $s \in S$.

For the 'if'-direction, assume that $\mathsf{SN}^\infty_R(S)$ does not hold. Then there exists a rewrite sequence $\sigma : s \twoheadrightarrow$ for some $s \in S$ which is not strongly convergent. Then for some depth $d \in \mathbb{N}$ there are infinitely many rewrite steps at depth d in σ; let d be minimal with this property. There are only finitely many steps above depth d and therefore σ factors into $\sigma : s \twoheadrightarrow s' \twoheadrightarrow$ such that after s' there are no rewrite steps above depth d (but infinitely many steps at depth d). The term s' has only finitely many subterms at depth d, and by the Pigeonhole Principle one of these subterms admits a rewrite sequence containing infinitely many root steps. Hence there exists a term $t \in \mathcal{F}^\infty(S)$ which is not root terminating.

The proof for $\mathsf{SN}^\omega_R(S)$ proceeds analogously. □

3 Characterizations of Local SN^ω and Local SN^∞

We give a complete characterization of the local version of SN^∞, based on an extension of the monotone algebra approach of [9].

Definition 3.1. A Σ-algebra $\langle A, [\cdot]\rangle$ consists of a non-empty set A and for each n-ary $f \in \Sigma$ a function $[f] : A^n \to A$, the *interpretation of* f.

Let $\mathcal{A} = \langle A, [\cdot]\rangle$ be a Σ-algebra, and $\alpha : \mathcal{X} \to A$ be an assignment of variables. The interpretation of finite terms $t \in \mathit{Ter}(\Sigma)$ is inductively defined as follows:

$$[x]^\alpha := \alpha(x) \qquad\qquad [f(t_1, \ldots, t_n)]^\alpha := [f]([t_1]^\alpha, \ldots, [t_n]^\alpha)$$

For ground terms $t \in \mathit{Ter}(\Sigma, \varnothing)$ we write $[t]$ for short, since the interpretation does not depend on α. We define the interpretation $[t]$ of infinite terms t as the limit of the interpretations of finite terms converging towards t. In the sequel we assume (without loss of generality) that the signature Σ contains at least one constant symbol; in case it does not, we add one. This ensures that every infinite term is indeed the limit of a sequence of finite terms.

Let A_i, A be sets equipped with metrics. A function $f : A_1 \times \ldots \times A_n \to A$ is *continuous* if whenever for $i = 1, \ldots, n$ the sequence $a_{i,1}, a_{i,2}, \ldots$ in A_i converges with limit a_i, then $\lim_{j\to\infty} f(a_{1,j}, \ldots, a_{n,j})$ exists and is equal to $f(a_1, \ldots, a_n)$.

Definition 3.2. A Σ-algebra $\langle A, [\cdot], d\rangle$ equipped with a metric $d : A \times A \to \mathbb{R}^+_0$ is called *continuous* if:

(i) for every $f \in \Sigma$ the function $[f]$ is continuous, and
(ii) for every sequence $\{t_i\}_{i\in\mathbb{N}}$ of finite ground terms $t_i \in \mathit{Ter}(\Sigma, \varnothing)$ that is convergent in $\mathit{Ter}^\infty(\Sigma, \varnothing)$, the sequence $\{[t_i]\}_{i\in\mathbb{N}}$ is convergent.

Note that clause (ii) of Definition 3.2 is a necessary and sufficient condition for the existence of a unique continuous extension $[\cdot] : \mathit{Ter}^\infty(\Sigma) \to \mathcal{A}$ to (possibly) infinite terms of the interpretation $[\cdot] : \mathit{Ter}(\Sigma) \to \mathcal{A}$. As a matter of fact this observation motivates the definition.

Lemma 3.3. *Let $\mathcal{A} = \langle A, [\cdot]\rangle$ be a continuous Σ-algebra. Let $t \in Ter(\Sigma, \mathcal{X})$ be a finite term, and $\sigma : \mathcal{X} \to Ter^{\infty}(\Sigma, \varnothing)$ a ground substitution. We define the map $\alpha : \mathcal{X} \to A$ for all $x \in \mathcal{X}$ by $\alpha(x) = [\sigma(x)]$. Then we have $[t\sigma] = [t]^{\alpha}$.*

Proof. We use induction on the term structure of t. The case of t being a variable is trivial, hence assume $t = f(t_1, \dots, t_n)$. For $i = 1, \dots, n$ let $\{t_{i,j}\}_{j \in \mathbb{N}}$ be a sequence of finite terms converging towards $t_i\sigma$. Then we have:

$$\begin{aligned}
[t\sigma] &= \lim_{j\to\infty}[f(t_{1,j}, \dots, t_{n,j})] && \text{by continuity of } [\cdot] \\
&= [f](\lim_{j\to\infty}[t_{1,j}], \dots, \lim_{j\to\infty}[t_{n,j}]) && \text{by continuity of } f \\
&= [f]([t_1\sigma], \dots, [t_n\sigma]) = [f]([t_1]^{\alpha}, \dots, [t_n]^{\alpha}) = [t]^{\alpha} && \text{by IH}
\end{aligned}$$

□

Let R be a binary relation on A. A function $f : A^n \to A$ is *monotone* with respect to R if $a \mathrel{R} b$ implies $f(\dots, a, \dots) \mathrel{R} f(\dots, b, \dots)$ for every $a, b \in A$.

Definition 3.4. A *weakly monotone Σ-algebra* $\mathcal{A} = \langle A, [\cdot], \succ, \sqsupseteq\rangle$ is a Σ-algebra $\langle A, [\cdot]\rangle$ where $\succ$ is a strict partial order, and $\sqsupseteq$ a quasi-order, on A such that:

(i) $\succ$ is well-founded,
(ii) $\forall xyz. (x \succ y \sqsupseteq z \Rightarrow x \succ z)$ and $\forall xy. (x \succ y \Rightarrow x \sqsupseteq y)$ (compatibility), and
(iii) for every symbol $f \in \Sigma$ the function $[f]$ is monotone with respect to $\sqsupseteq$.

A *weakly monotone Σ-algebra with undefined elements* is a weakly monotone Σ-algebra $\mathcal{A} = \langle A, [\cdot], \succ, \sqsupseteq\rangle$ with a set $\Omega \subseteq A$ of *undefined elements* for which:

(iv) for every $b \in \Omega$ and $a \in A \setminus \Omega$ we have $b \succ a$ (maximality), and
(v) for every $f \in \Sigma$ and $b \in \Omega$ we have $[f](\dots, b, \dots) \in \Omega$ (strictness).

All of the results in this paper remain valid if instead of requiring $\succ$ to be a strict partial order and $\sqsupseteq$ a quasi-order we allow arbitrary binary relations fulfilling conditions (i)–(v) of Definition 3.4.

Remark 3.5. The reason to consider weakly monotone algebras with more than just one undefined element is the following. For every TRS R, we want to be able to build a continuous weakly monotone algebra from the term algebra with carrier-set $Ter^{\infty}(\Sigma)$ by interpreting the terms t with $\mathsf{SN}^{\infty}_{R}(\{t\})$ by themselves, and the other terms by suitably chosen undefined objects. However, by just dropping the terms t that are not SN^{∞}_{R}, and replacing them by a single undefined element usually a continuous algebra is not obtained.

For example, let $\Sigma = \{\mathsf{I}, \mathsf{J}, \mathsf{c}\}$, where I, J are unary function symbols and c a constant. Let R be the (orthogonal) TRS over Σ with the rules $\mathsf{I}(x) \to x$ and $\mathsf{J}(x) \to x$. Here the terms $t \in Ter^{\infty}(\Sigma)$ with $\mathsf{SN}^{\infty}_{R}(\{t\})$ are precisely the finite terms, the terms $t \in Ter(\Sigma)$. Now suppose that $\mathcal{A} = \langle A, [\cdot], d_A, \succ, \sqsupseteq\rangle$ is a continuous, weakly monotone algebra with $A \supseteq Ter(\Sigma)$, an interpretation $[\cdot] : \Sigma \to A$ with the property that $[f]([t_1], \dots, [t_n]) = [f(t_1, \dots, t_n)]$ for all $f \in Ter(\Sigma)$, and d_A an extension of the metric in Definition 2.2. Then we find that $A \setminus Ter(\Sigma)$ contains more than one element (and in fact uncountably many elements). Note that for the induced interpretation function

$[\cdot] : Ter^{\infty}(\Sigma) \to A$ it holds that $[t] = t$ for all $t \in Ter(\Sigma)$. We find that $[\mathsf{I}^{\omega}] = [\mathsf{I}(\mathsf{I}(\mathsf{I}(\ldots)))] = [\lim \mathsf{I}^{n}(x)] = \lim[\mathsf{I}^{n}(x)] = \lim \mathsf{I}^{n}(x) \in A \setminus Ter(\Sigma)$, and similarly, $[\mathsf{J}^{\omega}] = \lim \mathsf{J}^{n}(x) \in A \setminus Ter(\Sigma)$. From this we conclude that the interpretations $[\mathsf{I}^{\omega}]$ and $[\mathsf{J}^{\omega}]$ of the infinite terms I^{ω} and J^{ω} are different elements in $A \setminus Ter^{\infty}(\Sigma)$: $[\mathsf{I}^{\omega}] \neq [\mathsf{J}^{\omega}]$ follows from $d_A([\mathsf{I}^{\omega}], [\mathsf{J}^{\omega}]) = d_A(\lim \mathsf{I}^{n}(x), \lim \mathsf{J}^{n}(x)) = \lim d_A(\mathsf{I}^{n}(x), \mathsf{J}^{n}(x)) = \lim d(\mathsf{I}^{n}(x), \mathsf{J}^{n}(x)) = 1$.

Definition 3.6. Let $\mathcal{A} = \langle A, [\cdot], \succ, \sqsupseteq \rangle$ be a weakly monotone Σ-algebra with undefined elements Ω.

(i) A set $S \subseteq Ter^{\infty}(\Sigma, \varnothing)$ is called *defined w.r.t.* Ω if, for all $s \in S$, $[s] \notin \Omega$.
(ii) A TRS R over Σ is called *(weakly) decreasing w.r.t.* Ω if for all $\ell \to r \in R$ and every assignment $\alpha : \mathcal{X} \to A$, $[\ell]^{\alpha} \notin \Omega$ implies $[\ell]^{\alpha} \succ [r]^{\alpha}$ ($[\ell]^{\alpha} \sqsupseteq [r]^{\alpha}$).

Theorem 3.7. *Let R be a TRS over Σ, and $S \subseteq Ter^{\infty}(\Sigma, \varnothing)$. Then the following statements are equivalent:*

(i) $\mathsf{SN}^{\omega}_{R}(S)$.
(ii) There exists a continuous weakly monotone Σ-algebra $\mathcal{A} = \langle A, [\cdot], d, \succ, \sqsupseteq \rangle$ with a set Ω of undefined elements such that S is defined w.r.t. Ω, and R is decreasing with respect to Ω.

Proof. For (i) ⇒ (ii) assume that $\mathsf{SN}^{\omega}_{R}(S)$ holds. We define $\mathcal{A} := \langle A, [\cdot], d, \succ, \sqsupseteq \rangle$ with $A := Ter^{\infty}(\Sigma, \varnothing)$, equipped with the metric d on A from Definition 2.2, and let $\Omega := A \setminus \mathcal{F}^{\omega}(S)$ be the set of undefined elements. We define the relations $\succ := (\to_{R,\epsilon} \cdot \to^{*}) \cap (\mathcal{F}^{\omega}(S) \times \mathcal{F}^{\omega}(S))$ and $\sqsupseteq := \to^{*}$, extended by $s \succ t$ for all $s \in \Omega$, $t \in \mathcal{F}^{\omega}(S)$ and $s \sqsupseteq t$ for all $s \in \Omega$, $t \in \mathcal{A}$. The interpretation $[\cdot]$ is defined for all $f \in \Sigma$ by $[f](t_1, \ldots, t_n) = f(t_1, \ldots, t_n)$.

Clearly $\mathcal{A}$ is a continuous Σ-algebra; we check that $\mathcal{A}$ is a weakly monotone Σ-algebra with undefined elements Ω. Assume that $\succ$ would not be well-founded. Then there exists a term $t \in \mathcal{F}^{\omega}(S)$ admitting an ω-rewrite sequence containing infinitely many root steps, contradicting $\mathsf{SN}^{\omega}_{R}(S)$. The compatibility $\succ \cdot \sqsupseteq \subseteq \succ$ and $\succ \subseteq \sqsupseteq$ holds by definition. For every $b \in \Omega$ and $a \in A \setminus \Omega$ we have $b \succ a$ by definition. Furthermore $b \in \Omega$ implies $[f](\ldots, b, \ldots) = f(\ldots, b, \ldots) \in \Omega$, since the family $\mathcal{F}^{\omega}(S)$ is closed under subterms. For monotonicity with respect to $\sqsupseteq$, we consider $f \in \Sigma$ and $s, t \in \mathcal{A}$ with $s \sqsupseteq t$. If $s \in \Omega$ then $[f](\ldots, s, \ldots) \in \Omega \sqsupseteq [f](\ldots, t, \ldots)$. If $s \in \mathcal{F}^{\omega}(S)$, then $[f](\ldots, s, \ldots) \sqsupseteq [f](\ldots, t, \ldots)$ as a consequence of the closure of rewriting $\to^{*}$ under contexts.

We check the remaining requirements of the theorem. For all $s \in S$ we have $[s] \notin \Omega$ by definition. Consider $\ell \to r \in R$ and $\alpha : \mathcal{X} \to A_I$ such that $[\ell]^{\alpha} \notin \Omega$. Then $[\ell]^{\alpha} \in \mathcal{F}^{\omega}(S)$ and hence $\alpha(x) \in \mathcal{F}^{\omega}(S)$ for all $x \in Var(\ell)$. Therefore we obtain $[\ell]^{\alpha} \equiv \ell\alpha \to_{R,\epsilon} r\alpha \equiv [r]^{\alpha}$ and $[r]^{\alpha} \in \mathcal{F}^{\omega}(S)$, hence $[\ell]^{\alpha} \succ [r]^{\alpha}$.

For (ii) ⇒ (i) assume that $\mathcal{A} := \langle A, [\cdot], \succ, \sqsupseteq \rangle$ and Ω fulfilling the requirements of the theorem are given. We show the following auxiliary lemmas:

$$\forall s, t \in Ter^{\infty}(\Sigma).\ [s] \notin \Omega \wedge s \to t \Rightarrow [t] \notin \Omega \wedge [s] \sqsupseteq [t] \qquad (*)$$

$$\forall s.\ [s] \notin \Omega \Rightarrow \forall t \in \mathcal{F}^{\omega}(s).\ [t] \notin \Omega \qquad (**)$$

Let $s, t \in Ter^{\infty}(\Sigma)$ with $[s] \notin \Omega$ and $s \to t$. There exist a context C, a rule $\ell \to r \in R$ and a substitution σ such that $s \equiv C[\ell\sigma] \to C[r\sigma] \equiv t$. By Lemma 3.3 together with the assumptions we obtain $[\ell\sigma] = [\ell]^{\alpha} \succ [r]^{\alpha} = [r\sigma]$ where the map $\alpha : \mathcal{X} \to A$ is defined by $\alpha(x) = [\sigma(x)]$ for all $x \in \mathcal{X}$. Since $\succ \subseteq \sqsupseteq$ and $[f]$ is monotone with respect to $\sqsupseteq$ for $f \in \Sigma$, we obtain $[s] \sqsupseteq [t]$. Furthermore $[t] \notin \Omega$, otherwise $[t] \in \Omega \succ [s] \sqsupseteq [t]$ and hence $[t] \succ [t]$, contradicting well-foundedness of $\succ$. We obtain $(**)$ by induction together with 'monotonicity' of Ω.

Assume $\mathsf{SN}^{\omega}_{R}(S)$ would not hold. By Lemma 2.8 there exists a term $t_0 \in \mathcal{F}^{\omega}(S)$ which admits an ω-reduction $t_0 \to t_1 \to \ldots$ containing infinitely many root steps. Then $t_0 \in \mathcal{F}^{\omega}(s)$ for some $s \in S$ and by assumption $[s] \notin \Omega$, hence by $(**)$ we obtain $t_i \notin \Omega$ for all $i \in \mathbb{N}$. Furthermore by $(*)$ if follows $[t_i] \sqsupseteq [t_{i+1}]$ for all $i \in \mathbb{N}$. Moreover for root steps $t_i \to_{R,\epsilon} t_{i+1}$ we get $[t_i] \succ [t_{i+1}]$ since then the context C in the proof of $(*)$ is empty. As a consequence we have infinitely often a strict decrease $\succ$ in the sequence $[t_0] \sqsupseteq [t_1] \ldots$, and by applying $\succ \cdot \sqsupseteq \subseteq \succ$ we can remove all $\sqsupseteq$ between them; giving rise to an infinite decreasing $\succ$-sequence, contradicting well-foundedness of $\succ$. □

Remark 3.8. A close inspection of the above proof yields that for Theorem 3.7 the requirement on the algebra to be continuous can be weakened. It suffices to require that for every infinite ground term t the sequence $[\mathsf{trunc}(t,n)]$ converges for $n \to \infty$. Here $\mathsf{trunc}(t,n)$ stands for the *truncation of t at depth n* defined for all $p \in \mathbb{N}^*$ by $\mathsf{trunc}(t,n)(p)$ is $t(p)$ if $|p| < n$, $\bot$ if $|p| = n$, and undefined, otherwise; where $\bot$ is an arbitrary, fixed constant symbol from the signature Σ.

However, we emphasise that for the characterization of $\mathsf{SN}^{\infty}_{R}(S)$ this weaker condition is *not* sufficient. Continuity of $[\cdot] : Ter^{\infty}(\Sigma) \to A$ is essential for the correctness of Theorem 3.10. It guarantees that for the limit steps in transfinite rewrite sequences, the limit of the interpretations coincides with the interpretation of the limit term.

We note that the weaker continuity condition used in [9, Theorem 3] does not suffice; see Example 3.9. Strengthening the condition to full continuity of the interpretation mapping would validate the theorem.

Example 3.9. We consider a TRS R which is SN^{ω} but not SN^{∞}. Interestingly, although the TRS is SN^{ω}, we display a term of which a normal form cannot be reached in ω many steps. Let R be the TRS consisting of the following rules:

$$f(x,x) \to f(A,B) \qquad A \to s(A) \qquad B \to s(B)\,.$$

It is not difficult to verify that R is indeed SN^{ω}, but SN^{∞} does not hold:

$$f(A,B) \to f(s(A),B) \to f(s(A),s(B)) \twoheadrightarrow f(s^{\omega},s^{\omega}) \to f(A,B) \to \ldots\,.$$

Note that the TRS R forms a counterexample to [9, Theorem 3], as the following Σ-algebra $\mathcal{A}$ fulfills all requirements of the theorem, but SN^{∞} does not hold. We choose the Σ-algebra $\mathcal{A} = \{A, B, F, a, b, f\}$ with $A \succ a$, $B \succ b$, $F \succ f$ and $\sqsupseteq := \succ \cup =$. The interpretation $[\cdot]$ is defined as follows:

$$
\begin{aligned}
[A] &= A \\
[B] &= B \\
[s](A \mid a) &= a \qquad\qquad [s](B \mid b) = b \qquad\qquad [s](F \mid f) = f \\
[f](A \mid a, B \mid b) &= f \qquad\qquad [f](\text{otherwise}) = F
\end{aligned}
$$

where | denotes ' or' and as truncation symbol c we chose $c := A$. Furthermore, for the metric we choose $d(x, y) = 0$ if $x = y$ and 1 otherwise. Then for all variable interpretations $\alpha : \mathcal{X} \to \mathcal{A}$ we have:

$$
\begin{aligned}
[f(x,x)]^\alpha &= F > f = [f(A,B)]^\alpha \\
[A]^\alpha &= A > a = [s(A)]^\alpha \\
[B]^\alpha &= B > b = [s(B)]^\alpha .
\end{aligned}
$$

Thus all rules are strictly decreasing. It is straightforward to verify that all functions $[g]$ are continuous, for every infinite ground term t the sequence $[\mathsf{trunc}(t, n)]$ converges (with limit in $\mathcal{A}$) for $n \to \infty$, and for every descending sequence $a_1 \sqsupseteq a_2 \sqsupseteq \cdots$ for which $\lim_{n\to\infty} a_i$ exists we have $a_1 \sqsupseteq \lim_{n\to\infty} a_i$.

Let A be a set equipped with a metric d and let $\sqsupseteq$ be a binary relation on A. We call the relation $\sqsupseteq$ *compatible with limits* if for every converging sequence $\{a_i\}_{i\in\mathbb{N}}$ with $a_0 \sqsupseteq a_1 \sqsupseteq \ldots$ we have $a_0 \sqsupseteq \lim_{i\to\infty} a_i$.

Theorem 3.10. *Let R be a TRS over Σ and $S \subseteq Ter^\infty(\Sigma, \varnothing)$. Then the following statements are equivalent:*

(i) $\mathsf{SN}^\infty_R(S)$.
(ii) There exists a continuous weakly monotone Σ-algebra $\mathcal{A} = \langle A, [\cdot], d, \succ, \sqsupseteq\rangle$ with a set Ω of undefined elements such that S is defined w.r.t. Ω, R is decreasing with respect to Ω, and $\sqsupseteq$ is compatible with limits.

Proof. We give the crucial steps for both directions. The remainder of the proof proceeds analogously to the proof of Theorem 3.7.

For (i) $\Rightarrow$ (ii) assume that $\mathsf{SN}^\infty_R(S)$ holds. We define $\mathcal{A} := \langle A, [\cdot], d, \succ, \sqsupseteq\rangle$ with $A := Ter^\infty(\Sigma, \varnothing)$, d the metric from Definition 2.2, and $\Omega := A \setminus \mathcal{F}^\infty(S)$; we define the relations $\succ := (\to_{R,\epsilon} \cdot \twoheadrightarrow) \cap (\mathcal{F}^\infty(S) \times \mathcal{F}^\infty(S))$, $\sqsupseteq := \twoheadrightarrow \cap (\mathcal{F}^\infty(S) \times \mathcal{F}^\infty(S))$, extended by $s \succ t$ for all $s \in \Omega$, $t \in \mathcal{F}^\infty(S)$ and $s \sqsupseteq t$ for all $s \in \Omega$, $t \in \mathcal{A}$. The interpretation $[\cdot]$ is defined for all $f \in \Sigma$ by $[f](t_1, \ldots, t_n) = f(t_1, \ldots, t_n)$. Consider a sequence $a_0 \sqsupseteq a_1 \sqsupseteq \ldots$ with $a_0 \in \mathcal{F}^\infty(S)$. Then $a_0 \twoheadrightarrow a_1 \twoheadrightarrow \ldots$ by definition and by $\mathsf{SN}^\infty_R(S)$ we obtain that $a := \lim_{i\to\infty} a_i$ exists, $a_0 \twoheadrightarrow a$ and $a_0 \sqsupseteq a$. Hence $\sqsupseteq$ is compatible with limits.

For the implication (ii) $\Rightarrow$ (i), the crucial step is to show that $s \twoheadrightarrow t$ implies $s \sqsupseteq t$. We use induction on the length of the rewrite sequence $s \to^\alpha t$. Note that the length α of a reduction is a countable ordinal, c.f. [6]. For $\alpha = \beta + 1$ we obtain $s \sqsupseteq t$ by induction hypothesis together with $(*)$ from the proof of Theorem 3.7. Assume that α is a (countable) limit ordinal. Then there exists a non-decreasing sequence $\{\beta_i\}_{i\in\mathbb{N}}$ of ordinals $\beta_i < \alpha$ such that $\alpha = \lim_{i\to\infty} \beta_i$. Let s_γ denote the term before the γ-th rewrite step in $s \to^\alpha t$. Then $s \twoheadrightarrow s_{\beta_1} \twoheadrightarrow s_{\beta_2} \cdots$

and $t = \lim_{i\to\infty} s_{\beta_i}$. Hence by induction hypothesis $s \sqsupseteq s_{\beta_1} \sqsupseteq s_{\beta_2} \ldots$; and by compatibility of $\sqsupseteq$ with limits we obtain $s \sqsupseteq t$. This gives us a handle for limit steps; the rest of the proof is analogous to the proof of Theorem 3.7. □

Finally, we generalize the Theorems 3.7 and 3.10 together with the concept of 'root termination' allowing for simpler, stepwise proofs of $\mathsf{SN}^\infty_R(S)$. This facility is incorporated in our tool. The following definition and theorem allow for modular proofs of SN^∞ and root termination of infinite terms. This is reminiscent to modular proofs of finitary root termination [1] (the dependency pairs method).

Definition 3.11. Let R_1 and R_2 be TRS over Σ, and $S \subseteq \mathit{Ter}^\infty(\Sigma)$. We say that R_1 *is ∞-root terminating relative to R_2 on S*, denoted $\mathsf{RT}^\infty_{R_1/R_2}(S)$, if no $s \in S$ admits a $\to_{R_1,\epsilon} \cup \to_{R_2}$-reduction containing infinitely many $\to_{R_1,\epsilon}$-steps.

We say R_1 *is root terminating relative to R_2 on S*, denoted $\mathsf{RT}^\omega_{R_1/R_2}(S)$, if the condition holds for rewrite sequences of length $\leq \omega$.

The following lemma is a direct consequence of Lemma 2.8 and Definition 3.11.

Lemma 3.12. *(i)* $\mathsf{SN}^\infty_R(S) \Leftrightarrow \mathsf{RT}^\infty_{R/R}(\mathcal{F}^\infty(S))$; *(ii)* $\mathsf{SN}^\omega_R(S) \Leftrightarrow \mathsf{RT}^\omega_{R/R}(\mathcal{F}^\omega(S))$.

For proving $\mathsf{SN}^\infty_R(S)$ using Theorem 3.10 we have to make all rules in R decreasing at once. For practical purposes it is often desirable to prove $\mathsf{SN}^\infty_R(S)$ stepwise, by repeatedly removing rules until no top-rules remain, that is, $\mathsf{RT}^\infty_{\varnothing/R}(\mathcal{F}^\infty(S))$ trivially holds. The following theorem enables us to do this, we can remove all decreasing rules, as long as the remaining rules are weakly decreasing.

Theorem 3.13. *Let $R_1 \subseteq R_2$, $R_1' \subseteq R_2$ be TRS over Σ, and $S \subseteq \mathit{Ter}^\infty(\Sigma, \varnothing)$. Let $\mathcal{A} = \langle A, [\cdot], d, \succ, \sqsupseteq\rangle$ be a continuous weakly monotone Σ-algebra with a set Ω of undefined elements such that S is defined w.r.t. Ω and it holds:*

(i) $R_1 \cup R_2$ is weakly decreasing with respect to Ω, and
(ii) R_1' is decreasing with respect to Ω.

Then $\mathsf{RT}^\omega_{R_1/R_2}(\mathcal{F}^\omega_{R_2}(S))$ implies $\mathsf{RT}^\omega_{(R_1\cup R_1')/R_2}(\mathcal{F}^\omega_{R_2}(S))$. If additionally $\sqsupseteq$ is compatible with limits, then $\mathsf{RT}^\infty_{R_1/R_2}(\mathcal{F}^\infty_{R_2}(S))$ implies $\mathsf{RT}^\infty_{(R_1\cup R_1')/R_2}(\mathcal{F}^\infty_{R_2}(S))$.

Proof. Minor modification of the proofs of Theorem 3.7 and 3.10, respectively. □

4 Tree Automata

We now come to the second contribution of our note, consisting of an application of tree automata to prove infinitary strong normalization, SN^∞, and a connection of tree automata with the algebraic framework treated above. For the notion of tree automata the reader is referred to [2]. We repeat the main definitions, for the sake of completeness, and to fix notations.

Definition 4.1. A *(finite nondeterministic top-down) tree automaton $\mathcal{T}$ over a signature Σ* is a tuple $\mathcal{T} = \langle Q, \Sigma, I, \Delta\rangle$ where Q is a finite set of *states*, disjoint from Σ; $I \subseteq Q$ is a set of *initial states*, and $\Delta \subseteq Ter(\Sigma \cup Q, \varnothing)^2$ is a ground term rewriting system over $\Sigma \cup Q$ with rules, or *transitions*, of the form:

$$q \to f(q_1, \ldots, q_n)$$

for n-ary $f \in \Sigma$, $n \geq 0$, and $q, q_1, \ldots, q_n \in Q$.

We define the notion of 'run' of an automaton on a term. For terms containing variables, we assume that a map $\alpha : \mathcal{X} \to \mathbf{2}^Q$ is given, so that each variable $x \in \mathcal{X}$ can be generated by any state from $\alpha(x)$.

Definition 4.2. Let $\mathcal{T} = \langle Q, \Sigma, I, \Delta \rangle$ be a tree automaton. Let $t \in Ter^{\infty}(\Sigma, \mathcal{X})$ be a term, $\alpha : Var(t) \to \mathbf{2}^Q$ a map from variables to sets of states, and $q \in Q$. Then a *q-run of $\mathcal{T}$ on t with respect to α* is a tree $\rho : \mathcal{P}os(t) \to Q$ such that:

(i) $\rho(\epsilon) = q$, and
(ii) $\rho(p) \to t(p)(\rho(p1), \ldots, \rho(pn)) \in \Delta$ for all $p \in \mathcal{P}os(t)$ with $t(p) \in \Sigma_n$, and
(iii) $\rho(p) \in \alpha(t(p))$ for all $p \in \mathcal{P}os(t)$ with $t(p) \in \mathcal{X}$.

We define $Q_\alpha(t) := \{q \in Q \mid \text{there exists a } q\text{-run of } \mathcal{T} \text{ on } t \text{ with respect to } \alpha\}$.

For ground terms t the above notions are independent of α. Then we say $\mathcal{T}$ *has a q-run on a term t* and write $Q(t)$ in place of $Q_\alpha(t)$. Moreover, we say that an automaton $\mathcal{T}$ *generates* a ground term t if $\mathcal{T}$ has a q-run on t such that $q \in I$. The language of an automaton is the set of ground terms it generates.

Definition 4.3. The *language* $\mathcal{L}(\mathcal{T})$ of a tree automaton $\mathcal{T}$ is defined by:

$$\mathcal{L}(\mathcal{T}) := \{t \in Ter^{\infty}(\Sigma, \varnothing) \mid Q(t) \cap I \neq \varnothing\}.$$

$\mathcal{T}$ is called *complete* if it generates all ground terms, i.e. if $\mathcal{L}(\mathcal{T}) = Ter^{\infty}(\Sigma, \varnothing)$.

Example 4.4. Consider the tree automaton $\mathcal{T} = \langle Q, \Sigma, I, \Delta \rangle$ with $Q := \{0, 1\}$, $I := \{0\}$, and with Δ consisting of the rules:

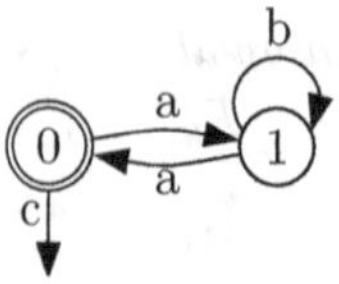

$$0 \to \mathsf{a}(1) \mid \mathsf{c} \qquad\qquad 1 \to \mathsf{a}(0) \mid \mathsf{b}(1)$$

where $\ell \to r_1 \mid \ldots \mid r_n$ is shorthand for rules $(\ell \to r_i)_{1 \leq i \leq n}$. The language of $\mathcal{T}$ is $\mathcal{L}(\mathcal{T}) = (\mathsf{a}\,\mathsf{b}^*\mathsf{a})^*\mathsf{c} \mid (\mathsf{a}\,\mathsf{b}^*\mathsf{a})^\omega \mid (\mathsf{a}\,\mathsf{b}^*\mathsf{a})^*\mathsf{a}\,\mathsf{b}^\omega$.

The following lemma states a continuity property of tree automata.

Lemma 4.5. *Let $\mathcal{T} = \langle Q, \Sigma, I, \Delta \rangle$ be a tree automaton, $q \in Q$, and $t \in Ter^{\infty}(\Sigma)$. Then $q \in Q(t)$ if and only if for all $n \in \mathbb{N}$ exists t_n with $q \in Q(t_n)$ and $t \equiv_{\leq n} t_n$.*

Proof. The 'only if'-direction is trivial, take $t_n := t$ for all $n \in \mathbb{N}$.

For the 'if'-direction, we prove $q \in Q(t)$ by constructing a q-run $\rho : \mathcal{P}os(t) \to Q$ of $\mathcal{T}$ on t. For ever $i \in \mathbb{N}$ there exists a q-run ρ_{t_i} of $\mathcal{T}$ on t_i by assumption. Define $T_0 := \{t_i \mid i \in \mathbb{N}\}$. In case T_0 is finite, then it follows that $t \in T_0$ and $q \in Q(t)$. Hence assume that T_0 is infinite.

First we define a decreasing sequence $T_0 \supseteq T_1 \supseteq T_2 \supseteq \ldots$ of infinite subsets of T_0 by induction as follows. Assume that T_i has already been obtained. By the Pigeonhole Principle there exists an infinite subset $T_{i+1} \subseteq T_i$ such that for all $v_1, v_2 \in T_{i+1}$ we have $v_1 \equiv_{\leq i} v_2$ and $\rho_{v_1} \equiv_{\leq i} \rho_{v_2}$.

We define the q-run ρ on t as follows. For each $i \in \mathbb{N}$ we pick a term $s_i \in T_{i+1}$ and define $\rho(p) := \rho_{s_i}(p)$ for all $p \in \mathcal{P}os(t)$ with $|p| = i$. Note that the definition of ρ does not depend no the choice of s_i. Furthermore note that for every $i \in \mathbb{N}$ the term s_i coincides with the term s_{i+1} on all positions $p \in \mathcal{P}os(t)$ with $|p| = i + 1$. Therefore the condition $\rho(p) \to t(p)(\rho(p1), \ldots, \rho(pn)) \in \Delta$ for every $p \in \mathcal{P}os(t)$ follows from $s_{|p|}$ fulfilling this condition. Hence ρ is a q-run on t and $q \in Q(t)$. □

Lemma 4.6. *Each of the following properties imply completeness of a tree automaton $\mathcal{T} = \langle Q, \Sigma, I, \Delta \rangle$:*

(i) there exists a single core state $q_c \in I$ *such that:*

$$\forall n \in \mathbb{N}. \forall f \in \Sigma_n. q_c \to f(q_c, \ldots, q_c) \in \Delta ;$$

(ii) there exists a set of core states $Q_c \cap I \neq \varnothing$ *such that for all* core inputs $q \in Q_c$ *there exist a tuple of* core outputs $q_1, \ldots, q_n \in Q_c$*:*

$$\forall n \in \mathbb{N}. \forall f \in \Sigma_n. \forall q \in Q_c. \exists q_1, \ldots, q_n \in Q_c. q \to f(q_1, \ldots, q_n) \in \Delta ;$$

(iii) there exists a set of core states $Q_c \subseteq I$ *such that for all tuples of* core outputs $q_1, \ldots, q_n \in Q_c$ *there exists a* core input $q \in Q_c$*:*

$$\forall n \in \mathbb{N}. \forall f \in \Sigma_n. \forall q_1, \ldots, q_n \in Q_c. \exists q \in Q_c. q \to f(q_1, \ldots, q_n) \in \Delta .$$

Proof. Note that (i) is an instance of (ii). For (ii) let $\Delta' \subseteq \Delta$ be such that the set Δ' contains for every $q \in Q$ exactly one transition of the form $\langle q, f(q_1, \ldots, q_n) \rangle$. We define $\rho(t, q)$ coinductively: $\rho(f(t_1, \ldots, t_n), q) := q(\rho(t_1, q_1), \ldots, \rho(t_n, q_n))$ where $\langle q, f(q_1, \ldots, q_n) \rangle \in \Delta'$. By construction $\rho(t, q)$ is a q-run on t. For (iii) it follows by induction that for every finite term $t \in \mathit{Ter}(\Sigma, \varnothing)$ has a q-run for some $q \in Q_c$. For infinite terms t take a sequence $\{t_i\}_{i \in \mathbb{N}}$ of finite terms converging towards t. By the Pigeonhole Principle there exists $q \in Q_c$ and a subsequence $\{s_i\}_{i \in \mathbb{N}}$ of $\{t_i\}_{i \in \mathbb{N}}$ such that every s_i has a q-run. Then by Lemma 4.5 we conclude that t has a q-run. □

5 Tree Automata as Certificates for SN^∞

We are now ready to use tree automata as 'certificates' for SN^∞.

Definition 5.1. Let R be a TRS over Σ, and let $S \subseteq \mathit{Ter}^\infty(\Sigma)$. A *certificate for* $\mathsf{SN}^\infty_R(S)$ is a tree automaton $\mathcal{T} = \langle Q, \Sigma, I, \Delta \rangle$ such that:

(i) $\mathcal{T}$ generates S, i.e. $S \subseteq \mathcal{L}(\mathcal{T})$, and
(ii) $Q_\alpha(\ell) \subsetneq Q_\alpha(r)$ if $Q_\alpha(\ell) \neq \varnothing$, for all $\ell \to r \in R$, and $\alpha : \mathit{Var}(\ell) \to \mathbf{2}^Q$.

Theorem 5.2. *Let R be a TRS over Σ, and $S \subseteq Ter^{\infty}(\Sigma)$. Then $\mathsf{SN}_R^{\infty}(S)$ holds if there exists a certificate for $\mathsf{SN}_R^{\infty}(S)$.*

The proof will be based on Theorem 3.10, the characterization of SN^{∞} in terms of interpretability in a continuous algebra. For this purpose we establish a bridge between tree automata certificates and continuous algebras. This bridge may need some intuitive explanation first. This concerns our use of tree automata states q decorated with a real numbers $r \in [0,1] = \{r \in \mathbb{R} \mid 0 \leq r \leq 1\}$, to be perceived as the degree of accuracy with which q can generate a certain term. Here 'accuracy' refers to the distance d in Definition 2.2. An example may be helpful.

Example 5.3. Consider the tree automaton $\mathcal{T}$ with the transitions

$$0 \to \mathsf{a} \qquad 1 \to \mathsf{b} \qquad 0 \to \mathsf{c}(0) \qquad 1 \to \mathsf{c}(1)$$

First we consider the 'run'-semantics $Q(\cdot)$ from Definition 4.1. Then for all $n \in \mathbb{N}$ we have $Q(\mathsf{c}^n(\mathsf{a})) = \{0\}$, meaning that $\mathsf{c}^n(\mathsf{a})$ can be generated by state 0, and likewise $Q(\mathsf{c}^n(\mathsf{b})) = \{1\}$. However, $Q(\mathsf{c}^{\omega}) = \{0,1\}$, and since c^{ω} is both the limit of $\mathsf{c}^n(\mathsf{a})$ and $\mathsf{c}^n(\mathsf{b})$, we face a problem if we aim at a continuous interpretation.

We redo this example, now with the accuracies r mentioned as superscripts of states 0, 1. More precisely, we use the continuous Σ-algebra $\mathcal{A}_{\mathcal{T}}$ defined below. Then $[\mathsf{c}^n(\mathsf{a})] = \{0^1, 1^{1-2^{-n}}\}$, meaning that $\mathsf{c}^n(\mathsf{a})$ can be generated from state 1 with accuracy 1, and also from state 0 but only with accuracy $1 - 2^{-n}$. Likewise, $[\mathsf{c}^n(\mathsf{b})] = \{0^{1-2^{-n}}, 1^1\}$. Furthermore $[\mathsf{c}^{\omega}] = \{0^1, 1^1\}$, which is indeed the limit of both $\{0^1, 1^{1-2^{-n}}\}$ and $\{0^{1-2^{-n}}, 1^1\}$, thereby resolving the clash with the continuity requirement.

Definition 5.4. Let $\mathcal{T} = \langle Q, \Sigma, I, \Delta\rangle$ be a tree automaton. We define a continuous weakly monotone Σ-algebra $\mathcal{A}_{\mathcal{T}} = \langle A, [\cdot], d, \succ, \sqsupseteq\rangle$ as follows. We let $A := \{\gamma \mid \gamma : Q \to [0,1]\}$ with undefined elements $\Omega_{\mathcal{T}} := \{\gamma \in A \mid \forall q \in Q.\, \gamma(q) < 1\}$.

For every $f \in \Sigma$ with arity n we define the interpretation $[f]$ by:

$$[f](\gamma_1, \ldots, \gamma_n) := \lambda q.\, \sup\big\{0.5 + 0.5 \cdot \min(\gamma_1(q_1), \ldots, \gamma_n(q_n)) \mid q \to f(q_1, \ldots, q_n) \in \Delta\big\}$$

where $\sup \varnothing := 0$.

For $\gamma \in A$ define $Q(\gamma) := \{q \in Q \mid \gamma(q) = 1\}$. Then $\succ$ and $\sqsupseteq$ on A are defined by: $\gamma_1 \succ \gamma_2 := Q(\gamma_1) \subsetneq Q(\gamma_2)$ and $\gamma_1 \sqsupseteq \gamma_2 := Q(\gamma_1) \subseteq Q(\gamma_2)$. As the metric d on A we choose $d(\gamma_1, \gamma_2) := \max\{|\gamma_1(q) - \gamma_2(q)| \mid q \in Q\}$.

The definition gives rise to a natural, continuous semantics associated with tree automata.

Lemma 5.5. *The algebra $\mathcal{A}_{\mathcal{T}}$ from Definition 5.4 is a continuous weakly monotone Σ-algebra with undefined elements Ω.*

Proof. We have $\succ \cdot \sqsupseteq \subseteq \succ$, and $\succ$ is well-founded since Q is finite. Consider a state $q \in Q$ for which $[f](\gamma_1, \ldots, \gamma_n)(q) = 1$, then there is $q \to f(q_1, \ldots, q_n) \in \Delta$

such that $\gamma_1(q_1) = 1,\ldots,\gamma_n(q_n) = 1$. Whenever additionally $\gamma_j \sqsupseteq \gamma'_j$ for some $1 \leq j \leq n$, then $\gamma'_j(q_j) = 1$ and therefore $[f](\ldots,\gamma'_j,\ldots)(q) = 1$. Hence $[f]$ is monotone with respect to $\sqsupseteq$ for all $f \in \Sigma$. Using the same reasoning it follows that Ω fulfills both requirements imposed on undefined elements. Hence $\mathcal{A}_\mathcal{T}$ is a weakly monotone Σ-algebra with undefine elements Ω.

For every $f \in \Sigma$ with arity n and every $\gamma_1, \gamma'_1, \ldots, \gamma_n, \gamma'_n \in A$ we have

$$d([f](\gamma_1,\ldots,\gamma_n), [f](\gamma'_1,\ldots,\gamma'_n)) \leq 0.5 \cdot \max\{d(\gamma_i,\gamma'_i) \mid 1 \leq i \leq n\}.$$

As a consequence, for the interpretation $[\cdot] : \mathit{Ter}(\Sigma,\varnothing) \to A$ of finite terms we have $d([s],[t]) \leq d(s,t)$ for all $s,t \in \mathit{Ter}(\Sigma,\varnothing)$. As a uniformly continuous map on the metric space $\langle \mathit{Ter}(\Sigma,\varnothing), d\rangle$, this interpretation can be extended to a continuous function $[\cdot] : \mathit{Ter}^\infty(\Sigma,\varnothing) \to A$ on the completion space $\langle \mathit{Ter}^\infty(\Sigma,\varnothing), d\rangle$. Hence $\mathcal{A}_\mathcal{T}$ is a continuous Σ-algebra. □

The following lemma connects the standard semantics of tree automata with the continuous algebra $\mathcal{A}_\mathcal{T}$. Roughly, in the continuous algebra the automaton can be found back, when considering only states with 'accuracy' 1 ($\gamma(q) = 1$).

Lemma 5.6. *Let $\mathcal{A}_\mathcal{T} = \langle A, [\cdot], d, \succ, \sqsupseteq\rangle$ be the Σ-algebra as in Definition 5.4. Then for all $t \in \mathit{Ter}^\infty(\Sigma,\varnothing)$, and $\alpha : \mathit{Var}(t) \to \mathbf{2}^Q$, $\beta : \mathit{Var}(t) \to A$ such that $\forall x \in \mathit{Var}(t).\ \alpha(x) = Q(\beta(x))$, it holds $Q_\alpha(t) = Q([t]^\beta)$.*

Proof. For the case $t \in \mathcal{X}$, there is nothing to be shown. Thus let $t \equiv f(t_1,\ldots,t_n)$. For '$\supseteq$', assume $q \in Q([t]^\beta)$. Then there exists $q \to f(q_1,\ldots,q_n) \in \Delta$ such that for $i = 1,\ldots,n$ we have $q_i \in Q([t_i]^\beta)$. Applying this argument (coinductively) to the subterms t_i we obtain a q-run $\rho := q(\rho_1,\ldots,\rho_n)$ of $\mathcal{T}$ on t (with respect to α) where ρ_i is a q_i-run of $\mathcal{T}$ on t_i for $i = 1,\ldots,n$. For '$\subseteq$', we show that $[t]^\beta(q) \geq 1 - 0.5^d$ for all $t \in \mathit{Ter}^\infty(\Sigma)$, $d \in \mathbb{N}$ and $q \in Q$ with $q \in Q_\alpha(t)$. Assume contrary this claim would not hold. Consider a counterexample with minimal $d \in \mathbb{N}$. Since $q \in Q_\alpha(t)$ there exists $q \to f(q_1,\ldots,q_n) \in \Delta$ such that $q_i \in Q_\alpha(t_i)$ for $i = 1,\ldots,n$. This implies $d \geq 1$ and from minimality of d we obtain $\forall i.\ [t_i]^\beta(q_i) \geq 1 - 0.5^{d-1}$. But then $[t]^\beta(q) \geq 0.5 + 0.5 \cdot \min([t_i]^\beta(q_i)) \geq 1 - 0.5^d$, contradicting the assumption. Hence $[t]^\beta(q) = 1$, and $q \in Q([t]^\beta)$. □

Using $\mathcal{A}_\mathcal{T}$ we now give the proof of Theorem 5.2.

Proof (Theorem 5.2). Let $\mathcal{T} = \langle Q, \Sigma, I, \Delta\rangle$ be a certificate for $\mathsf{SN}^\infty_R(S)$. Let $\mathcal{A}_\mathcal{T} = \langle A, [\cdot], d, \succ, \sqsupseteq\rangle$ and Ω as defined in Definition 5.4. According to Lemma 5.5 $\mathcal{A}_\mathcal{T}$ is a continuous weakly monotone Σ-algebra with undefined elements Ω. We prove that $\mathcal{A}_\mathcal{T}$ fulfills the requirements of Theorem 3.10.

As a consequence of Lemma 5.6 we obtain that $[s] \notin \Omega$ for all $s \in S$, since by assumption $S \subseteq \mathcal{L}(\mathcal{T})$; and $[\ell]^\alpha \notin \Omega$ implies $[\ell]^\alpha \succ [r]^\alpha$, for all rules $\ell \to r \in R$ and every $\alpha : \mathcal{X} \to A$. Finally, we check compatibility of $\sqsupseteq$ with limits. Let $\{\gamma_i\}_{i\in\mathbb{N}}$ be a converging sequence with $\gamma_0 \sqsupseteq \gamma_1 \sqsupseteq \ldots$, and define $\gamma := \lim_{i\to\infty} \gamma_i$. Note that $Q(\gamma_i) \subseteq Q(\gamma_{i+1})$ for all $i \in \mathbb{N}$. For every $q \in Q$ with $\gamma_0(q) = 1$ we have $\gamma_i(q) = 1$ for all $i \in \mathbb{N}$ and therefore $\gamma(q) = 1$. Hence $\gamma_0 \sqsupseteq \gamma$.

The algebra $\mathcal{A}_\mathcal{T}$ fulfills all requirements of Theorem 3.10, hence $\mathsf{SN}^\infty_R(S)$ holds. □

Example 5.7. Let $\Sigma := \{\mathsf{a}, \mathsf{b}, \mathsf{c}\}$ and $R := \{\mathsf{a}(\mathsf{c}) \to \mathsf{a}(\mathsf{b}(\mathsf{c})),\ \mathsf{b}(\mathsf{b}(\mathsf{c})) \to \mathsf{c}\}$ where a and b are unary symbols, and c is a constant. We are interested in SN^∞_R, that is, in infinitary normalization of R on the set of all (possibly infinite) terms. Consider the tree automaton $\mathcal{T} = \langle Q, \Sigma, I, \Delta\rangle$ depicted below:

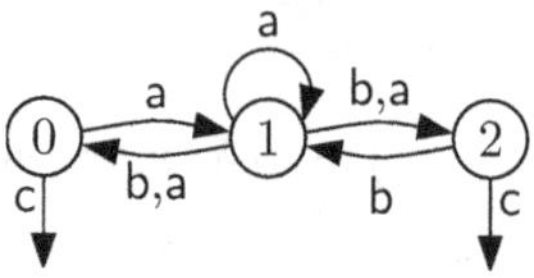

where $Q := \{0, 1, 2\}$, $I := Q$ and Δ consists of the following rules:

$$0 \to \mathsf{a}(1) \mid \mathsf{c} \qquad 1 \to \mathsf{a}(0) \mid \mathsf{a}(1) \mid \mathsf{a}(2) \mid \mathsf{b}(0) \mid \mathsf{b}(2) \qquad 2 \to \mathsf{b}(1) \mid \mathsf{c}$$

We show that $\mathcal{T}$ is a certificate for SN^∞_R, by checking the conditions of Definition 5.1. Completeness of $\mathcal{T}$ follows from Lemma 4.6 (iii), take $Q_\mathsf{c} = Q$. Second, as both rules of R have no variables, we do not have to consider assignments α. We verify that $Q(\ell) \subsetneq Q(r)$ for both rules. For the rule $\mathsf{a}(\mathsf{c}) \to \mathsf{a}(\mathsf{b}(\mathsf{c}))$ we compute $Q(\mathsf{a}(\mathsf{c})) = \{1\}$, for only from state 1 we can generate $\mathsf{a}(\mathsf{c})$: $1 \to \mathsf{a}(2) \to \mathsf{a}(\mathsf{c})$ (or $1 \to \mathsf{a}(0) \to \mathsf{a}(\mathsf{c})$). From state 2 there is no 'a-transition', and from state 0 we get stuck at $\mathsf{a}(1)$, for there is no rule $1 \to \mathsf{c}$. Similarly we find $Q(\mathsf{a}(\mathsf{b}(\mathsf{c}))) = \{0, 1\}$, hence $Q(\mathsf{a}(\mathsf{c})) \subsetneq Q(\mathsf{a}(\mathsf{b}(\mathsf{c})))$. For the second rule of R we find $Q(\mathsf{b}(\mathsf{b}(\mathsf{c}))) = \{2\} \subsetneq \{0, 2\} = Q(\mathsf{c})$. Thus we have shown $\mathcal{T}$ to be a certificate, and by Theorem 5.2 we may conclude SN^∞_R.

6 Improving Efficiency: Strict Certificates

The second requirement for an automaton to be a certificate for SN^∞ (item (ii) of Definition 5.1) is computationally expensive to check, since there are $2^{|Q|\cdot|\mathit{Var}(\ell)|}$ different maps $\alpha : \mathit{Var}(\ell) \to \mathbf{2}^Q$, leading to an exponential explosion in the number of states when searching for such an automaton.

Remark 6.1. For Theorem 5.2 it is not sufficient to check that the second condition holds for maps from variables to single states, that is, maps $\alpha : \mathit{Var}(\ell) \to \mathbf{2}^Q$ with $|\alpha(x)| = 1$ for all $x \in \mathcal{X}$.

To see this, consider the TRS $R := \{\mathsf{f}(x) \to \mathsf{f}(\mathsf{a}(x))\}$ with the tree automaton $\mathcal{T} = \langle Q, \Sigma, I, \Delta\rangle$ where $Q := I := \{0, 1\}$ and Δ consists of $0 \to \mathsf{f}(0)$, $1 \to \mathsf{f}(1)$, $0 \to \mathsf{a}(0)$, $0 \to \mathsf{a}(1)$, $1 \to \mathsf{a}(0)$, and $1 \to \mathsf{a}(1)$. Then $\mathcal{L}(\mathcal{T}) = \mathit{Ter}^\infty(\Sigma)$ and for every map $\alpha := x \mapsto \{q\}$ with $q \in Q$ we get $Q_\alpha(\ell) = \{q\} \subsetneq Q = Q_\alpha(r)$. Both conditions seem to be fulfilled, however SN^∞_R does not hold, since R admits an infinite root rewrite sequence $\mathsf{f}(\mathsf{a}^\omega) \to_{R,\epsilon} \mathsf{f}(\mathsf{a}^\omega) \to_{R,\epsilon} \dots$.

For the purpose of efficient implementations and the envisaged SAT encoding, we define the notion of 'strict certificates', and show that they have the same theoretical strength while being easier to check.

Definition 6.2. Let R be a TRS over Σ, and $S \subseteq \mathit{Ter}^\infty(\Sigma)$. A *strict certificate for* $\mathsf{SN}^\infty_R(S)$ is a tree automaton $\mathcal{T} = \langle Q, \Sigma, I, \Delta\rangle$ with a strict total order $< \subseteq Q \times Q$ such that:

(i) $S \subseteq \mathcal{L}(\mathcal{T})$, and
(ii) for every $\ell \to r \in R$ and $\alpha : Var(\ell) \to \mathbf{2}^Q$ with $1 \leq |\alpha(x)| \leq \#_x(\ell)$, for all $x \in Var(\ell)$, where $\#_x(\ell) \in \mathbb{N}$ the number of occurrences of x in ℓ, it holds:

$$Q_\alpha(\ell) \neq \varnothing \Longrightarrow Q_\alpha(\ell) \subseteq Q_\alpha(r) \text{ and } \\ \forall q \in Q_\alpha(\ell).\, \exists q' \in Q_\alpha(r).\, q' < q\,.$$

That strict certificates are certificates, the next theorem, will be proved below.

Theorem 6.3. *Let R be a TRS over Σ, and $S \subseteq Ter^\infty(\Sigma)$. Then every strict certificate for $\mathsf{SN}_R^\infty(S)$ is a certificate for $\mathsf{SN}_R^\infty(S)$.*

In the search for certificates, the computational complexity is improved when restricting the search to strict certicates, because the number of maps α which have to be considered is reduced to:

$$\prod_{x \in Var(\ell)} \left(\sum_{i=1}^{\#_x(\ell)} \binom{|Q|}{i}\right)$$

which is polynomial in the number of states $|Q|$. In particular if ℓ is linear then we need to consider $|Q|^{|Var(\ell)|}$ maps α.

Remark 6.4. Note that, in the definition of strict certificates, we cannot replace the condition $1 \leq |\alpha(x)| \leq \#_x(\ell)$ by $|\alpha(x)| = 1$. To see this, we consider the non-left-linear TRS $R := \{\mathsf{f}(x, x) \to \mathsf{f}(\mathsf{a}(x), \mathsf{a}(x))\}$ together with the tree automaton $\mathcal{T} = \langle Q, \Sigma, I, \Delta \rangle$ where $Q := I := \{0, 1\}$ and Δ consists of $1 \to \mathsf{f}(q, q)$, $0 \to \mathsf{f}(q, \overline{q})$ and $q \to \mathsf{a}(q')$ for all $q, q' \in Q$ where $\overline{q} = 1 - q$. Then $\mathcal{L}(\mathcal{T}) = Ter^\infty(\Sigma)$ and for every map $\alpha := x \mapsto \{q\}$ with $q \in Q$ we get $Q_\alpha(\ell) = \{1\}$ and $Q_\alpha(r) = \{0, 1\}$; thus $Q_\alpha(\ell) \subseteq Q_\alpha(r)$ and $0 < 1$ with $0 \in Q_\alpha(r)$. However R admits an infinite root rewrite sequence $\mathsf{f}(\mathsf{a}^\omega, \mathsf{a}^\omega) \to_{R,\epsilon} \mathsf{f}(\mathsf{a}^\omega, \mathsf{a}^\omega) \to_{R,\epsilon} \ldots$.

Note that the theorem holds even if one allows a partial order $<$ in the definition of strict certificates. However, that would not make the notion of strict certificates more general, because such a partial order can always be extended to a total order. The advantage of the definition as it stands is that we get the order for free. For every strict certificate with n states there exists an isomorphic automaton with states $Q := \{1, \ldots, n\}$ and $<$ being the natural order on integers. Thus, we can narrow the search for certificates to such automata.

Lemma 6.5. *Let $\mathcal{T} = \langle Q, \Sigma, I, \Delta \rangle$ be a tree automaton, $s \in Ter^\infty(\Sigma)$ and $\alpha : Var(s) \to \mathbf{2}^Q$. Let $\mathcal{B}$ consist of all maps $\beta : Var(s) \to \mathbf{2}^Q$ with $\beta(x) \subseteq \alpha(x)$ and $1 \leq |\beta(x)| \leq \#_x(s)$ for all $x \in Var(s)$. Then $Q_\alpha(s) = \bigcup_{\beta \in \mathcal{B}} Q_\beta(s)$.*

Proof. The part '$\supseteq$' is trivial, all maps $\beta \in \mathcal{B}$ are a restriction of α. For '$\subseteq$' let ρ be a q-run with respect to α on s. Let $\beta := \lambda x.\{\rho(p) \mid p \in \mathcal{P}os(s) \text{ with } s(p) = x\}$, then ρ is also a q-run with respect to β and $\forall x \in Var(s).\, 1 \leq |\beta(x)| \leq \#_x(s)$. □

Now we prove Theorem 6.3.

Proof (Theorem 6.3). Let R be a TRS over Σ, $S \subseteq \mathit{Ter}^\infty(\Sigma)$ a set of terms, and $\mathcal{T} = \langle Q, \Sigma, I, \Delta \rangle$ a strict certificate for $\mathsf{SN}^\infty_R(S)$ with a strict total order $<$ on the states. We show that $\mathcal{T}$ satisfies the conditions of Definition 5.1. Let $\ell \to r \in R$ and $\alpha : \mathit{Var}(\ell) \to \mathbf{2}^Q$ with $Q_\alpha(\ell) \neq \varnothing$. Let $\mathcal{B}$ consist of all maps $\beta : \mathit{Var}(\ell) \to \mathbf{2}^Q$ with $\beta(x) \subseteq \alpha(x)$ and $1 \leq |\beta(x)| \leq \#_x(\ell)$ for all $x \in \mathit{Var}(\ell)$. Then $Q_\alpha(\ell) = \bigcup_{\beta \in \mathcal{B}} Q_\beta(\ell)$ and $Q_\alpha(r) = \bigcup_{\beta \in \mathcal{B}} Q_\beta(r)$ by Lemma 6.5. Note that we have $Q_\beta(\ell) \subseteq Q_\beta(r)$ for all $\beta \in \mathcal{B}$ by assumption, hence $Q_\alpha(\ell) \subseteq Q_\alpha(r)$. Take the least $q \in Q_\alpha(\ell)$ with respect to $<$. Then there exists $\beta \in \mathcal{B}$ with $q \in Q_\beta(\ell)$ and by assumption $\exists q' \in Q_\beta(r).\, q' < q$. Hence $q' \in Q_\alpha(r)$ and $Q_\alpha(\ell) \subsetneq Q_\alpha(r)$. □

The additional requirement of an ordering $<$ on the states is not a weakening. Indeed, we can show that any certificate can be transformed into a strict one.

Lemma 6.6. *Let R be a TRS over Σ, and $S \subseteq \mathit{Ter}^\infty(\Sigma)$. If there is a certificate for $\mathsf{SN}^\infty_R(S)$ then there is a strict certificate for $\mathsf{SN}^\infty_R(S)$.*

Proof. Let R be a TRS over Σ, $S \subseteq \mathit{Ter}^\infty(\Sigma)$, and $\mathcal{T} = \langle Q, \Sigma, I, \Delta \rangle$ a certificate for $\mathsf{SN}^\infty_R(S)$. We construct a tree automaton $\mathcal{T}' = \langle Q', \Sigma, I', \Delta' \rangle$ and show that it meets the requirements of Definition 6.2. Let $Q' := \mathbf{2}^Q$, and $I' := \{Q_I \subseteq Q \mid Q_I \cap I \neq \varnothing\}$. We define Δ' to consist of all transitions of the form $Q_0 \to f(Q_1, \ldots, Q_n)$ with $f \in \Sigma$, $Q_0, \ldots, Q_n \subseteq Q$ such that $\varnothing \neq Q_0 \subseteq Q_0'$ where

$$Q_0' := \{q \in Q \mid \text{exists } q \to f(q_1, \ldots, q_n) \in \Delta \text{ such that } \forall i.\, q_i \in Q_i\}\,.$$

Note that the construction is similar to the construction for making tree automata deterministic [2]. The main difference concerns the set Q_0, which is not uniquely defined as $Q_0 := Q_0'$ in our setting (we allow subsets $Q_0 \subseteq Q_0'$). Therefore the automaton $\mathcal{T}'$ will in general not be deterministic. For all terms $s \in \mathit{Ter}(\Sigma)$ and maps $\alpha' : \mathit{Var}(s) \to \mathbf{2}^{Q'}$ we have:

$$Q'_{\alpha'}(s) = \{Q_0' \subseteq Q_\alpha(s) \mid \alpha : \mathit{Var}(s) \to \mathbf{2}^Q \text{ with } \forall x.\, \alpha(x) \in \alpha'(x)\} \qquad (*)$$

This follows from the above-mentioned analogy; we refer to [2] for a proof. From $(*)$ it immediately follows that $\mathcal{L}(\mathcal{T}) = \mathcal{L}(\mathcal{T}')$.

We define the strict order $>$ on Q' as $\subsetneq$, arbitrarily extended to a total order. Let $\ell \to r \in R$ and $\alpha' : \mathit{Var}(\ell) \to \mathbf{2}^{Q'}$ such that $Q'_{\alpha'}(\ell) \neq \varnothing$. We know that for every $\alpha : \mathit{Var}(s) \to \mathbf{2}^Q$ it holds $Q_\alpha(\ell) \subsetneq Q_\alpha(r)$ by assumption. Then together with $(*)$ it follows that $Q'_{\alpha'}(\ell) \subseteq Q'_{\alpha'}(r)$. Finally let Q_0' be the least element with respect to $>$ from $Q'_{\alpha'}(\ell)$. Then there exists a map $\alpha : \mathit{Var}(s) \to \mathbf{2}^Q$ such that $\forall x.\, \alpha(x) \in \alpha'(x)$ and $Q_0' \subseteq Q_\alpha(\ell)$, even $Q_0' = Q_\alpha(\ell)$, since otherwise $Q_0' > Q_\alpha(\ell)$ would contradict minimality of Q_0'. Then we have $Q_\alpha(\ell) \subsetneq Q_\alpha(r)$ and therefore $Q_\alpha(r) \in Q'_{\alpha'}(r)$ with $\forall q' \in Q'_{\alpha'}(\ell).\, Q_\alpha(r) < q'$. □

7 Examples and Tool

Here we consider a few illustrating examples. We have implemented our method into a tool that aims at proving $\mathsf{SN}^\infty_R(S)$ automatically. Actually, all certificates in this section have been found fully automatically by our tool. The program is available via `http://infinity.few.vu.nl/sni/`, it may be used to try examples online. The tool shows the interpretation of all symbols and rules (with respect to all variable assignments) in the form of transition tables such that decreasingness can be recognized easily. The start language S can be specified by providing a tree automaton $\mathcal{T}$ that generates S; the program then searches an extension of $\mathcal{T}$ which fulfills the requirements of Theorem 6.3.

Example 7.1. Consider the following TRS R defining the sequence morse:

$$\begin{aligned}
\mathsf{morse} &\to \mathsf{cons}(0, \mathsf{zip}(\mathsf{inv}(\mathsf{morse}), \mathsf{tail}(\mathsf{morse})))\\
\mathsf{zip}(\mathsf{cons}(x, y), z) &\to \mathsf{cons}(x, \mathsf{zip}(z, y))\\
\mathsf{inv}(\mathsf{cons}(0, x)) &\to \mathsf{cons}(1, \mathsf{inv}(x))\\
\mathsf{inv}(\mathsf{cons}(1, x)) &\to \mathsf{cons}(0, \mathsf{inv}(x))\\
\mathsf{tail}(\mathsf{cons}(x, y)) &\to y
\end{aligned}$$

Our tool proves $\mathsf{SN}^{\infty}_{R}(\{\mathsf{morse}\})$ fully automatically. First it instantiates y in the rule $\mathsf{tail}(\mathsf{cons}(x, y)) \to y$ with non-variable terms covering all ground instances, and then it finds the tree automaton $\mathcal{T} = \langle Q, \Sigma, I, \Delta\rangle$ with $I = Q = \{0, 1, 2\}$ where the set Δ consists of: $2 \to \mathsf{morse}$, $1 \mid 2 \to 0$, $1 \mid 2 \to 1$, $2 \to \mathsf{tail}(0 \mid 2)$, $1 \mid 2 \to \mathsf{inv}(1 \mid 2)$, $0 \mid 1 \mid 2 \to \mathsf{cons}(1, 1)$, $1 \mid 2 \to \mathsf{zip}(1 \mid 2, 1)$, and $1 \mid 2 \to \mathsf{zip}(1, 2)$. Note that with the productivity tool of [3] we could already prove productivity of this specification fully automatically.

Example 7.2. Consider the term rewriting system R consisting of the rules:

$$\mathsf{c} \to \mathsf{f}(\mathsf{a}(\mathsf{b}(\mathsf{c}))) \qquad \mathsf{f}(\mathsf{a}(x)) \to \mathsf{f}(x) \qquad \mathsf{f}(\mathsf{b}(x)) \to \mathsf{b}(\mathsf{f}(x))$$

and the tree automaton $\mathcal{T} = \langle Q, \Sigma, I, \Delta\rangle$ with (initial) states $I = Q = \{0, 1, 2, 3\}$ over the signature $\Sigma = \{\mathsf{c}, \mathsf{a}, \mathsf{b}, \mathsf{f}\}$ where the set Δ of transition rules is given by:

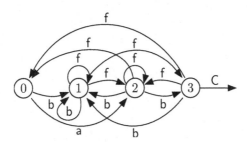

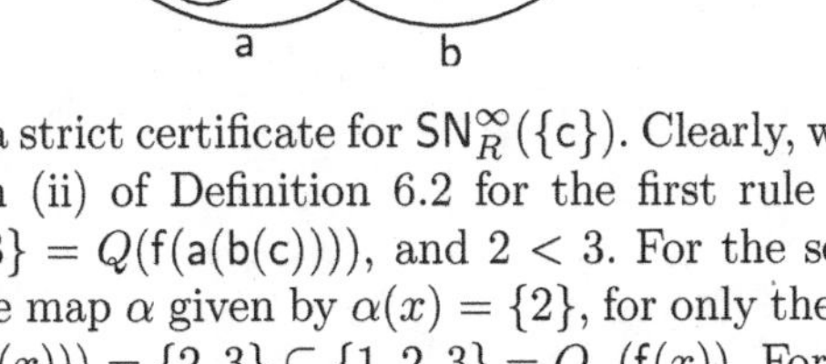

We show that $\mathcal{T}$ is a strict certificate for $\mathsf{SN}^{\infty}_{R}(\{\mathsf{c}\})$. Clearly, we have $\{\mathsf{c}\} \subseteq \mathcal{L}(\mathcal{T})$. To verify condition (ii) of Definition 6.2 for the first rule of R, observe that $Q(\mathsf{c}) = \{3\} \subsetneq \{2, 3\} = Q(\mathsf{f}(\mathsf{a}(\mathsf{b}(\mathsf{c}))))$, and $2 < 3$. For the second rule, we only have to consider the map α given by $\alpha(x) = \{2\}$, for only then $Q_\alpha(\mathsf{f}(\mathsf{a}(x))) \neq \varnothing$. We observe $Q_\alpha(\mathsf{f}(\mathsf{a}(x))) = \{2, 3\} \subsetneq \{1, 2, 3\} = Q_\alpha(\mathsf{f}(x))$. For the third rule of R we have to consider two assignments: α_1 that maps x to $\{1\}$, and α_3 that maps x to $\{3\}$. We get that $Q_{\alpha_1}(\mathsf{f}(\mathsf{b}(x))) = \{1, 2, 3\} \subsetneq Q = Q_{\alpha_1}(\mathsf{b}(\mathsf{f}(x)))$ (and $0 < q$ for all $q \in \{1, 2, 3\}$), and $Q_{\alpha_3}(\mathsf{f}(\mathsf{b}(x))) = \{1, 2\} \subsetneq Q = Q_{\alpha_3}(\mathsf{b}(\mathsf{f}(x)))$ (and $0 < 1, 2$).

References

1. Arts, T., Giesl, J.: Termination of term rewriting using dependency pairs. Theoretical Computer Science 236, 133–178 (2000)
2. Comon, H., Dauchet, M., Gilleron, R., Löding, C., Jacquemard, F., Lugiez, D., Tison, S., Tommasi, M.: Tree Automata Techniques and Applications (2007), `http://www.grappa.univ-lille3.fr/tata`
3. Endrullis, J., Grabmayer, C., Hendriks, D., Isihara, A., Klop, J.W.: Productivity of Stream Definitions. In: Csuhaj-Varjú, E., Ésik, Z. (eds.) FCT 2007. LNCS, vol. 4639, pp. 274–287. Springer, Heidelberg (2007)
4. Endrullis, J., Waldmann, J., Zantema, H.: Matrix Interpretations for Proving Termination of Term Rewriting. In: Furbach, U., Shankar, N. (eds.) IJCAR 2006. LNCS (LNAI), vol. 4130, pp. 574–588. Springer, Heidelberg (2006)
5. Goguen, J.A., Thatcher, J.W., Wagner, E.G., Wright, J.B.: Initial Algebra Semantics and Continuous Algebras. JACM 24(1), 68–95 (1977)
6. Klop, J.W., de Vrijer, R.C.: Infinitary Normalization. In: Artemov, S., Barringer, H., d'Avila Garcez, A.S., Lamb, L.C., Woods, J. (eds.) We Will Show Them: Essays in Honour of Dov Gabbay, vol. 2, pp. 169–192. College Publ. (2005)
7. Terese: Term Rewriting Systems. Cambridge Tracts in Theoretical Computer Science, vol. 55. Cambridge University Press, Cambridge (2003)
8. Zantema, H.: Termination of Term Rewriting: Interpretation and Type Elimination. Journal of Symbolic Computation 17, 23–50 (1994)
9. Zantema, H.: Normalization of Infinite Terms. In: Voronkov, A. (ed.) RTA 2008. LNCS, vol. 5117, pp. 441–455. Springer, Heidelberg (2008)

First-Class Object Sets

Erik Ernst

University of Aarhus, Denmark
eernst@cs.au.dk

Abstract. Typically, an object is a monolithic entity with a fixed interface. To increase flexibility in this area, this paper presents first-class object sets as a language construct. An object set offers an interface which is a disjoint union of the interfaces of its member objects. It may also be used for a special kind of method invocation involving multiple objects in a dynamic lookup process. With support for feature access and late-bound method calls, object sets are similar to ordinary objects, only more flexible. Object sets are particularly convenient as a lightweight primitive which may be added to a mainstream virtual machine in order to improve on the support for family polymorphism. The approach is made precise by means of a small calculus, and the soundness of its type system has been shown by a mechanically checked proof in Coq.

Keywords: Object sets, composition, multi-object method calls, types.

1 Introduction

In an object-oriented setting, the main concept is the object. It is typically a monolithic entity with a fixed interface, such as a fixed set of messages that the object accepts. This paper presents a language design where sets of objects are first-class entities, equipped with operations that enable such object sets to work in a way similar to monolithic objects. An object set offers an interface to the environment which is a disjoint union of the interfaces of its members, and it supports cross-member operations, known as *object set method calls*, which are similar in nature to late-bound method calls on monolithic objects. The object set thus behaves in a way which resembles the behavior of a monolithic instance of a 'large' class that combines all the classes of the members of the object set, e.g., by mixin composition or multiple inheritance.

Object sets are useful because they are more flexible than such a monolithic instance of a large class: There are no restrictions on which classes may be put together in the creation of an object set, and there is no need to declare a large, composite class and refer to that class by name everywhere. In fact, object set types are structural with member class granularity—the type of an object set is a set of classes, and every subset is a supertype. Moreover, object sets could be modified during their lifetime, which would correspond to a dynamic change of class in the monolithic case.

On the other hand, access to a feature of an object set requires explicit selection of the member class which provides this feature, and the object set method

S. Berardi, F. Damiani, and U. de'Liguoro (Eds.): TYPES 2008, LNCS 5497, pp. 83–99, 2009.

call mechanism is quite simple rather than convenient. This is because the emphasis in this language design has been put on expressing the required primitives in order to support typed sets of objects with cross-object features, rather than giving a proposal for convenient and pragmatic surface programming language design.

In fact, ongoing work on the implementation of the language gbeta [1,2,3] on .NET served as our starting point for the design of object sets. At the core of the semantics and typing of this language is the feature known as *family polymorphism* [4], which involves dependent types in the sense that run-time object identities are significant parts of the types of classes and objects. Mainstream platforms like .NET and the Java Virtual Machine do not support dependent types, which causes the insertion of many dynamic casts in bytecode in order to support family polymorphism. Object sets maintain information which makes it possible to avoid an entire category of such dynamic casts.

Objects in gbeta have a semantics which may be represented by collections of instances of mixins, i.e., as multi-entity phenomena rather than monolithic entities. Like object sets, they provide an interface which is a disjoint union of the interfaces of the included mixins, but unlike object sets there is no need to specify explicitly which mixin to use when accessing a feature. Like object sets, gbeta objects can have cross-entity features (such as methods or inner classes, which are then known as virtual), but, unlike object sets, these features are accessed in exactly the same way as single-mixin features. Similarly, since all gbeta objects are conceptually object sets there is no distinction (syntactically or otherwise) between the usage of object sets and "ordinary objects."

The language gbeta also supports dynamic change of class for existing objects (to a subclass), and this corresponds to the replacement of the contents of the gbeta object by a (larger) object set. In the context of the features included in this paper, this operation is simple and safe, though of course it would require addition of mutable references to object sets to make it work as a dynamic change of class. In the context of gbeta it is considerably more complex, because dynamic specialization of an object may have effects that correspond to a dynamic replacement of actual type arguments of the class of the gbeta object by some subtypes, which may cause a run-time error, e.g., because the value of an instance variable may thus become type incorrect. Because of this, dynamic object specialization in gbeta has been extended with restricted versions that are safe, but it is beyond the scope of this paper to model these refinements of the concept. Nevertheless, it is worth noting that it is possible to embody the object set primitive presented in this paper in a full-fledged programming language in such a way that it is convenient to use.

The contributions of this paper are the concept of object sets, and the precise definition of their semantics and typing in a formal calculus, FJ_{set}. Moreover, a mechanically checked proof of soundness [5] for this calculus has been constructed using the Coq [6] proof assistant, and the experience of doing this is reflected by a number of remarks throughout the paper.

The rest of the paper is organized as follows: Section 2 presents the calculus informally and discusses the design. Next, Sect. 3 gives the formal definitions, and Sect. 4 describes the soundness result. Finally, Sect. 5 describes related work, and Sect. 6 concludes.

2 An Informal Look at the $\mathbf{FJ_{set}}$ Calculus

The FJ_{set} calculus is derived from the Featherweight Java calculus [7] by adding the object set related operations, allowing covariant method return types, and removing casts. Covariant method return types are included because they are useful and standard today, and casts are left out because they do not provide extra benefits in this context.

The crux of this calculus is of course the ability to express and use object sets. An object set is a set of objects collected into a single, typed entity. An object set may be decomposed in order to use individual members of the set, and used as a whole in a special kind of method call, the *object set method call.* Each object set is associated with a set of classes, and each object in the set is uniquely associated with one particular class in the set of classes. Another way to describe this would be to say that each object in the set is *labeled* by a class. The object is an instance of that class or a subclass thereof. This makes it possible to access the object set members and to use each one of them according to an interface that it is known to support.

The correspondence between objects and classes in an object set is maintained by considering the set of objects and the set of classes as lists and pairing up the lists element by element. This is possible for an object set creation expression (a variant of the well-known `new` expression for monolithic objects) because such an expression contains the two lists syntactically, and this ensures that every object set from its creation has a *built-in definition of the mapping* from classes to member objects. It also equips the members of the object set with an ordering. This ordering is insignificant with respect to typing, but it is significant with respect to the dynamic semantics, because it determines which method implementation is most specific during an object set method call. In other words, the ordering of the members of an object set is a server side issue rather than a client side issue—crucial in the definition of its internal structure and behavior, but encapsulated and invisible at the level of types.

An expression denoting an object set may by subsumption be typed with an arbitrary subset of the associated classes, and they may be listed in an arbitrary order in the type. It is therefore possible to forget some of the objects and also to ignore the ordering of the objects. However, the dynamic semantics only operates on an object set when it has been evaluated to such an extent that it is an object set creation at top level. This ensures that the object set operations are consistent because they are based on the built-in mapping.

Two operations are provided to decompose an object set. They both rely on addressing a specific member of the set via its associated class. One operation provides access to the object associated with the given class, and the other operation deletes that object and class from the object set, thus producing a smaller object set.

The object set method call operation is provided in order to gather contributions from all suitable objects in an object set, in a process that resembles a fold operation on a list. The call is based on an ordinary method whose signature must follow a particular pattern, namely that it takes a positive number of arguments and that the type of the first argument is identical to the return type of the method. This makes it possible for the method to accept an arbitrary list of "ordinary" arguments—the arguments number two and up—and also to accept and return a value which plays the role as an accumulator of the final result. In this sense the object set method call supports iteration over the selected members of the object set and collection of contributions to the final result, not unlike a folding operator applied to a list. Note that an object set method call does not require static knowledge about the type of any of the objects in the set.

The design of the object set method call mechanism was chosen to enable iteration over a subset of the members of the object set supporting a specific interface, without adding extra language mechanisms. Pragmatically, it might be more natural to use actual iteration in an imperative setting, or to return a data structure like an array containing the eligible object set members. But in this context we prefer a minimal design, and hence we ended up choosing the programmer convention driven approach based on ordinary nested method calls.

Figure 1 shows a small example program in FJ_{set}. This program shows how to create objects and object sets, how to perform an object set method call, and how to decompose an object set in order to use a feature of one of its members; finally it indicates the result of the computation. In order to make the example compact and readable, we use an extension of the calculus that includes a `String` type, string literals, and concatenation of strings with the '`+`' operator.

Lines 1–13 define three classes to support modeling a human being from two different points of view in an object set; the only difference from standard Java code is that there are no constructors, but the constructors in FJ style calculi are trivial and somewhat of an anomaly, so we left them out. Note that the signature of the `print` method is such that it can be used for object set method calls: Its return type is also the type of the first (and only) argument.

The class `Main` has an instance variable (line 15) whose type is an object set, { `Printable` }, which means that it is guaranteed that there is an object labeled `Printable` in this object set, but there may be other objects as well. The `doPrint` method (line 16–20) makes two object set method calls (line 17 and 19) and one ordinary method call (line 18), and returns the concatenation of the results. The object set method call on line 17 involves only one member of `p`, because only the first one is labeled by `Agent` or a subclass thereof. The call on line 19 involves both objects in `p`. The expression `p@Printable` on line 18 extracts the object labeled as `Printable` in `p`, which is the `Person` object, and calls its `separator` method, which by ordinary late binding returns `" -- "`.

Finally, note that subsumption makes it possible for the instance variable `p` to refer to an object set of type { `Agent`, `Printable` }, and also that the usage of

```
class Printable extends Object {
    String print(String s) { return "Plain Printable"; }
    String separator() { return ", "; }
}
class Agent extends Printable {
    String id;
    String print(String s) { return s+" "+id; }
}
class Person extends Printable {
    String name;
    String print(String s) { return name+" "+s; }
    String separator() { return " -- "; }
}
class Main extends Object {
    {Printable} p;
    String doPrint() {
        return p.print@Agent("The name is") +
               p@Printable.separator() +
               p.print@Printable("");
    }
}

// the following yields "The name is Bond -- James Bond"
new Main(new {Agent, Printable}
    (new Agent("Bond"), new Person("James"))).doPrint()
```

Fig. 1. A small example program in FJ_{set}

different classes in the object set method call can be used to filter the contributors to such a call in various ways.

3 The FJ_{set} Calculus

We now proceed to present the syntax, the dynamic semantics, and the type system of the FJ_{set} calculus, interspersed with short discussions about why the calculus is designed the way it is. We also give some remarks on how the presentation in this paper and the accompanying Coq proof fit together, reflecting the process of learning to use Coq, and based on the assumption that this kind of knowledge is useful for the development of a strong culture of using proof assistant software.

3.1 Syntax and Notation

A program is a class table and a main expression, and the semantics of a program is to execute the main expression in the context of the given classes. As is common, we assume the existence of a fixed, globally accessible class table, CT, which lists all the class definitions in the program.

Q	::=	$\texttt{class C extends D}\,\{\,\overline{\texttt{T f}};\,\overline{\texttt{M}}\,\}$	*class declarations*
M	::=	$\texttt{T m}(\overline{\texttt{T x}})\ \{\,\texttt{return e;}\,\}$	*method declarations*
e	::=	$\texttt{x} \mid \texttt{e.f} \mid \texttt{e.m}(\overline{\texttt{e}}) \mid \texttt{e@C} \mid \texttt{e}\backslash\texttt{C} \mid$ $\texttt{new C}(\overline{\texttt{e}}) \mid \texttt{new}\,\{\overline{\texttt{C}}\}\,(\overline{\texttt{e}}) \mid \texttt{e.m@C}(\overline{\texttt{e}})$	*expressions*
v	::=	$\texttt{new C}(\overline{\texttt{v}}) \mid \texttt{new}\,\{\overline{\texttt{C}}\}\,(\overline{\texttt{v}})$	*values*
T,U	::=	$\texttt{C} \mid \{\overline{\texttt{C}}\}$	*types*
Object,this			*predefined names*
C,D			*class names*
f,g			*field names*
x			*variable names*
m			*method names*
N			*any kind of name*

Fig. 2. Syntax of FJ_{set}

$$fields(\texttt{Object}) = \bullet \qquad\qquad \frac{\texttt{class C extends D}\ \{\overline{\texttt{T f};}\ \overline{\texttt{M}}\} \quad fields(\texttt{D}) = \overline{\texttt{U g}}}{fields(\texttt{C}) = \overline{\texttt{U g}}, \overline{\texttt{T f}}}$$

$$\frac{\texttt{class C extends D}\ \{\overline{\texttt{U f};}\ \overline{\texttt{M}}\} \quad \texttt{T m}(\overline{\texttt{T x}})\ \{\texttt{return e;}\} \in \overline{\texttt{M}}}{mBody(\texttt{m}, \texttt{C}) = \overline{\texttt{x}}.\texttt{e}} \qquad \frac{\texttt{class C extends D}\ \{\overline{\texttt{U f};}\ \overline{\texttt{M}}\} \quad \texttt{m} \notin \overline{\texttt{M}} \quad mBody(\texttt{m}, \texttt{D}) = \overline{\texttt{x}}.\texttt{e}}{mBody(\texttt{m}, \texttt{C}) = \overline{\texttt{x}}.\texttt{e}}$$

$$\frac{\texttt{class C extends D}\ \{\overline{\texttt{U f};}\ \overline{\texttt{M}}\} \quad \texttt{T m}(\overline{\texttt{T x}})\ \{\texttt{return e;}\} \in \overline{\texttt{M}}}{mType(\texttt{m}, \texttt{C}) = (\overline{\texttt{T}} \to \texttt{T})} \qquad \frac{\texttt{class C extends D}\ \{\overline{\texttt{U f};}\ \overline{\texttt{M}}\} \quad \texttt{m} \notin \overline{\texttt{M}} \quad mType(\texttt{m}, \texttt{D}) = (\overline{\texttt{T}} \to \texttt{T})}{mType(\texttt{m}, \texttt{C}) = (\overline{\texttt{T}} \to \texttt{T})}$$

$$distinct(\bullet) \qquad\qquad \frac{\texttt{N} \notin \overline{\texttt{N}} \quad distinct(\overline{\texttt{N}})}{distinct(\texttt{N}\ \overline{\texttt{N}})}$$

Fig. 3. Auxiliary functions for FJ_{set}

The syntax of the calculus from the level of classes and down is shown in Fig. 2. Notationally, we use overbars to denote lists of terms, so $\overline{\texttt{C}}$ stands for the list $\texttt{C}_1\,\texttt{C}_2 \ldots \texttt{C}_n$ for some natural number n; n=0 is allowed and yields the empty list, '$\bullet$'. There may be separators such as commas or semicolons between the elements of such a list, but they are implicit and implied by the context.

Several constructs in the syntax are identical to the ones known from Featherweight Java. Class and method definitions are standard, using the variant of

$$\frac{\texttt{class C extends D}\ \{\,\overline{\texttt{T}}\,\overline{\texttt{f}};\ \overline{\texttt{M}}\,\}}{\vdash \texttt{C} \sqsubset: \texttt{D}} \qquad \frac{\vdash \texttt{C} \sqsubset: \texttt{C}'' \quad \vdash \texttt{C}'' \sqsubset: \texttt{C}'}{\vdash \texttt{C} \sqsubset: \texttt{C}'}$$

$$\vdash \texttt{C} \sqsubset: \texttt{C} \qquad \frac{\overline{\texttt{D}} \subseteq \overline{\texttt{C}} \quad \mathit{distinct}(\overline{\texttt{D}})}{\vdash \{\overline{\texttt{C}}\} \subset: \{\overline{\texttt{D}}\}}$$

$$\frac{\vdash \texttt{T} \sqsubset: \texttt{U}}{\vdash \texttt{T} <: \texttt{U}} \qquad \frac{\vdash \texttt{T} \subset: \texttt{U}}{\vdash \texttt{T} <: \texttt{U}} \qquad \frac{\vdash \texttt{C}_i \sqsubset: \texttt{C}}{\texttt{C} \in: \{\overline{\texttt{C}}\}}$$

Fig. 4. Subclassing and subtyping for $\mathrm{FJ}_{\mathrm{set}}$

$$\vdash \texttt{Object}\ \text{OK} \qquad \frac{\texttt{C} \in \texttt{CT}}{\vdash \texttt{C}\ \text{OK}} \qquad \frac{\mathit{distinct}(\overline{\texttt{C}})}{\vdash \{\overline{\texttt{C}}\}\ \text{OK}}$$

Fig. 5. Wellformedness rules for $\mathrm{FJ}_{\mathrm{set}}$

Featherweight Java that omits explicit constructors. The standard expressions are variables, field lookups, method calls, and `new` expressions.

The remaining expressions are concerned with object sets. A class selection expression, `e@C`, provides the object labeled with the class `C` from the object set `e`. A class exclusion expression, `e\C`, provides an object set from which the object labeled with `C` as well as `C` itself has been deleted. The expression `new` $\{\overline{\texttt{C}}\}$ $(\overline{\texttt{e}})$ denotes creation of an object set which contains each of the objects denoted by the expression list $\overline{\texttt{e}}$, labeled by the list of classes $\overline{\texttt{C}}$.

Finally, the expression $\texttt{e.m@C}(\overline{\texttt{e}})$ denotes an object set method call, which selects all objects from the object set `e` which are labeled with the class `C` or a subclass thereof and calls a method `m` on each of them in the order they appear in the class list of the built-in mapping of the object set `e`. The method `m` must be defined in or inherited by the class `C`, and it must take a non-zero number of arguments where the first argument has the same type as the method return type, in order to enable the nested method call process mentioned in Sect. 1.

3.2 Auxiliary Methods, Subtyping, and Wellformedness

Figure 3 defines the auxiliary functions used for field lookup and similar tasks. They are standard except for the function *distinct*, which simply expresses that a given list of names (of any kind such as class names, method names, etc.) are distinct. As is common, quoting a class definition as a premise of a rule indicates the requirement that `CT` must contain that class definition.

The rules in Fig. 4 show subclassing ($\vdash \texttt{C} \sqsubset: \texttt{D}$), which is standard; subtyping for object sets ($\vdash \texttt{T} \subset: \texttt{U}$), which corresponds to the superset relation among the sets; and subtyping, which combines the two. Furthermore the judgement $\texttt{C} \in: \{\overline{\texttt{C}}\}$ holds whenever `C` is a superclass of one of the classes $\overline{\texttt{C}}$; this is used in the dynamic semantics of object set method calls.

In the Coq formalization of the calculus, transitivity for subclassing includes the requirement that the two pairs of classes are distinct, i.e., that $\mathtt{C} \neq \mathtt{C}''$ and $\mathtt{C}'' \neq \mathtt{C}'$. An easy induction shows that each of the two definitions of subclassing is able to derive all the subclass judgements of the other. However, in order to show in Coq that subclassing is decidable, the addition of these requirements solves a problem because they make it easy to see that a subtype judgment derivation tree must have a limited size if it exists.

The requirement in the rule for object set subtyping that the classes $\overline{\mathtt{D}}$ are distinct is necessary in order to prevent duplicates in the class list of object set types, which would be unsound. This is only required for the supertype, $\{\overline{\mathtt{D}}\}$, because distinctness for the subtype is ensured by other rules, in particular in the typing of object set creation expressions shown below in Fig. 7.

The type wellformedness requirements are shown in Fig. 5. They state that a class name is well-formed if there is a class of that name in the class table, and that an object set type, $\{\overline{\mathtt{C}}\}$, is well-formed if it consists of distinct class names. Finally, CT must satisfy $\mathtt{Object} \notin \mathtt{CT}$. Note that the class names in an object set type are not explicitly required to be defined in CT, because this requirement is a consequence of other rules. In general, the wellformedness requirements in this calculus are sufficient to enable the proof of soundness, but they are also minimal in the sense that removing any of them invalidates the proof. We believe that the use of proof assistant software may tighten the specification of well-formedness requirements in calculi, which is an area that otherwise easily gets a slightly imprecise treatment.

3.3 Expression Evaluation

The dynamic semantics of $\mathrm{FJ}_{\mathrm{set}}$ is presented in Fig. 6. Selection of a redex in a larger expression is defined in terms of evaluation contexts rather than congruence rules; they are listed at the bottom of the figure, where E denotes an expression with exactly one hole and $\mathtt{E}^+$ denotes a (non-empty) list of expressions with exactly one hole. With respect to the evaluation order, this calculus follows the tradition from FJ, whereby the evaluation order is restricted as little as possible, and particular strategies like call-by-value are available as one of the possible choices.

The rules for field lookup and method invocation are standard. The rule for class selection, (R-Select), selects the member of the given object set labeled with the specified class. This rule serves as an example of the evaluation order issue mentioned above: evaluation has to proceed until the top level expression is an object set creation expression ($\mathtt{new}\,\{\overline{\mathtt{C}}\}\,(\ldots)$) in order to reveal $\{\overline{\mathtt{C}}\}$ and thus the built-in mapping of the object set, but the arguments need not be fully evaluated.

We stated earlier that an object set offers an interface which is a disjoint union of the interfaces of its members. The class selection operation fulfills this promise as follows: For a given object set, the classes used to label some members of the object set are made explicit in its type (others may have been lost by subsumption). The interface of the object set is the union of the interfaces of

$$\frac{fields(\texttt{C}) = \overline{\texttt{T f}}}{\texttt{new C}(\overline{\texttt{e}})\texttt{.f}_i \leadsto \texttt{e}_i}$$
(R-Field)

$$(\texttt{new }\{\overline{\texttt{C}}\}\ (\overline{\texttt{e}}))\backslash\texttt{C}_i \leadsto \texttt{new }\{\overline{\texttt{C}}\backslash\texttt{\#i}\}\ (\overline{\texttt{e}}\backslash\texttt{\#i})$$
(R-Drop)

$$\frac{mBody(\texttt{m}, \texttt{C}) = \overline{\texttt{x}}.\texttt{e}_0}{(\texttt{new C}(\overline{\texttt{e}}))\texttt{.m}(\overline{\texttt{e}'}) \leadsto [\texttt{this/new C}(\overline{\texttt{e}}),\overline{\texttt{x/e}'}]\texttt{e}_0}$$
(R-Invk)

$$\frac{\vdash \texttt{C}_i \sqsubset: \texttt{C} \quad i = min\{\, j \mid \vdash \texttt{C}_j \sqsubset: \texttt{C}\,\} \quad \texttt{v}_i = \texttt{new D}(\overline{\texttt{v}'}) \quad mBody(\texttt{m}, \texttt{D}) = \overline{\texttt{x}}.\texttt{e}_0 \quad \texttt{e}_0' = [\texttt{this/new D}(\overline{\texttt{v}'}),\overline{\texttt{x}}/\texttt{e}\overline{\texttt{e}}]\,\texttt{e}_0}{(\texttt{new }\{\overline{\texttt{C}}\}\ (\overline{\texttt{v}}))\texttt{.m@C}(\texttt{e}\overline{\texttt{e}}) \leadsto (\texttt{new }\{\overline{\texttt{C}}\backslash\texttt{\#i}\}\ (\overline{\texttt{v}}\backslash\texttt{\#i}))\texttt{.m@C}(\texttt{e}_0'\overline{\texttt{e}})}$$
(R-SInvk)

$$(\texttt{new }\{\overline{\texttt{C}}\}\ (\overline{\texttt{e}}))\texttt{@C}_i \leadsto \texttt{e}_i$$
(R-Select)

$$\frac{\texttt{C} \not\in: \{\overline{\texttt{C}}\}}{(\texttt{new }\{\overline{\texttt{C}}\}\ (\overline{\texttt{v}}))\texttt{.m@C}(\texttt{e}\overline{\texttt{e}}) \leadsto \texttt{e}}$$
(R-SInvk-Done)

$$\begin{array}{lcl} \texttt{E} & ::= & [\_] \mid \texttt{E.f} \mid \texttt{E.m}(\overline{\texttt{e}}) \mid \texttt{e.m}(\texttt{E}^+) \mid \texttt{new C}(\texttt{E}^+) \mid \texttt{new }\{\overline{\texttt{C}}\}\ (\texttt{E}^+) \mid \texttt{E@C} \mid \\ & & \texttt{E}\backslash\texttt{C} \mid \texttt{E.m@C}(\overline{\texttt{e}}) \mid \texttt{e.m@C}(\texttt{E}^+) \\ \texttt{E}^+ & ::= & \overline{\texttt{e}}\,\texttt{E}\,\overline{\texttt{e}'} \end{array}$$

Fig. 6. Evaluation rules and evaluation contexts for FJ_{set}

these classes, and thus the object set supports access to all these features of all those members. There are no naming conflicts because the choice of class is made explicit, i.e., it is a disjoint union. In a full-fledged language it is much more convenient for the programmer if the explicit class selection is avoided, but this is trivial in the cases where there is no naming conflict, and it should be handled explicitly when a conflict exists; the language gbeta uses such an implicit approach.

The rule for class exclusion, (R-Drop), deletes the requested class and the corresponding member from the object set. This rule introduces notation for a simple function that deletes the i'th element from a list, namely $\overline{\texttt{t}}\backslash\texttt{\#i}$, where $\overline{\texttt{t}}$ are terms of any kind, e.g., class names or expressions. Usage of this notation implies that the list is long enough to contain the position to delete.

In the Coq formalization of this calculus the (R-Drop) rule zips the list of classes and the list of expressions together to a list of pairs, then deletes the pair which contains the specified class, and then unzips the shortened list of pairs to get the resulting list of classes and list of expressions. The reason for this choice is that it provides a direct correspondence between the classes and expressions, whereas an alternative approach based on looking up the i'th element in both lists causes a large number of extra conditions regarding the upper bound of the index i. This is a typical situation where the convenient formalization in Coq does not correspond exactly to a well-known or convenient notation for presentation in a paper, but the deletion-by-position notation $\overline{\texttt{t}}\backslash\texttt{\#i}$ is a simple and relatively readable way to bridge the gap.

The object set method call semantics is specified by two rules, (R-SInvk) and (R-SInvk-Done). As mentioned, an object set method call amounts to a composite operation which includes a method call on each of the members of the set labeled by a class supporting that method. The rule (R-SInvk) specifies what to do when the object set contains a member supporting the requested method `m`, and the rule (R-SInvk-Done) specifies what to do in the end when all such objects have been processed.

Whether an object set member supports `m` is determined by requiring that the member is associated with a subclass of the class `C` specified in the object set call. This means that each selected object will be an instance of `C` or one of its subclasses, and the method `m` will be defined for that object, with a signature which is identical to the signature of `m` in `C`, except for possible covariance in the return type. Objects supporting unrelated methods with the same name `m` are ignored.

The rule (R-SInvk) specifies how to call one method `m` and provide the results produced by this method call to the next method call. It requires that the list of classes $\overline{\mathtt{C}}$ associated with the object set contains a subclass $\mathtt{C}_i$ of the class requested in the call, `C`, and selects the 'first' one (the one with the smallest index i). It then removes the selected object from the receiver object set and repeats the object set method call with the result of the invocation of `m` on the selected object as its first argument. Note that the minimality of i is not needed for soundness, it is needed in order to ensure that object set method calls have a predictable semantics: it should accumulate the results from its members according to their built-in ordering.

However, the first argument does not look like a method call, it is actually given as $[\mathtt{this/new\ D(\overline{v'}),\overline{x}/e\overline{e}}]\mathtt{e}_0$, but inspection of the rule for method call, (R-Invk), reveals that this is the result of taking one evaluation step after the method invocation $\mathtt{new\ D(\overline{v'}).m(e\overline{e})}$. It is necessary to express the rule in this form in order to maintain the property that all rules are compositional.

A similar investigation shows that the receiver of the object set method call after the step in (R-SInvk) is the result of taking one step after excluding the selected class $\mathtt{C}_i$ from the receiver object set before the step. Compositionality again forces the rule to take that step rather than expressing the result in terms of an explicit class exclusion operation.

The semantics of an object set method call may thus seem to be expressible in terms of other operations, but this is not the case because there is no way to select the class $\mathtt{C}_i$ appropriately without this operation. A primitive could be provided in order to make such a selection, but we have not found any such primitive which enables the same functionality without requiring strictly more static knowledge about the contents of object sets.

Finally, the rule (R-SInvk-Done) yields the first argument of the object set method call in the situation where no object in the object set can be selected.

3.4 Typing

The type rules for $\mathrm{FJ}_{\mathrm{set}}$ are shown in Fig. 7. The rules for the typing of variables, field lookups, ordinary method invocations, and ordinary object creation are standard.

$$\frac{\Gamma(\mathtt{x}) = \mathtt{T}}{\Gamma \vdash \mathtt{x} : \mathtt{T}} \quad \text{(T-VAR)}$$

$$\frac{\Gamma \vdash \mathtt{e} : \{\overline{\mathtt{C}}\} \qquad \vdash \mathtt{C}_i \text{ OK}}{\Gamma \vdash \mathtt{e@C}_i : \mathtt{C}_i} \quad \text{(T-SELECT)}$$

$$\frac{\Gamma \vdash \mathtt{e} : \mathtt{C} \qquad \mathit{fields}(\mathtt{C}) = \overline{\mathtt{T}\,\mathtt{f}}}{\Gamma \vdash \mathtt{e.f}_i : \mathtt{T}_i} \quad \text{(T-FIELD)}$$

$$\frac{\Gamma \vdash \mathtt{e} : \{\overline{\mathtt{C}}\}}{\Gamma \vdash \mathtt{e}\backslash\mathtt{C}_i : \{\overline{\mathtt{C}}\backslash\mathtt{\#i}\}} \quad \text{(T-DROP)}$$

$$\frac{\Gamma \vdash \mathtt{e} : \mathtt{C} \qquad \mathit{mType}(\mathtt{m}, \mathtt{C}) = (\overline{\mathtt{T}} \to \mathtt{T}) \qquad \Gamma \vdash \overline{\mathtt{e}} : \overline{\mathtt{U}} \qquad \vdash \overline{\mathtt{U}} <: \overline{\mathtt{T}}}{\Gamma \vdash \mathtt{e.m}(\overline{\mathtt{e}}) : \mathtt{T}} \quad \text{(T-INVK)}$$

$$\frac{\Gamma \vdash \mathtt{e} : \{\overline{\mathtt{C}}\} \qquad \Gamma \vdash \overline{\mathtt{e}} : \mathtt{T}\,\overline{\mathtt{T}} \qquad \mathit{mType}(\mathtt{m}, \mathtt{C}) = (\mathtt{T}\,\overline{\mathtt{T}} \to \mathtt{T})}{\Gamma \vdash \mathtt{e.m@C}(\overline{\mathtt{e}}) : \mathtt{T}} \quad \text{(T-SINVK)}$$

$$\frac{\mathit{fields}(\mathtt{C}) = \overline{\mathtt{T}\,\mathtt{f}} \qquad \Gamma \vdash \overline{\mathtt{e}} : \overline{\mathtt{U}} \qquad \vdash \overline{\mathtt{U}} <: \overline{\mathtt{T}}}{\Gamma \vdash \mathtt{new}\ \mathtt{C}(\overline{\mathtt{e}}) : \mathtt{C}} \quad \text{(T-NEW)}$$

$$\frac{\mathit{distinct}(\overline{\mathtt{C}}) \qquad \Gamma \vdash \overline{\mathtt{e}} : \overline{\mathtt{U}} \qquad \vdash \overline{\mathtt{U}} <: \overline{\mathtt{C}}}{\Gamma \vdash \mathtt{new}\ \{\overline{\mathtt{C}}\}\ (\overline{\mathtt{e}}) : \{\overline{\mathtt{C}}\}} \quad \text{(T-SNEW)}$$

Fig. 7. Type rules for $\mathrm{FJ}_{\mathrm{set}}$

The rule (T-SELECT) specifies that the target must be typable as an object set containing the requested class, and the resulting type is then that class. It would be easy to change this rule and (R-SELECT) to select a subclass, i.e., to allow for the selection of a class C as long as C $\in:$ $\{\overline{\mathtt{C}}\}$, but this could prevent the selection of a class C′ from an object set that is also associated with some subclass C″ of C′ or make the operation ambiguous, and since there is no depth subtyping for object set types it would not enhance the expressive power or the flexibility of the language.

The rule (T-DROP) specifies that the target must be typable as an object set that includes the class to exclude, which is then removed from the type of the object set to produce the result type. For the same reasons as above it would not be useful to allow the requested class to be a superclass of the excluded class. The rule (T-SINVK) specifies how to type object set method calls. It requires that the receiver is typable as an object set, but does not require anything about the set of classes associated with this object set. On the other hand, the method m must be defined or inherited in the class C, it must take at least one argument, and the type of the first argument must be identical to the return type, which is also the type of the entire object set method call. Finally, the rule (T-SNEW) specifies the typing of object set creations. It simply requires that the classes used as labels are distinct and that each member has a subtype of its associated class.

It would be very easy to change the (T-SINVK) rule to require C $\in:$ $\{\overline{\mathtt{C}}\}$ and adapt the soundness proof accordingly, which would guarantee that the object set method call would include at least one actual method call, but this is not required for soundness. Similarly, it would be easy to relax the rule such that

$$\frac{\mathit{override}(\mathtt{m},\mathtt{D},\mathtt{T},\overline{\mathtt{T}}) \quad \Gamma;\ \mathtt{this}:\mathtt{C} \vdash \mathtt{e} : \mathtt{U} \quad \vdash \mathtt{U} <: \mathtt{T} \quad \vdash \mathtt{T}, \overline{\mathtt{T}}\ \text{OK} \quad \mathit{distinct}(\overline{\mathtt{x}})}{\vdash \mathtt{T}\ \ \mathtt{m}(\overline{\mathtt{T}}\,\overline{\mathtt{x}})\{\ \mathtt{return}\ \mathtt{e};\ \}\ \text{OK in}\ \mathtt{C},\mathtt{D}}$$

(T-Method)

$$\frac{\vdash \mathtt{D}\ \text{OK} \quad \vdash \overline{\mathtt{T}}\ \text{OK} \quad \vdash \overline{\mathtt{M}}\ \text{OK in}\ \mathtt{C},\mathtt{D} \quad \vdash \mathtt{D} \not<: \mathtt{C} \quad \vdash \mathtt{D} <: \mathtt{Object} \quad \mathit{distinct}(\mathit{fields}(\mathtt{D})\ \overline{\mathtt{f}}) \quad \mathit{distinct}(\mathit{names}(\overline{\mathtt{M}}))}{\vdash \mathtt{class}\ \mathtt{C}\ \mathtt{extends}\ \mathtt{D}\ \{\ \overline{\mathtt{T}}\,\overline{\mathtt{f}};\ \overline{\mathtt{M}}\ \}\ \text{OK}}$$

(T-Class)

$$\frac{\mathit{mType}(\mathtt{m},\mathtt{D})\ \mathit{is\ undefined}}{\mathit{override}(\mathtt{m},\mathtt{D},\mathtt{T},\overline{\mathtt{T}})} \qquad \frac{\mathit{mType}(\mathtt{m},\mathtt{D}) = (\overline{\mathtt{T}} \to \mathtt{T}') \quad \vdash \mathtt{T} <: \mathtt{T}'}{\mathit{override}(\mathtt{m},\mathtt{D},\mathtt{T},\overline{\mathtt{T}})}$$

Fig. 8. Class and method typing for $\mathrm{FJ}_{\mathrm{set}}$

the return type only has to be a subtype of the type of the first argument rather than being identical to it, but it would be hard to exploit this information unless the rule were modified to enforce that there is at least one actual method call. Even then, the accumulation of contributions from several members of the object set would have to start "from scratch" at each member, because the type of the first argument is fixed. Hence, these variations do not seem to be worthwhile.

Finally, Fig. 8 shows the rules for class and method typing, i.e., rules that apply type checking to the entire program. As opposed to the traditional treatment, these rules include all the requirements needed for programs to be well-formed—for instance in order to avoid cyclic inheritance graphs.

The rule (T-Method) specifies that a method `m` defined in a class `C` with superclass `D` must correctly override any definitions of `m` available in the superclass, it must have a body whose type is a subtype of the declared return type, it must have distinct argument names, and the specified types must be well-formed. The only non-standard element here is the requirement that argument names must be distinct.

The rules for *override* are given at the bottom of the figure; they are used to specify when a definition of a method `m` with argument types $\overline{\mathtt{T}}$ and return type `T` is correct in relation to definitions available in the superclass `D`. It is standard except that it allows for covariance in the return type, just like the Java language of today.

The rule (T-Class) specifies the standard requirements that the superclass `D`, all field types, and all methods must be well-formed. Moreover, the superclass cannot be a subclass of `C` itself, which prevents cycles in the inheritance graph; and the superclass must be a subclass of `Object`, which ensures that all inheritance chains are finite. This finiteness ensures that subclassing is decidable, which is used in the progress proof. Finally, there are distinctness requirements for field and method names.

All in all, this is not much more involved or verbose than the usual class and method typing rules, but it is complete in the sense that there are no additional (informal and maybe even implicit) well-formedness rules about programs to worry about. We think that it would be useful to make program well-formedness fully explicit, as we have done it here. It is, of course, a consequence of using proof assistant software, because the proofs cannot be completed unless such things are made precise and included in the specification.

4 Soundness

The $\mathrm{FJ}_{\mathrm{set}}$ calculus is sound, which is shown via the standard preservation and progress results:

Theorem 1 (Preservation). *If the expression* $\mathtt{e}$ *in the environment* Γ *is typable by* $\Gamma \vdash \mathtt{e} : \mathtt{T}$ *and it can take the step* $\mathtt{e} \leadsto \mathtt{e}'$, *then* $\Gamma \vdash \mathtt{e}' : \mathtt{U}$ *for some type* $\mathtt{U}$ *such that* $\vdash \mathtt{U} <: \mathtt{T}$.

Theorem 2 (Progress). *If* $\mathtt{e}$ *is an expression typeable by* $\emptyset \vdash \mathtt{e} : \mathtt{T}$ *then either* $\mathtt{e}$ *is a value or there exists an expression* $\mathtt{e}'$ *such that* $\mathtt{e} \leadsto \mathtt{e}'$.

A complete proof of these properties which has been mechanically checked by the proof assistant Coq is is available for download [5]. It consists of approximately 6500 lines of Gallina code, divided into approximately 3500 lines specifically on the calculus and approximately 3000 lines of standard language metatheory facilities from the Coq tutorial given at POPL 2008 [8].

5 Related Work

Mixins [9] and traits [10,11] are language mechanisms which improve on the flexibility of ordinary object-oriented inheritance. Usually a subclass is created by extending one or more existing classes with a new class body, and this class body cannot be reused in the extension of other classes; but mixins and traits promote class bodies to a first class status, such that they can be reused. Mixins are typically as general as class bodies, whereas traits leave out state and access control in return for a more robust and flexible symmetric composition mechanism with renaming, exclusion and similar operators. Object sets were conceived as a useful primitive mechanism that supports a version of mixins that includes cross-mixin-instance features (virtuals, in the family polymorphism sense) directly, thus enabling these features to be computed at runtime while maintaining more precise type information than standard object-oriented platforms are capable of.

Multiple inheritance, e.g., as in C++ [12] or Eiffel [13], is so semantically different and so much more tied to compile-time that such a mechanism as object sets is unlikely to provide any benefits as an implementation device. Object sets might still be supported as a surface language feature, but in this case they should be redesigned to emphasize convenient usage for programmers.

Dynamic languages like Self [14,15] support a very general and flexible style of composite objects by means of parent slots and genuine delegation. Object sets are less flexible, but in return they are statically typed. Moreover, objects sets provide features based on either explicit disambiguation of which member to use, or by object set method calls which are used to enable collaboration among the members of the object set; both of these are more low-level than delegation in a language like Self, but they fit more smoothly as new primitives in a standard, typed object-oriented context, and they are specifically optimized for maintaining a kind of type information which is needed in order to support family polymorphism.

The concept of roles [16] is related to inheritance mechanisms like mixins, but it adds a dynamic element in that roles can be added and removed for existing objects; the role and the role player may or may not have distinct object identities. Gottlob et al. [17] describe a delegation based approach to support roles in Smalltalk, i.e., in a dynamically typed context. Handling delegation or roles in a statically typed context is a much greater challenge, but Kniesel [18] shows how both static and dynamic delegation can be handled in a type safe manner and used to support a kind of roles, and the support for dynamic reclassification in Fickle [19] may be used similarly, especially in order to change object behavior. Object sets fit into this context in the same way as they do for delegation in general.

Featherweight Wrap Java [20] is an extension of Featherweight Java with mutable state and the ability to add extra behavior before or after the main behavior of an object by means of a wrapper object which uses delegation or consultation to the wrappee. It is statically typed and proved sound. It avoids much of the complexity of, e.g., Kniesel's approach by letting the wrapper be a subtype of the declared wrappee interface of the wrapper class rather than a subtype of the actual wrappee type, which means that interfaces cannot be not accumulated from multiple extensions. Object sets are quite different from this approach because they do not support traditional delegation, but on the other hand they enable accumulation of arbitrary types, require explicit disambiguation, and add in object set method calls to allow cross-object behavior that includes objects without requiring them to know about each other's types.

Object sets are similar to extensible records in some ways. For instance, Gaster and Jones [21] define polymorphic, extensible records and unions based on row variables, i.e., mappings from labels to types. With object sets, the associated classes work as labels and types combined; this reduces the flexibility because there cannot be two labels with the same type, but given that object sets are intended to model composite objects it would correspond to repeated inheritance to have more than one member associated with the same class, and this would preclude a surface level syntax where class selection is implicit, due to name clashes. Object sets as presented here do not support extension; this is because we consider a '`lacks C`' construct which promises that there is no class `C` in this object set to be unmanageable in real-world software development, and this would probably affect languages built on object sets even if they are confined to the virtual machine layer. On the other hand, extensible records do not have a late-bound operation that corresponds to our object set method calls; they only use statically known components.

A well-established approach to extensible records is the Haskell `HList` library [22], where Kiselyov, Lämmel and Schupke use type level natural numbers and a number of layers on top of that to support type safe heterogeneous lists. Such lists are actually nested tuples, and the approach relies heavily on being able to use large type expressions which are inferred and never show up in the source code. If explicit typing is considered a valuable source of documentation then object sets are more manageable because they abstract away from the ordering of elements, and they may provide access to an arbitary set of members without depending on the internal structure, i.e., the order of known members and the presence of unknown members.

Finally, other languages that include support for family polymorphism such as Scala [23], CaesarJ [24], and ObjectTeams [25] could benefit from having support for object sets in an underlying virtual machine, especially because this might allow them to deviate from the implementation strategy whereby one surface language object corresponds to exactly one virtual machine object. This could increase the runtime flexibility of classes and objects, at a cost in performance but with a more direct and complete support for typing at the virtual machine level.

6 Conclusion

We have presented the concept of object sets as a first class language construct which is capable of emulating the main features of traditional, monolithic objects: access to the disjoint union of the features of all object set members in the type, and support for a kind of method calls whereby the choice of methods to call is made dynamically, corresponding to feature access and method calls for ordinary objects. However, object sets are more flexible than ordinary objects, because they combine the features of several classes (like mixins or multiple inheritance, but without the name clashes), and they provide the machinery needed in order to support dynamic metamorphosis of object sets. The mechanism is useful in its own right, but it is likely to benefit from a pragmatic layer on top of the operations shown in this paper, because this makes the syntax more compact and convenient. An example of a language which does this is gbeta. The mechanisms of this paper might then provide good service as primitives on main-stream platforms such as .Net or JVM, which would make these platforms capable of handling languages supporting family polymorphism, such as gbeta, with a significantly reduced need for compiler generated dynamic casts.

Acknowledgments. The design of the object set method call mechanism owes some very useful insights to Kim Birkelund. The Coq proof was developed from a starting point created by Bruno de Fraine which was a proof of soundness for Featherweight Java without casts, which again used a number of files from the POPL 2008 Coq tutorial. Finally, the anonymous reviewers provided very good feed-back.

References

1. Ernst, E.: gbeta – A Language with Virtual Attributes, Block Structure, and Propagating, Dynamic Inheritance. PhD thesis, Devise, Department of Computer Science, University of Aarhus, Aarhus, Denmark (June 1999)
2. Ernst, E.: Higher-order hierarchies. In: Cardelli, L. (ed.) ECOOP 2003. LNCS, vol. 2743, pp. 303–329. Springer, Heidelberg (2003)
3. Ernst, E., Ostermann, K., Cook, W.R.: A virtual class calculus. In: Proceedings POPL 2006, Charleston, SC, USA, pp. 270–282. ACM, New York (2006)
4. Ernst, E.: Family polymorphism. In: Knudsen, J.L. (ed.) ECOOP 2001. LNCS, vol. 2072, pp. 303–326. Springer, Heidelberg (2001)
5. Ernst, E.: Coq proof of soundness for the FJ_{set} calculus (October 2008), `http://www.cs.au.dk/~eernst/papers/objsetproof.tgz`
6. Bertot, Y., Castéran, P.: Interactive Theorem Proving and Program Development. Coq'Art: The Calculus of Inductive Constructions. Texts in Theoretical Computer Science. Springer, Heidelberg (2004)
7. Igarashi, A., Pierce, B., Wadler, P.: Featherweight Java: A minimal core calculus for Java and GJ. TOPLAS 23(3), 396–459 (2001)
8. Aydemir, B., Bohannon, A., Pierce, B., Vaughan, J., Vytiniotis, D., Weirich, S., Zdancewic, S.: Using proof assistants for programming language research (January 2008), `http://www.cis.upenn.edu/~plclub/popl08-tutorial/`
9. Bracha, G., Cook, W.R.: Mixin-based inheritance. In: Proceedings OOPSLA/ECOOP, pp. 303–311 (1990)
10. Schärli, N., Ducasse, S., Nierstrasz, O., Black, A.P.: Traits: Composable units of behaviour. In: Cardelli, L. (ed.) ECOOP 2003. LNCS, vol. 2743, pp. 248–274. Springer, Heidelberg (2003)
11. Ducasse, S., Nierstrasz, O., Schärli, N., Wuyts, R., Black, A.P.: Traits: A mechanism for fine-grained reuse. ACM Transactions on Programming Languages and Systems 28(2), 331–388 (2006)
12. Stroustrup, B.: The C++ Programming Language: Special edn. Addison Wesley, Reading (2000)
13. Meyer, B.: Eiffel: The Language. Prentice-Hall, Englewood Cliffs (1991)
14. Ungar, D., Smith, R.B.: Self: The power of simplicity. In: Proceedings OOPSLA 1987, Orlando, FL, October 1987, pp. 227–242 (1987)
15. Agesen, O., Bak, L., Chambers, C., Chang, B.W., Hölzle, U., Maloney, J., Smith, R.B., Ungar, D., Wolczko, M.: The Self 4.0 Programmer's Reference Manual. Sun Microsystems, Inc., Mountain View (1995)
16. Kristensen, B.B., Østerbye, K.: Roles: Conceptual abstraction theory and practical language issues. TAPOS 2(3), 143–160 (1996)
17. Gottlob, G., Schrefl, M., Röck, B.: Extending object-oriented systems with roles. ACM Transactions on Information Systems 14(3), 268–296 (1996)
18. Kniesel, G.: Type-safe delegation for dynamic component adaptation. In: Demeyer, S., Bosch, J. (eds.) ECOOP 1998 Workshops. LNCS, vol. 1543, pp. 136–137. Springer, Heidelberg (1998)
19. Drossopoulou, S., Damiani, F., Dezani-Ciancaglini, M., Giannini, P.: Fickle: Dynamic object re-classification. In: Knudsen, J.L. (ed.) ECOOP 2001. LNCS, vol. 2072, pp. 130–149. Springer, Heidelberg (2001)
20. Bettini, L., Capecchi, S., Giachino, E.: Featherweight wrap java: Wrapping objects and methods. Journal of Object Technology 7(2), 5–29 (2008)

21. Gaster, B.R., Jones, M.P.: A polymorphic type system for extensible records and variants. Technical Report NOTTCS-TR-96-3, Department of Computer Science, University of Nottingham (November 1996)
22. Kiselyov, O., Lammel, R., Schupke, K.: Strongly typed heterogeneous collections. In: Haskell Workshop, pp. 96–107 (2004)
23. Odersky, M.: The Scala Language Specification. EPFL, Switzerland. Version 2.7 edn. (January 2009), `http://www.scala-lang.org/docu/files/ScalaReference.pdf`
24. Aracic, I., Gasiunas, V., Mezini, M., Ostermann, K.: An overview of CaesarJ. In: Rashid, A., Aksit, M. (eds.) Transactions on Aspect-Oriented Software Development I. LNCS, vol. 3880, pp. 135–173. Springer, Heidelberg (2006)
25. Herrmann, S.: A precise model for contextual roles: The programming language ObjectTeams/Java. Applied Ontology 2(2), 181–207 (2007)

Monadic Translation of Intuitionistic Sequent Calculus

José Espírito Santo[1], Ralph Matthes[2], and Luís Pinto[1]

[1] Departamento de Matemática, Universidade do Minho, Portugal
[2] I.R.I.T. (C.N.R.S. and University of Toulouse III), France
{jes,luis}@math.uminho.pt, matthes@irit.fr

Abstract. This paper proposes and analyses a monadic translation of an intuitionistic sequent calculus. The source of the translation is a typed λ-calculus previously introduced by the authors, corresponding to the intuitionistic fragment of the call-by-name variant of $\overline{\lambda}\mu\tilde{\mu}$ of Curien and Herbelin, and the target is a variant of Moggi's monadic meta-language, where the rewrite relation includes extra permutation rules that may be seen as variations of the "associativity" of bind (the Kleisli extension operation of the monad).

The main result is that the monadic translation simulates reduction strictly, so that strong normalisation (which is enjoyed at the target, as we show) can be lifted from the target to the source. A variant translation, obtained by adding an extra monad application in the translation of types, still enjoys strict simulation, while requiring one fewer extra permutation rule from the target.

Finally we instantiate, for the cases of the identity monad and the continuations monad, the meta-language into the simply-typed λ-calculus. By this means, we give a generic account of translations of sequent calculus into natural deduction, which encompasses the traditional mapping studied by Zucker and Pottinger, and CPS translations of intuitionistic sequent calculus.

1 Introduction

This paper is about a monadic translation of intuitionistic sequent calculus. By the latter we mean the intuitionistic, call-by-name fragment of Curien-Herbelin's system for classical logic [1]. In the spirit of the Curry-Howard correspondence, such a system is handled as an extension of the simply-typed λ-calculus, identified by the authors in [5], and named $\lambda\mathbf{J}^{\mathbf{m}se}$.

The target of the monadic translation is a variant of Moggi's monadic meta-language [12], named λ_{M} here. To recall, this is an extension of the simply-typed λ-calculus where the type system includes a monad M, and the term language includes constructions for the unit and the Kleisli extension (a. k. a. bind) operation of the monad. The main point is that the set of reduction rules of the meta-language is extended by two new rules, which can be seen as variations

S. Berardi, F. Damiani, and U. de'Liguoro (Eds.): TYPES 2008, LNCS 5497, pp. 100–116, 2009.

of the usual "associativity" rule for bind, and which together with this "associativity" rule can be seen as forming a variation of one single principle in the ordinary λ-calculus, that we name assoc.

The monadic translation we introduce generalizes the ordinary monadic translation of the (call-by-name) λ-calculus [7], and, in particular, is based on the principle that functions from A to B are interpreted as functions from MA (computations of type A) to MB (computations of type B). The main result we obtain is a strict simulation theorem (one reduction step in the sequent calculus is mapped to one or more reduction steps in the monadic target). A variant of the monadic translation, based on the interpretation of functions from A to B as functions from MA to MMB, also enjoys strict simulation, and requires one less of the new reduction rules from the target system.

One of the uses of the above results is in obtaining *strong normalisation* for sequent calculus, i. e., the absence of an infinite sequence of proof transformations starting with a well-formed proof. Indeed, strong normalisation follows immediately from strict simulation, since the target system is itself strongly normalising. This fact, in turn, rests on the strong normalisability of the extension of λ-calculus with the assoc reduction rule [3]. This emphasis on strict simulation and strong normalisation follows the line of [5,6], but is in contrast with the uses of the monadic language in the study of programming languages semantics and compilation, where other kinds of relationship between source and target calculi, like equational correspondence, or reflection, are often obtained [7,16].

On the other hand, we may regard the monadic translations and their properties, not as a goal in itself, but as a parametric means to analyse a family of situations, via instantiation of the monad of the meta-language. In fact, we study two such instantiations, one for the identity monad, the other for the continuations monad, where by "instantiation" we mean composition of the monadic translation with an interpretation of the monadic language into the λ-calculus.

Through this method we obtain a *generic account* of translations of sequent calculus into natural deduction. The identity monad gives an analysis of what in our framework is the traditional mapping studied by Zucker and Pottinger [17,14], together with some of its variants. The continuations monad obtains an analysis of a CPS translation of $\lambda\mathbf{J}^{\mathbf{m}se}$ similar to the one at the basis of [5].

The methodology of this generic account should be contrasted with that of [7]. There, it is the monadic translation that varies, in order to capture a family of situations (in the case of [7], several CPS translations), while the monad remains instantiated to the continuations monad. Here, the monadic translation remains fixed, while, by varying the monad, we uncover a common root to seemingly unrelated translations of sequent calculus into natural deduction.

The paper is organised as follows. Section 2 presents sequent calculus $\lambda\mathbf{J}^{\mathbf{m}se}$. Section 3 presents our version λ_{M} of the monadic meta-language. Section 4 defines and proves the properties of the monadic translation and its optimized variant, and strong normalisation for $\lambda\mathbf{J}^{\mathbf{m}se}$ is obtained. Section 5 gives the generic account of translation into natural deduction. Finally, Section 6 concludes with some remarks.

2 Intuitionistic Sequent Calculus

The calculus $\lambda\mathbf{J}^{\mathbf{m}se}$ that is used here has been proposed in [5] (whose journal version is [6]). It corresponds to the intuitionistic fragment of the call-by-name variant of $\overline{\lambda}\mu\tilde{\mu}$-calculus of Curien and Herbelin [1]. We quite closely follow the presentation of the definition of $\lambda\mathbf{J}^{\mathbf{m}se}$ in [6].

There are three classes of expressions in $\lambda\mathbf{J}^{\mathbf{m}se}$:

$$\begin{array}{ll} \text{(Terms)} & t, u ::= x \mid \lambda x.t \mid \{c\} \\ \text{(Co-terms)} & l ::= [] \mid u :: l \mid (x)c \\ \text{(Commands)} & c ::= tl \end{array}$$

Terms can be variables (of which we assume a denumerable set ranged over by letters x, y, z), lambda-abstractions $\lambda x.t$ or coercions $\{c\}$ from commands to terms.

Co-terms provide means of forming lists of arguments, generalised arguments, or explicit substitutions. A co-term of the form $(x)c$, binds variable x in c and provides the generalised application facility. Operationally it can be thought of as "substitute for x in c". A co-term of the form $[]$ or $u :: l$ is called an *evaluation context* and is denoted by E. Evaluation contexts of the form $u :: l$ allow for multiary applications and, when passed to a term, indicate that, after consumption of argument u, computation should carry on with arguments in l. The co-term $[]$ marks the end of an evaluation context, while the expression $(x)x$ is just ill-formed and, in particular, not a co-term.

A command tl has a double role: if l is of the form $(x)c$, tl is an explicit substitution; otherwise, tl is a general form of application.

In writing expressions, sometimes we add parentheses to help their parsing. Also, we assume that the scope of binders λx and (x) extends as far as possible. We follow usual practise in that names of bound variables are considered as immaterial and that the binding occurrences on display are meant to be well-chosen so that no unwanted effects arise. It is then straightforward to define what it means to replace every free occurrence of variable x in a capture-avoiding way by a term t in a term u, co-term l or command c, yielding term $[t/x]u$, co-term $[t/x]l$ and command $[t/x]c$, respectively.

The calculus $\lambda\mathbf{J}^{\mathbf{m}se}$ has a form of sequent for each class of expressions:

$$\Gamma \vdash t : A \qquad \Gamma | l : A \vdash B \qquad \Gamma \xrightarrow{c} B$$

Letters A, B, C are used to range over the set of types (=formulas), built from a base set of type variables (ranged over by X) using the function type (that we write $A \supset B$). In sequents, contexts Γ are viewed as finite sets of declarations $x : A$, where no variable x occurs twice. The context $\Gamma, x : A$ is obtained from Γ by adding the declaration $x : A$, and will only be written if this yields again a valid context, i.e., if x is not declared in Γ. We can think of a term (resp. co-term) as an annotation for a selected formula in the *rhs* (resp. *lhs*). Commands annotate sequents generated as a result of logical cuts, where there is no selected formula on the *rhs* or *lhs*; as such we write them on top of the sequent arrow.

$$\frac{}{\Gamma|[] : A \vdash A}\,LAx \qquad \frac{}{\Gamma, x : A \vdash x : A}\,RAx$$

$$\frac{\Gamma \vdash u : A \quad \Gamma|l : B \vdash C}{\Gamma|u :: l : A \supset B \vdash C}\,LIntro \qquad \frac{\Gamma, x : A \vdash t : B}{\Gamma \vdash \lambda x.t : A \supset B}\,RIntro$$

$$\frac{\Gamma, x : A \xrightarrow{c} B}{\Gamma|(x)c : A \vdash B}\,LSel \qquad \frac{\Gamma \xrightarrow{c} A}{\Gamma \vdash \{c\} : A}\,RSel$$

$$\frac{\Gamma \vdash t : A \quad \Gamma|l : A \vdash B}{\Gamma \xrightarrow{tl} B}\,Cut$$

Fig. 1. Typing rules of $\lambda\mathbf{J}^{\mathbf{mse}}$

The typing rules of $\lambda\mathbf{J}^{\mathbf{mse}}$ are presented in Figure 1, stressing the parallel between left and right rules.

The standard typing rules for substitution for each syntactic class are admissible: replacing a variable of declared type A by a term of type A does not change the type. We also have the usual weakening rules: If a sequent with context Γ is derivable and Γ is replaced by a context Γ' that is a superset of Γ, then also this sequent is derivable.

We consider the following base reduction rules on expressions:[1]

$$\begin{array}{llll} (\beta) & (\lambda x.t)(u :: l) \to u((x)tl) & (\mu) & (x)xl \to l, \text{ if } x \notin l \\ (\pi) & \{tl\}E \to t\,(l@E) & (\epsilon) & \{t[]\} \to t \\ (\sigma) & t(x)c \to [t/x]c, & & \end{array}$$

where, in general, $l@l'$ is a co-term that represents an "eager" concatenation of l and l', viewed as lists, and is defined as follows[2]:

$$[]@l' = l' \qquad (u :: l)@l' = u :: (l@l') \qquad ((x)tl)@l' = (x)t\,(l@l')$$

Concatenation obeys to the following further admissible form of cut rule:

$$\frac{\Gamma|l : A \vdash B \quad \Gamma|l' : B \vdash C}{\Gamma|l@l' : A \vdash C}$$

The one-step reduction relation $\to$ is inductively defined as the term closure of the reduction rules.

For detailed comments on the reduction rules, the subject reduction property (that holds true), an analysis of normal forms and critical pairs (yielding local

---

1 Naming practise for binding occurrences excludes x as a free variable in u or l in the left-hand side of rule β. The widening of the binding scope of x in the right-hand side is noteworthy, but it is only meant to correspond to weakening.

2 Concatenation is "eager" in the sense that, in the last case, the right-hand side is not $(x)\{tl\}l'$ but, in the only important case that l' is an evaluation context E, its π-reduct. One immediately verifies $l@[] = l$ and $(l@l')@l'' = l@(l'@l'')$ by induction on l. Associativity would not hold with the lazy version of @.

confluence) and the identification of $\lambda\mathbf{J}^{\mathbf{m}se}$ as the intuitionistic fragment of CBN $\overline{\lambda}\mu\tilde{\mu}$, see [5,6].

We stress that the rule β does not execute any substitution. This makes a simulation of $\lambda\mathbf{J}^{\mathbf{m}se}$ in another system more difficult, not only because substitution is delayed, but also because the scope of the bound variable is enlarged.

3 Monadic Lambda-Calculus

The main result of [5,6] is a proof of strong normalization of $\lambda\mathbf{J}^{\mathbf{m}se}$ that does not refer to the strong normalization results by Lengrand [10] and Polonovski [13] about $\overline{\lambda}\mu\tilde{\mu}$, but by a syntactic transformation to simply-typed λ-calculus that strictly simulates reduction. The technique is a variation of continuation-passing style, called continuation-and-garbage-passing style [8]. CPS translations alone do not suffice for a strict simulation of all reductions. In the present article, we move from CPS translations to monadic translations, whose target we call monadic lambda-calculus. Strong normalisation of the monadic lambda-calculus itself does not rest any longer on simply-typed λ-calculus with only β-reduction; instead the following rule has to be added:

$$s((\lambda x.t)r) \to (\lambda x.st)r \ ,$$

where x is not free in s and s is a λ-abstraction. We call this rule assoc and the extension of λ-calculus obtained by adding it $\lambda[\beta, \mathsf{assoc}]$.

Proposition 1. *The calculus $\lambda[\beta, \mathsf{assoc}]$ is strongly normalizing, i. e., there is no infinite reduction sequence $t = t_0 \to t_1 \to t_2 \to \ldots$ with a typable term t.* [3]

Proof. A proof by Lengrand may be found in [11]. A stronger result was stated in [3], concerning the addition to the λ-calculus, not only of assoc (even without the abstraction proviso), but also of another permutation rule, due to Regnier [15], and named here perm. The "proof" of the strong result given in [3] was incomplete. A complete proof may be found in [4]. Strong normalisation of $\lambda[\beta, \mathsf{assoc}, \mathsf{perm}]$ will be needed below in Section 5.1 for translation F. □

Although our first aim is to give an alternative syntactic proof of strong normalization of $\lambda\mathbf{J}^{\mathbf{m}se}$, we want to be able to interpret $\lambda\mathbf{J}^{\mathbf{m}se}$ in as many monads as possible, and not just the identity monad. Hence, we take as target calculus the extension of simply-typed λ-calculus where the type system includes a monad—a type transformation called M as the single unary constant for building types—and the term language includes constructions for the unit and the Kleisli extension (a.k.a. bind) operation of the monad M, as follows: the term language is extended by the following clauses: If s is a term then ηs is a term, and if r and t are terms, then $\mathsf{bind}(r, x.t)$ is a term. The variable x is considered as bound by "$x.$" in t.

[3] A term t is typable if there is a context Γ and a type A such that $\Gamma \vdash t : A$.

$$\begin{array}{lrcl}
(\beta_\lambda) & (\lambda x.t)s & \to & [s/x]t \\
(\beta_{\mathsf{bind}}) & \mathsf{bind}(\eta s, x.t) & \to & [s/x]t \\
(\pi_{\lambda,\lambda}) & (\lambda y.u)((\lambda x.t)r) & \to & (\lambda x.(\lambda y.u)t)r \\
(\pi_{\mathsf{bind},\lambda}) & \mathsf{bind}((\lambda x.t)r, y.u) & \to & (\lambda x.\mathsf{bind}(t, y.u))r \\
(\pi_{\mathsf{bind},\mathsf{bind}}) & \mathsf{bind}(\mathsf{bind}(r, x.t), y.u) & \to & \mathsf{bind}(r, x.\mathsf{bind}(t, y.u))
\end{array}$$

Fig. 2. Base reduction rules of λ_{M}

The usual typing rules of simply-typed λ-calculus are extended as follows:

$$\frac{\Gamma \vdash s : A}{\Gamma \vdash \eta s : MA}\,\eta \qquad \frac{\Gamma \vdash r : MA \quad \Gamma, x : A \vdash t : MB}{\Gamma \vdash \mathsf{bind}(r, x.t) : MB}\,\mathsf{bind}$$

The monadic language was introduced by Moggi [12] as an equational theory and was used to interpret the computational lambda-calculus. Its corresponding reduction theory is considered in [7] and [16] and includes rules for the 3 monadic laws. Our monadic λ-calculus λ_{M} brings into play two more permutation rules. The base reduction rules of λ_{M} are shown in Figure 2. The implicit proviso for the three latter rules – the permutation rules – is that x is not free in $\lambda y.u$. Again, we write $\to$ for the term closure of the base reduction rules.

While β_{bind} and $\pi_{\mathsf{bind},\mathsf{bind}}$ correspond to two of the three monad laws, we do not need the eta rule of the monad $\mathsf{bind}(r, x.\eta x) \to r$.

Note that the rule $\pi_{\mathsf{bind},\lambda}$ orients the direct equational consequence of β_λ,

$$\mathsf{bind}((\lambda x.t)r, y.u) =_{\beta_\lambda} \mathsf{bind}([r/x]t, y.u) =_{\beta_\lambda} (\lambda x.\mathsf{bind}(t, y.u))r\ ,$$

in a specific way. Likewise, $\pi_{\lambda,\lambda}$ – which is just a different presentation of rule assoc – directs an equational consequence of β_λ. So, from a purely equational point of view, our notion of λ_{M} is not stronger than the ordinary one that only reflects the monad laws. Moreover, we even omitted the eta rule.

To the best of our knowledge, rules $(\pi_{\lambda,\lambda})$ and $(\pi_{\mathsf{bind},\lambda})$ have not been considered before in combination with the traditional monad rules. However as we show below, the enriched system λ_{M} enjoys good properties, which would hold even in presence of the monadic eta rule.

The λ_{M}-calculus can be interpreted in $\lambda[\beta, \mathsf{assoc}]$ so that strict simulation of reduction is obtained. The translation corresponds to defining the *identity monad* in $\lambda[\beta, \mathsf{assoc}]$. The translation $|\_| : \lambda_{\mathsf{M}} \to \lambda[\beta, \mathsf{assoc}]$ is defined on types by $|X| := X$, $|A \supset B| := |A| \supset |B|$ and $|MA| := A$, and is defined on terms by $|x| := x$, $|\lambda x.t| := \lambda x.|t|$, $|tu| := |t||u|$, $|\eta s| := |s|$ and $|\mathsf{bind}(r, x.t)| := (\lambda x.|t|)|r|$. Evidently, this respects the typing rules.

Lemma 1. *If $\Gamma \vdash s : A$ is derivable in λ_{M}, $|\Gamma| \vdash |s| : |A|$ is derivable in $\lambda[\beta, \mathsf{assoc}]$, where $|\Gamma|$ is the result of replacing each declaration $x : A$ in Γ by $x : |A|$.*

Under these definitions, β_λ and β_{bind} become β (the usual rule of λ-calculus that is β_λ, but quantified over a different set of terms), and all three permutation rules become the assoc rule. (The ordinary eta rule of the monad would be just mapped to one step of β.) Thus we have the strongest possible simulation result.[4]

Lemma 2. *If* $t \to u$ *in* λ_{M}, $|t| \to |u|$ *in* $\lambda[\beta, \mathsf{assoc}]$.

From the above result and strong normalization of $\lambda[\beta, \mathsf{assoc}]$, we immediately get the following result.

Corollary 1. *The calculus* λ_{M} *is strongly normalizing.*

Now, given that all critical pairs for the rules of λ_{M} are joinable, we also obtain a confluence result.

Corollary 2. $\to$ *is confluent for the typable terms of* λ_{M}.

4 Translations of $\lambda \mathbf{J}^{\mathbf{m}se}$ into Monadic λ-Calculus

Here, we show how to translate $\lambda \mathbf{J}^{\mathbf{m}se}$ into λ_{M} such that one obtains strict simulation and thus can infer strong normalization of $\lambda \mathbf{J}^{\mathbf{m}se}$ from Corollary 1. Hence, this is an alternative syntactic proof of strong normalization of $\lambda \mathbf{J}^{\mathbf{m}se}$. While the translation in the following section works on the types in usual CBN fashion [7], a more complicated type translation in Section 4.2 even yields strict simulation within λ_{M} without the rule $\pi_{\lambda,\lambda}$.

4.1 Main Monadic Translation

A type A of $\lambda \mathbf{J}^{\mathbf{m}se}$ is translated to $\overline{A} = MA^*$ of λ_{M}, with the type A^* defined by recursion on A (where the definition of $\overline{A}$ is used as an abbreviation):

$$X^* = X \qquad \text{and} \qquad (A \supset B)^* = \overline{A} \supset \overline{B}$$

Note that, for the identity monad, this trivializes to $\overline{A} = A^* = A$. Any term t of $\lambda \mathbf{J}^{\mathbf{m}se}$ is translated into a term $\bar{t}$ of λ_{M}, any command c of $\lambda \mathbf{J}^{\mathbf{m}se}$ into a term $\bar{c}$ and any pair of a co-term l of $\lambda \mathbf{J}^{\mathbf{m}se}$ and a variable w of λ_{M}, with w not free in l, into a term l_w of λ_{M}.[5] This is done so that the typing rules in Figure 3 are derivable, where $\overline{\Gamma}$ is derived from Γ by replacing every $x : C$ in Γ by $x : \overline{C}$.

The definitions are in Figure 4, where it is understood that f, v and w are fresh variable names. The definition of $[]_w$ is given with the extra $(\lambda k.k)$ so as to form an (administrative) redex which will guarantee strict simulation of ϵ and of the initial cases of π, see the proofs of Lemma 4 and Theorem 1. Also $(\lambda v.l_v)(f\overline{u})$ is a redex for strict simulation purposes, and we will "monadically" abstract away from it in the optimized translation in Section 4.2.

---

[4] Strict simulation would just mean that one step in the source calculus is mapped to at least one step of the target calculus, which would be sufficient to inherit strong normalization of the source calculus from the target calculus.

[5] Whenever we write l_w (or E_w), it will be understood that w does not occur free in the co-term l (or E).

$$\frac{\Gamma \vdash t : A}{\overline{\Gamma} \vdash \overline{t} : \overline{A}} \qquad \frac{\Gamma \xrightarrow{c} A}{\overline{\Gamma} \vdash \overline{c} : \overline{A}} \qquad \frac{\Gamma | l : A \vdash B}{\overline{\Gamma}, w : \overline{A} \vdash l_w : \overline{B}}$$

Fig. 3. Derived typing rules for monadic translation of $\lambda\mathbf{J}^{\mathbf{mse}}$

$$\begin{array}{rclcrclcrcl} \overline{x} &=& x & \quad & []_w &=& (\lambda k.k)w & \quad & & & \\ \overline{\lambda x.t} &=& \eta(\lambda x.\overline{t}) & & (u :: l)_w &=& \mathsf{bind}(w, f.(\lambda v.l_v)(f\overline{u})) & & \overline{tl} &=& [\overline{t}/w]l_w \\ \overline{\{c\}} &=& \overline{c} & & ((x)c)_w &=& (\lambda x.\overline{c})w & & & & \end{array}$$

Fig. 4. Monadic translation of $\lambda\mathbf{J}^{mse}$

Lemma 3. *The translation satisfies* $\overline{[t/x]u} = [\overline{t}/x]\overline{u}$, $([t/x]l)_w = [\overline{t}/x](l_w)$ *and* $\overline{[t/x]c} = [\overline{t}/x]\overline{c}$. *The proviso for the second equation is that x is not w.* □

Lemma 4. *For $w \notin E$*[6]*, one has* $[l_w/v]E_v \to^+ (l@E)_w$.

Proof. For $E = []$, we calculate

$$[l_w/v]([]_v) = [l_w/v]((\lambda k.k)v) = (\lambda k.k)l_w \to_{\beta_\lambda} l_w = (l@[])_w \ .$$

For $E = u :: l'$, do induction on l.

Case $[]$: $[(\lambda k.k)w/v]E_v \to_{\beta_\lambda} [w/v]E_v = E_w$ (v once in E_v + renaming)

Case $u' :: l$:

$$\begin{array}{rcl} [(u' :: l)_w/v]E_v &=& \mathsf{bind}(\mathsf{bind}(w, g.(\lambda v'.l_{v'})(g\overline{u'})), f.(\lambda v.l'_v)(f\overline{u})) \\ &\to_{\pi_{\mathsf{bind},\mathsf{bind}}}& \mathsf{bind}(w, g.\mathsf{bind}((\lambda v'.l_{v'})(g\overline{u'}), f.(\lambda v.l'_v)(f\overline{u}))) \\ &\to_{\pi_{\mathsf{bind},\lambda}}& \mathsf{bind}(w, g.(\lambda v'.\mathsf{bind}(l_{v'}, f.(\lambda v.l'_v)(f\overline{u})))(g\overline{u'})) \\ &=& \mathsf{bind}(w, g.(\lambda v'.[l_{v'}/w]E_w)(g\overline{u'})) \\ &\to^+& \mathsf{bind}(w, g.(\lambda v'.(l@E)_{v'})(g\overline{u'})) \qquad \text{by IH for } l \\ &=& (u' :: (l@E))_w = ((u' :: l)@E)_w \end{array}$$

Case $(y)c$ with $c = t_1 l_1$:

$$\begin{array}{rcl} [((y)c)_w/v]E_v &=& \mathsf{bind}((\lambda y.\overline{c})w, f.(\lambda v.l'_v)(f\overline{u})) \\ &\to_{\pi_{\mathsf{bind},\lambda}}& (\lambda y.\mathsf{bind}(\overline{c}, f.(\lambda v.l'_v)(f\overline{u})))w \\ &=& (\lambda y.[\overline{c}/v]E_v)w \\ &=& (\lambda y.[[\overline{t_1}/v'](l_1)_{v'}/v]E_v)w \\ &=& (\lambda y.[\overline{t_1}/v'][(l_1)_{v'}/v]E_v)w \\ &\to^+& (\lambda y.[\overline{t_1}/v'](l_1@E)_{v'})w \qquad \text{by IH for } l_1 \\ &=& ((y)t_1(l_1@E))_w = (((y)c)@E)_w \end{array}$$

□

Theorem 1 (Simulation). *If $t \to t'$ in $\lambda\mathbf{J}^{\mathbf{mse}}$, then $\overline{t} \to^+ \overline{t'}$ in λ_{M}. If $l \to l'$ in $\lambda\mathbf{J}^{\mathbf{mse}}$, then $l_w \to^+ l'_w$ in λ_{M}. If $c \to c'$ in $\lambda\mathbf{J}^{\mathbf{mse}}$, then $\overline{c} \to^+ \overline{c'}$ in λ_{M}.*

[6] By writing $(l@E)_w$, we already implicitly assume that $w \notin E$, but this condition is not visible in the left-hand side of the statement, hence we indicate it.

Proof. We only have to consider a rewrite step at the root since the cases corresponding to the closure rules follow by routine induction. This is so because w has one free occurrence in l_w (it has only one occurrence), and so the definition of $\overline{tl}$ is uncritical ($\overline{t}$ cannot be lost as a subterm through substitution into l_w).

Case β: $(\lambda x.t)(u :: l) \to u(x)tl$.

$$\begin{array}{rl} \overline{(\lambda x.t)(u :: l)} & = \mathsf{bind}(\eta(\lambda x.\overline{t}), f.(\lambda v.l_v)(f\overline{u})) \\ & \to_{\beta_{\mathsf{bind}}} (\lambda v.l_v)((\lambda x.\overline{t})\overline{u}) \\ & \to_{\pi_{\lambda,\lambda}} (\lambda x.(\lambda v.l_v)\overline{t})\overline{u} \\ & \to_{\beta_\lambda} (\lambda x.[\overline{t}/v]l_v)\overline{u} = (\lambda x.\overline{tl})\overline{u} = \overline{u(x)tl} \end{array}$$

Case σ: $t(x)c \to [t/x]c$:

$$\overline{t(x)c} = (\lambda x.\overline{c})\overline{t} \to_{\beta_\lambda} [\overline{t}/x]\overline{c} = \overline{[t/x]c} \quad \text{(Lemma 3)}$$

Case ϵ: $\{t[]\} \to t$: $\quad \overline{\{t[]\}} = (\lambda k.k)\overline{t} \to_{\beta_\lambda} \overline{t}$

Case μ: $(x)xl \to l$, if $x \notin l$.

$$((x)xl)_w = (\lambda x.\overline{xl})w = (\lambda x.[x/w]l_w)w \to_{\beta_\lambda} [w/x][x/w]l_w = l_w$$

Case π: $\{tl\}E \to t\,(l@E)$. Apply substitution $[\overline{t}/w]$ to Lemma 4:

$$\overline{\{tl\}E} = [\overline{tl}/v]E_v = [[\overline{t}/w]l_w/v]E_v = [\overline{t}/w][l_w/v]E_v \to^+ [\overline{t}/w](l@E)_w = \overline{t(l@E)} \ ,$$

using the usual substitution lemmas. □

Corollary 3. *The calculus $\lambda\mathbf{J}^{\mathbf{mse}}$ is strongly normalizing.*

Proof. Use the previous theorem, the preservation of typability expressed in Figure 3 and Corollary 1. □

We remark that $\pi_{\lambda,\lambda}$ would not have been necessary if rule β of $\lambda\mathbf{J}^{\mathbf{mse}}$ were already σ-reduced on the right-hand side, thus with $[u/x]tl$. The calculation would be as follows:

$$\begin{array}{rll} \overline{(\lambda x.t)(u :: l)} & = \mathsf{bind}(\eta(\lambda x.\overline{t}), f.(\lambda v.l_v)(f\overline{u})) & \\ & \to_{\beta_{\mathsf{bind}}} (\lambda v.l_v)((\lambda x.\overline{t})\overline{u}) & \\ & \to_{\beta_\lambda} (\lambda v.l_v)([\overline{u}/x]\overline{t}) & \\ & = (\lambda v.l_v)\overline{[u/x]t} & \text{(Lemma 3)} \\ & \to_{\beta_\lambda} [\overline{[u/x]t}/v]l_v = \overline{[u/x]tl} & \end{array}$$

Our monadic translation when restricted to λ-calculus essentially captures the usual CBN monadic translation [7], call it $(\_)^\circ$. This translation for variables and λ-abstraction behaves as our translation, and for applications does $(tu)^\circ := \mathsf{bind}(t^\circ, f.fu^\circ)$. Our translation of a λ-calculus application tu, encoded in $\lambda\mathbf{J}^{\mathbf{mse}}$ as $t(u :: [])$, reaches the expected term after two β_λ-steps:

$$\overline{t(u :: [])} = \mathsf{bind}(\overline{t}, f.(\lambda v.[]_v)(f\overline{u})) \to^2_{\beta_\lambda} \mathsf{bind}(\overline{t}, f.f\overline{u})$$

We also notice that the property "$t \to_\beta u$ in the λ-calculus $\Rightarrow t^\circ \to_{\beta_{\mathsf{bind}},\beta_\lambda} u^\circ$ in the λ_{M}-calculus", that holds of mapping $(\_)^\circ$ (an easy, perhaps new result), is also shared by our translation.

$$\overline{x} = x \qquad []_w = (\lambda k.k)w$$
$$\overline{\lambda x.t} = \eta(\lambda x.\eta\overline{t}) \qquad (u :: l)_w = \mathsf{bind}(w, f.\mathsf{bind}(f\overline{u}, v.l_v)) \qquad \overline{tl} = [\overline{t}/w]l_w$$
$$\overline{\{c\}} = \overline{c} \qquad ((x)c)_w = (\lambda x.\overline{c})w$$

Fig. 5. Optimized monadic translation of $\lambda\mathbf{J}^{\mathbf{m}se}$

4.2 Optimized Translation

Now, a translation is given that allows simulation of $\lambda\mathbf{J}^{\mathbf{m}se}$ even in λ_{M}^{-} that is obtained from λ_{M} by omitting the rule $\pi_{\lambda,\lambda}$. The symbols of the previous subsection will be reused, but their definition will be changed.

A type A of $\lambda\mathbf{J}^{\mathbf{m}se}$ is translated to $\overline{A} = MA^*$ of λ_{M}, with the type A^* defined by recursion on A (where the definition of $\overline{A}$ is used as an abbreviation):

$$X^* = X \qquad \text{and} \qquad (A \supset B)^* = \overline{A} \supset M\overline{B}$$

Note that, for the identity monad, this again trivializes to $\overline{A} = A^* = A$. But the crucial change is that an extra M is inserted on top of $\overline{B}$ in the translation of $A \supset B$. For the special case of $MA = \neg\neg A$, this is logically equivalent to the translation used in [5,6].

Any term t of $\lambda\mathbf{J}^{\mathbf{m}se}$ is translated into a term $\overline{t}$ of λ_{M}, any command c of $\lambda\mathbf{J}^{\mathbf{m}se}$ into a term $\overline{c}$ and any pair of a co-term l of $\lambda\mathbf{J}^{\mathbf{m}se}$ and a variable w of λ_{M}, with w not free in l, into a term l_w of λ_{M}. This is done so that the typing rules in Figure 3 are again derivable, where, obviously, all symbols have to be interpreted according to the current definitions.

The definitions are in Figure 5, where the usual freshness assumptions are understood. Changes with respect to Figure 4 concern λ-abstraction with an extra η and $(u :: l)_w$ where bind replaces the β redex. In fact, $\mathsf{bind}(f\overline{u}, v.l_v)$ is just the monadic version of $(\lambda v.l_v)(f\overline{u})$ that was used formerly. For the identity monad, the translation thus agrees with that of Section 4.1. However, in the general case, $\mathsf{bind}(f\overline{u}, v.l_v)$ would not be well-typed with the definitions of Section 4.1. For $\Gamma|u :: l : A \supset B \vdash C$, one would have $w : \overline{A \supset B}$ and hence $f : (A \supset B)^*$. Therefore, $f\overline{u}$ would have type $\overline{B}$ and finally $v : B^*$, which is not enough. We remark that one can base an alternative translation with A^* as in Section 4.1 on the idea of enforcing the admissible rule $\Gamma|l : A \vdash B \Rightarrow \overline{\Gamma}, w : A^* \vdash l_w : \overline{B}$. Simulation results for this alternative translation needed extensions of the η rule $\mathsf{bind}(t, x.\eta x) \to t$ that did not seem to be well justified.

Lemma 3 also holds for the definitions of the present section.

Theorem 2 (Simulation). *If $t \to t'$ in $\lambda\mathbf{J}^{\mathbf{m}se}$, then $\overline{t} \to^+ \overline{t'}$ in λ_{M}^{-}. If $l \to l'$ in $\lambda\mathbf{J}^{\mathbf{m}se}$, then $l_w \to^+ l'_w$ in λ_{M}^{-}. If $c \to c'$ in $\lambda\mathbf{J}^{\mathbf{m}se}$, then $\overline{c} \to^+ \overline{c'}$ in λ_{M}^{-}.*

Proof. As in the proof of Theorem 1, it suffices to consider the base cases of reduction at the root. The cases σ, ϵ and μ can be copied verbatim from the proof of Theorem 1. For β, one calculates that

$$\overline{(\lambda x.t)(u :: l)} \to_{\beta_{\mathsf{bind}}} \mathsf{bind}((\lambda x.\eta\overline{t})\overline{u}, v.l_v) \to_{\pi_{\mathsf{bind},\lambda}} (\lambda x.\mathsf{bind}(\eta\overline{t}, v.l_v))\overline{u} \to_{\beta_{\mathsf{bind}}} \overline{u(x)tl}$$

Case π: $\{tl\}E \to t\,(l@E)$. The treatment of $E = []$ is immediate due to the extra redex in the definition of $[]_w$.

Sub-case $E = u :: l'$. We have to show $\overline{\{tl\}E} \to^+ \overline{t(l@E)}$, which is done by induction on l, simultaneously for all t. □

5 Generic Account of Translation into Natural Deduction

An instantiation of the monadic translation with a particular monad gives an interpretation of the intuitionistic sequent calculus $\lambda\mathbf{J}^{\mathbf{m}se}$ into natural deduction. In this section we show that two such instantiations relate to known interpretations, namely variants of both the Zucker-Pottinger translation and a CPS translation. These interpretations receive, thus, a generic account through the monadic translation.

5.1 Direct Translations

In this subsection we study certain "direct" translations of $\lambda\mathbf{J}^{\mathbf{m}se}$ into the λ-calculus. One of these, named N here, implements the traditional interpretation of sequent calculus into natural deduction studied by Zucker [17] and Pottinger [14]. The directness comes from the fact no translation of types is involved, and also because these translations give a straightforward expression in terms of the λ-calculus of the computational interpretations of $\lambda\mathbf{J}^{\mathbf{m}se}$-expressions. The direct translations, as we will see, turn out to be related to the monadic translation, when the latter is instantiated with the identity monad.

A direct translation. Let F be the mapping from $\lambda\mathbf{J}^{\mathbf{m}se}$ to λ, based on the idea of mapping, say, $t(u_1 :: u_2 :: [])$ and $t(u_1 :: u_2 :: (x)c)$ to

$$(\lambda x.x)(rs_1s_2) \text{ and } (\lambda x.s)(rs_1s_2) \ ,$$

where r, s_i, and s are the translations of t, u_i, and c, respectively. Formally, F is given by

$$\begin{array}{rclrclrcl} F(x) &=& x & F(r,[]) &=& (\lambda x.x)r & & & \\ F(\lambda x.t) &=& \lambda x.F(t) & F(r,u :: l) &=& F(rF(u),l) & F(tl) &=& F(F(t),l) \\ F(\{c\}) &=& F(c) & F(r,(x)c) &=& (\lambda x.F(c))r & & & \end{array}$$

We will need the target of F to be equipped not only with the assoc reduction rule, but also with

$$(\lambda x.t)rs \to (\lambda x.ts)r \ ,$$

for x not free in s (a proviso that, as for assoc, already follows from the variable convention). This is a well-known permutation rule [15,9], which we name here perm. Let $\lambda[\beta, \mathsf{assoc}, \mathsf{perm}]$ be the λ-calculus equipped with both assoc and perm.

As mentioned in the proof of Proposition 1, normalisation of $\lambda[\beta, \mathsf{assoc}, \mathsf{perm}]$ holds as a consequence of a result stated in [3] and fully proved in [4].

Proposition 2. *If* $t \to u$ *in* $\lambda\mathbf{J}^{\mathbf{mse}}$ *then* $F(t) \to^{+} F(u)$ *in* $\lambda[\beta, \mathsf{assoc}, \mathsf{perm}]$.

Proof. A by-hand proof would be possible, but we give an indirect proof, joining scattered results from the literature. The point is that F is the composition of the following mappings

$$\lambda\mathbf{J}^{\mathbf{mse}} \xrightarrow{(\_)^{\circ}} \lambda\mathbf{J}^{\mathbf{ms}} \subset \lambda^{\mathbf{G}tz} \xrightarrow{(\_)^{*}} \lambda\mathbf{s} \xrightarrow{(\_)^{\sharp}} \lambda[\beta\pi]$$

where $(\_)^{\circ} : \lambda\mathbf{J}^{\mathbf{mse}} \to \lambda\mathbf{J}^{\mathbf{ms}}$ comes from [6], $(\_)^{*} : \lambda^{\mathbf{G}tz} \to \lambda\mathbf{s}$ comes from [2], and $(\_)^{\sharp} : \lambda\mathbf{s} \to \lambda[\beta\pi]$ comes from [3]. $\lambda\mathbf{J}^{\mathbf{ms}}$ is the system preceding $\lambda\mathbf{J}^{\mathbf{mse}}$ in the "spectrum" of intuitionistic systems studied in [5,6]. The difference relatively to $\lambda\mathbf{J}^{\mathbf{mse}}$ is that there is neither a separate class of commands, nor co-terms []; instead, selection has the general form $(x)t$. $\lambda^{\mathbf{G}tz}$ is identical to $\lambda\mathbf{J}^{\mathbf{ms}}$, except that it has a more general π reduction rule, in that the call-by-name restriction is not imposed, and the concatenation operator is lazy; so each π step in $\lambda\mathbf{J}^{\mathbf{ms}}$ corresponds to one or more π steps in $\lambda^{\mathbf{G}tz}$. $\lambda\mathbf{s}$ is λ plus a substitution construction, equipped with rules for generating (β), executing (σ), and delaying (π) substitution. $\lambda[\beta\pi]$ is identical to $\lambda[\beta, \mathsf{assoc}, \mathsf{perm}]$, except that in $\lambda[\beta\pi]$ the abstraction proviso in the assoc rule is not imposed. [7] Mapping $(\_)^{\circ}$ erases the coercion $\{-\}$ and encodes [] as $(x)x$. Mapping $(\_)^{*}$ has the same spirit as F, except that tl is mapped to a substitution, instead of a β-redex. Finally mapping $(\_)^{\sharp}$ "raises" substitutions to β-redexes. The present proposition is corollary of three simulation results: Proposition 3.6 of [6] concerning $(\_)^{\circ}$, Proposition 1 of [2] concerning $(\_)^{*}$, and Proposition 7 of [3] concerning $(\_)^{\sharp}$. All three state that each reduction step of the source generates one or more reduction steps of the target, except in one case: $(\_)^{\sharp}$ collapses β steps of $\lambda\mathbf{s}$. So, one has to supplement Proposition 1 of [2] with the remark - useless for the purposes of [2], but needed now - that $(\_)^{*}$ always generates at least one reduction step different from β in the target, when translating a reduction step of its source. Finally we observe that the simulation property of $(\_)^{\sharp}$ still holds when one takes the assoc rule of $\lambda[\beta\pi]$ with the abstraction proviso, and therefore the target of $(\_)^{\sharp}$ can be taken as $\lambda[\beta, \mathsf{assoc}, \mathsf{perm}]$. □

Identity-monadic translations. Let G be the composition of the monadic translation with the mapping $|\_| : \lambda_{\mathsf{M}} \to \lambda[\beta, \mathsf{assoc}]$ from the end of Section 3. G maps, say, $t(u_1 :: u_2 :: [])$ and $t(u_1 :: u_2 :: (x)c)$ respectively to

$$(\lambda f.(\lambda z.(\lambda f'.(\lambda z'.(\lambda x.x)z)(f's_2))z')(fs_1))r$$
$$(\lambda f.(\lambda z.(\lambda f'.(\lambda z'.(\lambda x.s)z)(f's_2))z')(fs_1))r$$

if we let again r, s_i, and s be the translations of t, u_i, and c, respectively. A recursive definition of G is:

$$\begin{array}{rlrlr} G(x) = x & & []_w = (\lambda k.k)w & & \\ G(\lambda x.t) = \lambda x.G(t) & & (u :: l)_w = (\lambda f.(\lambda v.l_v)(fG(u)))w & & G(tl) = [G(t)/w]l_w \\ G(\{c\}) = G(c) & & ((x)c)_w = (\lambda x.G(c))w & & \end{array}$$

[7] The idea is that perm and the relaxed assoc (called π_1 and π_2 in [3] respectively) form a coherent set of rules for "delaying" a "substitution" $(\lambda x.-)r$ surrounding a term t, whenever this t occurs in the function or argument positions of an application.

Proposition 3. *If $t \to u$ in $\lambda\mathbf{J}^{\mathbf{mse}}$ then $G(t) \to^+ G(u)$ in $\lambda[\beta, \mathsf{assoc}]$.*

Proof. Immediate consequence of Theorem 2, and the 1-1 mapping of reduction steps given by $|\_|$. □

Comparison of translations. The previous proposition guarantees that, by changing from the encoding of commands of F to the encoding of commands of G, we dispense with perm in the target. For instance, consider

$$c_1 = \{t(u_1 :: (x)c)\}(u_1' :: (y)c') \qquad \text{and} \qquad c_2 = t(u_1 :: (x)v(u_1' :: (y)c'))$$

with $c = v[]$. Then $s = F(c) = (\lambda z.z)F(v)$ and $c_1 \to_\pi c_2$ in $\lambda\mathbf{J}^{\mathbf{mse}}$. In the target we have (if $s_1 = F(u_1)$, $s' = F(c')$, and $s_1' = F(u_1')$)

$$F(c_1) = (\lambda y.s')\big((\lambda x.s)(rs_1)s_1'\big) \qquad \text{and} \qquad F(c_2) = \big(\lambda x.(\lambda y.s')(F(v)s_1')\big)(rs_1)$$

After reducing s to $F(v)$ and performing one perm step, one obtains from $F(c_1)$ the term $(\lambda y.s')((\lambda x.F(v)s_1')(rs_1))$, which in turn reaches $F(c_2)$ after one assoc step. On the other hand,

$$G(c_1) = \Big(\lambda f'.(\lambda z'.(\lambda y.s')z')(f's_1')\Big)\Big(((\lambda f.(\lambda z.(\lambda x.s)z)(fs_1))r\Big)$$

$$G(c_2) = \Big(\lambda f.((\lambda z.(\lambda x.(\lambda f'.(\lambda z'.(\lambda y.s')z')(f's_1'))G(v))z)\,(fs_1)\Big)r$$

with $s = G(c) = (\lambda z.z)G(v)$, $s_1 = G(u_1)$, $s' = G(c')$, and $s_1' = G(u_1')$. Now $G(c_1)$ reaches $G(c_2)$ after 3 assoc steps, provided s is first reduced to $G(v)$.

The proximity between F and G becomes clearer if we give the definition of F in the following style:

$$\begin{array}{rclrclrcl} F(x) &=& x & []_w &=& (\lambda k.k)w & & & \\ F(\lambda x.t) &=& \lambda x.F(t) & (u :: l)_w &=& [wF(u)/v]l_v & F(tl) &=& [F(t)/w]l_w \\ F(\{c\}) &=& F(c) & ((x)c)_w &=& (\lambda x.F(c))w & & & \end{array}$$

So, the only difference between F and G is in the clause for $(u :: l)_w$, where the two β-redexes appearing in the clause for G are contracted in the clause for F.

F translates the left introduction $u :: l$ as the traditional map between sequent calculus and natural deduction does - through a combination of application and substitution; but, because of the definitions of $[]_w$ and $((x)c)_w$, F's translation of a cut generates a β-redex whose contractum (a certain substitution) would be the translation of that cut by the traditional mapping. So, if we let N denote the traditional mapping, N is defined as F, except that now we put $[]_w = w$ and $((x)c)_w = [w/x]N(c)$. Notice that N also corresponds to taking the definition of G and uniformly contracting all β-redexes in the clauses defining l_w.

Proposition 4. *If $t \to u$ in $\lambda\mathbf{J}^{\mathbf{mse}}$, then $Nt \to_\beta^* Nu$ in the λ-calculus.*[8]

[8] In fact, all but β-steps are identified by N, meaning that only the reduction rule corresponding to the key step of cut-elimination has a non-trivial translation. This agrees with the properties of the map $(\_)^\flat : LJ \to \lambda$ studied in [3], a map implementing the traditional translation of another sequent calculus, named LJ there. Each β-step in the source of N is guaranteed to generate exactly one step in the target only when it happens at root position. The same applies to $(\_)^\flat$, but is not acknowledged in Proposition 10 of [3].

Let us sum up in the following diagram, where double-headed arrows denote 0 or more steps of reduction, except for the two central, vertical arrows, which denote 1 or more steps (a fact signaled by a little black triangle).

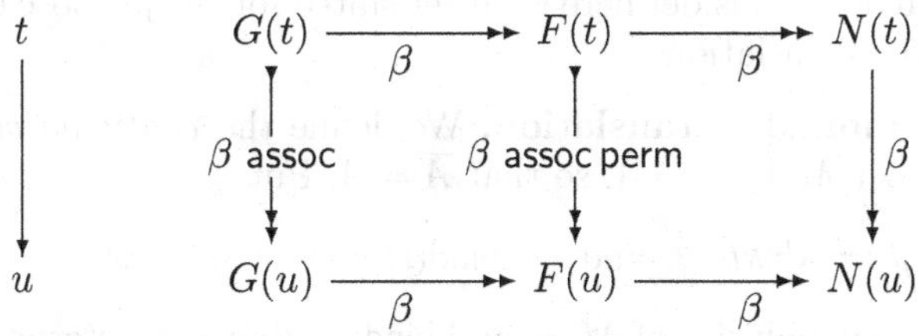

The map G, obtained by instantiating the monadic map, has a sharper simulation property than the previously known maps.

5.2 CPS Translation

In this subsection we introduce a CPS translation of $\lambda\mathbf{J}^{\mathbf{m}se}$, and compare it with the monadic translation instantiated with the continuations monad.

A CPS translation. Let A^* be given by $X^* = X$ and $(A \supset B)^* = \widehat{A} \supset \widehat{B}$, where $\widehat{A} = \neg\neg A^*$. The CPS translation of $t \in \lambda\mathbf{J}^{\mathbf{m}se}$ - denoted $\widehat{t}$ - is defined as $\lambda k.(t : k)$, where the so-called colon-operation is given as follows:

$$\begin{array}{rclcrcl}
(x : K) &=& xK & \quad & ([] : K) &=& \lambda w.wK \\
(\lambda x.t : K) &=& K(\lambda x.\widehat{t}) & & (u :: l : K) &=& \lambda w.w(\lambda f.(l : K)(f\widehat{u})) \\
(\{c\} : K) &=& (c : K) & & ((x)c : K) &=& \lambda x.(c : K) \\
& & & & (t[] : K) &=& (t : K) \\
& & & & (t(u :: l) : K) &=& (t : \lambda f.(l : K)(f\widehat{u})) \\
& & & & (t(x)c : K) &=& ((x)c : K)\widehat{t}
\end{array}$$

This CPS translation is considered in [6]. In [5,6] a different CPS translation is given, based on the definition $(A \supset B)^* = \neg\widehat{B} \supset \neg\widehat{A}$, and a weak simulation result for it is proved, stating that each reduction step in the source $\lambda\mathbf{J}^{\mathbf{m}se}$ is mapped to 0 or more β-reduction steps in the λ-calculus. A variant of the proof sketched in [5,6] gives an even weaker result for the present CPS translation.

Proposition 5. *If $t \to u$ in $\lambda\mathbf{J}^{\mathbf{m}se}$, then $\widehat{t} \to^* \widehat{u}$ in $\lambda[\beta, \mathsf{assoc}]$.*

Indeed one needs assoc in the target, precisely for the simulation of β:

$$\begin{array}{rcl}
((\lambda x.t)(u :: []) : K) &=& (\lambda f.([] : K)(f\widehat{u}))(\lambda x.\widehat{t}) \\
&\to_\beta& ([] : K)((\lambda x.\widehat{t})\widehat{u}) \\
&\to_{\mathsf{assoc}}& (\lambda x.([] : K)\widehat{t})\widehat{u} \\
&\to_\beta^2& (\lambda x.(t : K))\widehat{u} = (u(x)(t[]) : K)
\end{array}$$

For this special case of β, the LHS also reduces to the RHS using β and perm. However, it is no longer possible to replace assoc by perm if instead of the empty list we have another list format.

Despite its weaker simulation properties, the CPS translation satisfying $(A \supset B)^* = \widehat{A} \supset \widehat{B}$ that we consider here is better suited for the purpose of comparing with the monadic translation.

Continuation-monadic translation. We define the *continuations monad* in the λ-calculus. Let $MA := \neg\neg A$, so that $\overline{A} = \widehat{A}$. Put

$$\eta t := \lambda k.kt \qquad \text{and} \qquad \mathsf{bind}(r, x.s) := \lambda k.r(\lambda x.sk)$$

We may see these definitions of M, η, and bind as giving an interpretation of λ_{M} into λ. Under this interpretation, the reduction rules of λ_{M} hold as $\beta\eta$-equalities in the λ-calculus.

The instantiation of the monadic translation (that is, the composition of the monadic translation with the present interpretation of λ_{M} into λ) gives:

$$\begin{array}{rlrl} C(x) &= x & []_w &= (\lambda k.k)w \\ C(\lambda x.t) &= \lambda k.k(\lambda x.C(t)) & (u :: l)_w &= \lambda k.w(\lambda f.(\lambda v.l_v)(fC(u))k) \\ C(\{c\}) &= C(c) & ((x)c)_w &= (\lambda x.C(c))w \\ & & C(tl) &= [C(t)/w]l_w \end{array}$$

One immediately obtains that C maps each reduction step in $\lambda\mathbf{J}^{\mathbf{ms}}$ to a $\beta\eta$-equality in the λ-calculus.

Comparison with CPS. $C(t)$ is close to the CPS translation $\widehat{t}$. To see this, we introduce the "colon-free" translation $\widetilde{t}$, an intermediate point between $C(t)$ and $\widehat{t}$. $\widetilde{t}$ is defined as $\lambda k.(t : -)k$, where $(t : -)$ is given by:

$$\begin{array}{rlrlrl} (x : -) &= x & []^w &= w & & \\ (\lambda x.t : -) &= \lambda k.k(\lambda x.\widetilde{t}) & (u :: l)^w &= \lambda k.w(\lambda f.(\lambda v.l^v k)(f\widetilde{u})) & (tl : -) &= [\widetilde{t}/w]l^w \\ (\{c\} : -) &= (c : -) & ((x)c)^w &= \lambda k.(\lambda x.(c : -)k)w & & \end{array}$$

Then one proves: i) $(t : -)K$ reduces to $(t : K)$ - whence $\widetilde{t}$ reduces to $\widehat{t}$; ii) $(c : -)K$ reduces to $(c : K)$; iii) $\lambda w.l^w K$ reduces to $(l : K)$. The proof is a simultaneous induction on t, c, and l. Here reduction means 0 or more *administrative* β-steps. An administrative step is of one of two forms:

1. reduction of redexes $(\lambda k.t)K$ - pushing continuations inside;
2. $\lambda w.(\lambda x.t)w \to_\beta \lambda x.t$, with w not in t - notice the implicit α-conversion.

If $t \to u$ in $\lambda\mathbf{J}^{\mathbf{mse}}$, then $\widetilde{t} =_\beta \widetilde{u}$. This follows from the remarks just made, together with Proposition 5 and the fact assoc $\subseteq =_\beta$.

After the transfiguration of the CPS translations, it is perspicuous that: i) $C(t)$ reduces to $(t : -)$ (which in turn η-expands to $\widetilde{t}$; we let eta $= \eta^{-1}$ and refer to η-expansion as eta-reduction); ii) $C(c)$ reduces to $(c : -)$; iii) l_w reduces to l^w. The proof is again a simultaneous induction on t, c, and l. Here reduction means 0 or more steps of one of the following forms:

1. β, for reducing the redex $[]_w$;
2. eta, needed to bridge $(t : -)$ and $\widetilde{t}$;
3. eta followed by perm, for reducing the generic $(\lambda x.C(c))w$ to the continuations-monad-specific $\lambda k.(\lambda x.C(c)k)w$.
4. perm for bridging the change in the placement of variable k, when moving from the clause for $(u :: l)_w$ to the clause for $(u :: l)^w$.

Let us sum up with this diagram.

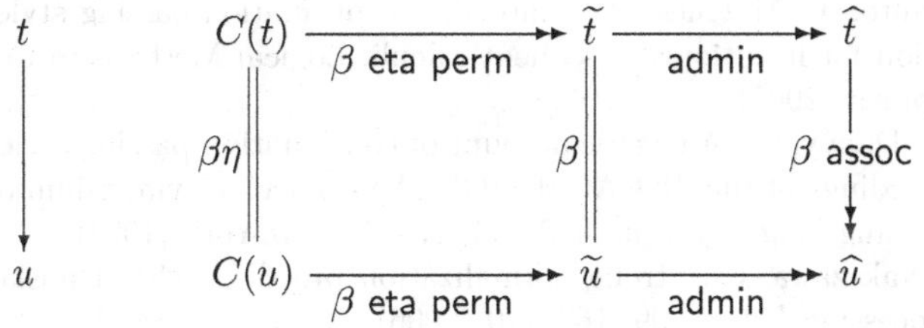

These results show how close is the CPS translation $\widehat{t}$ of being a mere instantiation of our monadic translation; and how the CPS translation helps explaining, in terms of reduction, the equational relationship existing between $C(t)$ and $C(u)$, when t and u are related by a reduction step in the source calculus.

6 Final Remarks

This paper raises two issues that deserve further consideration. The first issue is whether the technique of "garbage-passing", as used in the translation of λ-calculus with control operators in [8] and later for translation of intuitionistic sequent calculus [5,6], can be captured through some monad. Less ambitiously, one would hope for a precise comparison that allows to see why there is no need for extra rules such as assoc in the target of the garbage-passing translation. The second issue is the systematization of "associativity" principles in the monadic meta-language; indeed, it is conspicuous that one principle is missing, namely $\pi_{\lambda,\mathsf{bind}}$, which reads $(\lambda y.u)\mathsf{bind}(r, x.t) \to \mathsf{bind}(r, x.(\lambda y.u)t)$. The uses and properties of this rule are not yet entirely clear.

Acknowledgements. We are thankful to the referees for their constructive feedback. The first and third authors are supported by FCT through the Centro de Matemática da Universidade do Minho. The second author thanks for an invitation by that institution to Braga in May 2008. All authors were supported by the European Union FP6-2002-IST-C Coordination Action 510996 "Types for Proofs and Programs" and the first and third authors are also supported by RESCUE FCT project PTDC/EIA/65862/2006.

References

1. Curien, P.-L., Herbelin, H.: The duality of computation. In: Proc. of 5th ACM SIGPLAN Int. Conf. on Functional Programming (ICFP 2000), Montréal, pp. 233–243. IEEE, Los Alamitos (2000)

2. Espírito Santo, J.: Completing Herbelin's programme. In: Ronchi Della Rocca, S. (ed.) TLCA 2007. LNCS, vol. 4583, pp. 118–132. Springer, Heidelberg (2007)
3. Espírito Santo, J.: Delayed substitutions. In: Baader, F. (ed.) RTA 2007. LNCS, vol. 4533, pp. 169–183. Springer, Heidelberg (2007)
4. Espírito Santo, J.: Addenda to Delayed Substitutions (2008) (manuscript available from the author's web page)
5. Espírito Santo, J., Matthes, R., Pinto, L.: Continuation-passing style and strong normalisation for intuitionistic sequent calculi. In: Ronchi Della Rocca, S. (ed.) TLCA 2007. LNCS, vol. 4583, pp. 133–147. Springer, Heidelberg (2007)
6. Espírito Santo, J., Matthes, R., Pinto, L.: Continuation-passing style and strong normalisation for intuitionistic sequent calculi. Logical Methods in Computer Science (to appear, 2009)
7. Hatcliff, J., Danvy, O.: A generic account of continuation-passing styles. In: POPL 1994: Proceedings of the 21st ACM SIGPLAN-SIGACT symposium on Principles of programming languages, pp. 458–471. ACM, New York (1994)
8. Ikeda, S., Nakazawa, K.: Strong normalization proofs by CPS-translations. Information Processing Letters 99, 163–170 (2006)
9. Kfoury, A.J., Wells, J.B.: New notions of reduction and non-semantic proofs of beta-strong normalisation in typed lambda-calculi. In: Proceedings of LICS 1995, pp. 311–321 (1995)
10. Lengrand, S.: Call-by-value, call-by-name, and strong normalization for the classical sequent calculus. In: Gramlich, B., Lucas, S. (eds.) Post-proc. of the 3rd Workshop on Reduction Strategies in Rewriting and Programming (WRS 2003). Electronic Notes in Theoretical Computer Science, vol. 86, Elsevier, Amsterdam (2003)
11. Lengrand, S.: Temination of lambda-calculus with the extra call-by-value rule known as assoc. arXiv:0806.4859v2 (2007)
12. Moggi, E.: Notions of computation and monads. Inf. Comput. 93(1), 55–92 (1991)
13. Polonovski, E.: Strong normalization of $\overline{\lambda}\mu\tilde{\mu}$ with explicit substitutions. In: Walukiewicz, I. (ed.) FOSSACS 2004. LNCS, vol. 2987, pp. 423–437. Springer, Heidelberg (2004)
14. Pottinger, G.: Normalization as a homomorphic image of cut-elimination. Annals of Mathematical Logic 12(3), 323–357 (1977)
15. Regnier, L.: Une équivalence sur les lambda-termes. Theoretical Computer Science 126(2), 281–292 (1994)
16. Sabry, A., Wadler, P.: A reflection on call-by-value. ACM Trans. Program. Lang. Syst. 19(6), 916–941 (1997)
17. Zucker, J.: The correspondence between cut-elimination and normalization. Annals of Mathematical Logic 7(1), 1–112 (1974)

Towards a Type Discipline for Answer Set Programming

Camillo Fiorentini, Alberto Momigliano, and Mario Ornaghi

Dipartimento di Scienze dell'Informazione, Università degli Studi di Milano, Italy
{fiorenti,momiglia,ornaghi}@dsi.unimi.it

Abstract. We argue that it is high time that types had a beneficial impact in the field of Answer Set Programming and in particular Disjunctive Datalog as exemplified by the *DLV* system. Things become immediately more challenging, as we wish to present a type system for *DLV-Complex*, an extension of DLV with uninterpreted function symbols, external implemented predicates and types. Our type system owes to the seminal polymorphic type system for Prolog introduced by Mycroft and O'Keefe, in the formulation by Lakshman and Reddy. The most innovative part of the paper is developing a *declarative grounding* procedure which is at the same time appropriate for the operational semantics of ASP and able to handle the new features provided by DLV-Complex. We discuss the soundness of the procedure and evaluate informally its success in reducing, as expected, the set of ground terms. This yields an automatic reduction in size and numbers of (non isomorphic) models. Similar results could have only been achieved in the current untyped version by careful use of generator predicates in lieu of types.

Keywords: Answer set programming, type checking, grounding, many sorted interpretation.

1 Introduction

The advantages of static type checking for programming languages are almost universally recognized and well-understood, although types managed to make it into logic programming somewhat belatedly [21]. From the pioneering paper [18] advocating the introduction of types in Prolog, different approaches emerged, belonging roughly to two main camps: the descriptive and the prescriptive one. The former aims at capturing some aspects of a program behavior, for example its success set, up to much more complex static and dynamic properties as with CiaoPP [12].

We favor the *prescriptive* approach, where types are an integral part of the program's meaning, thus allowing the user to discard ill-formed elements at compile time. This helps the developer to proceed in a more disciplined way, being able to receive early feedback about mistakes that may be hard to find, especially in a purely declarative setting in the non-infrequent case where the answer would be merely "no".

The type theory of logic programming has significantly developed in the passing years, following the ever-increasing role of types in the general theory of programming languages – see the type system of Mercury [14] and λProlog [19] for a modern take to prescriptive typing. Still, for the sake of this paper, we will adapt to disjunctive

S. Berardi, F. Damiani, and U. de'Liguoro (Eds.): TYPES 2008, LNCS 5497, pp. 117–135, 2009.

logic programing a rather elementary SML-like polymorphic discipline; in addition we will enrich it semantically, following Lakshman & Reddy's approach [15], where typed logic programs are interpreted over first order many sorted structures, see also [13].

Our objective here is to advocate the usefulness of prescriptive typing for Answer Set Programming (ASP) [9] in general. ASP is a form of declarative programming based on the stable models semantics [10]. It has been proposed as a language for knowledge representation and as a tool for formalizing and solving hard search problems [1]. The main idea is to reduce a search problem encoded with a (disjunctive) logic program P to the problem of computing stable models of P using an answer set solver, namely a system for generating stable models, such as Smodels [22] and DLV [16]. In this sense ASP's operational semantics is very different from the usual SLD(NF)-resolution [17]. This is also why typing ASP is interesting and less banal that at first sight: the original paper by Mycroft and O'Keefe was entirely syntactical and so the vast majority of the ensuing research was intrinsically biased towards standard SLD(NF)-resolution.

Although some papers in the ASP literature use a many sorted language or deals with object-oriented features — we refer to Sect. 6 for a detailed discussion — it is safe to say that types have never been *integrated* into ASP; in fact the above papers do not consider a *many-sorted model theory*, nor any form of *parametric* polymorphism. The present paper introduces static type-checking in DLV-Complex [5], an extension of DLV with uninterpreted function symbols – hence we are leaving the comfort of Datalog – as well with external types and predicates requiring an oracle to be computable. In other terms, one may exploit external sources of knowledge. To take into account this possibility, we consider programs over *grounding structures*. The latter provide a declarative semantics for grounding, as they play the role of pre-interpretations [17] in presence of an external implementation of data types and related operations.

The paper is organized as follows: we begin in Section 2 by briefly illustrating the role of typing in ASP through examples. Section 3 and 4 give a static semantics to the language both in terms of types and many-sorted interpretations. In Section 5 we extend the above analysis to DLV-Complex's programs and offer a rational reconstruction of type-driven grounding. After discussing related work and summing up the results in Section 6, we add a short Appendix (A) explaining the basics of ASP for the reader unfamiliar with this topic.

2 An Informal Introduction to Typed DLV

For the sake of this paper, an ASP program P is a set of rules (or clauses) of the form:

$$A_1 \vee \cdots \vee A_n \leftarrow B_1, \ldots, B_k, not\ B_{k+1}, \ldots, not\ B_m$$

where $A_1, \ldots, A_n, B_1, \ldots, B_m$ are atoms, i.e., formulas of the form $p(\boldsymbol{t})$, where $\boldsymbol{t}$ stands for a sequence $t_1, \ldots, t_j$ of terms. Usually *not* is referred to as "negation as failure" or "default negation" and $\vee$ as "epistemic disjunction".

Example 1. The following `pencil problem` is a simple non-typed ASP program, which illustrates the meaning of default negation.

```
color(red). color(blue). pencil(p1). pencil(p2).        % facts
nice(X) :- pencil(X), color(Y), not ugly(Y).            % clause c1
ugly(Y) :- color(Y), pencil(X), not nice(X).            % clause c2
```

By rule c1, we can infer that a pencil X is nice if we do not have evidence for ugly(Y) (i.e., if we can assume not ugly(Y) as a default). Symmetrically, by rule c2, we can infer ugly(Y) if we do not have evidence for nice(X). Let us assume that "not ugly(Y)" holds. We infer that all the pencils are nice, i.e., we have the answer set {nice(p1),nice(p2)}. Symmetrically, assuming "not nice(X)" we obtain the answer set {ugly(red),ugly(blue)}.

As shown by the example, an ASP program P represents a problem that may have multiple solutions, represented by its answer sets. The answer set semantics is defined in terms of *grounding*: let $grnd(C)$ be the set of the ground instances of a clause C, obtained by substituting its variables with ground terms of the language underlying P. We define $grnd(P) = \cup_{C \in P} grnd(C)$. An answer set is a set of ground literals. The answer sets of P are the stable models of $grnd(P)$ (see the Appendix for the formal definitions).

An ASP solver computes the stable models in two stages: firstly it builds an optimized version of $grnd(P)$, namely a set $grnd^{\bullet}(P)$ of ground clauses with the same semantics of $grnd(P)$. Then it generates the answer sets of $grnd^{\bullet}(P)$.

Example 2. In DLV, the pencil program of Example 1 grounds to:

```
% EDB facts:
color(red).  color(blue).  pencil(p1).  pencil(p2).
% Residual ground instantiation of the program:
ugly(red)  :- not nice(p1).     ugly(red)  :- not nice(p2).
ugly(blue) :- not nice(p1).     ugly(blue) :- not nice(p2).
nice(p1) :- not ugly(red).      nice(p1)  :- not ugly(blue).
nice(p2)  :- not ugly(red).     nice(p2)  :- not ugly(blue).
```

We point out that only 12 simplified clauses of the 36 of *grnd*(pencil) are generated, called the *residual ground instantiation*. For example, ugly(red) :- color(red), pencil(p1), not nice(p1) is simplified to ugly(red) :- not nice(p1), because the facts color(red) and pencil(p1) are true in every model. In contrast, the clause ugly(p1) :- color(p1), pencil(p1), not nice(p1) has not been generated because the atom color(p1) in the body is false in every stable model. We get the answer sets:

```
{color(red), color(blue), pencil(p1), pencil(p2), nice(p1), nice(p2)}
{color(red), color(blue), pencil(p1), pencil(p2), ugly(red), ugly(blue)}
```

Now we come to the main issue of this paper. Looking at the example, one easily recognizes that the optimizations performed by the grounding algorithm, henceforth indicated by *GA*, are related to the implicit presence of two types of objects: pencils and colors. Such types are *interpreted in the same way in all stable models*. We believe that making typing explicit has several advantages. In particular, the programmer has a better control on the grounding process, because a type-driven grounder will not generate clauses corresponding to type information, thus yielding smaller $grnd^{\bullet}(P)$ and smaller answer sets.[1] Furthermore there is a clear-cut distinction between *data base facts*, which represent relevant pieces of information to be shown in answer sets, and *typing constraints*, which only have the role of narrowing grounding, as exemplified next.

Example 3. A typed program is (informally) declared as a unit. We use a concrete syntax inspired by [15]. The typed version of the program of Example 1 is:

```
unit pencils.
type pencil --> p1 ; p2.        % Cf. SML's datatype pencil = p1 | p2;
type color --> red ; blue.
pred nice(pencil), ugly(color).
prog
   nice(X) :- not ugly(Y).
   ugly(X) :- not nice(Y).
```

The *enumerated* type `color` is interpreted as the set of constants {`red, blue`}, and `pencil` as {`p1,p2`}. The `pred` keyword introduces a set of predicate symbols and their declared types. By `nice(pencil)` we indicate that `nice` requires an argument of type `pencil` to be a well-typed predicate. Finally, we have the program section, whose clauses are well-typed. Of course we do not need to declare the type of the variables, as they can be inferred. In the example, the rules for clauses in Sect. 5 would yield these well-typing judgments of the form $\Gamma \vdash H \leftarrow B : \mathrm{clause}$.

$$\begin{array}{lcl} X : pencil, Y : color & \vdash & nice(X) \leftarrow not\ ugly(Y) \ : \mathrm{clause} \\ X : color, Y : pencil & \vdash & ugly(X) \leftarrow not\ nice(Y) \ : \mathrm{clause} \end{array}$$

We will explicitly connect the concrete and abstract syntax in Example 6. A typed *GA* will use the variable contexts $X : color, Y : pencil$ instead of the predicates `pencil(X)` and `color(Y)` in the untyped program. The difference is that the untyped *GA* catalogues `color(red), color(blue),...` as "external data base" (`EDB`) facts and puts them in the answer sets, as shown in Example 2, since it has no mean to distinguish type and data-base information. In contrast, a typed *GA* would *generate* only the residual grounding instantiation of Example 2, which has the more concise answer sets {`nice(p1),nice(p2)`} and {`ugly(red),ugly(blue)`}.

We remark that our model-theory fix the interpretation of external types, functions as well as of equality, but not of program predicates. The latter depends on the program and may have a variety of models. In pure logic programming, types, functions and equality are interpreted on term models. Equality is axiomatized by Clark's Equality Theory (CET) [17]. For the predicates defined in our example, CET includes the standard equality axioms and the freeness axioms $\neg(\mathtt{red} = \mathtt{blue})$ and $\neg(\mathtt{p1} = \mathtt{p2})$. A set of equations has a solution in CET iff a unifier exists. Thus standard unification is a sound and complete decision method for CET. If we add non-free functions, we no longer have pure term models and the unification algorithm cannot be applied as such.

Example 4. The following program uses the externally implemented type `#int` of integers, where we use the notation `#` to indicate external types and predicates, as in DLV-Complex [5].

[1] We remark that constructions such as `#hide` in Smodels, may not show the type information, but the latter is still present during the generation of the model.

```
unit usr.
type #int.
type usr --> john ; ted.
func age: usr -> #int.
     age(john) = 12.
     age(ted) = 15.
pred older(usr,usr,#int).
prog
    older(P1,P2,A) :- age(P1) = age(P2)+A, A > 0.
```

The domain of the external type #int is predefined, together with the usual operations on it. In our example, the function age is defined by two equations, although it could be an arbitrarily complex computable function. Both age and the predefined $+, =, >$ are used in the body of the program rule. An ASP system should be able to solve constraint problems, preferably at the grounding stage. If this were the case, the user would not have to specify a #maxint and the older clause would not have to be grounded up to that.

The next example uses parametric polymorphism. We adopt the usual Prolog notation for lists [_|_] as well as using 'A for the (quantified) type variable α (following SML's notation).

Example 5. We assume that #list is an external type constructor, while rec is internal.

```
unit store.
type  usr --> john ; ted.
type  #int.
type  rec('A,'B) --> r('A,'B).
type  #list('A).
pred  is_user(usr), stored_at(#int,'A), store(#list(rec(#int,'A))).
prog
      is_user(U) :- stored_at(K,U).
      stored_at(K,E) :- store(S), #member(r(K,E),S).
      store([r(1,john),r(2,ted)]).
```

Note that for every instantiation [T/'A], the constructor #list(T) yields an external implementation of finite lists with elements of type T, together with a set of pre-defined operations such as #member('A,#list('A)).

Since the signature of a DLV-Complex program consists of two distinct parts, one containing functions and (external) data predicates, the other (internal) predicates, we present the type system, the abstract syntax and the intended model semantics in three steps over the next three Sections. We firstly introduce a many-sorted first order data signature Σ^* induced by a polymorphic type system. Then we define the *intended data models* as suitable Σ^*-interpretations. Finally, we introduce the program signature, DLV-Complex programs and their intended models.

3 The Type and Term Language

We start from the type system. Types are defined inductively applying a type constructors *c* (*#c*) to *internal* and *external* base types. Internal atomic types include a unit type and a constant “o” denoting the type of propositions. To stay inside first-order logic, “o” can only appear at the right of an arrow type. Terms are obtained from logical variables X and function symbols. Signatures give kinds to (any) type constructors and types to internal function symbols and external predicates. The latter two can be used polymorphically.

$$
\begin{array}{rrcl}
\text{Base types} & T_0 & ::= & \alpha \mid \mathrm{o} \mid \mathrm{unit} \mid \#int \mid \#string \ldots \\
\text{Compound types} & T_{n+1} & ::= & c(\boldsymbol{T_n}) \mid \#c(\boldsymbol{T_n}) \\
\text{Types} & T & ::= & \bigcup_{n\geq 0} T_n \\
\text{Signature} & \Sigma & ::= & \emptyset \mid \Sigma, c : \boldsymbol{type} \to \mathrm{type} \mid \Sigma, \#c : \boldsymbol{type} \to \mathrm{type} \\
 & & & \mid \Sigma, f : \forall \boldsymbol{\alpha}.\, \boldsymbol{T} \to T' \mid \Sigma, \#p : \forall \boldsymbol{\alpha}.\, \boldsymbol{T} \to \mathrm{o} \\
\text{Contexts} & \Gamma & ::= & \emptyset \mid \Gamma, \alpha : \mathrm{type} \mid \Gamma, X : T \\
\text{Terms} & t & ::= & X \mid f_{\boldsymbol{T}}(\boldsymbol{t}) \\
\text{Atoms} & A & ::= & \#p(\boldsymbol{t}) \mid t = t'
\end{array}
$$

Before we discuss type checking for the above syntactic categories (Fig. 1), some common notation and conventions: tokens in bold such as $\boldsymbol{T}$ stand for a sequence, i.e. $T_1,\ldots,T_n$. We view Σ and Γ as sets and we use Γ, e for $\Gamma \cup \{e\}$ with $e \notin \Gamma$. We will often write α instead of α : type. $\Gamma \vdash_\Sigma J_1,\ldots,J_n$ abbreviates $\Gamma \vdash_\Sigma J_1,\ldots,\Gamma \vdash_\Sigma J_n$. We often write $\Gamma \vdash J$, leaving the signature Σ understood. In the rule sf, f is new in Σ. In the rules for Declarations, Types, Terms and Atoms, Γ is assumed to be a well-formed context.

In the rules fd and pd, $[\boldsymbol{I}/\boldsymbol{\alpha}]$ indicates the substitution of the variables $\boldsymbol{\alpha}$ by the types $\boldsymbol{I}$. It is worth remarking that while in our type systems constants can be parameterized by type *schemata*, rather than closed (simple) types, we do not support *impredicative* polymorphism. Indeed, we can think of this still as form of prenex polymorphism, where type schema can be seen as axiom schema in first-order logic.

Example 6. The concrete syntax of Example 5 corresponds to the signature:

$$
\begin{array}{l}
usr : \mathrm{type},\ \#int : \mathrm{type},\ john : usr,\ ted : usr,\ 0 : \#int,\ + : \#int, \#int \to \#int, \ldots \\
rec : \mathrm{type}, \mathrm{type} \to \mathrm{type},\ r : \forall \alpha, \beta. \alpha, \beta \to rec(\alpha, \beta),\ \#list : \mathrm{type} \to \mathrm{type}, \\
nil : \forall \alpha. \#list(\alpha),\ cons : \forall \alpha. \alpha, \#list(\alpha) \to \#list(\alpha),\ \#member : \forall \alpha. \alpha, \#list(\alpha) \to o \ldots
\end{array}
$$

The system in Fig. 1 is an extension with rules for external sources of knowledge of the one presented in [15]. One difference however is in the judgments “$f_{\boldsymbol{I}}$ decl” and “$\#p_{\boldsymbol{I}}$ decl”, which introduce *indexed* function and predicate declarations. In [15] instead, indexes are only used at the semantic level, to provide a first-order interpretation of polymorphic declarations. Indexes can be reconstructed as we will discuss in Sect. 4.

Some relevant admissible properties of the type system are listed below, where $fv(J)$ indicates the set of the variables occurring free in J and $\Gamma | fv(J)$ the restriction of Γ to terms containing variables in $fv(J)$. Here we have J range over Decl, Typs, Trms and Atms.

uq) Uniqueness of typing: $\Gamma \vdash_\Sigma t : T$ and $\Gamma \vdash_\Sigma t : T'$ implies $T = T'$, modulo α-conversion.

str) Strengthening: if $\Gamma \vdash_\Sigma J$ then every variable free in J is declared in Γ. Moreover, $\Gamma | fv(J) \vdash_\Sigma J$.

Cntx

$$\frac{}{\emptyset \text{ cntx}}\,cn0 \qquad \frac{\Gamma \text{ cntx}}{\Gamma, \alpha \text{ cntx}}\,cn1 \qquad \frac{\Gamma \text{ cntx} \quad \Gamma \vdash_\Sigma T : \text{type}}{\Gamma, X : T \text{ cntx}}\,cn2$$

Sig

$$\frac{}{\emptyset \text{ sig}}\,s0 \qquad \frac{\Sigma \text{ sig}}{\Sigma,\ (c : \boldsymbol{type} \to \text{type}) \text{ sig}}\,sc \qquad \frac{\Sigma \text{ sig}}{\Sigma,\ (\#c : \boldsymbol{type} \to \text{type}) \text{ sig}}\,sc\#$$

$$\frac{\Sigma \text{ sig} \quad \boldsymbol{\alpha} \vdash_\Sigma \boldsymbol{T}, T : \boldsymbol{type}}{\Sigma,\ f : \forall \boldsymbol{\alpha}.\boldsymbol{T} \to T \text{ sig}}\,sf \qquad \frac{\Sigma \text{ sig} \quad \boldsymbol{\alpha} \vdash_\Sigma \boldsymbol{T} : \boldsymbol{type}}{\Sigma,\ \#p : \forall \boldsymbol{\alpha}.\boldsymbol{T} \to \text{o sig}}\,sp$$

Dcl

$$\frac{\Gamma \vdash_\Sigma \boldsymbol{I} : \boldsymbol{type} \quad f : \forall \boldsymbol{\alpha}.\boldsymbol{T} \to T \in \Sigma}{\Gamma \vdash_\Sigma f_{\boldsymbol{I}} : \boldsymbol{T}[\boldsymbol{I}/\boldsymbol{\alpha}] \to T[\boldsymbol{I}/\boldsymbol{\alpha}] \text{ decl}}\,fd \qquad \frac{\Gamma \vdash_\Sigma \boldsymbol{I} : \boldsymbol{type} \quad \#p : \forall \boldsymbol{\alpha}.\boldsymbol{T} \to \text{o} \in \Sigma}{\Gamma \vdash_\Sigma \#p_{\boldsymbol{I}} : \boldsymbol{T}[\boldsymbol{I}/\boldsymbol{\alpha}] \to \text{o decl}}\,pd$$

Typs

$$\frac{}{\Gamma \vdash_\Sigma \text{unit} : \text{type}}\,Tu \qquad \frac{}{\Gamma, \alpha \vdash_\Sigma \alpha : \text{type}}\,T0$$

$$\frac{\Gamma \vdash_\Sigma \boldsymbol{T} : \boldsymbol{type} \quad c : \boldsymbol{type} \to \text{type} \in \Sigma}{\Gamma \vdash_\Sigma c(\boldsymbol{T}) : \text{type}}\,Ti$$

Trms

$$\frac{}{\Gamma, X : T \vdash_\Sigma X : T}\,t0 \qquad \frac{\Gamma \vdash_\Sigma \boldsymbol{t} : \boldsymbol{T} \quad \Gamma \vdash_\Sigma f_{\boldsymbol{I}} : \boldsymbol{T} \to T \text{ decl}}{\Gamma \vdash_\Sigma f_{\boldsymbol{I}}(\boldsymbol{t}) : T}\,ti$$

Atms

$$\frac{\Gamma \vdash_\Sigma \boldsymbol{t} : \boldsymbol{T} \quad \Gamma \vdash_\Sigma \#p_{\boldsymbol{I}} : \boldsymbol{T} \to o \text{ decl}}{\Gamma \vdash_\Sigma \#p(\boldsymbol{t}) : \text{o}}\,pi \qquad \frac{\Gamma \vdash_\Sigma \boldsymbol{t} : \boldsymbol{T} \quad \Gamma \vdash_\Sigma \boldsymbol{t}' : \boldsymbol{T}}{\Gamma \vdash_\Sigma \boldsymbol{t} = \boldsymbol{t}' : \text{o}}\,eq$$

Fig. 1. Type checking rules

wk) Weakening: if $\Gamma \subseteq \Gamma'$ and $\Gamma \vdash_\Sigma J$, then $\Gamma' \vdash_\Sigma J$.

sbi) Substitution: we implicitly assume the α-conversions needed to avoid capture and, for $\theta \equiv [T/\alpha]$, $\Gamma\theta$ is defined as: $\emptyset\theta = \emptyset, (\Gamma, \alpha : \text{type})\theta = (\Gamma\theta),\ \alpha : \text{type}$ and $(\Gamma,\ X : T')\theta = (\Gamma\theta),\ X : (T'\theta)$.

$$\frac{\Gamma, \alpha \vdash_\Sigma T : \text{type} \qquad \Gamma \vdash_\Sigma T' : \text{type}}{\Gamma[T'/\alpha] \vdash_\Sigma T[T'/\alpha] : \text{type}}\,sb1 \qquad \frac{\Gamma, X : T' \vdash_\Sigma t : T \qquad \Gamma \vdash_\Sigma t' : T'}{\Gamma \vdash_\Sigma t[t'/X] : T}\,sb2$$

4 Grounding Structures

We firstly introduce a many-sorted first order signature Σ^*. Then we define grounding structures as suitable Σ^*-interpretations. A polymorphic signature Σ yields an infinite many sorted signature Σ^* in the following way:

- Every ground type T is a sort in Σ^*;
- for every $f : \forall \boldsymbol{\alpha}.\boldsymbol{T} \to T$ such that $\emptyset \vdash_\Sigma f_{\boldsymbol{I}} : \boldsymbol{T}[\boldsymbol{I}/\boldsymbol{\alpha}] \to T[\boldsymbol{I}/\boldsymbol{\alpha}]$ decl is provable, $f_{\boldsymbol{I}} : \boldsymbol{T}[\boldsymbol{I}/\boldsymbol{\alpha}] \to T[\boldsymbol{I}/\boldsymbol{\alpha}]$ is a function declaration in Σ^*;

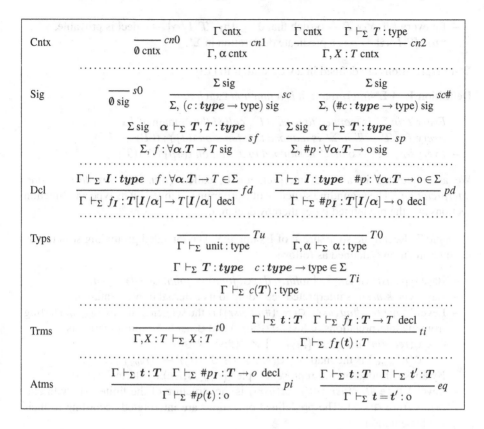

- for every $\#p : \forall\alpha.\boldsymbol{T} \to o$ such that $\emptyset \vdash_\Sigma \#p_{\boldsymbol{I}} : \boldsymbol{T}[\boldsymbol{I}/\alpha] \to o$ decl is provable, $\#p_{\boldsymbol{I}} : \boldsymbol{T}[\boldsymbol{I}/\alpha] \to o$ is a predicate declaration in Σ^*.

Σ^*-interpretations are defined in a way similar to [13]:

Definition 1. *A Σ^*-interpretation is a function ι such that:*

- *Every T in Σ^* is mapped to a set $\iota(T)$, called the* domain *of T.*
- *Every $f_{\boldsymbol{I}} : \boldsymbol{T} \to T$ in Σ^* is mapped to a function $\iota(f_{\boldsymbol{I}}) : \iota(\boldsymbol{T}) \to \iota(T)$.*
- *Every $\#p_{\boldsymbol{I}} : \boldsymbol{T} \to o$ in Σ^* is mapped to a relation $\iota(\#p_{\boldsymbol{I}}) \subseteq \iota(\boldsymbol{T})$.*

We are interested in a fixed interpretation G, called *grounding structure* or *pre-interpretation*.[2] The grounding structure of a program is defined by its type and function sections and the predefined types thereby declared.

Example 7. Recall the signature Σ of Example 5. The intended grounding structure is the interpretation G defined as follows.

- Base type *usr*: $\mathrm{G}(usr) = \{john, ted\}$, $\mathrm{G}(john) = john$, $\mathrm{G}(ted) = ted$.
- Base type *#int*: it is interpreted according to its external implementation.
- Level 1 type $rec(\#int, usr)$: $\mathrm{G}(rec(\#int, usr))$ is the set generated by $r_{\#int,usr}$ starting from the interpretations of level 0 $\mathrm{G}(\#int)$ and $\mathrm{G}(usr)$. A representation is:
  - $\mathrm{G}(rec(\#int, usr)) = \{r_{\#int,usr}\} \times \mathrm{G}(\#int) \times \mathrm{G}(usr)$;
  - $\mathrm{G}(r_{\#int,usr})$ is the map $\langle i \in \mathrm{G}(\#int), u \in \mathrm{G}(usr)\rangle \mapsto \langle r_{\#int,usr}, i, u\rangle$.

  By abuse of notation, we represent triples $\langle r_{\#int,usr}, i, u\rangle$ as "terms" $r(i,u)$.
- Level 1 type $\#list(usr)$: $\mathrm{G}(\#list(usr))$ is the structure of the finite lists with elements from $\mathrm{G}(usr)$. The predefined operations are interpreted according to their implementation.
- ... and so on. We remark that the set-theoretic representation applies at each level.

We now move to the evaluation of a term with respect to an *assignment* in an interpretation.

Definition 2. *Let ι be a Σ^*-interpretation and Γ a context. An* assignment *a for Γ in ι maps every β of Γ to a sort $a(\beta)$ of Σ^* and every $X : T$ of Γ to a value $v \in \iota(T\,a)$, where $T\,a$ is the sort of Σ^* obtained by grounding each variable β occurring in T by the sort $a(\beta)$.*

Given ι, we denote by $\mathrm{ass}(\Gamma,\iota)$ the set of all the assignments for Γ in ι. Take $\Gamma \vdash_\Sigma t : T$ and $a \in \mathrm{ass}(\Gamma,\iota)$. The evaluation of t with respect to a, denoted $t \leadsto_a v$, is performed according to the following rules:

$$\frac{}{X \leadsto_a a(X)}(ev1) \qquad \frac{\boldsymbol{t} \leadsto_a \boldsymbol{v} \qquad \Gamma_\Sigma \vdash f_{\boldsymbol{I}} : \boldsymbol{T} \to T \text{ decl}}{f_{\boldsymbol{I}}(\boldsymbol{t}) \leadsto_a \iota(f_{\boldsymbol{I}a})(\boldsymbol{v})}(ev2)$$

Theorem 1. *Let Σ be a signature, ι a Σ^*-interpretation, $a \in ass(\Gamma,\iota)$ and $\Gamma \vdash_\Sigma t : T$. Then there is a unique $v \in \iota(T\,a)$ such that $t \leadsto_a v$.*

[2] Our notion of pre-interpretation generalizes Lloyd's [17] interpreting #-predicates.

Theorem 1 follows by the existence and uniqueness of the $\Gamma \vdash_\Sigma f_I : \boldsymbol{T} \to T$ decl judgment and as such depends on the indexing of function symbols via a simple table look up "modulo matching" over Σ.

Example 8. Let Σ be the signature of Ex. 5 and $\Gamma = \{\alpha, \beta : \text{type}, X : \alpha, Y : \beta\}$. We have that $\Gamma \vdash_\Sigma r(X,Y) : rec(\alpha,\beta)$. Let G be the interpretation defined in Ex. 7 and $a \in \text{ass}(\Gamma, \mathrm{G})$ such that $a(\alpha) = \#int$, $a(\beta) = usr$, $a(X) = 5$ and $a(Y) = ted$. Then $r(X,Y) \rightsquigarrow_a \langle r_{\#int,usr}, 5, ted\rangle$.

Finally, the truth relation $\iota \models_a A$ for a well-typed atom A in Γ and $a \in \text{ass}(\Gamma, \iota)$ is defined by cases on A:

- $\iota \models_a \#p_I(\boldsymbol{t})$ iff $\boldsymbol{t} \rightsquigarrow_a \boldsymbol{v}$ and $\boldsymbol{v} \in \iota(p_I)$;
- $\iota \models_a t_1 = t_2$ iff $t_j \rightsquigarrow_a v$ for $j = 1, 2$.

5 Program Grounding

A *program* signature is of the form $\Sigma \cup \Pi$ where Π is a set of *program predicate* declarations. We remark that Σ still accounts for the external predicates $\#p$. As we saw in Section 2, the latter are interpreted on top of the grounding structure G defined in the type and function section of the program at hand. To formalize this, we resort to G-expansions. We start with $\Sigma \cup \Pi$ as above: the associated many sorted signature $(\Sigma \cup \Pi)^*$ has the form $\Sigma^* \cup \Pi^*$, where Π^* contains the ground predicate declarations $p_I : \boldsymbol{T} \to \mathrm{o}$. A grounding structure G is a Σ^*-interpretation and a G-expansion is a $(\Sigma^* \cup \Pi^*)$-interpretation ι such that $\iota|\Sigma^* = \mathrm{G}$, i.e. the *reduct* of ι w.r.t. Σ^*. We can represent ι as the set

$$H_\iota = \{ \langle p_I, \boldsymbol{v}\rangle \mid p_I : \boldsymbol{T} \to \mathrm{o} \in \Pi^* \text{ and } \boldsymbol{v} \in \iota(p_I) \}$$

By abuse of notation, we write $p(\boldsymbol{v})$ instead of $\langle p_I, \boldsymbol{v}\rangle$ and call it a G-*atom*. We dub the set of G-atoms the G-*base*, playing a role analogous to the Herbrand base. In this way, H_ι can be thought as a kind of "G-answer set".

Example 9. Back to program of Example 5 and the related grounding structure of Example 7. The following interpretation ι is the minimum model of the program (the unit "store"), seen as a set of G-atoms:

$$\{store([r(1,john), r(2,ted)]), stored\_at(1,john), stored\_at(2,ted), is\_user(john), is\_user(ted)\}$$

The loose notation $store([r(1,john), r(2,ted)])$, $stored\_at(1,john), \ldots$ means that

$$\langle cons, \langle r, 1, john, \rangle, \langle cons, \langle r, 2, ted\rangle, nil\rangle\rangle \in \mathrm{G}(store_{usr}), \ \langle 1, john\rangle \in \mathrm{G}(stored\_at_{usr}), \ldots$$

where $cons = cons_{rec(\#int,usr)}$ and $nil = nil_{rec(\#int,usr)}$.

Now we can formally introduce programs and enrich the syntax judgments of the previous section. The form of a clause is $A_1 \vee \cdots \vee A_n \leftarrow B_1, \ldots, B_m$, where $A_1, \ldots, A_n$ are program atoms and $B_1, \ldots, B_m$ are literals, i.e., atoms, external predicates, negated atoms or negated external predicates. For the sake of simplicity, we do not consider here classical negation, see the appendix.

- Atom A ::= ... (as before).
- Program Atom PA ::= $p(\boldsymbol{t})$, with p defined in Π.
- Negated Atom NA ::= $not\ A \mid not\ PA$.
- Literal B ::= $A \mid PA \mid NA$.
- Head HD ::= $A_1 \vee \cdots \vee A_n$; for $n = 0$, $HD \equiv false$.
- Body BD ::= $B_1, \cdots, B_m$; for $m = 0$, $BD \equiv true$.
- Clause C ::= $HD \leftarrow BD$.

To distinguish external and internal program predicates, we split the judgment for well-formed atoms $p(\boldsymbol{t}) : \mathrm{o}^+$ and $\#q(\boldsymbol{t}) : \mathrm{o}^\#$, where q is declared in Σ and p in Π. The corresponding rules are omitted, since they correspondingly split the rule pi in Fig. 1. We introduce the judgments $NA : o^-$, $HD : \mathrm{head}$, $BD : \mathrm{body}$ and $C : \mathrm{clause}$, where in $B_i : \mathrm{o}^{s_i}$, the superscript s_i may be $+, -, \#$.

$$\frac{\Gamma \vdash_{\Sigma\cup\Pi} A : \mathrm{o}^+}{\Gamma \vdash_{\Sigma\cup\Pi} not\ A : \mathrm{o}^-}\ at^- \qquad \frac{\Gamma \vdash_{\Sigma\cup\Pi} A : \mathrm{o}^\#}{\Gamma \vdash_{\Sigma\cup\Pi} not\ A : \mathrm{o}^-}\ at^-$$

$$\frac{\Gamma \vdash_{\Sigma\cup\Pi} A_1 : \mathrm{o}^+, \ldots, A_n : \mathrm{o}^+}{\Gamma \vdash_{\Sigma\cup\Pi} A_1 \vee \cdots \vee A_n : \mathrm{head}}\ h \qquad \frac{\Gamma \vdash_{\Sigma\cup\Pi} B_1 : \mathrm{o}^{s_1}, \ldots, B_m : \mathrm{o}^{s_m}}{\Gamma \vdash_{\Sigma\cup\Pi} B_1, \ldots, B_m : \mathrm{body}}\ b$$

$$\frac{\Gamma \vdash_{\Sigma\cup\Pi} HD : \mathrm{head} \quad \Gamma \vdash_{\Sigma\cup\Pi} BD : \mathrm{body}}{\Gamma \vdash_{\Sigma\cup\Pi} HD \leftarrow BD : \mathrm{clause}}\ cl$$

Let $\Sigma \cup \Pi$ be a program signature, G a grounding structure and ι be a G-expansion. The truth relation $\iota \models_a A$ is extended to clauses as usual. Similarly, the standard properties of the type system stated on Page 122 also hold.

Now, we aim to define the grounding of a program P with respect to the structure G, notation $\mathrm{grnd}_{\mathrm{G}}(P)$, where P consists of a set of clauses $\forall \boldsymbol{\alpha} : \mathrm{type}.\forall \boldsymbol{X} : \boldsymbol{T}.\ C$ such that $\boldsymbol{\alpha} : \mathrm{type}, \boldsymbol{X} : \boldsymbol{T} \vdash C : \mathrm{clause}$. Let $a \in \mathrm{ass}(\Gamma, \mathrm{G})$; if we write $\Gamma \vdash C$, for "$\Gamma \vdash_{\mathrm{G}} C : \mathrm{clause}$", we can define grounding on the structure of C:

$$\begin{aligned}
\mathrm{grnd}_{\mathrm{G}}(\Gamma \vdash p(\boldsymbol{t}), a) &= p(\boldsymbol{v}) \quad \text{where } \boldsymbol{t} \leadsto_a \boldsymbol{v} \\
\mathrm{grnd}_{\mathrm{G}}(\Gamma \vdash not\ A, a) &= not\ \mathrm{grnd}_{\mathrm{G}}(\Gamma \vdash A, a) \\
\mathrm{grnd}_{\mathrm{G}}(\Gamma \vdash B_1, \ldots, B_m, a) &= \mathrm{grnd}_{\mathrm{G}}(\Gamma \vdash B_1, a), \ldots, \mathrm{grnd}_{\mathrm{G}}(\Gamma \vdash B_m, a) \\
\mathrm{grnd}_{\mathrm{G}}(\Gamma \vdash A_1 \vee \cdots \vee A_n, a) &= \mathrm{grnd}_{\mathrm{G}}(\Gamma \vdash A_1, a) \vee \cdots \vee \mathrm{grnd}_{\mathrm{G}}(\Gamma \vdash A_n, a) \\
\mathrm{grnd}_{\mathrm{G}}(\Gamma \vdash H \leftarrow B, a) &= \mathrm{grnd}_{\mathrm{G}}(\Gamma \vdash H, a) \leftarrow \mathrm{grnd}_{\mathrm{G}}(\Gamma \vdash B, a) \\
\mathrm{grnd}_{\mathrm{G}}(\Gamma \vdash C) &= \textstyle\bigcup_{a \in ass(\Gamma, G)} \{\mathrm{grnd}_{\mathrm{G}}(\Gamma \vdash C, a)\} \\
\mathrm{grnd}_{\mathrm{G}}(P) &= \textstyle\bigcup_{\Gamma \vdash C \in P} \mathrm{grnd}_{\mathrm{G}}(\Gamma \vdash C)
\end{aligned}$$

We can see $\mathrm{grnd}_{\mathrm{G}}(P)$ as a set of propositional clauses over the G-base. We have

$$H_\iota \models_{\mathrm{prop}} \mathrm{grnd}_{\mathrm{G}}(P) \quad \text{iff} \quad \iota \models P \tag{1}$$

where $\models_{\mathrm{prop}}$ is propositional truth. By the equivalence (1), we can define the G-answer sets of P as those of $\mathrm{grnd}_{\mathrm{G}}(P)$. The latter are defined using the Gelfond-Lifschitz transformation, as explained in the appendix.

We now discuss the interaction between grounding and typing: this brings about some immediate problems. The first one is *ground type reconstruction*, as illustrated by the following example.

Example 10

```
pred  p(#list('A)), q(#list('A)), r(#list(#char)), t(#nat).
prog
      p([X|Y]) :- q(Y).
      q(Z) :- r(Z).
      r([]).
      t(3).
```

Now, given the ground instance $p([3]) \leftarrow q([\,])$, the atom $p([3])$ is in the minimal model of the program, which clearly violates the typing discipline. To overcome this problem, we have to assign to $[\,]$ the "right ground type" T. A solution is to assign an index to constants, apparently similar to how constants are declared in the simply-typed λ-calculus to recover a principal type, viz. $inl_B : A \rightarrow A \vee B$. In our case, $\emptyset \vdash_\Sigma [\,]_{\#int} : T$ has solution $T = \#list(\#int)$. Using indexed terms, the above clauses become $p([3]) \leftarrow q([\,]_{\#int})$, as well as $q([\,]_{\#char}) \leftarrow r([\,]_{\#char})$ and $r([\,]_{\#char})$. Now the standard ASP semantics correctly works.

The reader may wonder whether function indexing can be avoided and thus reconstructed. Interestingly enough, the dual of the property of *transparency* [11], for which the type declaration of every (internal) function is such that every type variable occurring in the domain also occurs in the range, ensures, in the absence of overloading, index reconstruction. Space prevents us from providing a formal proof.

Any non-trivial type signature is a source of infinity. We draw our inspiration from the way in which [5] deals with computable functions. We view type instantiation and grounding of individual variables as two separate phases. The intuition is that polymorphic predicate definitions behave as *open units*, i.e., groups of clauses *abstracting* type variables. During grounding an open unit is instantiated by grounding the type variables over a finite number of ground types of its signature ($tinst(P)$). Subsequently $grnd^\bullet(tinst(P))$ grounds individual variables. In the absence of a satisfactory notion of modules and their instantiation in the ASP framework yet, we will not formally define $tinst(P)$. Rather, we directly illustrate a case where modularity and type instantiation do not work properly. In this situation, the user has to provide the type instantiation with a so-called "`@with`-directive".

Example 11

```
unit choices.
#maxint=1.
type pencil --> p1 ; p2.
type color --> red ; blue.
func  col : pencil -> color.
      col(p1) = red.
      col(p2) = blue.
pred  choose(pencil), ok(#int), nice('A), ugly('B).
prog
```

```
nice(X) v ugly(Y).
choose(X) :- nice(X), not ugly(col(X)).
@with ugly(#int):
ok(X+1) :- nice(X), X < 1.
```

Type inference yields:

$$\text{poly:} \quad \alpha,\beta : \text{type}, X : \alpha, Y : \beta \vdash \mathit{nice}_{\alpha}(X) \vee \mathit{ugly}_{\beta}(Y) : \text{head}$$
$$\text{used by: i) } X : \mathit{pencil} \vdash \mathit{choose}(X) \leftarrow \mathit{nice}_{\mathit{pencil}}(X), \mathit{not}\ \mathit{ugly}_{\mathit{color}}(\mathit{col}(X)) : \text{clause}$$
$$\text{ii)} X : \#\mathit{int} \vdash \mathit{ok}(X+1) \leftarrow \mathit{nice}_{\#\mathit{int}}(X), X < 1 : \text{clause}$$

The type instantiation inferred from i) is $[\mathit{pencil}/\alpha, \mathit{color}/\beta]$, while ii) gives the partial instantiation $[\#\mathit{int}/\alpha]$. The `@with`-directive provides $[\#\mathit{int}/\beta]$. The result of the type instantiation is:

$$\begin{array}{lrl} c_1 & X : \mathit{pencil}, Y : \mathit{color} \vdash & \mathit{nice}(X) \vee \mathit{ugly}(Y) : \text{head} \\ c_2 & X, Y : \#\mathit{int} \vdash & \mathit{nice}(X) \vee \mathit{ugly}(Y) : \text{head} \\ c_3 & X : \mathit{pencil} \vdash & \mathit{choose}(X) \leftarrow \mathit{nice}(X), \mathit{not}\ \mathit{ugly}(\mathit{col}(X)) : \text{clause} \\ c_4 & X : \#\mathit{int} \vdash & \mathit{ok}(X+1) \leftarrow \mathit{nice}(X), X < 1 : \text{clause} \end{array}$$

The inferred contexts will be relevant next, in Example 13. With the directive `@with ugly(#int)`, the instantiations of *nice* and of *ugly* in c_1 and c_2 do not interfere, i.e. composition is modular. However, the directive `@with ugly(color)` would have set $[\mathit{color}/\beta]$. Then the clauses c_1 and c_2 would have shared the same $\mathit{ugly}_{\texttt{color}}$ predicate.

A sufficient condition to avoid interference is to impose that the head predicates have different indexed names in different type instantiations. In our example, we could obtain this by declaring a unit `allNiceOrAllUgly`(α,β) with $\mathit{nice}, \mathit{ugly} : \forall \alpha, \beta. \alpha \to o$.

Now we explain the grounding phase. We extend the notion of context so that it contains everything related to the external predicates and operations, while the right-hand side of $\vdash$ contains only the program and the free functions. Namely we transform a clause $\Gamma \vdash C$ into an *extended clause* $\Gamma' | \Delta \vdash C'$, where $\Gamma \subseteq \Gamma'$, Δ is a set of Σ-formulas, which we call *constraints*, and C' now contains only predicates from Π and function symbols that are internal. This can be done by moving external predicates in the constraint context Δ and by introducing suitable equalities in the usual CLP's way.

Example 12. The clauses $c_1, \ldots, c_4$ of Example 11 are transformed into the following extended clauses $e_1, \ldots, e_4$:

$$\begin{array}{lrl} e_1 & X : \mathit{pencil}, Y : \mathit{color} \mid \top \vdash & \mathit{nice}(X) \vee \mathit{ugly}(Y) \\ e_2 & X, Y : \#\mathit{int} \mid \top \vdash & \mathit{nice}(X) \vee \mathit{ugly}(Y) \\ e_3 & X : \mathit{pencil}, U : \mathit{color} \mid U = \mathit{col}(X) \vdash & \mathit{choose}(X) \leftarrow \mathit{nice}(X), \mathit{not}\ \mathit{ugly}(U) \\ e_4 & X, J : \#\mathit{int} \mid J = X+1, X < 1 \vdash & \mathit{ok}(J) \leftarrow \mathit{nice}(X) \end{array}$$

In e_3, the term $\mathit{col}(X)$ occurring in c_3 is replaced by a new special variable $U : \mathit{color}$ and $U = \mathit{col}(X)$ is inserted in the constraints context. Similarly for $J = X+1$ in e_4. Furthermore, the external predicate $X < 1$ has been moved to the constraints context.

Finally, we explain how GA uses extended contexts. Take the signature $\Sigma \cup \Pi$ and extended clause $\Gamma|\Delta \vdash C$. Clearly, GA may access both the signature and the grounding structure G. Take an assignment a such that $a \in \mathrm{ass}(\Gamma, \mathrm{G})$. Since Δ contains Σ-formulas and G is a Σ-interpretation, GA can evaluate the truth relation $\mathrm{G} \models_a \Delta$.

$$\mathrm{egrnd}_{\mathrm{G}}(\Gamma \mid \Delta \vdash C, a) = \begin{cases} \mathrm{grnd}_{\mathrm{G}}(\Gamma \vdash C, a) & \text{if } \mathrm{G} \models_a \Delta \\ \top & \text{otherwise} \end{cases}$$

$$\mathrm{egrnd}_{\mathrm{G}}(\Gamma \mid \Delta \vdash C) = \bigcup_{a \in ass(\Gamma, G)} \{ \mathrm{egrnd}_{\mathrm{G}}(\Gamma \mid \Delta \vdash C, a) \}$$

It has the following property. Let P be a $(\Sigma \cup \Pi)$-program and P_e be the corresponding extended program. The grounding $\mathrm{egrnd}_{\mathrm{G}}(P_e)$ is the union of $\mathrm{egrnd}_{\mathrm{G}}(\Gamma|\Delta \vdash C)$, for $\Gamma|\Delta \vdash C \in P_e$. One can prove that the stable models of $\mathrm{grnd}_{\mathrm{G}}(P)$ coincide with those of $\mathrm{egrnd}_{\mathrm{G}}(P_e)$.

Example 13. Take the clause e_3 of Example 12. An assignment for its context is $a_1 = [p1/X, red/U]$. Since $\mathrm{G} \models_{a_1} U = col(X)$, we have

$$\mathrm{egrnd}_{\mathrm{G}}(e_3, a_1) = choose(p1) \leftarrow nice(p1), not\ ugly(red)$$

In contrast, for the assignment $a_2 = [p1/X, blue/U]$, we have $\mathrm{G} \not\models_{a_2} U = col(X)$. Thus $\mathrm{egrnd}_{\mathrm{G}}(e_3, a_2) = \top$, which can be ignored. Since we have finite types, we generate all the possible assignments and, at the end of the process, we obtain:

$$\begin{array}{ll} nice(p1) \lor ugly(red) & nice(p1) \lor ugly(blue) \\ nice(p2) \lor ugly(red) & nice(p2) \lor ugly(blue) \\ nice(0) \lor ugly(0) & nice(0) \lor ugly(1) \\ nice(1) \lor ugly(0) & nice(1) \lor ugly(1) \\ \multicolumn{2}{l}{choose(p1) \leftarrow nice(p1), not\ ugly(red)} \\ \multicolumn{2}{l}{choose(p2) \leftarrow nice(p2), not\ ugly(blue)} \\ \multicolumn{2}{l}{ok(1) \leftarrow nice(0)} \end{array}$$

As the example shows, a type-driven grounder GA has the following features:

- In the grounding process, only well-typed instances are generated, viz. in Example 13, $nice(p1) \lor ugly(p1), nice(p1) \lor ugly(p2), \ldots$ are excluded.
- If a variable X functionally depends on other variables $\boldsymbol{Y}$, one value is generated for it, depending on the values generated for $\boldsymbol{Y}$. For instance, in clause e_3 of Ex. 12 the variable U functionally depends on X, thus in Ex. 13 only the clauses $choose(p1) \leftarrow nice(p1), not\ ugly(red)$ and $choose(p2) \leftarrow nice(p2), not\ ugly(blue)$ are added.

This should facilitate the generation of small ground instances. More importantly, types give the user a better control of the grounding process.

In the previous example, the grounding structure is finite. To complete the picture, we should consider the general case of possibly infinite grounding structures. We will not introduce new ideas, but consider how types could be managed by existing approaches. Specifically, we consider intelligent grounding w.r.t. the class of finitely ground programs [6], denoted by $\mathscr{FG}$. Here we explain the idea informally. Intelligent grounding is performed starting from a modular decomposition $\langle M_1, M_2, \ldots, M_n \rangle$ of the program P. The module M_1 does not depend on any other modules and grounds to a G_1 having the same stable models as M_1. We use G_1 to build a $\mathrm{G}_2 = ig(M_2, \mathrm{G}_1)$, having the same stable models of $M_1 \cup M_2$. We proceed in this way until we obtain $\mathrm{G}_n = ig(M_n, \mathrm{G}_{n-1})$, which has the same models of P. If G_n is finite, one says that P is an $\mathcal{FG}$-program.

Example 14. This contrived example shows a program that can be divided in two modules (we use `a,b,...` as constants of type `#char`).

```
unit show.
type #int, #char, #list('A).
pred select(#list('A)), show(#list('A)).
prog
  select([a]) v select([4]).                          %module m1
  select([b,a]).
  select([X]) :- select([X+X]).
  show(M):- select(L),select(M), #reverse(L,M).       %module m2
```

The module decomposition is $\langle m1, m2 \rangle$. Note that $m2$ is polymorphic. Type inferences yield the following extended clauses:

$$
\begin{array}{llll}
l_1 & & \vdash & select([a]) \vee select([4]) \\
l_2 & & \vdash & select([b,a]) \\
l_3 & X, I : \#int \mid I = X + X & \vdash & select([X]) \leftarrow select([I]) \\
l_4 & L, M : \#list(\#char) \mid \#reverse(L, M) & \vdash & show(M) \leftarrow select(L), select(M) \\
l_5 & L, M : \#list(\#int) \mid \#reverse(L, M) & \vdash & show(M) \leftarrow select(L), select(M)
\end{array}
$$

Now we proceed to the intelligent grounding phase, by applying the instantiation algorithm given in [6] and using extended clauses. We firstly ground the module $m1$. We assume that the grounding algorithm uses $a, b, c, \ldots$ as the internal representation for characters, $0, 1, 2, \ldots$ for numbers and $[v_1, \ldots, v_n]$ for lists, so that we do not need to distinguish between values and ground terms. Following [6], we start from the already ground clauses:

$$\vdash select([a]) \vee select([4]) \qquad\qquad \vdash select([b,a])$$

We consider the set $\mathcal{G}$ of atoms $select([a])$, $select([4])$, $select([b,a])$ occurring in the head of the ground clauses of $m1$. We use $\mathcal{G}$ to instantiate l_3 by looking for all the grounding substitutions that unify $select([I])$ with one of the atoms in it. We get:

$$X : \#int \mid 4 = X + X \ \vdash \ select([X]) \leftarrow select([4])$$

Differently from Example 13, the *GA* tries to instantiate clauses looking for body atoms that match with atoms in the current $\mathcal{G}$. In this way, the variable X is not instantiated. It has to be grounded by solving the constraint $4 = X + X$. The solution is of course $X = 2$. The result of this step is:

$$
\begin{array}{rcl}
 & \vdash & select([a]) \vee select([4]) \\
 & \vdash & select([b,a]) \\
 & \vdash & select([2]) \leftarrow select([4]) \\
X, I : \#int \mid I = X + X & \vdash & select([X]) \leftarrow select([I])
\end{array}
$$

and the new $\mathcal{G}$ contains also $select([2])$. We iteratively apply the same process, which yields the instantiation $I = 2$ and the equation $2 = X + X$. We get:

$$\begin{array}{rl} & \vdash\ select([a]) \vee select([4]) \\ & \vdash\ select([b,a]) \\ & \vdash\ select([2]) \leftarrow select([4]) \\ & \vdash\ select([1]) \leftarrow select([2]) \\ X, I : \#int \mid I = X + X & \vdash\ select([X]) \leftarrow select([I]) \end{array}$$

Now no new atom is added to $\mathcal{G}$ and we stop the grounding process for $m1$. The next step is to use $m1$ to ground $m2$. We just consider one of the grounding operations activated in this phase, that is the instantiation and simplification of $select(L), select(M)$ by $select([b,a])$ in the first clause of $m2$. We get $\cdots \mid \#reverse([b,a],[b,a]) \vdash show([b,a])$, where the body has been simplified because $select([b,a])$ is true. The evaluation of $\#reverse([b,a],[b,a])$ fails and the result is $\top$, i.e., no new clause is generated.

The main differences with respect to [6] are:

- Typing does not even consider ill typed substitutions.
- The treatment of external functions is shifted to the context. The context also contains the external atoms. Thus, the problem of grounding a clause in presence of predicates and operations defined in the grounding structure G is reduced to a constraint solving problem over G. The class of the groundable programs depends both on the grounding strategy and on the class of the constraints that the grounder can solve. For example, we could reconstruct value invention as treated in [5] in terms of programs that guarantee the solvability of a special class of constraints.

6 Related Work and Conclusions

The introduction of polymorphism in logic programming has required some care: there are predicate definitions that are semantically non-problematic, yet may lead to run-time errors [15,11]. A sufficient condition is *definitional genericity*, which requires that the type of a defining occurrence of a predicate must be a renaming of the assigned type signature of the predicate. This ensures the type soundness result of [18], stating that if a program and a goal are well-typed, then at each resolution step variables can only be instantiated to terms consistent with their typing.

Although our system owes its static semantics to Mycroft and O'Keefe's seminal paper, the very different operational semantics of ASP allows us to dispense of those restrictions – as a matter of fact a dual property to transparency is instrumental in inferring type indexes that ensure ground type reconstruction.

Some papers in the ASP literature, e.g. [9,2] use a many sorted language or deals with object-oriented features. Sorts, however, are merely syntactic sugar, to be translated back into first-order logic, hence they are not given a semantics in terms of many sorted first order signatures, not (polymorphic) type checking is addressed. W.r.t. object orientation OntoDLV [8] has recently extended DLV-Complex, catering to the specification and reasoning on enterprise ontologies; it extends ASP with object-oriented constructs, such as classes, objects, multiple inheritance, sets and lists. The semantics is based on a notion of well-defined interpretation of a program, which is defined over the Herbrand Universe of the program. Differently from our approach, it lacks a formal notion of "external" structure where to interpret external predicates, moreover "generics" are not considered.

Modern prescriptive types system for logic programming have significantly evolved since the Eighties: for example, Mercury's type system [14] is a fairly complex extension of [18], supporting higher order types as λProlog, but also type classes similar to Haskell, in addition with a form of existential types. Moreover its mode system [20] provides support for subtypes as well as for uniqueness, similar to linear types. We plan to investigate how modes can be used in our approach to deal with grounding, as we have touched upon in Example 13.

Finally we mention the extensive research about polymorphic type inference in relation algebra and (object-oriented) databases, see [7,4] for examples. However, this is only loosely related to our efforts, as there the aim is to extend a ML-like type system to capture the *principal* type of a program involving database operations or more in general of an expression in relational algebra, where a type is seen as a set of attribute names for a given schema.

In conclusion, we have presented a polymorphic types system for DLV-Complex and developed a declarative semantics for grounding, while studied its impact on model generation. Although our research is at an early stage, we believe that it shows the potential of types in ASP. A fortiori this should hold true for ASP systems such as Smodels [22] that implements a simpler logic. Beside the well-known advantages of prescriptive typing, in the ASP setting one may reap significant benefits w.r.t. the size of grounding; finally, we have shown that certain constraints can be directly codified in terms of typing, further reducing the size of the search space. This fits with current research aimed at finding a tighter relationship between ASP and constraint solving [2]. Directions for future research include:

- quantitatively evaluate the efficiency of type-driven grounding in terms of size and number of non-isomorphic models;
- extend the type system with other constructs, such as aggregate types, subtyping, etc., in order to account of other features of ASP, e.g. *preference* and *cardinality* constraints [16]. In particular, we speculate that we can elegantly define aggregate types using concrete data as in [3];
- From a practical standpoint, evaluate whether it is feasible to decompile the many-sorted language into untyped first order logic so that the present systems can readily be used or it is more effective to redesign grounding to take into full account the typing discipline.

Acknowledgments. This research was partially supported by the MIUR project "Potenziamento e Applicazioni della Programmazione Logica Disgiuntiva". We wish to thanks the anonymous referees for having appreciated the potential of the paper, notwithstanding several lacunae and imperfections. Those, we have done our best to fix.

References

1. Baral, C.: Knowledge Representation, Reasoning and Declarative Problem Solving. CUP (2003)
2. Baselice, S., Bonatti, P.A., Gelfond, M.: Towards an integration of answer set and constraint solving. In: Gabbrielli, M., Gupta, G. (eds.) ICLP 2005. LNCS, vol. 3668, pp. 52–66. Springer, Heidelberg (2005)
3. Bertoni, A., Mauri, G., Miglioli, P.: A characterization of abstract data as model-theoretic invariants. In: Maurer, H.A. (ed.) ICALP 1979. LNCS, vol. 71, pp. 26–37. Springer, Heidelberg (1979)
4. Buneman, P., Ohori, A.: Polymorphism and type inference in database programming. ACM Trans. Database Syst. 21(1), 30–76 (1996)
5. Calimeri, F., Cozza, S., Ianni, G.: External sources of knowledge and value invention in logic programming. Ann. Math. Artif. Intell. 50(3-4), 333–361 (2007)
6. Calimeri, F., Cozza, S., Ianni, G., Leone, N.: Computable functions in ASP: Theory and implementation. In: de la Banda, M.G., Pontelli, E. (eds.) ICLP 2008. LNCS, vol. 5366, pp. 407–424. Springer, Heidelberg (2008)
7. Van den Bussche, J., Waller, E.: Polymorphic type inference for the relational algebra. J. Comput. Syst. Sci. 64(3), 694–718 (2002)
8. Ricca, F., Gallucci, L., Schindlauer, R., Dell'Armi, T., Grasso, G., Leone, N.: OntoDLV: an ASP-based system for enterprise ontologies. Journal of Logic and Computation (to appear, 2009)
9. Gelfond, M.: Answer sets. In: van Harmelen, F., Lifschitz, V., Porter, B. (eds.) Handbook of knowledge representation, vol. 7. Elsevier, Amsterdam (2007)
10. Gelfond, M., Lifschitz, V.: The stable model semantics for logic programming. In: ICLP/SLP, pp. 1070–1080 (1988)
11. Hanus, M.: Horn clause programs with polymorphic types: Semantics and resolution. Theor. Comput. Sci. 89(1), 63–106 (1991)
12. Hermenegildo, M.V., Puebla, G., Bueno, F., López-García, P.: Integrated program debugging, verification, and optimization using abstract interpretation (and the Ciao system preprocessor). Sci. Comput. Programming 58(1-2), 115–140 (2005)
13. Hill, P.M., Topor, R.W.: A semantics for typed logic programs. In: Types in Logic Programming, pp. 1–62. MIT Press, Cambridge (1992)
14. Jefferey, D.: Expressive Type Systems for Logic Programming Languages. PhD thesis, The University of Melbourne (2002)
15. Lakshman, T.L., Reddy, U.S.: Typed PROLOG: A semantic reconstruction of the Mycroft-O'Keefe type system. In: ISLP, pp. 202–217 (1991)
16. Leone, N., et al.: The DLV system for knowledge representation and reasoning. ACM TOCL 7(3), 499–562 (2006)
17. Lloyd, J.W.: Foundations of logic programming (2nd extended ed.). Springer, Heidelberg (1987)
18. Mycroft, A., O'Keefe, R.A.: A polymorphic type system for PROLOG. Artif. Intell. 23(3), 295–307 (1984)
19. Nadathur, G., Qi, X.: Optimizing the runtime processing of types in polymorphic logic programming languages. In: Sutcliffe, G., Voronkov, A. (eds.) LPAR 2005. LNCS, vol. 3835, pp. 110–124. Springer, Heidelberg (2005)
20. Overton, D., Somogyi, Z., Stuckey, P.J.: Constraint-based mode analysis of Mercury. In: PPDP, pp. 109–120. ACM, New York (2002)
21. Pfenning, F.: Types in logic programming. MIT Press, Cambridge (1992)
22. Simons, P., Niemelä, I., Soininen, T.: Extending and implementing the stable model semantics. Artif. Intell. 138(1-2), 181–234 (2002)

A Syntax and Semantics of DLV

In this section we provide the bare minimum of definitions of the syntax and semantics of DLV [16] (disjunctive Datalog) and we briefly describe its extension with functions [5].

A *term* is either a variable or a constant. An *atom* a is an expression $p(t_1,\ldots,t_n)$, where p is a predicate of arity n and $t_1,\ldots,t_n$ are terms. A *classical literal* is an atom of the form a or $\neg a$, where $\neg$ is the classical negation. A *literal* l has the form l or *not* l, where l is a classical literal and *not* represents the *negation as failure* [17]. A *disjunctive rule* r is a formula of the form

$$A_1 \vee \cdots \vee A_n \leftarrow B_1,\ldots,B_k, not\ B_{k+1},\ldots,not\ B_m$$

where $A_1,\ldots,A_n,B_1,\ldots,B_m$ are classical literals, $n \geq 0$ and $m \geq k \geq 0$. The disjunction $A_1 \vee \cdots \vee A_n$ is the *head* of r, the conjunction $B_1,\ldots,B_k, not\ B_{k+1},\ldots,not\ B_m$ is the *body* of r. We use the following notations: $HD(r) = \{A_1,\ldots,A_n\}$ (the *head literals*), $BD^+(r) = \{B_1,\ldots,B_k\}$ (the *positive body*), $BD^-(r) = \{B_{k+1},\ldots,B_m\}$ (the *negative body*). A rule with empty head (i.e, $n = 0$) is called a *constraint*, a rule with empty body (i.e., $k = m = 0$) is called a *fact*. A rule r is *safe* if any variable occurring in r also appears in $BD^+(r)$. A *DLV program* is a finite set of safe rules. We recall some basics of logic programming: The Herbrand Universe U_P of P is the set of all constants occurring in P (we assume $U_P \neq \emptyset$). The Herbrand Base B_P of P is the set of all the literals constructible from the predicate symbols of P and the constants of U_P. For every rule r of P, $grnd(r)$ denotes the set of rules obtained by applying all possible substitutions from the variables in r to elements of U_P; $grnd(P)$ is the union of the sets $grnd(r)$, for every r of P. An *interpretation* I of P is a consistent subset of B_P, i.e. I does not contain pairs of classical literals of the form a and $\neg a$. Let P be a positive program (namely, for every rule r of P, $BD^-(r) = \emptyset$) and let $I \subseteq B_P$ be an interpretation.

- I is *closed under* P if, for every $r \in grnd(P)$, $BD^+(r) \subseteq I$ implies $HD(r) \cap I \neq \emptyset$.
- I is an *answer set* for P if I is minimal under set inclusion among all the interpretations closed under P.

The *reduct* or *Gelfond-Lifschitz* transformation of a ground program P with respect to an interpretation $I \subseteq B_P$ is the positive ground program P^I obtained from P by:

- deleting all rules $r \in P$ such that $BD^-(r) \cap I \neq \emptyset$;
- deleting the negative body from the remaining rules.

An *answer set* for a program P is an interpretation $I \subseteq B_P$ such that I is an answer set of $grnd(P)^I$.

Recently, DLV has been extended to DLV-Complex by introducing functions, sets and lists. This is obtained through the concept of *value invention* [5], which is based on the possibility of using externally defined predicates in the body of clauses. External predicates must be (well) moded, e.g. `#p(i,o)` denotes a call to an external predicates p(c,X) with ground input c that will "invent" an output value for X. The semantics of DLV programs has to be extended in order to treat external predicates, which are interpreted by means of external oracles. A DLV-Complex program with external predicates is required to be safe and *VI-restricted* [5]. In this case, it is guaranteed that the grounding process will halt with a finite ground program whose answer sets are the expected ones. A function symbol f(X) is introduced in DLV-Complex by associating with it a constructor predicate #f(o,i) and a destructor predicate #f(i,o). Given a value c, #f(V,c) invents a value v representing the ground term f(c), while #f(v,X) reconstructs the value c.

Type Inference for a Polynomial Lambda Calculus*

Marco Gaboardi and Simona Ronchi Della Rocca

Dipartimento di Informatica
Università degli Studi di Torino
Corso Svizzera 185, 10149 Torino, Italy
{gaboardi,ronchi}@di.unito.it

Abstract. We study the type inference problem for the Soft Type Assignment system (STA) for λ-calculus introduced in [1], which is correct and complete for polynomial time computations. In particular we design an algorithm which, given in input a λ-term, provides all the constraints that need to be satisfied in order to type it. For the propositional fragment of STA, the satisfiability of the constraints is decidable. We conjecture that, for the whole system, the type inference is undecidable, but our algorithm can be used for checking the typability of some particular terms.

1 Introduction

In [1], we have introduced a type assignment system for λ-calculus, named STA (Soft Type Assignment), inspired by the Soft Linear Logic of Lafont [2], which characterizes the polynomial time computations, in the sense that a well typed term can be reduced to normal form in a number of β-reduction steps which is polynomial in its size, and moreover all polynomial time functions can be represented by well typed terms, through an appropriate coding. In this paper we approach the problem of type inference in STA. In the simple types setting, type inference is decidable, and it corresponds to the property of having a principal typing, i.e., a typing for a term from which all (and only) the types derivable for the term itself can be built, through a substitution. STA has both modal and second order types, so the type inference is more difficult to be studied in this setting. We approach the problem in two steps, first for the propositional fragment and then for the full system. In both cases we need the notion of *type scheme*, which is an abstract representation of a set of types. Namely types can be obtained from type schemes through an operation of substitution. A notion of type scheme, for reasoning about type inference, was introduced first in [3] in the setting of intersection types, and it has been used, in different forms, for second order type inference [4], and for modal type inference [5]. We prove that, in propositional case, the type inference for STA is decidable. We introduce an algorithm which, given a term M, generates a triple $\mathbf{\Pi}(\mathtt{M}) = \langle \Psi, U, \mathcal{H} \rangle$, where U is a

* Paper partially supported by MIUR-Cofin'07 CONCERTO Project.

S. Berardi, F. Damiani, and U. de'Liguoro (Eds.): TYPES 2008, LNCS 5497, pp. 136–152, 2009.

type scheme, Ψ is a context assigning type schemes to the free variables of M, and $\mathcal{H} = \langle \mathcal{P}, \mathcal{C} \rangle$ is a pair of constraint sets. The constraint set $\mathcal{P}$ is a unification set of type schemes while $\mathcal{C}$ is a set of (in)equalities between exponentials. Informally $\mathcal{P}$ represents the conditions on the terms functionality, while $\mathcal{C}$ represents the conditions on the modalities. A pair of constraint sets is *satisfied* if the unification in $\mathcal{P}$ succeeds and moreover there is a substitution replacing exponentials by natural numbers in $\mathcal{C}$, in such a way that the (in)equalities become true. The algorithm is correct and complete, in the sense that M can be typed only in case the sets of constraints can be satisfied, and moreover all the typings for M can be built from Ψ and U through substitutions satisfying them. Since the satisfiability of the constraints is decidable in polynomial time, the type inference is decidable in polynomial time too.

Then we extend our study to second order types. We define an algorithm showing all the conditions that must be satisfied in order to type a term in the system. Namely, when applied to a term M, the algorithm produces as output a type scheme, a type scheme context, and five sets of constraints $\mathcal{G}, \mathcal{F}, \mathcal{Q}, \mathcal{P}$ and $\mathcal{C}$, where, $\mathcal{P}, \mathcal{C}$ are as in the propositional case, $\mathcal{G}$ is a semi-unification set of type schemes, and $\mathcal{F}$ and $\mathcal{Q}$ represent the conditions on the quantified abstracted variables. Also in this case the algorithm is correct and complete, but we conjecture that the satisfiability of the second order constraints is undecidable. We think the proof of Wells of undecidability of typability in System F adapts also in this case [6]. In any case, the algorithm is quite useful for checking the typability in particular cases, and in fact we use it for building two terms, the first one typable in System F but untypable in STA, and the second one typable in STA and not typable in DLAL [7], which is an alternative polynomial type assignment inspired by Light Affine Logic [8].

The paper is organized as follows. In Section 2 we introduce the type assignment system STA, and we recall its properties. In Section 3 we present the type inference algorithm for the propositional fragment and we prove it correct and complete. Moreover in Section 4 we discuss its complexity. In Section 5 we extend the analysis to second order types. Finally Section 6 contains a short conclusion.

2 The System STA

In this section we introduce the type assignment system STA, and we show its properties. STA is presented in a version which is slightly different from the presentation given in [1]. The difference is only in the management of contexts, in [1] contexts were sets of type assignments, here instead they are multisets of type assignments. This version is clearly equivalent to the original one [9], preserving the complexity properties, but it makes easier the design of the type inference algorithm.

Definition 1

i) The set $\mathbf{T}$ *of* soft types *is defined as follows:*

$$A, B, C ::= \alpha \mid \sigma \multimap A \mid \forall\alpha.A \text{ (Linear Types)} \qquad \sigma, \tau ::= A \mid !\sigma$$

Table 1. STA in the multiset version

$$\frac{}{\mathtt{x} : A \vdash \mathtt{x} : A}\ (Ax) \qquad \frac{\mathfrak{A} \vdash \mathtt{M} : \sigma}{\mathfrak{A}, \mathtt{x} : A \vdash \mathtt{M} : \sigma}\ (w) \qquad \frac{\mathfrak{A}, (\mathtt{x} : \tau)^{(r)} \vdash \mathtt{M} : \sigma}{\mathfrak{A}, \mathtt{x} :!\tau, \vdash \mathtt{M} : \sigma}\ (m)$$

$$\frac{\mathfrak{A} \vdash \mathtt{M} : \sigma \multimap A \quad \mathfrak{B} \vdash \mathtt{N} : \sigma \quad \mathfrak{A} \approx \mathfrak{B}}{\mathfrak{A}, \mathfrak{B} \vdash \mathtt{MN} : A}\ (\multimap E) \qquad \frac{\mathfrak{A} \vdash \mathtt{M} : \forall \alpha.A}{\mathfrak{A} \vdash \mathtt{M} : A[B/\alpha]}\ (\forall E)$$

$$\frac{\mathfrak{A}, \mathtt{x} : \sigma \vdash \mathtt{M} : A \quad \mathtt{x} \notin \mathrm{dom}(\mathfrak{A})}{\mathfrak{A} \vdash \lambda \mathtt{x}.\mathtt{M} : \sigma \multimap A}\ (\multimap I) \qquad \frac{\mathfrak{A} \vdash \mathtt{M} : A \quad \alpha \notin \mathrm{FTV}(\mathfrak{A})}{\mathfrak{A} \vdash \mathtt{M} : \forall \alpha.A}\ (\forall I) \qquad \frac{\mathfrak{A} \vdash \mathtt{M} : \sigma}{!\mathfrak{A} \vdash \mathtt{M} :!\sigma}\ (sp)$$

where α, β range over a countable set of type variables. $\equiv$ denotes the syntactical identity between types.

ii) *A context $\mathfrak{A}$ is a finite multiset of type assignments of the shape $\mathtt{x} : \sigma$, such that if $\mathtt{x} : \sigma_1 \in \mathfrak{A}$ and $\mathtt{x} : \sigma_2 \in \mathfrak{A}$ then there exists $A \in \mathbf{T}$ and $n, m \in \mathbb{N}$ such that $\sigma_1 \equiv \underbrace{!...!}_{n} A$ and $\sigma_2 \equiv \underbrace{!...!}_{m} A$. Contexts are ranged over by $\mathfrak{A}, \mathfrak{B}, \mathfrak{C}$. When a context is a set we denote it by Γ, Δ.*

iii) STA *proves statements of the shape $\mathfrak{A} \vdash \mathtt{M} : \sigma$ where $\mathfrak{A}$ is a context, $\mathtt{M}$ is a term of λ-calculus, and σ is a type. The rules of the system are given in Table 1. The term $\mathtt{M}$ is* typable *in* STA *if there is a context $\mathfrak{A}$ and a type σ such that $\mathfrak{A} \vdash \mathtt{M} : \sigma$.*

As usual $\multimap$ associates to the right and has precedence on $\forall$, while ! has precedence on everything else. $\mathrm{FTV}(\sigma)$ denotes the set of free type variables of the type σ. $B[A/\alpha]$ denotes the capture free substitution of all occurrences of the type variable α by the linear type A: note that this kind of substitution preserves the correct syntax of types. $\forall \boldsymbol{\alpha}.A$ shortens $\forall \alpha_1 ... \alpha_n.A$ for $n \geq 0$. Two contexts $\mathfrak{A}$ and $\mathfrak{B}$ are *coherent*, denoted $\mathfrak{A} \approx \mathfrak{B}$, if and only if their multiset union $\mathfrak{A}, \mathfrak{B}$ is a context. Let $\mathfrak{A} = \{\mathtt{x}_1 : \sigma_1, \ldots, \mathtt{x}_n : \sigma_n\}$ then $\mathrm{dom}(\mathfrak{A}) = \{\mathtt{x}_1, \ldots, \mathtt{x}_n\}$, $!\mathfrak{A} = \{\mathtt{x}_1 :!\sigma_1, \ldots, \mathtt{x}_n :!\sigma_n\}$ and $\mathrm{FTV}(\mathfrak{A}) = \{\alpha \in \mathrm{FTV}(\sigma) \mid \mathtt{x} : \sigma \in \mathfrak{A}\}$. $\Sigma \rhd \mathfrak{A} \vdash \mathtt{M} : \sigma$ denotes that there is a derivation Σ proving $\mathfrak{A} \vdash \mathtt{M} : \sigma$. $|\mathtt{M}|$ denotes the number of sybols in $\mathtt{M}$.

Hygiene condition. We assume that free and bound type variables have different names, and also type variables bounded by different quantifiers are named differently.

Theorem 2 (Complexity of STA **[1])**

i) (Soundness) *If $\Pi \rhd \Gamma \vdash \mathtt{M} : \sigma$, then $\mathtt{M}$ can be evaluated to normal form in a number of β-reduction steps $O(|\mathtt{M}|^{(\mathtt{d}(\Pi)+1)})$, where $\mathtt{d}(\Pi)$ is the maximum nesting of rules (sp) in Π.*

ii) (Completeness) *Every polynomial time function can be encoded by a term typable in* STA.

3 Type Inference for the Propositional Fragment

As we said in the introduction, if we restrict ourselves to consider just the propositional fragment, the type inference is decidable. In this section we will show the type inference algorithm, which is based on the notion of type scheme and a unification procedure for type schemes.

Table 2. Unification Algorithm

$$\frac{}{\mathtt{U}(a,a)=\langle\emptyset,\emptyset\rangle}\ (\mathtt{U}_1) \qquad \frac{\mathtt{U}(\phi,\psi)=\langle\mathcal{P}_1,\mathcal{C}_1\rangle \quad \mathtt{U}(U,V)=\langle\mathcal{P}_2,\mathcal{C}_2\rangle}{\mathtt{U}(\phi\multimap U,\psi\multimap V)=\langle\mathcal{P}_1\cup\mathcal{P}_2,\mathcal{C}_1\cup\mathcal{C}_2\rangle}\ (\mathtt{U}_4)$$

$$\frac{a\notin \mathrm{FV}(U)}{\mathtt{U}(a,U)=\langle\{a=U\},\emptyset\rangle}\ (\mathtt{U}_2) \qquad \frac{\mathtt{U}(\phi\multimap U,V)=\langle\mathcal{P},\mathcal{C}\rangle}{\mathtt{U}(\phi\multimap U,!^pV)=\langle\mathcal{P},\mathcal{C}\cup\{p=0\}\rangle}\ (\mathtt{U}_5)$$

$$\frac{\mathtt{U}(a,V)=\langle\mathcal{P},\mathcal{C}\rangle}{\mathtt{U}(a,!^pV)=\langle\mathcal{P},\mathcal{C}\cup\{p=0\}\rangle}\ (\mathtt{U}_3) \qquad \frac{\mathtt{U}(\psi,\phi)=\langle\mathcal{P},\mathcal{C}\rangle}{\mathtt{U}(\phi,\psi)=\langle\mathcal{P},\mathcal{C}\rangle}\ (\mathtt{U}_6) \qquad \frac{\mathtt{U}(U,V)=\langle\mathcal{P},\mathcal{C}\rangle}{\mathtt{U}(!^pU,!^qV)=\langle\mathcal{P},\mathcal{C}\cup\{p=q\}\rangle}\ (\mathtt{U}_7)$$

3.1 Type Schemes, Substitutions and Constraints

Definition 3. Linear type schemes *and* type schemes *are respectively defined by the grammars*

$$U,V,Z ::= a \mid \phi\multimap U \qquad\qquad \phi,\psi,\xi ::= U \mid !^pU$$

where the exponential p,q,r *belong to a countable set,* a,b,c,d *belong to a countable set of* linear scheme variables. $\mathcal{T}$ *denotes the set of type schemes.*

$\mathrm{FV}(\phi)$ is the set of all linear scheme variables and exponentials occurring in ϕ. Two type schemes ϕ,ψ are *disjoint* if $\mathrm{FV}(\phi)\cap\mathrm{FV}(\psi)=\emptyset$.

A *scheme substitution* s is a total function mapping linear scheme variables to linear types and exponentials to natural numbers. So a scheme substitution maps type schemes to types. The application of s to a type scheme is defined as

$$s(a)=A \text{ if } [a\mapsto A]\in s \qquad s(\phi\multimap U)=s(\phi)\multimap s(U)$$

$$s(!^pU)=\underbrace{!...!}_{n}\,s(U) \text{ if } [p\mapsto n]\in s$$

In what follows, $s[a_1\mapsto\tau_1,\dots,a_n\mapsto\tau_n]$ denotes the scheme substitution defined as s except on variables $a_1,\dots,a_n$ to which it assigns $\tau_1,\dots\tau_n$.

A type scheme can be seen as an abstract representation of the set of types that can be obtained from it through a scheme substitution. For example, a represents the set of all linear types, while $!^p(a\multimap b)$ represents the set of types $\{\underbrace{!...!}_{n}(A\multimap B)\mid A,B \text{ are linear types and } n\geq 0\}$.

Two type schemes ϕ and ψ can be *unified* if there is a scheme substitution s such that $s(\phi)\equiv s(\psi)$.

A *type scheme context* is a multiset of *variable type scheme assignments* of the shape $\mathtt{x} : \phi$ where $\mathtt{x}$ is a variable and ϕ is a type scheme. Type scheme contexts are ranged over by Ψ, Φ. $\mathrm{dom}(\Psi)$ denotes the set of variables $\{\mathtt{x} \mid \mathtt{x} : \phi \in \Psi\}$. Multiset union of type scheme contexts is denoted by $\sqcup$. The expression $\Phi = \Phi' \odot \Psi$ denotes the fact that $\Phi = \Phi' \sqcup \Psi$ and $\mathrm{dom}(\Phi') \cap \mathrm{dom}(\Psi) = \emptyset$. Scheme substitutions are easily extended to type scheme contexts, *i.e.* if $\Psi = \mathtt{x}_1 : \phi_1, \ldots, \mathtt{x}_n : \phi_n$ then $s(\Psi) = \mathtt{x}_1 : s(\phi_1), \ldots, \mathtt{x}_n : s(\phi_n)$.

A *constraints sequence* $\mathcal{H}$ is a couple $\langle \mathcal{P}, \mathcal{C} \rangle$ of constraints sets. The set of *scheme variable constraints* $\mathcal{P}$ is a set of constraints of the shape $a = U$ where a is a linear scheme variable and U is a linear type scheme such that $a \notin \mathrm{FV}(U)$. The set of *exponentials constraints* $\mathcal{C}$ is a set of linear (in)equations of the shape $p = q, p \geq q, p \geq q_1 + q_2, p > q$ or $p = 0$. $\mathcal{H}_1 \uplus \mathcal{H}_2$ denotes the component-wise union of the constraints sequences $\mathcal{H}_1$ and $\mathcal{H}_2$. Sometimes we omit the empty set of a constraints sequence, i.e. $\mathcal{H} \uplus \{p = q\}$ denotes $\mathcal{H} \uplus \langle \emptyset, \{p = q\} \rangle$.

A scheme substitution s satisfies $\mathcal{H}$ if and only if $s(a) \equiv s(U)$, for every equation $a = U$ in $\mathcal{P}$, and $s(p)\ \mathtt{op}\ s(q)$, for every $p\ \mathtt{op}\ q$ in $\mathcal{C}$.

3.2 Unification Algorithm

In Table 2, we introduce the algorithm U, which allows to unify type schemes under some assumptions. U proves judgments of the shape

$$\mathtt{U}(\phi, \psi) = \mathcal{H}$$

where ϕ and ψ are two type schemes and $\mathcal{H} = \langle \mathcal{P}, \mathcal{C} \rangle$ is a constraint sequence representing the constraints under which ϕ and ψ can be unified. Namely $\mathcal{P}$ represents the constraints on the structure of the type schemes, while $\mathcal{C}$ the constraints on the number of modalities.

Note that rule ($\mathtt{U}_6$) keeps down the number of rules, nevertheless it can be cause of non termination (infinite derivations). It is easy (but boring) to give a different definition of the algorithm without the rule ($\mathtt{U}_6$), by making explicit a symmetric version of all the rules. In what follows we assume to use such an extended version of the algorithm. Note that some inputs does not admit a derivation, by rule ($\mathtt{U}_2$), in such a cases the unification fails. The following easy theorem assures a weak form of successful termination which will be useful in the sequel.

Theorem 4 (U Termination). *Let $\phi, \psi \in \mathcal{T}$ be disjoint. Then, there exists $\mathcal{H}$ such that $\mathtt{U}(\phi, \psi) = \mathcal{H}$.*

The algorithm U is correct and complete, as shown in the following.

Theorem 5 (U Correctness). *Let $\phi, \psi \in \mathcal{T}$. If $\mathtt{U}(\phi, \psi) = \mathcal{H}$ then for every scheme substitution s satisfying $\mathcal{H}$*

$$s(\phi) \equiv s(\psi)$$

Proof. By induction on the derivation of $\mathtt{U}(\phi, \psi) = \langle \mathcal{P}, \mathcal{C} \rangle$. Note that the existence of a scheme substitution satisfying the constraints in $\langle \mathcal{P}, \mathcal{C} \rangle$ is decidable. □

Theorem 6 (U Completeness). *If $s(\phi) \equiv s(\psi)$ then there exists $\mathcal{H}$ such that $\mathtt{U}(\phi, \psi) = \mathcal{H}$ and s satisfies $\mathcal{H}$.*

Proof. By induction on the shape of ϕ and ψ. □

3.3 The Algorithm

The type inference algorithm defined in Table 3 proves statement of the shape

$$\mathbf{\Pi}(\mathtt{M}) = \langle \Psi, U, \mathcal{H} \rangle$$

where Ψ is a type scheme context, U is a linear type scheme and $\mathcal{H}$ is a constraints sequence. The type inference algorithm uses the procedure `Unify`, defined in Table 4, that is just an extension of the unification algorithm U to type scheme contexts and type schemes.

It is worth noticing the difference between this algorithm and the type inference algorithm for simple types. The latter generates a principal typing, which is a typing for the input term, and for which all and only the typings derivable for the same term are derivable, through substitutions. If the input term cannot be typed, then the algorithm fails. In the current setting, our algorithm generates a sort of an abstract representation of all the typings for the input term M, in the sense that, if the constraint sequence $\mathcal{H}$ can be satisfied by a scheme substitution s, then $s(\Psi) \vdash \mathtt{M} : s(U)$ is a typing for M, and moreover all typings for M can be built from $\mathbf{\Pi}(\mathtt{M})$ by a scheme substitution satisfying $\mathcal{H}$, plus some applications of rules dealing with the modality. If the constraints are not satisfiable, then M cannot be typed.

Table 3. Type Inference Algorithm

$\mathbf{\Pi}(\mathtt{x}) = \mathtt{let}\ a, p\ \mathtt{be\ fresh\ in}\ \langle \{\mathtt{x} :!^p a\}, a, \{\emptyset, \emptyset\} \rangle$
$\mathbf{\Pi}(\lambda\mathtt{x.M}) = \mathtt{let}\ \mathbf{\Pi}(\mathtt{M}) = \langle \Psi, U, \mathcal{H} \rangle\ \mathtt{in}$ $\mathtt{let}\ \Psi = \Psi' \odot \{\mathtt{x} :!^{s_1} V_1, \ldots, \mathtt{x} :!^{s_n} V_n\}\mathtt{in\ let}\ a, r\ \mathtt{be\ fresh\ in}$ $\mathtt{if}\ n = 0\ \mathtt{then}\ \langle \Psi', !^r a \multimap U, \mathcal{H} \rangle$ $\mathtt{if}\ n = 1\ \mathtt{then}\ \langle \Psi', !^r V_1 \multimap U, \mathcal{H} \uplus \{r \geq s_1\} \rangle$ $\mathtt{if}\ n > 1\ \mathtt{then}\ \langle \Psi', !^r V_1 \multimap U, \mathcal{H} \uplus \{r > s_1, \ldots, r > s_n\} \rangle$
$\mathbf{\Pi}(\mathtt{MN}) = \mathtt{let}\ \mathbf{\Pi}(\mathtt{M}) = \langle \Psi_{\mathtt{M}}, U, \mathcal{H}_{\mathtt{M}} \rangle\ \mathtt{and}\ \mathbf{\Pi}(\mathtt{N}) = \langle \Psi_{\mathtt{N}}, V, \mathcal{H}_{\mathtt{N}} \rangle\ \mathtt{be\ disjoint\ in}$ $\mathtt{let}\ a, q_i, p\ \mathtt{be\ fresh\ in\ let}\ \Psi'_{\mathtt{N}} = \{\mathtt{z} :!^{q_i} V_i \mid \exists \mathtt{z} :!^{p_i} V_i \in \Psi_{\mathtt{N}}\},$ $\mathcal{H} = \mathtt{Unify}(\Psi_{\mathtt{M}}, \Psi'_{\mathtt{N}}, U, !^p V \multimap a)\ \mathtt{in}\ \langle \Psi_{\mathtt{M}} \sqcup \Psi'_{\mathtt{N}}, a, \mathcal{H}_{\mathtt{M}} \uplus \mathcal{H}_{\mathtt{N}} \uplus \mathcal{H} \uplus \{q_i \geq p_i + p\} \rangle$

We need to prove that the type inference algorithm is well defined.

Theorem 7 (Π Termination). *Let $\mathtt{M} \in \Lambda$. Then there exist Ψ, U and $\mathcal{H}$ such that $\mathbf{\Pi}(\mathtt{M}) = \langle \Psi, U, \mathcal{H} \rangle$.*

Proof. By induction on the structure of M, using Theorem 4.

The use of multisets instead of sets as contexts in STA helps in the design of the algorithm, maintaining the correcteness of typing. Note that in the definition of $\mathbf{\Pi}$, in the abstraction case we can freely take only the type scheme of the first occurrence (if any) of the variable to be abstracted since all the type schemes have already been unified. The same holds for the `Unify` procedure. We can now finally prove the main theorems of this section.

Theorem 8 ($\mathbf{\Pi}$ Correctness). *Let $\mathbf{\Pi}(\mathtt{M}) = \langle \Psi, U, \mathcal{H}\rangle$. Then, for each scheme substitution s satisfying $\mathcal{H}$,*

$$s(\Psi) \vdash \mathtt{M} : s(U)$$

Proof. By induction on the derivation proving $\mathbf{\Pi}(\mathtt{M}) = \langle \Psi, U, \mathcal{H}\rangle$. We will show just the most difficult case, when the term is of the shape PN. Consider the case $\mathbf{\Pi}(\mathtt{PN}) = \langle \Psi, U, \mathcal{H}\rangle$. By definition U is a scheme variable a, $\mathbf{\Pi}(\mathtt{P}) = \langle \Psi_{\mathtt{P}}, U_{\mathtt{P}}, \mathcal{H}_{\mathtt{P}}\rangle$, $\mathbf{\Pi}(\mathtt{N}) = \langle \Psi_{\mathtt{N}}, U_{\mathtt{N}}, \mathcal{H}_{\mathtt{N}}\rangle$ and they are all disjoint. Let s be a scheme substitution satisfying $\mathcal{H}$. Since s clearly satisfies $\mathcal{H}_{\mathtt{P}}$ and $\mathcal{H}_{\mathtt{N}}$ then by induction we have both $s(\Psi_{\mathtt{P}}) \vdash \mathtt{P} : s(U_{\mathtt{P}})$ and $s(\Psi_{\mathtt{N}}) \vdash \mathtt{N} : s(U_{\mathtt{N}})$. By definition $\mathtt{U}(U_{\mathtt{P}}, !^p U_{\mathtt{N}} \multimap a) = \mathcal{H}'$ with $\mathcal{H}' \subseteq \mathcal{H}$, so since s satisfies $\mathcal{H}$, by Theorem 5: $s(U_{\mathtt{P}}) \equiv s(!^p U_{\mathtt{N}}) \multimap s(a)$. Let $\Psi'_{\mathtt{N}} = \{\mathtt{z} :!^{q_i} V_i \mid \exists \mathtt{z} :!^{p_i} V_i \in \Psi_{\mathtt{N}}\}$. Then clearly $s(\Psi) = s(\Psi_{\mathtt{P}} \sqcup \Psi'_{\mathtt{N}}) = s(\Psi_{\mathtt{P}}), s(\Psi'_{\mathtt{N}})$. So, let $s(p) = k$. Then, the following derivation can be built

$$\dfrac{s(\Psi_{\mathtt{P}}) \vdash \mathtt{P} : s(!^p U_{\mathtt{N}}) \multimap s(a) \qquad \dfrac{\dfrac{s(\Psi_{\mathtt{N}}) \vdash \mathtt{N} : s(U_{\mathtt{N}})}{!^k s(\Psi_{\mathtt{N}}) \vdash \mathtt{N} :!^k s(U_{\mathtt{N}})}\,(sp)^k}{s(\Psi'_{\mathtt{N}}) \vdash \mathtt{N} :!^k s(U_{\mathtt{N}})}\,(m)^*}{s(\Psi_{\mathtt{P}}), s(\Psi'_{\mathtt{N}}) \vdash \mathtt{PN} : s(a)}\,(\multimap E)$$

□

Table 4. `Unify` procedure

$$\begin{array}{l} \mathtt{Unify}(\Phi, \Psi, \phi, \psi) = \mathtt{let}\ \mathtt{x}_1, \dots, \mathtt{x}_m = \mathrm{dom}(\Phi) \cap \mathrm{dom}(\Psi),\ \forall 1 \le i \le m \\ \qquad \Phi(\mathtt{x}_i) = \{!^{s_1} a_1, \dots, !^{s_n} a_n\},\ \Psi(\mathtt{x}_i) = \{!^{r_1} b_1, \dots, !^{r_k} b_k\}, \\ \qquad \mathtt{U}(\phi, \psi) = \langle \mathcal{P}_0, \mathcal{C}_0\rangle,\ \mathtt{U}(a_1, b_1) = \langle \mathcal{P}_i, \mathcal{C}_i\rangle \\ \qquad \mathtt{in}\langle \bigcup_{j=0}^{m} \mathcal{P}_j, \bigcup_{j=0}^{m} \mathcal{C}_j\rangle, \end{array}$$

Theorem 9 ($\mathbf{\Pi}$ Completeness). *Let $\mathbf{\Pi}(\mathtt{M}) = \langle \Psi, U, \mathcal{H}\rangle$. If $\mathfrak{A} \vdash \mathtt{M} : \sigma$, then there exists a scheme substitution s satisfying $\mathcal{H}$ such that*

$$\Sigma \rhd s(\Psi) \vdash \mathtt{M} : s(U)$$

Moreover, the sequent $\mathfrak{A} \vdash \mathtt{M} : \sigma$ can be obtained from Σ by a (maybe empty) sequence of applications of the rules (w), (m) and (sp).

Proof. By induction on the derivation Π proving $\mathfrak{A} \vdash \mathtt{M} : \sigma$. We will show just the case where Π ends as

$$\dfrac{\Sigma' \rhd \mathfrak{A} \vdash \mathtt{N} : \sigma \multimap A \qquad \Theta' \rhd \mathfrak{B} \vdash \mathtt{P} : \sigma \qquad \mathfrak{A} \approx \mathfrak{B}}{\mathfrak{A}, \mathfrak{B} \vdash \mathtt{NP} : A}\,(\multimap E)$$

Let $\mathbf{\Pi}(\mathtt{NP}) = \langle \Psi, U, \mathcal{H}\rangle$. Then, there are disjoint $\mathbf{\Pi}(\mathtt{N}) = \langle \Psi_{\mathtt{N}}, U_{\mathtt{N}}, \mathcal{H}_{\mathtt{N}}\rangle$, $\mathbf{\Pi}(\mathtt{P}) = \langle \Psi_{\mathtt{P}}, U_{\mathtt{P}}, \mathcal{H}_{\mathtt{P}}\rangle$. By induction, there are scheme substitutions $s_{\mathtt{N}}$ and $s_{\mathtt{P}}$ satisfying respectively $\mathcal{H}_{\mathtt{N}}$ and $\mathcal{H}_{\mathtt{P}}$, such that $\Sigma'' \rhd s_{\mathtt{N}}(\Psi_{\mathtt{N}}) \vdash \mathtt{N} : s_{\mathtt{N}}(U_{\mathtt{N}})$ and $\Theta'' \rhd s_{\mathtt{P}}(\Psi_{\mathtt{P}}) \vdash \mathtt{P} : s_{\mathtt{P}}(U_{\mathtt{P}})$ and Σ' and Θ' can be obtained respectively from Σ'' and Θ'' by some applications of the rules (w), (m) and/or (sp).

Since $\mathcal{H}_{\mathtt{N}}$ and $\mathcal{H}_{\mathtt{P}}$ are disjoint, we can build a scheme substitution s' satisfying both, just acting as each one of the previous substitutions on the corresponding domain. By definition of $\mathbf{\Pi}$, $\Psi = \Psi_{\mathtt{N}} \sqcup \Psi'_{\mathtt{P}}$ where, if $\Psi_{\mathtt{P}} = \mathtt{x}_1 :!^{p_1} V_1, \ldots, \mathtt{x}_n :!^{p_n} V_n$, then $\Psi'_{\mathtt{P}} = \mathtt{x}_1 :!^{q_1} V_1, \ldots, \mathtt{x}_n :!^{q_n} V_n$ for fresh $q_1, \ldots, q_n$. Moreover, for fresh a and p, if $\mathtt{Unify}(\Psi_{\mathtt{N}}, \Psi'_{\mathtt{P}}, U_{\mathtt{N}}, !^p U_{\mathtt{P}} \multimap a) = \mathcal{H}'$ then $U \equiv a$, and $\mathcal{H} = \mathcal{H}_{\mathtt{N}} \uplus \mathcal{H}_{\mathtt{P}} \uplus \mathcal{H}' \uplus \{q_i \geq p_i + p\}$. Since a and p are fresh, we can choose s' satisfying also $s'(U_{\mathtt{N}}) \equiv \sigma \multimap A \equiv s'(!^p U_{\mathtt{P}} \multimap a)$. Hence in particular by Theorem 6 s' satisfies $\mathcal{H}'$. Moreover, since $q_1, \ldots, q_n$ are fresh, it is easy to extend s' to a scheme substitution $s = s'[q_1 \mapsto s(p_1) + s(p), \ldots, q_n \mapsto s(p_n) + s(p)]$. Clearly s satisfies $\mathcal{H}$. Let $s(p) = k$. Then we can build the following derivation

$$\dfrac{\Sigma' \rhd s(\Psi_{\mathtt{N}}) \vdash \mathtt{N} : s(U_{\mathtt{N}}) \qquad \dfrac{\Theta' \rhd s(\Psi_{\mathtt{P}}) \vdash \mathtt{P} : s(U_{\mathtt{P}})}{!^k s(\Psi_{\mathtt{P}}) \vdash \mathtt{P} :!^k s(U_{\mathtt{P}})}\,(sp)^k}{s(\Psi_{\mathtt{N}}), !^k s(\Psi_{\mathtt{P}}) \vdash \mathtt{NP} : s(\forall t.a)}\,(\multimap E)$$

and $\mathfrak{A}, \mathfrak{B} \vdash \mathtt{NP} : A$ can be obtained from it by a sequence of applications of the rules (w), (m) and (sp). □

In the following we will give some examples, and in the next section we will discuss the constraints resolution in them. These example are useful both to understand the behaviour of the algorithm and to compare the typability power of STA and other type assignment systems. Namely the first term (2) is typable in STA and in simple type assignment system, the second term (222) is typable in the simple type assignment system but untypable in STA, and the third one (2(yz)) is typable in STA but untypable in the propositional fragment of DLAL.

Example 10

1. Let $2 \equiv \lambda\mathtt{s}.\lambda\mathtt{z}.\mathtt{s}(\mathtt{sz})$. Then $\mathbf{\Pi}(2) = \langle \emptyset, U, \langle \mathcal{P}, \mathcal{C}\rangle\rangle$ where

$$U = !^{r_5} a_2 \multimap (!^{r_4} a_1 \multimap b_2)$$
$$\mathcal{P} = \{a_2 = !^{q_1} a_1 \multimap b_1, a_3 = !^{q_2} b_1 \multimap b_2, a_2 = a_3\}$$
$$\mathcal{C} = \{r_1 \geq p_1 + q_1, r_2 = p_3, r_2 \geq p_2 + q_2, r_3 \geq r_1 + q_2, r_4 \geq r_3, r_5 > p_3\}\rangle$$

2. A more involved example is related to the term 222. Then, we obtain $\mathbf{\Pi}(222) = \langle \emptyset, U, \langle \mathcal{P}, \mathcal{C}\rangle\rangle$ where

$$U = a''$$
$$\mathcal{P} = \mathcal{P}^0 \cup \mathcal{P}^1 \cup \mathcal{P}^2 \cup \{a_2^1 = !^{r_5^2} a_2^2 \multimap (!^{r_4^2} a_1^2 \multimap b_2^2), a' = !^{r_4^1} a_1^1 \multimap b_2^1,$$
$$a' = !^{p''}(!^{r_5^0} a_2^0 \multimap (!^{r_4^0} a_1^0 \multimap b_2^0)) \multimap a''\ \}$$
$$\mathcal{C} = \mathcal{C}^0 \cup \mathcal{C}^1 \cup \mathcal{C}^2 \cup \{p' = r_5^1\}$$

and $\mathcal{P}^i = \{a_2^i = !^{q_1^i} a_1^i \multimap b_1^i, a_3^i = !^{q_2^i} b_1^i \multimap b_2^i, a_2^i = a_3^i\}$ while $\mathcal{C}^i = \{r_1^i \geq p_1^i + q_1^i, r_2^i = p_3^i, r_2^i \geq p_2^i + q_2^i, r_3^i \geq r_1^i + q_2^i, r_4^i \geq r_3^i, r_5^i > p_3^i\}$.

3. Let us consider now the term 2(yz). The application of the algorithm produces: $\mathbf{\Pi}(2(\mathtt{yz})) = \langle \{\mathtt{y} :!^r c, \mathtt{z} :!^s d\}, U, \langle \mathcal{P}^*, \mathcal{C}^* \rangle \rangle$, where:

$$\begin{aligned} U &= f \\ \mathcal{P}^* &= \mathcal{P} \cup \{a_2 = e, c =!^t d \multimap e, f =!^{r_4} a_1 \multimap b_2\} \\ \mathcal{C}^* &= \mathcal{C} \cup \{r_5 = t', r \geq r' + t', s \geq s' + t', s' \geq t + s''\} \end{aligned}$$

where $\mathcal{P}$ and $\mathcal{C}$ are defined as in point 1 of this example.

4 Constraints Resolution

Let $\mathbf{\Pi}(\mathtt{M}) = \langle \Psi, U, \mathcal{H} \rangle$, where $\mathcal{H} = \langle \mathcal{P}, \mathcal{C} \rangle$. The resolution of the constraints in $\mathcal{H}$ is splitted in two phases. The first one is the application of the standard Robinson resolution [10] to $\mathcal{P}$, so obtaining a new set of constraints, that can be in its turn splitted in a set $\mathcal{P}'$ of constraints on schemes, and $\mathcal{C}'$ of constraints on exponentials. Then the second phase is to find a scheme substitution satisfying the constraints $\mathcal{P}'$ and $\mathcal{C} \cup \mathcal{C}'$. Some examples can clarify the procedure.

Example 11

1. Let us continue Example 10.1, i.e., $\mathbf{\Pi}(2)$. Then, the application of the Robinson resolution to the set $\mathcal{P}$ and $\mathcal{C}$ generates $\mathcal{P}' = \{a_2 =!^{q_1} a_1 \multimap b_1, a_1 = b_1, b_1 = b_2, a_2 = a_3\}$ and $\mathcal{C}' = \{q_1 = q_2\}$ respectively. The substitution

$$s = s'[a_1, b_1, b_2 \mapsto \alpha; a_2, a_3 \mapsto \alpha \multimap \alpha; p_1, p_2, p_3, q_1, q_2, r_1, r_2, r_3, r_4 \mapsto 0; r_5 \mapsto 1]$$

satisfies the constraints $\mathcal{P}'$, $\mathcal{C}$, $\mathcal{C}'$ for all s', and generates the typing $\emptyset \vdash 2 : !(\alpha \multimap \alpha) \multimap \alpha \multimap \alpha$. Hence the term is typable.

2. Let us continue Example 10.2, i.e., $\mathbf{\Pi}(222)$. The application of the Robinson resolution is boring but easy, applied to $\mathcal{P}$ it produces a solvable set of constraints on type schemes and the final type is defined through the type scheme equation

$$a'' =!^{q_1^2} (!^{q_1^0} a_1^0 \multimap a_1^0) \multimap (!^{q_1^0} a_1^0 \multimap a_1^0).$$

But Robinson algorithm changes also the set of constraints on exponentials $\mathcal{C}$ into the set $\mathcal{C}' = \mathcal{C} \cup \{q_1^0 = r_4^0, q_1^2 = r_5^0, r_4^1 = p'', q_1^2 = r_4^2, q_1^2 = q_2^2, q_1^1 = q_2^1, q_1^0 = q_2^0, q_1^1 = r_5^2\}$ which can be simplified in $\mathcal{C}'' = \{r_5^2 > r_2^2 \geq p_2^2, r_1^1 \geq p_1^1 + q_1^1, p' > r_2^1 \geq p_2^1 + q_1^1, r_4^1 \geq r_3^1 \geq r_1^1 + q_1^1, q_1^2 > r_2^0, q_1^2 = 0\}$. This set is clearly not satisfiable since the last two constraints are contradictory, and so the term is not typable.

3. Let us continue Example 10.3, i.e. $\mathbf{\Pi}(2(\mathtt{yz}))$. The application of Robinson resolution to $\mathcal{P}^*$ gives a set $\mathcal{P}^{**} = \mathcal{P}' \cup \{c =!^t d \multimap e, a_2 = e, f =!^{r_4} a_1 \multimap b_2\}$ and the set of exponential constraints becomes $\mathcal{C}^{**} = \mathcal{C}' \cup \mathcal{C} \cup \{r_5 = t', r \geq r' + t', s \geq s' + t', s' \geq t + s''\}$, where $\mathcal{P}'$and $\mathcal{C}'$ are defined as at point 1 of this example while $\mathcal{C}$ is defined as in Example 10.1. Let s be the substitution at point 1 of this example; then the substitution:

$$s^* = s[c \mapsto (\alpha \multimap \alpha) \multimap \alpha \multimap \alpha; d \mapsto \alpha \multimap \alpha; r', s', t, s'' \mapsto 0; r, s, t' \mapsto 1]$$

satisfies the constraints in $\mathcal{P}^{**}$ and $\mathcal{C}^{**}$ and generates the typing:

$$\mathtt{y} :!((\alpha \multimap \alpha) \multimap \alpha \multimap \alpha), \mathtt{z} :!(\alpha \multimap \alpha) \vdash 2(\mathtt{yz}) : \alpha \multimap \alpha.$$

Note that the term $2(\mathtt{yz})$ is not typable in DLAL due to the presence of two free variables that must be duplicated.

4.1 Type Inference Complexity

It can be shown that our algorithm works in polynomial time. In particular it is easy to verify that the construction of $\mathbf{\Pi}(\mathtt{M}) = \langle \Psi, U, \mathcal{H} \rangle$ can be done in time polynomial in $|\mathtt{M}|$.

Let $\mathcal{H} = \langle \mathcal{P}, \mathcal{C} \rangle$. The application of Robinson resolution to $\mathcal{P}$, generating $\mathcal{P}'$ and $\mathcal{C}'$, is polynomial in the number of both the scheme variables and exponentials in $\mathcal{P}$. The solution of the constraints in $\mathcal{P}'$ can be done through the standard algorithm working on the dag representation of schemes, and so it is polynomial in the number of scheme variables in $\mathcal{P}'$, which coincides with the number of scheme variables in $\mathcal{P}$.

As far as the exponential resolution task, i.e., the problem of solving the constraints in $\mathcal{C} \cup \mathcal{C}'$, is concerned, apparently it seems more difficult, since the problem of solving integer inequalities is in general NP-complete [11]. Nevertheless, following the method shown in [12], we can solve the problem over rational, which takes time polynomial in the number of exponentials. Clearly the set of solutions is closed under multiplication by positive integers. Now an integer solution can be obtained simply multiplying a rational solution by a suitable integer.

It is easy to check that the number of symbols in the constraints generated by Π is polynomial in $|\mathtt{M}|$. So the type inference problem for the propositional fragment can be decided in polynomial time in the size of the term.

5 Type Inference for the Full System

5.1 Schemes, Substitutions and Constraints

Definition 12. *The grammar of* type schemes $\mathcal{T}$, *given in Definition 1, is extended as follows*

$$U, V, Z ::= a \mid \phi \multimap U \mid [t].a \mid [t].\phi \multimap U \text{ (Linear type schemes)} \quad \phi, \psi ::= U \mid !^{p}U$$

where t, u, v belong to a countable set of sequence variables.

The notation $[t]$ does not introduce bound variables. Note that schemes of the shape $[t].[u].U$ are not allowed. $\mathrm{FV}(\phi)$ now denotes the set of linear scheme variables, exponentials and sequence variables occurring in ϕ. Two type schemes ϕ, ψ are *disjoint* if $\mathrm{FV}(\phi) \cap \mathrm{FV}(\psi) = \emptyset$.

A *scheme substitution* s is extended to map sequence variables to sequences of type variables. Namely the application of s to a type scheme is extended by the following rule

$$s([t].U) = \begin{cases} s(U) & \text{if } [t \mapsto \varepsilon] \in s \\ \forall \boldsymbol{\alpha}.s(U) & \text{if } [t \mapsto \boldsymbol{\alpha}] \in s \end{cases}$$

As in the propositional case, a type scheme is an abstract representation of all the types that can be obtained from it by a scheme substitution., e.g., the type scheme $[t].([u].b) \multimap a$ represents the set $\{\forall\boldsymbol{\alpha}.(\forall\boldsymbol{\beta}.A) \multimap B, (\forall\boldsymbol{\beta}.A) \multimap B, \forall\boldsymbol{\alpha}.A \multimap B \mid A, B \in \mathbf{T}\}$. The notion of *type scheme context* and its notation can be straightforwardly adapted from the one for the propositional fragment.

A *constraints sequence* $\mathcal{H}$ is a triple $\langle \mathcal{P}, \mathcal{C}, \mathcal{Q}\rangle$ of constraints sets, where $\mathcal{P}$ and $\mathcal{C}$ are as in Subsection 3.1, and $\mathcal{Q}$ is a set of equations of the shape $t = u$ or $t = \varepsilon$, where t, u are sequence variables. $\mathcal{Q}$ is satisfied by a scheme substitution s if $s(t) = s(u)$ $(s(t) = \varepsilon)$, for every $t = u$ $(t = \varepsilon)$ in it.

Table 5. Unification Algorithm

$$\frac{}{\mathtt{U}(a,a) = \langle\emptyset,\emptyset,\emptyset\rangle}\,(\mathtt{U}_0) \qquad \frac{}{\mathtt{U}(a,b) = \langle\{a=b\},\emptyset,\emptyset\rangle}\,(\mathtt{U}_1) \qquad \frac{a \notin \mathrm{FV}(\phi \multimap U)}{\mathtt{U}(a,\phi \multimap U) = \langle\{a = \phi \multimap U\},\emptyset,\emptyset\rangle}\,(\mathtt{U}_2)$$

$$\frac{\mathtt{U}(a,U) = \langle\mathcal{P},\mathcal{C},\mathcal{Q}\rangle}{\mathtt{U}(a,[t].U) = \langle\mathcal{P},\mathcal{C},\mathcal{Q}\cup\{t=\epsilon\}\rangle}\,(\mathtt{U}_3) \qquad \frac{\mathtt{U}(a,V) = \langle\mathcal{P},\mathcal{C},\mathcal{Q}\rangle}{\mathtt{U}(a,!^pV) = \langle\mathcal{P},\mathcal{C}\cup\{p=0\},\mathcal{Q}\rangle}\,(\mathtt{U}_4)$$

$$\frac{\mathtt{U}(\phi,\psi) = \langle\mathcal{P}_1,\mathcal{C}_1,\mathcal{Q}_1\rangle \quad \mathtt{U}(U,V) = \langle\mathcal{P}_2,\mathcal{C}_2,\mathcal{Q}_2\rangle}{\mathtt{U}(\phi \multimap U,\psi \multimap V) = \langle\mathcal{P}_1\cup\mathcal{P}_2,\mathcal{C}_1\cup\mathcal{C}_2,\mathcal{Q}_1\cup\mathcal{Q}_2\rangle}\,(\mathtt{U}_5)$$

$$\frac{\mathtt{U}(\phi \multimap U,V) = \langle\mathcal{P},\mathcal{C},\mathcal{Q}\rangle}{\mathtt{U}(\phi \multimap U,[t].V) = \langle\mathcal{P},\mathcal{C},\mathcal{Q}\cup\{t=\varepsilon\}\rangle}\,(\mathtt{U}_6) \qquad \frac{\mathtt{U}(\phi \multimap U,V) = \langle\mathcal{P},\mathcal{C},\mathcal{Q}\rangle}{\mathtt{U}(\phi \multimap U,!^pV) = \langle\mathcal{P},\mathcal{C}\cup\{p=0\},\mathcal{Q}\rangle}\,(\mathtt{U}_7)$$

$$\frac{\mathtt{U}(U,V) = \langle\mathcal{P},\mathcal{C},\mathcal{Q}\rangle}{\mathtt{U}([t].U,[u].V) = \langle\mathcal{P},\mathcal{C},\mathcal{Q}\cup\{t=u\}\rangle}\,(\mathtt{U}_8) \qquad \frac{\mathtt{U}([t].U,V) = \langle\mathcal{P},\mathcal{C},\mathcal{Q}\rangle}{\mathtt{U}([t].U,!^pV) = \langle\mathcal{P},\mathcal{C}\cup\{p=0\},\mathcal{Q}\rangle}\,(\mathtt{U}_9)$$

$$\frac{\mathtt{U}(\psi,\phi) = \langle\mathcal{P},\mathcal{C},\mathcal{Q}\rangle}{\mathtt{U}(\phi,\psi) = \langle\mathcal{P},\mathcal{C},\mathcal{Q}\rangle}\,(\mathtt{U}_{10}) \qquad \frac{\mathtt{U}(U,V) = \langle\mathcal{P},\mathcal{C},\mathcal{Q}\rangle}{\mathtt{U}(!^pU,!^qV) = \langle\mathcal{P},\mathcal{C}\cup\{p=q\},\mathcal{Q}\rangle}\,(\mathtt{U}_{11})$$

5.2 Unification Algorithm

In Table 5 we present a unification algorithm U extending the one presented in the propositional case. U proves judgments of the shape

$$\mathtt{U}(\phi,\psi) = \langle\mathcal{P},\mathcal{C},\mathcal{Q}\rangle$$

where ϕ and ψ are the two schemes that must be unified and $\langle\mathcal{P},\mathcal{C},\mathcal{Q}\rangle$ is a constraint sequence. Since the notation $[t]$ does not introduce bound variables in type schemes, we can consider it as a first order symbol. Then the unification problem we are considering is an instance of first order unification. As in the propositional case we have the following easy results.

Theorem 13 (U **Termination).** *Let* $\phi, \psi \in \mathcal{T}$ *be disjoint. Then, there exist* $\mathcal{P}, \mathcal{C}$ *and* $\mathcal{Q}$ *such that* $\mathtt{U}(\phi, \psi) = \langle \mathcal{P}, \mathcal{C}, \mathcal{Q} \rangle$*.*

Theorem 14 (U **Correctness).** *Let* $\phi, \psi \in \mathcal{T}$*. If* $\mathtt{U}(\phi, \psi) = \mathcal{H}$ *then, for every substitution* s *satisfying* $\mathcal{H}$

$$s(\phi) \equiv s(\psi)$$

Proof. By induction on the derivation of $\mathtt{U}(\phi, \psi) = \langle \mathcal{P}, \mathcal{C}, \mathcal{Q} \rangle$ noting that $\mathcal{Q}$ contains equalities of the shape $t = u$ or $t = \varepsilon$, hence the existence of a substitution satisfying this kind of constraints is decidable. □

We need now to prove that the algorithm U is also complete. The design of the type inference algorithm will be such that we need just to prove the completeness for the $\equiv$ relation of types. This agrees with the fact proved in [13] that typing in System F does not need the explicit use of α-rule.

Theorem 15 (U **Completeness).** *If* $s(\phi) \equiv s(\psi)$ *then there exists* $\mathcal{H}$ *such that* $\mathtt{U}(\phi, \psi) = \mathcal{H}$ *and* s *satisfies* $\mathcal{H}$*.*

Proof. By straighforward induction on the shape of ϕ and ψ. We will show just the case when $\phi \equiv [t].U$ and $\psi \equiv [u].V$. Let $s(\psi) \equiv s(\phi) = \forall \boldsymbol{\alpha}.\sigma$. Then $s(U) \equiv s(V)$, and by induction $\mathtt{U}(U, V) = \mathcal{H}'$. By rule $\mathtt{U}_8$, $\mathtt{U}([t].U, [u].V) = \langle \mathcal{H} \cup \{t = u\} \rangle$. So $s' = s[t \mapsto \boldsymbol{\alpha}, u \mapsto \boldsymbol{\alpha}]$ is the desired substitution. □

Remark. Note that a stronger completeness property holds for U, namely if $s(\phi)$ and $s(\psi)$ are α-equivalent, then there exists $\mathcal{H}$ such that $\mathtt{U}(\phi, \psi) = \mathcal{H}$ and there is a scheme substitution s' satisfying $\mathcal{H}$ such that $s'(\phi) \equiv s'(\psi)$, and $s'(\phi)$ is α-equivalent to both $s(\phi)$, $s(\psi)$. In fact, if $s(\phi)$ and $s(\psi)$ are α-equivalent, it is always possible to build a substitution s' such that $s'(\phi) \equiv s'(\psi)$, by renaming the bound variables, and then Theorem 15 can be applied.

Table 6. Type Inference Algorithm

$\mathbf{\Pi}(\mathtt{x}) = \mathtt{let}\ u, t, a, b, p\ \mathtt{be\ fresh\ in}\ \langle \{\mathtt{x} :!^p [t].a\}, [u].b, \{([t].a, b)\}, [u \mapsto \{[t].a\}], \{\emptyset, \emptyset, \emptyset\} \rangle$
$\mathbf{\Pi}(\lambda\mathtt{x.M}) = \mathtt{let}\ \mathbf{\Pi}(\mathtt{M}) = \langle \Psi, U, \mathcal{G}, \mathcal{F}, \mathcal{H} \rangle\ \mathtt{in}$ $\mathtt{let}\ \Psi = \Psi' \odot \{\mathtt{x} :!^{s_1} V_1, \ldots, \mathtt{x} :!^{s_n} V_n\}\ , \mathcal{I} = \textsc{range}(\Psi')\ \mathtt{in\ let}\ u, t, a, r\ \mathtt{be\ fresh\ in}$ $\mathtt{if}\ n = 0\ \mathtt{then}\ \langle \Psi', [u].!^r([t].a) \multimap U, \mathcal{G}, \mathcal{F} + [u \mapsto \mathcal{I}], \mathcal{H} \rangle$ $\mathtt{else\ if}\ n = 1\ \mathtt{then}\ \langle \Psi', [u].!^r V_1 \multimap U, \mathcal{G}, \mathcal{F} + [u \mapsto \mathcal{I}], \mathcal{H} \uplus \{r \geq s_1, \} \rangle$ $\mathtt{else\ if}\ n > 1\ \mathtt{then}\ \langle \Psi', [u].!^r V_1 \multimap U, \mathcal{G}, \mathcal{F} + [u \mapsto \mathcal{I}], \mathcal{H} \uplus \{r > s_1, \ldots, r > s_n\} \rangle$
$\mathbf{\Pi}(\mathtt{MN}) = \mathtt{let}\ \mathbf{\Pi}(\mathtt{M}) = \langle \Psi_{\mathtt{M}}, U, \mathcal{G}_{\mathtt{M}}, \mathcal{F}_{\mathtt{M}}, \mathcal{H}_{\mathtt{M}} \rangle\ \mathtt{and}\ \mathbf{\Pi}(\mathtt{N}) = \langle \Psi_{\mathtt{N}}, V, \mathcal{G}_{\mathtt{N}}, \mathcal{F}_{\mathtt{N}}, \mathcal{H}_{\mathtt{N}} \rangle\ \mathtt{be\ disjoint\ in}$ $\mathtt{let}\ u, t, a, b, q_i, p\ \mathtt{be\ fresh\ in}\ \mathtt{let}\ \Psi'_{\mathtt{N}} = \{\mathtt{z} :!^{q_i} V_i \mid \exists \mathtt{z} :!^{p_i} V_i \in \Psi_{\mathtt{N}}\},$ $\mathcal{I} = \textsc{range}(\Psi_{\mathtt{M}} \sqcup \Psi_{\mathtt{N}}),\ \mathcal{H} = \mathtt{Unify}(\Psi_{\mathtt{M}}, \Psi'_{\mathtt{N}}, U, !^p V \multimap [t].a)\ \mathtt{in}$ $\langle \Psi_{\mathtt{M}} \sqcup \Psi'_{\mathtt{N}}, [u].b, \mathcal{G}_{\mathtt{M}} \cup \mathcal{G}_{\mathtt{N}} \cup \{([t].a, b)\}, \mathcal{F}_{\mathtt{M}} + \mathcal{F}_{\mathtt{N}} + [u \mapsto \mathcal{I}], \mathcal{H}_{\mathtt{M}} \uplus \mathcal{H}_{\mathtt{N}} \uplus \mathcal{H} \uplus \{q_i \geq p_i + p\} \rangle$

5.3 The Algorithm

The Type Inference Algorithm follows the same lines of the type inference algorithm for System F designed by Ronchi Della Rocca and Giannini in [4]. In order to define it, we need to introduce some further notions.

Definition 16. *The* containment relation $\leq$ *between soft types is the relation defined as follows* $\forall\boldsymbol{\alpha}.A \leq A[\boldsymbol{B}/\boldsymbol{\alpha}]$*, for some* B.

Note that $\sigma \leq \tau$ corresponds to the fact that to a term $\mathtt{M}$ of type σ we can assign also the type τ by some applications of the rule $(\forall E)$. The relation $\leq$ is clearly decidable. Remembering that $\boldsymbol{\alpha}$ could be an empty sequence, $\leq$ is obviously reflexive. Moreover, it is transitive, hence a preorder. Note that $\forall\boldsymbol{\alpha}.\tau \multimap \sigma \leq \tau_1 \multimap \sigma_1$ implies $\forall\boldsymbol{\alpha}.\tau \leq \tau_1$ and $\forall\boldsymbol{\alpha}.\sigma \leq \sigma_1$, while in general the converse does not hold.

A *scheme system* $\mathcal{G}$ is a set of pairs of type schemes. A set of *binding constraints* $\mathcal{F}$ is a function from sequence variables to finite sets of schemes.

Definition 17. *Let* s *be a scheme substitution.*

- s *satisfies a scheme system* $\mathcal{G} = \{(U_1, V_1), \ldots, (U_n, V_n)\}$ *if and only if* $s(U_i) \leq s(V_i)$*,* $(1 \leq i \leq n)$*.*
- s *satisfies a binding constraints* $\mathcal{F} = \{u_1 \mapsto \Gamma_1, \ldots, u_n \mapsto \Gamma_n\}$ *if and only if* $\forall i \leq n, \forall\alpha \in s(u_i), \forall U \in \Gamma_i : \alpha \notin \mathrm{FV}(s(U))$

The type inference algorithm defined in Table 6 proves statement of the shape

$$\mathbf{\Pi}(\mathtt{M}) = \langle \Psi, U, \mathcal{G}, \mathcal{F}, \mathcal{H} \rangle$$

where Ψ is a type scheme assignment context, U is a linear type scheme, $\mathcal{G}$ is a scheme system, $\mathcal{F}$ is a set of binding constraints and $\mathcal{H}$ is a constraints sequence. The type inference algorithm call the `Unify` procedure, defined in Table 7, on contexts and schemes which need to be unified through the unification algorithm.

Theorem 18 ($\mathbf{\Pi}$ Termination). *Let* $\mathtt{M} \in \Lambda$*. Then* $\mathbf{\Pi}(\mathtt{M}) = \langle \Psi, U, \mathcal{G}, \mathcal{F}, \mathcal{H} \rangle$

Proof. By induction on the structure of $\mathtt{M}$. It is easy to verify that the schemes which need to be unified by the algorithm are always disjoint, so Theorem 14 applies. □

Table 7. `Unify` procedure

$$\begin{array}{ll}\mathtt{Unify}(\Phi, \Psi, \phi, \psi) = & \mathtt{let}\ \mathtt{x}_1, \ldots, \mathtt{x}_m = \mathrm{dom}(\Phi) \cap \mathrm{dom}(\Psi),\ \forall 1 \leq i \leq m \\ & \Phi(\mathtt{x}_i) = \{!^{s_1} V_1, \ldots, !^{s_n} V_n\},\ \Psi(\mathtt{x}_i) = \{!^{r_1} U_1, \ldots, !^{r_k} U_k\}, \\ & \mathtt{U}(\phi, \psi) = \langle \mathcal{P}_0, \mathcal{C}_0, \mathcal{Q}_0 \rangle,\ \mathtt{U}(V_1, U_1) = \langle \mathcal{P}_i, \mathcal{C}_i, \mathcal{Q}_i \rangle \\ & \mathtt{in}\langle \bigcup_{j=0}^{m} \mathcal{P}_j, \bigcup_{j=0}^{m} \mathcal{C}_j, \bigcup_{j=0}^{m} \mathcal{Q}_j \rangle\end{array}$$

Finally we can now prove the main theorems of this section.

Theorem 19 ($\mathbf{\Pi}$ Correctness). *Let $\mathbf{\Pi}(\mathtt{M}) = \langle \Psi, U, \mathcal{G}, \mathcal{F}, \mathcal{H}\rangle$. Then, for each substitution s satisfying $\mathcal{G}$, $\mathcal{F}$ and $\mathcal{H}$*

$$s(\Psi) \vdash \mathtt{M} : s(U)$$

Proof. By induction on the derivation proving $\mathbf{\Pi}(\mathtt{M}) = \langle \Psi, U, \mathcal{G}, \mathcal{F}, \mathcal{H}\rangle$. We will show just the most difficult case, when the term $\mathtt{M}$ is of the shape $\mathtt{PN}$.

Consider the case $\mathbf{\Pi}(\mathtt{PN}) = \langle \Psi, U, \mathcal{G}, \mathcal{F}, \mathcal{H}\rangle$. By hypothesis $\mathbf{\Pi}(\mathtt{P}) = \langle \Psi_{\mathtt{P}}, U_{\mathtt{P}}, \mathcal{G}_{\mathtt{P}}, \mathcal{F}_{\mathtt{P}}, \mathcal{H}_{\mathtt{P}}\rangle$, $\mathbf{\Pi}(\mathtt{N}) = \langle \Psi_{\mathtt{N}}, U_{\mathtt{N}}, \mathcal{G}, \mathcal{F}_{\mathtt{N}}, \mathcal{H}_{\mathtt{N}}\rangle$ and $U \equiv [u].b$. Let s be a substitution satisfying $\mathcal{G}$, $\mathcal{F}$ and $\mathcal{H}$. By induction hypothesis since s clearly satisfies $\mathcal{G}_{\mathtt{P}}$, $\mathcal{G}_{\mathtt{N}}$, $\mathcal{F}_{\mathtt{P}}$, $\mathcal{F}_{\mathtt{N}}$, $\mathcal{H}_{\mathtt{P}}$ and $\mathcal{H}_{\mathtt{N}}$, then we have both $s(\Psi_{\mathtt{P}}) \vdash \mathtt{P} : s(U_{\mathtt{P}})$ and $s(\Psi_{\mathtt{N}}) \vdash \mathtt{N} : s(U_{\mathtt{N}})$.

Moreover by definition $\mathtt{U}(U_{\mathtt{P}}, !^p U_{\mathtt{N}} \multimap [t].a) = \mathcal{H}'$ with $\mathcal{H}' \subseteq \mathcal{H}$, so since s satisfies $\mathcal{H}$ by Theorem 14: $s(U_{\mathtt{P}}) \equiv s(!^p U_{\mathtt{N}}) \multimap s([t].a)$. Let $\Psi'_{\mathtt{N}} = \{\mathtt{z} :!^{q_i} V_i \mid \exists \mathtt{z} : !^{p_i} V_i \in \Psi_{\mathtt{N}}\}$. Then clearly $s(\Psi) = s(\Psi_{\mathtt{P}} \sqcup \Psi'_{\mathtt{N}}) = s(\Psi_{\mathtt{P}}), s(\Psi'_{\mathtt{N}})$. Moreover since by hypothesis s satisfies $\mathcal{G}$, then in particular $s([t].a) \leq s(b)$. So, let $s(u) = \boldsymbol{\alpha}$ and $s(p) = k$. Then, the conclusion follows by the derivation

$$\dfrac{\dfrac{s(\Psi_{\mathtt{P}}) \vdash \mathtt{P} : s(!^p U_{\mathtt{N}}) \multimap s([t].a) \qquad \dfrac{\dfrac{s(\Psi_{\mathtt{N}}) \vdash \mathtt{N} : s(U_{\mathtt{N}})}{!^k s(\Psi_{\mathtt{N}}) \vdash \mathtt{N} :!^k s(U_{\mathtt{N}})}\,(sp)^k}{s(\Psi'_{\mathtt{N}}) \vdash \mathtt{N} :!^k s(U_{\mathtt{N}})}\,(m)^*}{\dfrac{s(\Psi_{\mathtt{P}}), s(\Psi'_{\mathtt{N}}) \vdash \mathtt{PN} : s([t].a)}{s(\Psi_{\mathtt{P}}), s(\Psi'_{\mathtt{N}}) \vdash \mathtt{PN} : s(b)}\,(\forall E)^*}\,(\multimap E)}{s(\Psi_{\mathtt{P}}), s(\Psi'_{\mathtt{N}}) \vdash \mathtt{PN} : \forall \boldsymbol{\alpha}.s(b)}\,(\forall I)^*$$

Note that we have freely applied the $(\forall I)$ rule over variables in $\boldsymbol{\alpha}$ since s satisfies the binding constraints $\mathcal{F}$. □

Theorem 20 ($\mathbf{\Pi}$ Completeness). *Let $\mathbf{\Pi}(\mathtt{M}) = \langle \Psi, U, \mathcal{G}, \mathcal{F}, \mathcal{H}\rangle$. If $\mathfrak{A} \vdash \mathtt{M} : \sigma$ then there exists a substitution s satisfying $\mathcal{G}, \mathcal{F}$ and $\mathcal{H}$ such that*

$$\Sigma \rhd s(\Psi) \vdash \mathtt{M} : s(U)$$

Moreover, the sequent $\mathfrak{A} \vdash \mathtt{M} : \sigma$ can be obtained from Σ by a (maybe empty) sequence of applications of the rules (w), (m) and (sp).

Proof. By induction on the derivation Π proving $\mathfrak{A} \vdash \mathtt{M} : \sigma$. We consider here the two most difficult cases. Let Π ends as

$$\dfrac{\Sigma \rhd \mathfrak{A} \vdash \mathtt{N} : \sigma \multimap A \qquad \Theta \rhd \mathfrak{B} \vdash \mathtt{P} : \sigma \qquad \mathfrak{A} \approx \mathfrak{B}}{\mathfrak{A}, \mathfrak{B} \vdash \mathtt{NP} : A}\,(\multimap E)$$

Let $\mathbf{\Pi}(\mathtt{NP}) = \langle \Psi, U, \mathcal{G}, \mathcal{F}, \mathcal{H}\rangle$, $\mathbf{\Pi}(\mathtt{N}) = \langle \Psi_{\mathtt{N}}, U_{\mathtt{N}}, \mathcal{G}_{\mathtt{N}}, \mathcal{F}_{\mathtt{N}}, \mathcal{H}_{\mathtt{N}}\rangle$ and $\mathbf{\Pi}(\mathtt{P}) = \langle \Psi_{\mathtt{P}}, U_{\mathtt{P}}, \mathcal{G}_{\mathtt{P}}, \mathcal{F}_{\mathtt{P}}, \mathcal{H}_{\mathtt{P}}\rangle$. By definition of $\mathbf{\Pi}$, $\Psi = \Psi_{\mathtt{N}} \sqcup \Psi'_{\mathtt{P}}$ where, if $\Psi_{\mathtt{P}} = \mathtt{x}_1 : !^{p_1} V_1, \ldots, \mathtt{x}_n :!^{p_n} V_n$, then $\Psi'_{\mathtt{P}} = \mathtt{x}_1 :!^{q_1} V_1, \ldots, \mathtt{x}_n :!^{q_n} V_n$ for fresh $q_1, \ldots, q_n$. Moreover, for fresh u, t, a, b and p, if $\mathcal{I} = \mathrm{RANGE}(\Psi)$ and $\mathtt{Unify}(\Psi_{\mathtt{N}}, \Psi'_{\mathtt{P}}, U_{\mathtt{N}}, !^p U_{\mathtt{P}} \multimap [t].a) = \mathcal{H}'$ then $U \equiv [u].b$, $\mathcal{G} = \mathcal{G}_{\mathtt{N}} \cup \mathcal{G}_{\mathtt{P}} \cup \{([t].a, b)\}$, $\mathcal{F} = \mathcal{F}_{\mathtt{N}} + \mathcal{F}_{\mathtt{P}} + [u \mapsto \mathcal{I}]$ and $\mathcal{H} = \mathcal{H}_{\mathtt{N}} \uplus \mathcal{H}_{\mathtt{P}} \uplus \mathcal{H}' \uplus \{q_i \geq p_i + p\}$.

By induction hypothesis there exists a scheme substitution $s_{\mathtt{N}}$ satisfying $\mathcal{G}_{\mathtt{N}}, \mathcal{F}_{\mathtt{N}}$ and $\mathcal{H}_{\mathtt{N}}$ such that $\Sigma' \rhd s_{\mathtt{N}}(\Psi_{\mathtt{N}}) \vdash \mathtt{N} : s_{\mathtt{N}}(U_{\mathtt{N}})$ and a substitution $s_{\mathtt{P}}$ satisfying $\mathcal{G}_{\mathtt{P}}, \mathcal{F}_{\mathtt{P}}$

and $\mathcal{H}_{\mathtt{P}}$ such that $\Theta' \rhd s_{\mathtt{P}}(\Psi_{\mathtt{P}}) \vdash \mathtt{P} : s_{\mathtt{P}}(U_{\mathtt{P}})$ and Σ and Θ can be obtained from Σ' and Θ' by a sequence of applications of the rules (w), (m) and (sp). This implies that $U_{\mathtt{N}}$ and $!^p U_{\mathtt{P}} \multimap [t].a$ are unifiable from Theorem 15.

Since $\mathbf{\Pi}(\mathtt{N})$ and $\mathbf{\Pi}(\mathtt{P})$ are disjoint we can build a substitution s' acting as $s_{\mathtt{N}}$ on schemes in $\mathbf{\Pi}(\mathtt{N})$ and as $s_{\mathtt{P}}$ on schemes in $\mathbf{\Pi}(\mathtt{P})$. Note that $s'(U_{\mathtt{N}}) \equiv \sigma \multimap A \equiv s'(!^p U_{\mathtt{P}} \multimap [t].a)$, where t and a are fresh. Hence in particular s' satisfies $\mathcal{H}'$.

Since $u, b, q_1, \ldots, q_n$ are fresh, it is easy to extend s' to a substitution $s = s'[b \mapsto s([t].a), u \mapsto \epsilon, q_1 \mapsto s(p_1) + s(p), \ldots, q_n \mapsto s(p_n) + s(p)]$. Clearly s satisfies $\mathcal{G}, \mathcal{F}$ and $\mathcal{H}$. If $s(p) = k$, then the following derivation can be built

$$\dfrac{\Sigma' \rhd s(\Psi_{\mathtt{N}}) \vdash \mathtt{N} : s(U_{\mathtt{N}}) \qquad \dfrac{\Theta' \rhd s(\Psi_{\mathtt{P}}) \vdash \mathtt{P} : s(U_{\mathtt{P}})}{!^k s(\Psi_{\mathtt{P}}) \vdash \mathtt{P} : !^k s(U_{\mathtt{P}})}\,(sp)^k}{s(\Psi_{\mathtt{N}}), !^k s(\Psi_{\mathtt{P}}) \vdash \mathtt{NP} : s([t].a)}\,(\multimap E)$$

and $\mathfrak{A}, \mathfrak{B} \vdash \mathtt{NP} : A$ can be obtained from it by a sequence of applications of the rules (w), (m) and (sp).

Consider the case where Π ends as

$$\dfrac{\Sigma \rhd \mathfrak{A} \vdash \mathtt{N} : \forall\alpha.A}{\mathfrak{A} \vdash \mathtt{N} : A[B/\alpha]}\,(\forall E)$$

Let $\mathbf{\Pi}(\mathtt{N}) = \langle \Psi, U, \mathcal{G}, \mathcal{F}, \mathcal{H} \rangle$. By induction hypothesis there is s satisfying $\mathcal{G}, \mathcal{F}$ and $\mathcal{H}$ such that $\Theta \rhd s(\Psi) \vdash \mathtt{N} : s(U)$ and Σ is derivable from Θ by applying a sequence of rule $(w), (m)$ and (sp). So in particular we have $s(U) \equiv \forall\alpha.A$ and by an inspection of the rules it is easy to verify that $U \equiv [u].V$ for some V and fresh u. Moreover $A \equiv \forall\beta.C$ for some C. Hence in particular $s = s'[u \mapsto \alpha\beta]$ for some substitution s'. Let $a_1, \ldots, a_n$ be such that $s'(a_i) = C_i[\alpha]$, where $C_i[\alpha]$ denotes a type C_i, where α occurs free $(1 \leq i \leq n)$. Then $s_1 = s'[u \mapsto \beta, a_1 \mapsto C_1[B], \ldots, a_n \mapsto C_n[B]]$ and s_1 does the intended work, since the Hygiene Condition. Moreover since u is fresh it is easy to verify that s_1 satisfies $\mathcal{G}, \mathcal{F}$ and $\mathcal{H}$. □

5.4 Examples

Example 21

1. It is easy to verify that $\mathbf{\Pi}(\lambda\mathrm{x.xx}) = \langle \emptyset, U, \mathcal{G}, \mathcal{F}, \mathcal{H} \rangle$ where

$$U = [w].!^r([t_1].a_1) \multimap [v].c \qquad \mathcal{G} = \{([t_1].a_1, b_1), ([t_2].a_2, b_2), ([t].a, c)\}$$
$$\mathcal{F} = \{u_1 \mapsto \{[t_1].a_1\}, u_2 \mapsto \{[t_2].a_2\}, v \mapsto \{[t_1].a_1, [t_2].a_2\}\}$$
$$\mathcal{H} = \langle \{b_1 = !^q([u_2].b_2) \multimap [t].a, a_1 = a_2\}, \{u_1 = \epsilon, t_1 = t_2\}, \{p_1 = p_3, p_3 \geq p_2 + q, r > p_1, r > p_3\} \rangle$$

   The substitution $s = s'[a_1 \mapsto \alpha, a_2 \mapsto \alpha, b_1 \mapsto (\forall\beta.\beta) \multimap \gamma, b_2 \mapsto \beta, c \mapsto \gamma, a \mapsto \gamma, t_1 \mapsto \alpha, t_2 \mapsto \alpha, u_1 \mapsto \epsilon, u_2 \mapsto \beta, v \mapsto \gamma, w \mapsto \epsilon, t \mapsto \epsilon, p_1 \mapsto 0, p_2 \mapsto 0, p_3 \mapsto 0, q \mapsto 0, r \mapsto 1]$ satisfies $\mathcal{G}$, $\mathcal{F}$ and $\mathcal{H}$. Hence the term is typable.
2. It is boring but easy to obtain the constraints in $\mathbf{\Pi}((\lambda\mathrm{x.xx})2) = \langle \emptyset, U, \mathcal{G}, \mathcal{F}, \langle \mathcal{P}, \mathcal{C}, \mathcal{Q} \rangle \rangle$. Making the substitutions in $\mathcal{P}$ and $\mathcal{Q}$ we obtain

$U = [z].d \qquad \mathcal{G} = \{([t_2].a_2, c), ([t_z].a_z, b_z), ([t_{sz}].a_{sz}, b_{sz}), ([t_{s^2z}].a_{s^2z}, b_{s^2z}),$
$([t_s].a_s, !^{p_{sz}}([u_z].b_z) \multimap [t_{sz}].a_{sz}), ([z_1].c, d),$
$([t_1].(!^s([t_s].a_s) \multimap ([v].!^r([t_z].a_z) \multimap [u_{s^2z}].b_{s^2z})), b_2), \quad (1)$
$([t_1].(!^s([t_s].a_s) \multimap ([v].!^r([t_z].a_z) \multimap [u_{s^2z}].b_{s^2z})), !^{q_1}([u_2].b_2) \multimap [t_2].a_2), \quad (2)$
$([t_s].a_s, !^{p_2}([u_{sz}].b_{sz}) \multimap [t_{s^2z}].a_{s^2z}) \quad (3)\}$
$\mathcal{F} = \{u_1, u_2, z_1 \mapsto \{[t_1].!^s([t_s].a_s) \multimap ([v].!^r([t_z].a_z) \multimap [u_{s^2z}].b_{s^2z})\},$
$u_z \mapsto \{[t_z].a_z\}, u_{sz}, u_{s^2z} \mapsto \{[t_z].a_z, [t_s].a_s\}, v \mapsto \{[t_s].a_s\}\}$
$\mathcal{C} = \{r_1 > q_1, r \geq p_{sz} + p_2, s > p_2\}$

The equation (1) implies that b_2 is of the shape

$$!^{s_1}[w_1].b_2^1 \multimap [w_2].!^{s_2}[w_2']b_2^2 \multimap [w_3].b_2^3$$

Moreover it implies that each substitution s satisfying the constraints must be such that $s(s_1) = s(s), s(s_2) = s(r)$ while equation (2) implies $s(s) = s(q_1)$. Remembering that $\mathcal{G}$ is a semi-unification set, equations (1), (2) and (3) imply that $s(t_s) = \epsilon$ and $s(a_s) =!^{s(p_2)}A \multimap B$. Substituting this in equation (2) we have $s(s) = s(p_2)$ but this is in contrast with the constraints in $\mathcal{H}$. Note that this term is typable in System F.

3. Note that the term 2(yz) of Example 10.3 is also typable in the full STA system and in System F but it is still not typable in DLAL due again to the presence of the two free variables.

6 Conclusion

We proved that the type inference problem for STA is decidable in polynomial time in the length of the input term if we restrict ourselves to consider just the propositional fragment. For the whole system we conjecture that the problem is undecidable since the presence of the second order quantifier. Nevertheless we showed an algorithm generating all the constraints that need to be satisfied in order to type a given term. It would be possible to follow the same method as in [4] for System F. Namely, for every $n \in \mathbb{N}$ we can define a bounded type containment relation $\leq_{\mathcal{J}}^n$ such that $\forall \boldsymbol{a}.A \leq_{\mathcal{J}}^n C$ if and only if $C \equiv A[\boldsymbol{B}/\boldsymbol{\alpha}]$ and the variables in $\boldsymbol{\alpha}$ occur in the syntax tree of A at a depth less or equal to n.

Then, we can define a countable set of type assignment systems STA^n which is a complete stratification of the system STA. For each $n \in \mathbb{N}$, the system STA^n is obtained by replacing the $(\forall E)$ rule in Table 1 by the following rule:

$$\frac{\Gamma \vdash \mathtt{M} : A \quad A \leq_{\mathcal{J}}^n B}{\Gamma \vdash \mathtt{M} : B} \ (n\text{-}\forall E)$$

In every STA^n the type inference problem is decidable. We leave the checking of the undecidability of the conjecture and the design of the stratified system for future investigations.

References

1. Gaboardi, M., Ronchi Della Rocca, S.: A soft type assignment system for λ-calculus. In: Duparc, J., Henzinger, T.A. (eds.) CSL 2007. LNCS, vol. 4646, pp. 253–267. Springer, Heidelberg (2007)
2. Lafont, Y.: Soft linear logic and polynomial time. Theoretical Computer Science 318(1-2), 163–180 (2004)
3. Coppo, M., Dezani-Ciancaglini, M., Venneri, B.: Principal type schemes and lambda-calculus semantics. In: Seldin, J.P., Hindley, J.R. (eds.) To H. B. Curry: Essays on Combinatory Logic, Lambda Calculus and Formalism, pp. 535–560. Academic Press, Inc., New York (1980)
4. Giannini, P., Ronchi Della Rocca, S.: A type inference algorithm for a stratified polymorphic type discipline. Information and Computation 109(1/2), 115–173 (1994)
5. Coppola, P., Dal Lago, U., Ronchi Della Rocca, S.: Elementary affine logic and the call by value lambda calculus. In: Urzyczyn, P. (ed.) TLCA 2005. LNCS, vol. 3461, pp. 131–145. Springer, Heidelberg (2005)
6. Wells, J.B.: Typability and type checking in the second-order λ-calculus are equivalent and undecidable. In: Proceedings of the Ninth Annual IEEE Symposium on Logic in Computer Science (LICS 1994), pp. 176–185. IEEE Computer Society, Los Alamitos (1994)
7. Baillot, P., Terui, K.: Light types for polynomial time computation in lambda-calculus. In: Proceedings of the Nineteenth Annual IEEE Symposium on Logic in Computer Science (LICS 2004), pp. 266–275. IEEE Computer Society, Los Alamitos (2004)
8. Asperti, A.: Light affine logic. In: Proceedings of the Thirteenth Annual IEEE Symposium on Logic in Computer Science (LICS 1998), pp. 300–308. IEEE Computer Society, Los Alamitos (1998)
9. Gaboardi, M.: Linearity: an Analytic Tool in the study of Complexity and Semantics of Programming Languages. PhD thesis, Università degli Studi di Torino - Institut National Polytechnique de Lorraine (2007)
10. Robinson, J.A.: Machine-oriented logic based on resolution principle. Journal of the ACM 12, 23–41 (1965)
11. Karp, R.M.: Reducibility among combinatorial problems. In: Miller, R.E., Thatcher, J.W. (eds.) Complexity of Computer Computations, pp. 85–103. Plenum Press (1972)
12. Baillot, P., Terui, K.: A feasible algorithm for typing in elementary affine logic. In: Urzyczyn, P. (ed.) TLCA 2005. LNCS, vol. 3461, pp. 55–70. Springer, Heidelberg (2005)
13. Kfoury, A.J., Ronchi Della Rocca, S., Tiuryn, J., Urzyczyn, P.: Alpha-conversion and typability. Information and Computation 150(1), 1–21 (1999)

Local Theory Specifications in Isabelle/Isar

Florian Haftmann[*] and Makarius Wenzel[**]

Technische Universität München
Institut für Informatik, Boltzmannstraße 3, 85748 Garching, Germany
http://www.in.tum.de/~haftmann/,
http://www.in.tum.de/~wenzelm/

Abstract. The proof assistant Isabelle has recently acquired a "local theory" concept that integrates a variety of mechanisms for structured specifications into a common framework. We explicitly separate a local theory "target", i.e. a fixed axiomatic specification consisting of parameters and assumptions, from its "body" consisting of arbitrary definitional extensions. Body elements may be added incrementally, and admit local polymorphism according to Hindley-Milner. The foundations of our local theories rest firmly on existing Isabelle/Isar principles, without having to invent new logics or module calculi.

Specific target contexts and body elements may be implemented within the generic infrastructure. This results in a large combinatorial space of specification idioms available to the user. Here we introduce targets for locales, type-classes, and class instantiations. The available selection of body elements covers primitive definitions and theorems, inductive predicates and sets, and recursive functions. Porting such existing definitional packages is reasonably simple, and allows to re-use sophisticated tools in a variety of target contexts. For example, a recursive function may be defined depending on locale parameters and assumptions, or an inductive predicate definition may provide the witness in a type-class instantiation.

1 Introduction

Many years ago, Isabelle locales were introduced [12] as a mechanism to organize formal reasoning in a modular fashion: after defining a locale as a context of fixed parameters (**fixes**) and assumptions (**assumes**), theorems could be proved within that scope, while an exported result (with additional premises) would be provided at the same time. Such "**theorem** (*in locale*)" statements have become popular means to organize formal theory developments. A natural extension of results within a local context is "**definition** (*in locale*)". Traditional locales would support **defines** elements within the axiomatic specification, essentially simulating definitions by equational assumptions, but this turned out to be unsatisfactory. It is not possible to add further definitions without extending the locale, and there is no support for polymorphism. Moreover, Isabelle/HOL users

[*] Supported by DFG project NI 491/10-1.
[**] Supported by BMBF project Verisoft XT (grant 01 IS 07008).

S. Berardi, F. Damiani, and U. de'Liguoro (Eds.): TYPES 2008, LNCS 5497, pp. 153–168, 2009.

rightly expect to have the full toolbox of definitional packages available (e.g. inductive predicates and recursive functions), not just primitive equations.

These needs are addressed by our *local theory* concept in Isabelle/Isar. A local theory provides an abstract interface to manage definitions and theorems relative to a context of fixed parameters and assumptions. This generic infrastructure is able to support the requirements of existing module concepts in Isabelle, notably locales and type-classes. Thus we integrate and extend the capabilities of structured specifications significantly, while opening a much broader scope for alternative module mechanisms.

Implementing such local theory targets is a delicate task, but only experts in module systems need to do it. In contrast, it is reasonably easy to produce definitional packages for use in the body of any local theory. Here we have been able to rationalize the traditional theory specification primitives of Higher-Order Logic considerably, such that the local versions are both simpler and more general than their global counterparts.

Overview. We examine the flexibility of the local theory concept by an example of type class specification and instantiation (§2). After a careful exposition of the relevant foundations of Isabelle/Pure and Isabelle/Isar (§3), we introduce the main local theory architecture (§4) and describe some concrete target mechanisms (§5).

2 Example: Type Classes

The following example in Isabelle/HOL [14] uses type-classes to model general orders and orders that admit well-founded induction. Earlier [11] we integrated traditional axiomatic type-classes with locales, now both theory structuring concepts are also fitted into the bigger picture of local theories.

Basic Isabelle notation approximates usual mathematics, despite some bias towards λ-calculus and functional languages like Haskell. The general syntax for local theory specifications is "*target* **begin** *body* **end**", where *body* consists of a sequence of specification elements (definitions and theorems with proofs), and *target* determines a particular interpretation of the body elements relative to a local context (with parameters and assumptions).

The most common targets are **locale** and **class**. These targets are special in being explicitly named, and allow further body additions at any time. The syntax for this is "**context** *name* **begin** *body* **end**", with the abbreviation of "*specification* (**in** *name*)" for "**context** *name* **begin** *specification* **end**". The latter also integrates the existing "**theorem** (**in** *locale*)" into our framework.

Other targets, like the **instantiation** shown later, demand that the body is a closed unit that provides required specifications, finishes proof obligations etc.

General Orders. We define an abstract algebra over a binary relation *less-eq* that observes the partial ordering laws.

class *order* =
  fixes *less-eq* :: $\alpha \Rightarrow \alpha \Rightarrow$ *bool* (**infix** $\preceq$ 50)
  assumes *refl*: $x \preceq x$
    and *trans*: $x \preceq y \Longrightarrow y \preceq z \Longrightarrow x \preceq z$
    and *antisym*: $x \preceq y \Longrightarrow y \preceq x \Longrightarrow x = y$
begin

This class context provides a hybrid view on our abstract theory specification. The term *less-eq* :: $\alpha \Rightarrow \alpha \Rightarrow$ *bool* refers to a fixed parameter of a fixed type; the parameter *less-eq* also observes assumptions. At the same time, the canonical type-class interpretation [11] provides a polymorphic constant for arbitrary *order* types, i.e. any instance of *less-eq* :: β::*order* $\Rightarrow \beta \Rightarrow$ *bool*. Likewise, the locale assumptions are turned into theorems that work for arbitrary types β::*order*.

Our class target augments the usual Isabelle type-inference by a separate *type improvement* stage, which identifies sufficiently general occurrences of *less-eq* with the locale parameter, while leaving more specific instances as constants. By handling the choice of locale parameters vs. class constants within the type-checking phase, we also avoid extra syntactic ambiguities: the above mixfix annotation (**infix** $\preceq$ 50) is associated with the class constant once and for all. See §5.3 for further details.

end

Back in the global context, *less-eq* :: α::*order* $\Rightarrow \alpha \Rightarrow$ *bool* refers to a global class operation for arbitrary *order* types α; the notation $x \preceq y$ also works as expected. Global class axioms are available as theorems *refl*, *trans*, *antisym*.

The old **axclass** [17] would have achieved a similar effect. At this point we could even continue with further definitions and proofs relative to this polymorphic constant only, e.g. *less* :: α::*order* $\Rightarrow \alpha \Rightarrow$ *bool* depending on the global *less-eq* :: α::*order* $\Rightarrow \alpha \Rightarrow$ *bool*. But then the resulting development would be more special than necessary, with the known limitations of type-classes of at most one instantiation per type constructor. So we now continue within the hybrid class/locale context, which provides type-class results as expected, but also admits general locale interpretations [2].

context *order*
begin

We now define *less* as the strict part of *less-eq*, and prove some simple lemmas.

definition *less* :: $\alpha \Rightarrow \alpha \Rightarrow$ *bool* (**infix** $\prec$ 50)
  where $x \prec y \leftrightarrow x \preceq y \wedge \neg\, y \preceq x$

lemma *irrefl*: $\neg\, x \prec x$ $\langle$*proof*$\rangle$
lemma *less-trans*: $x \prec y \Longrightarrow y \prec z \Longrightarrow x \prec z$ $\langle$*proof*$\rangle$
lemma *asym*: $x \prec y \Longrightarrow y \prec x \Longrightarrow C$ $\langle$*proof*$\rangle$

end

Again this produces a global constant *less* :: α::*order* $\Rightarrow \alpha \Rightarrow$ *bool*, whose definition depends on the original class operation *less-eq* :: α::*order* $\Rightarrow \alpha \Rightarrow$ *bool*. The additional variant *order.less* (*rel* :: $\alpha \Rightarrow \alpha \Rightarrow$ *bool*) (x :: α) (y :: α) stems from the associated locale context and makes this dependency explicit. The latter is more flexible, but also slightly more cumbersome to use.

Well-founded Induction and Recursion. Next we define well-founded orders by extending the specification of general orders.

class *wforder* = *order* +
  assumes *less-induct*: $(\bigwedge x{::}\alpha.\ (\bigwedge y.\ y \prec x \Longrightarrow P\ y) \Longrightarrow P\ x) \Longrightarrow P\ x$
begin

With this induction rule available, we can define a recursion combinator by means of an inductive relation that corresponds to the function's graph, see also [16].

inductive *wfrec-rel* :: $((\alpha \Rightarrow \beta) \Rightarrow \alpha \Rightarrow \beta) \Rightarrow \alpha \Rightarrow \beta \Rightarrow bool$
  for $F :: (\alpha \Rightarrow \beta) \Rightarrow \alpha \Rightarrow \beta$
  where *rec*: $(\bigwedge z.\ z \prec x \Longrightarrow \textit{wfrec-rel}\ F\ z\ (g\ z)) \Longrightarrow \textit{wfrec-rel}\ F\ x\ (F\ g\ x)$

definition *cut* :: $\alpha \Rightarrow (\alpha \Rightarrow \beta) \Rightarrow \alpha \Rightarrow \beta$
  where $\textit{cut}\ x\ f\ y = (\textit{if}\ y \prec x\ \textit{then}\ f\ y\ \textit{else undefined})$
lemma *cuts-eq*: $\textit{cut}\ x\ f = \textit{cut}\ x\ g \leftrightarrow (\forall y.\ y \prec x \longrightarrow f\ y = g\ y)$ ⟨*proof*⟩

definition *adm-wf* :: $((\alpha \Rightarrow \beta) \Rightarrow \alpha \Rightarrow \beta) \Rightarrow bool$
  where $\textit{adm-wf}\ F \leftrightarrow (\forall f\ g\ x.\ (\forall z.\ z \prec x \longrightarrow f\ z = g\ z) \longrightarrow F\ f\ x = F\ g\ x)$
lemma *adm-lemma*: $\textit{adm-wf}\ (\lambda f\ x.\ F\ (\textit{cut}\ x\ f)\ x)$ ⟨*proof*⟩

definition *wfrec* :: $((\alpha \Rightarrow \beta) \Rightarrow \alpha \Rightarrow \beta) \Rightarrow \alpha \Rightarrow \beta$
  where $\textit{wfrec}\ F = (\lambda x.\ (\textit{THE}\ y.\ \textit{wfrec-rel}\ (\lambda f\ x.\ F\ (\textit{cut}\ x\ f)\ x)\ x\ y))$

lemma *wfrec-unique*: $\textit{adm-wf}\ F \Longrightarrow \exists!y.\ \textit{wfrec-rel}\ F\ x\ y$ ⟨*proof*⟩
theorem *wfrec*: $\textit{wfrec}\ F\ x = F\ (\textit{cut}\ x\ (\textit{wfrec}\ F))\ x$ ⟨*proof*⟩

This characterizes a polymorphic combinator *wfrec* that works for arbitrary types β, relative to the locally fixed type parameter α. Thus $\textit{wfrec}\ (\lambda f :: \alpha \Rightarrow \textit{nat}.\ \textit{body})$ or $\textit{wfrec}\ (\lambda f :: \alpha \Rightarrow \textit{bool}.\ \textit{body})$ may be used in the current context. *THE* is the definite choice operator, sometimes written ι in literature; *undefined* is an unspecified constant.

end

Back in the global context, we may refer either to the exported locale operation $\textit{wforder.wfrec}\ (\textit{rel} :: \alpha \Rightarrow \alpha \Rightarrow \textit{bool})\ (F :: (\alpha \Rightarrow \beta) \Rightarrow \alpha \Rightarrow \beta)$ or the overloaded constant $\textit{wfrec}\ (F :: (\alpha{::}\textit{wforder} \Rightarrow \beta) \Rightarrow \alpha \Rightarrow \beta)$. Here α and β are again arbitrary, although the class constraint needs to be observed in the second case.

Lexicographic Products. The product $\alpha \times \beta$ of two *order* types is again an instance of the same algebraic structure, provided that the *less-eq* operation is defined in a suitable manner, such that the class assumptions can be proven. We shall establish this in the body of the following **instantiation** target.

instantiation $*$ :: (*order*, *order*) *order*
begin

We now define the lexicographic product relation by means of a (non-recursive) inductive definition, depending on hypothetical *less-eq* on fixed *order* types α and β.

inductive *less-eq-prod* :: $\alpha \times \beta \Rightarrow \alpha \times \beta \Rightarrow bool$
  where *less-eq-fst*: $x \prec v \Longrightarrow (x, y) \preceq (v, w)$
  | *less-eq-snd*: $x = v \Longrightarrow y \preceq w \Longrightarrow (x, y) \preceq (v, w)$

This definition effectively involves overloading of the polymorphic constant *less-eq* on product types, but the details are managed by our local theory target context. Here we have even used the derived definitional mechanism **inductive**, which did not support overloading in the past.

The above specification mentions various type instances of *less-eq*, for α, β, and $\alpha \times \beta$. All of these have been written uniformly with the $\preceq$ notation. This works smoothly due to an additional improvement stage in the type-inference process.

The outline of the corresponding instantiation proof follows. The **instance** element below initializes the class membership goal as in the existing global **instance** command [17], but the type arity statement is not repeated here.

instance
proof
  fix $p\ q\ r :: \alpha \times \beta$
  show $p \preceq p$ $\langle proof \rangle$
  { **assume** $p \preceq r$ **and** $r \preceq q$ **then show** $p \preceq q$ $\langle proof \rangle$ }
  { **assume** $p \preceq q$ **and** $q \preceq p$ **then show** $p = q$ $\langle proof \rangle$ }
qed

end

By concluding the instantiation, the new type arity becomes available in the theory, i.e. *less-eq* $(p :: \alpha{::}order \times \beta{::}order)\ (q :: \alpha \times \beta)$ can be used for arbitrary α, β in *order*.

3 Foundations

Isabelle consists of two main layers: the *logical framework* of Isabelle/Pure for higher-order Natural Deduction, and the *architectural framework* of Isabelle/Isar for organizing logical and extra-logical concepts in a structured manner.

Our local theory concepts rely only on existing foundations. We refrain from inventing new logics or module calculi, but merely observe pre-existent properties carefully to employ them according to our needs. This principle of leaving the logical basis unscathed is an already well-established Isabelle tradition. It enables implementation of sophisticated tools without endangering soundness. This is Milner's "LCF-approach" in its last consequence, cf. the discussion in [19, §3].

3.1 The Pure Logical Framework

The logic of Isabelle/Pure [15] is a reduced version of Higher-Order Logic according to Church [8] and Gordon [9]. This minimal version of HOL is used as a logical framework to represent object-logics, such as the practically important Isabelle/HOL [14].

Logical Entities. The Pure logic provides three main categories of formal entities: types, terms, and theorems (with implicit proofs).

Types τ are simple first-order structures, consisting of type variables α or type constructor applications $(\tau_1, \ldots, \tau_n)\ \kappa$, usually written postfix. Type *prop* represents framework propositions, and the infix type $\alpha \Rightarrow \beta$ functions.

Terms t are formed as simply-typed λ-terms, with variables $x :: \tau$, constants $c :: \tau$, abstraction $\lambda x :: \tau.\ t[x]$ and application $t_1\ t_2$. Types are usually left implicit, cf. Hindley-Milner type-inference [13]. Terms of type *prop* are called propositions. The logical structure of propositions is determined by quantification $\bigwedge x :: \alpha.\ B[x]$ or implication $A \Longrightarrow B$; these framework connectives express object-logic rules in Natural Deduction style. Isabelle/Pure also provides built-in equality $t_1 \equiv t_2$ with rules for $\alpha\beta\eta$-conversion.

Theorems *thm* are abstract containers for derivations within the logical environment. Primitive inferences of Pure operate on sequents $\Gamma \vdash \varphi$, where φ is the main conclusion and Γ its local context of hypotheses. There are standard introduction and elimination rules for $\bigwedge$ and $\Longrightarrow$ operating on sequents.

This low-level inference system is directly implemented in the Isabelle inference kernel; it corresponds to dependently-typed λ-calculus with propositions as types, although proof terms are usually omitted. It is useful to think of theorems as representing full proof terms, even though the implementation may omit them: the formal system can be categorized as "λHOL" within the general setting of Pure Type Systems (PTS) [3]. This provides a unified view of terms and derivations, with terms depending on terms $\lambda x :: \alpha.\ b[x]$, proofs depending on terms $\bigwedge x :: \alpha.\ B[x]$, and proofs depending on proofs $A \Longrightarrow B$.

Object-logic inferences are expressed at the level of Pure propositions, *not* Pure rules. For example, in Isabelle/HOL the modus ponens is represented as $(A \longrightarrow B) \Longrightarrow A \Longrightarrow B$, implication introduction as $(A \Longrightarrow B) \Longrightarrow A \longrightarrow B$. Isabelle provides convenient (derived) principles of *resolution* and *assumption* [15] to back-chain such object rules, or close branches within proofs, respectively.

This second level of Natural Deduction is encountered by Isabelle users most of the time, e.g. when doing "**apply** (*rule r*)" in a tactic script. Thus the first level of primitive inferences remains free for internal uses, to support local scopes of fixed variables and assumptions. Both Isar proof texts [18] and locales [12,1,2] operate on this primitive level of Pure, and the Isabelle/Isar framework ensures that local hypotheses are managed according to the block structure of the text, such that users never have to care about the Γ part of primitive sequents.

The notation $\langle\varphi\rangle$ shall refer to some theorem $\Gamma \vdash \varphi$, where Γ is clear from the context. The cIsabelle syntax 'φ' references facts the same way but uses ASCII back-quotes in the source instead of the funny parentheses.

Theories. Derivations in Isabelle/Pure depend on a global theory environment Θ, which holds declarations of type-constructors $(\alpha_1, \ldots, \alpha_n)\ \kappa$ (specifying the number of arguments), term constants $c :: \tau$ (specifying the most general type scheme), and axioms a: A (specifying the proposition). The following concrete syntax shall be used for these three theory declaration primitives:

type $\forall\overline{\alpha}.\ (\overline{\alpha})\ \kappa$	— type constructor κ
const $c :: \forall\overline{\alpha}.\ \tau[\overline{\alpha}]$	— term constant c
axiom $a: \forall\overline{\alpha}.\ A[\overline{\alpha}]$	— proof constant a

These primitives support *global schematic polymorphism*, which means that type variables given in the declaration may be instantiated by arbitrary types. The logic provides admissible inferences for this: moving from $\Gamma \vdash \varphi[\alpha]$ to $\Gamma \vdash \varphi[\tau]$ essentially instantiates whole proof trees.

We take the notational liberty of explicit type quantification $\forall\, \alpha.\ A[\alpha]$, even though the Pure logic is not really polymorphic. Type quantifiers may only occur in global theory declarations and theorems, but never in hypothetical statements, or the binding position of a λ-abstraction. This restricted type quantification behaves like schematic type variables, as indicated by question marks in Isabelle: results $\forall\, \alpha.\ \bigwedge x :: \alpha.\ x \equiv x$ and $\bigwedge x :: ?\alpha.\ x \equiv x$ are interchangeable.

Unrestricted declarations of types, terms, and axioms are rarely used in practice. Instead there are certain disciplined schemes that qualify as *definitional specifications* due to nice meta-theoretical properties. In the Pure framework, we can easily justify the well-known principle of *constant definition*, which relates a polymorphic term constant with an existing term:

$$\textbf{constdef}\ c :: \forall\, \overline{\alpha}.\ \tau[\overline{\alpha}]\ \textbf{where}\ \forall\, \overline{\alpha}.\ c[\overline{\alpha}] \equiv \mathit{rhs}[\overline{\alpha}]$$

Here the constant name c needs to be new, and *rhs* needs to be a closed term with all its type variables already included in its type τ. If these conditions hold, **constdef** expands to corresponding **const** $c :: \forall\, \overline{\alpha}.\ \tau[\overline{\alpha}]$ and **axiom** $\forall\, \overline{\alpha}.\ c[\overline{\alpha}] \equiv \mathit{rhs}[\overline{\alpha}]$.

This constant definition principle observes *parametric polymorphism.* Isabelle also supports *ad-hoc polymorphism* (overloading) which can be shaped into a disciplined version due to Haskell-style type-classes on top of the logic, see also [17] and [11]. The latter concept of "less ad-hoc polymorphism" allows us to reconstruct overloading-free definitions and proofs, via explicit dictionary terms.

We also introduce an explicit definition scheme for "proof constants", which gives proven facts an explicit formal status within the theory context:

$$\textbf{thmdef}\ \forall\, \overline{\alpha}.\ b = \langle B[\overline{\alpha}]\rangle$$

Moreover, the weaker variant of **thm** $b = \langle B\rangle$ shall serve the purpose of casual naming of facts, without any impact on the internal structure of derivations.

3.2 Isar Proof Contexts

The main purpose of the Isabelle/Isar infrastructure is to elevate the underlying logical framework to a scalable architecture that supports structured reasoning. This works by imposing certain Isar *policies* on the underlying Pure *primitives.* It is important to understand, that Isar is not another calculus, but an architecture to organize existing logical principles, and enrich them by non-logical support structure. The relation of Pure vs. Isar is a bit like that of a CPU (execution primitives) and an operating system (high-level abstractions via policies).

Isabelle/Isar was originally motivated by the demands for human-readable proofs [18]: the Isar proof language provides a structured walk through the text, maintaining local facts and goals, all relative to a *proof context* at each position. This idea of Isar proof context has turned out a useful abstraction to organize various advanced concepts in Isabelle, with locales [12,1,2] being the classic example. A more recent elaboration on the same theme are LCF-style proof tools that work relative to some local declarations and may be transformed in a concrete application context later; [7] covers a Gröbner Base procedure on abstract rings that may get used on concrete integers etc.

Subsequently we briefly review the main aspects of Isar proof contexts, as required for our local theory infrastructure.

Logical Context Elements. The idea is to turn the Γ part of the primitive calculus (§3.1) into an explicit environment, consisting of declarations for all three logical categories: type variables, term variables, and assumptions:

type α	— type variable α
fix $x :: \tau[\overline{\alpha}]$	— term variable x
assume a: $A[\overline{\alpha}][\overline{x}]$	— proof variable a

Strictly speaking there is no explicit **type** element in Isabelle/Isar, because type variables are handled implicitly according to Hindley-Milner discipline [13]: when entering a new term (proposition) into the context, its type variables are fixed; when exporting results from the context, type variables are generalized as far as possible, unless they occur in the types of term variables that are still fixed.

Exporting proven results from the scope of **fix** and **assume** corresponds to $\bigwedge/\Longrightarrow$ introduction rules. In other words, the logical part of an Isar context may be fully internalized into the Pure logic.

Isar also admits derived context elements, parameterized by a *discharge rule* that is invoked when leaving the corresponding scope. In particular, simple (non-polymorphic) definitions may be provided as follows:

$$\textbf{vardef}\ x :: \tau[\overline{\alpha}]\ \textbf{where}\ x \equiv \mathit{rhs}[\overline{\alpha}]$$

Here the variable name x needs to be new, and *rhs* needs to be a closed term, mentioning only previously fixed type variables. If these conditions hold, **vardef** expands to corresponding **fix** and **assume** elements, with a discharge rule that expands a local result $\langle B[x]\rangle$ to $\langle B[\mathit{rhs}]\rangle$, thanks to reflexivity of $\equiv$.

Although **vardef** resembles the global **constdef**, it only works for fixed types! In particular, **vardef** $\mathit{id} :: \alpha \Rightarrow \alpha$ **where** $\mathit{id} \equiv \lambda x :: \alpha.\ x$ merely results in context elements **type** α **fix** $\mathit{id} :: \alpha \Rightarrow \alpha$ **assume** $\mathit{id} \equiv \lambda x :: \alpha.\ x$, for the fixed (hypothetical) type α. There is no *let*-polymorphism at that stage, because the logic lacks type quantification.

Generic Context Data. The Isar proof context is able to assimilate arbitrary user-data in a type-safe fashion, using a functor interface in ML (see also [19, §3]). This means almost everything can be turned into context data. Common examples include type-inference information (constraints), concrete syntax (mixfix grammar), or hints for automated reasoning tools.

The global theory is only extended monotonically, but Isar contexts support opening and closing of local scopes. Moving between contexts requires replacing references to hypothetical types, terms, and proofs within user data accordingly. The Isar framework cannot operate on user data due to ML's type-safety, but a slightly different perspective allows us to transform arbitrary content, by passing through an explicit *morphism* in just the right spot, cf. [7, §3–4].

So instead of transforming fully abstract data directly, the framework transforms *data declarations*, i.e. implementation specific functions that maintain the data in the context. The interface for this is the generic context element "**declaration** d", where d: *morphism* $\to$ *context* $\to$ *context* is a data operation provided by some external module implementing context data. The morphism is provided by the framework at some later stage, it determines the differences of the present abstract context wrt. the concrete application environment, by providing mappings for the three logical categories of types, terms, and theorems. The implementation of d needs to apply this morphism wherever logical entities occur in the data; see [7, §2] for a simple example.

Using the Isabelle/Isar infrastructure on top of the raw logic, we can now introduce the concept of *constant abbreviations* that are type-checked like global polymorphic constants locally, but expanded before the logic ever sees them:

$$\textbf{abbrev}\ c :: \forall \overline{\beta}.\ \tau[\overline{\alpha}, \overline{\beta}]\ \textbf{where}\ \forall \overline{\beta}.\ c[\overline{\beta}] \equiv \mathit{rhs}[\overline{\alpha}, \overline{\beta}]$$

Here the type variables $\overline{\alpha}$ need to be fixed in the context, but $\overline{\beta}$ is arbitrary. In other words, we have conjured up proper *let*-polymorphism in the abstract syntax layer of Isabelle/Isar, without touching the Pure logic.

4 Local Theory Infrastructure

We are now ready to introduce the key concept of *local theory*, which models the general idea of interpreting definitional elements relatively to a local context.

Basic specification elements shall be explicitly separated into two categories: axiomatic **fix/assume** vs. definitional **define/note**. Together with our implicit treatment of types, this achieves an orthogonal arrangement of λ- and *let*-bindings for all three logical categories as follows:

	λ-binding	*let*-binding
types	fixed α	arbitrary β
terms	**fix** $x :: \tau$	**define** $c \equiv t$
theorems	**assume** a: A	**note** $b = \langle B \rangle$

A local theory specification is divided into a *target* part, which is derived from the background theory by adding axiomatic elements, and a *body* consisting of any number of definitional elements. The target also provides a particular interpretation of definitional primitives. Concrete body elements are produced by definitional packages invoked within corresponding **begin/end** blocks (cf. §2).

The key duality is that of background theory vs. target context, but there is also an *auxiliary context* that allows to hide the effect of the target interpretation internally. So the general structure of a local theory is a sandwich of three layers:

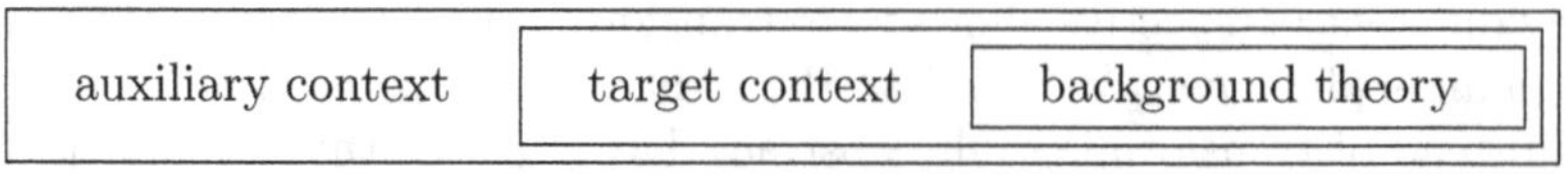

This allows one to make **define** appear like **vardef** and **note** like **thm** (cf. §3.2), while the main impact on the target context and background theory is exposed to the end-user only later. By fixing the body elements and their effect on the auxiliary context once and for all, we achieve a generic programming interface for definitional packages that work uniformly for arbitrary interpretations.

Canonical Interpretation via λ-Lifting. Subsequently we give a formal explanation of local theory interpretation by the blue-print model of λ-lifting over a fixed context **type** α **fix** $x :: \tau[\alpha]$ **assume** a: $A[\alpha][x]$. Restricting ourselves to a single variable of each category avoids cluttered notation; generalization to multiple parameters and assumptions is straightforward.

The idea is that **define** $a \equiv b[\alpha, \beta][x]$ relative to fixed α and $x :: \tau[\alpha]$ becomes a constant definition with explicit abstraction over the term parameter and generalization over the type parameters; the resulting theorem is re-imported into the target context by instantiation with the original parameters. The illusion of working fully locally is completed in the auxiliary context, by using hidden equational assumptions (see below).

The same principle works for **note** $a = \langle B[\alpha, \beta][x] \rangle$, but there is an additional dependency on assumption a: A, and the lifting over parameters is only partially visible, because proof terms are implicit.

The following λ-lifting scheme works for independent **define** and **note** elements, in an initial situation where the target and auxiliary context coincide:

specification	**define** $a \equiv b[\alpha, \beta][x]$
1. background theory	**constdef** $\forall \alpha\ \beta.\ thy.a \equiv \lambda x.\ b[\alpha, \beta][x]$
2. target context	**abbrev** $\forall \beta.\ loc.a \equiv thy.a[\alpha, \beta]\ x$
3. auxiliary context	**vardef** $a \equiv thy.a[\alpha, \beta]\ x$
local result	$\langle a \equiv b[\alpha, \beta][x] \rangle$

specification	**note** $a = \langle B[\alpha, \beta][x] \rangle$
1. background theory	**thmdef** $\forall \alpha\ \beta.\ thy.a = \langle \bigwedge x.\ A[\alpha][x] \Longrightarrow B[\alpha, \beta][x] \rangle$
2. target context	**thm** $\forall \beta.\ loc.a = \langle B[\alpha, \beta][x] \rangle$
3. auxiliary context	**thm** $a = \langle B[\alpha, \beta][x] \rangle$
local result	$\langle B[\alpha, \beta][x] \rangle$

This already illustrates the key steps of any local theory interpretation. Step (1) jumps from the auxiliary context right into the background theory to provide proper foundation, using global primitives of **constdef** and **thmdef**. Step (2) is where the particular target view is produced, here merely by applying fixed entities to revert the abstractions; other targets might perform additional transformations. Note that β is arbitrary in the target context. Step (3) bridges the distance of the target and auxiliary context, to make the local result appear literally as specified (with fixed β).

Extra care is required when mixing several **define** and **note** elements, as subsequent terms and facts may depend on the accumulated auxiliary parameters introduced by **vardef**. Export into the background theory now involves definitional expansion, and import into the auxiliary context folding of hypothetical equations. Here is the interpretation of **note** $a = \langle B[\alpha, \beta][c, x]\rangle$ depending on a previous **define** $c \equiv b[\alpha, \beta][x]$:

	note $a = \langle B[\alpha, \beta][c, x]\rangle$
1.	**thmdef** $\forall\, \alpha\ \beta.\ thy.a = \langle \bigwedge x.\ A[\alpha][x] \Longrightarrow B[\alpha, \beta][thy.c[\alpha, \beta]\ x, x]\rangle$
2.	**thm** $\forall\, \beta.\ loc.a = \langle B[\alpha, \beta][thy.c[\alpha, \beta]\ x, x]\rangle$
3.	**thm** $a = \langle B[\alpha, \beta][c, x]\rangle$
	$\langle B[\alpha, \beta][x]\rangle$

Each **define** element adds another **vardef** to the auxiliary context, to cater for internal term dependencies of any subsequent **define/note** within the same specification package (e.g. **inductive**). Thus package implementors need not care about term dependencies, but work directly with local variables (and with fixed types). Whenever a package concludes, our infrastructure resets the auxiliary context to the current target context, so the user will continue with polymorphic constant abbreviations standing for global terms. Only then, types appear in most general form according to the Hindley-Milner discipline, as expected by the end-user.

5 Common Local Theory Targets

5.1 Global Theories

A global theory is a trivial local theory, where the target context coincides with the background theory. The canonical interpretation (§4) is reduced as follows:

	define $a \equiv b[\beta]$
1.	**constdef** $\forall\, \beta.\ thy.a \equiv b[\beta]$
2.	(omitted)
3.	**vardef** $a \equiv thy.a[\beta]$
	$\langle a \equiv b[\beta]\rangle$

	note $a = \langle B[\beta]\rangle$
1.	**thmdef** $\forall\, \beta.\ thy.a = \langle B[\beta]\rangle$
2.	(omitted)
3.	**thm** $a = \langle B[\beta]\rangle$
	$\langle B[\beta]\rangle$

Here we trade a fixed term variable a (with fixed type) for a global constant $thy.a$ (with schematic type). The auxiliary context hides the difference in typing and name space details. This abstract view is a considerable advantage for package implementations, even without using the full potential of local theories yet.

The Isabelle/Isar toplevel ensures that local theory body elements occurring on the global level are wrapped into a proper target context as sketched above. Thus a local theory package like **inductive** may be invoked seamlessly in any situation.

5.2 Locales

Locales [12] essentially manage named chunks of Isabelle/Isar context elements (§3.2), such as **locale** *loc* = **fixes** x **assumes** $A[x]$, together with **declaration** elements associated with conclusions. Locale expressions [1] allow to combine atomic locales *loc*, either by renaming *loc* y or merge $loc_1 + loc_2$ of the underlying axiomatic specifications. Results stemming from a locale expression may be interpreted later, giving particular terms and proofs for the axiomatic part [2].

The key service provided by the locale mechanism is that of breaking up complex expressions into atomic locales. Down at that level, it hands over to the generic local theory infrastructure, by providing a target context that accepts **define** and **note** according to the canonical λ-lifting again (§4). There is only one modification: instead of working with primitive **fixes** and **assumes** of the original **locale** definition, there is an additional indirection through a global predicate definition **constdef** $thy.loc \equiv \lambda x.\ A[x]$.

The Isabelle/Isar toplevel initializes a locale target for "**context** *loc* **begin** *body* **end**" or the short version "*specification* (**in** *loc*)"; the latter generalizes the traditional "**theorem** (**in** *loc*)" form [12] towards arbitrary definitions within locales, including derived mechanisms like "**inductive** (**in** *loc*)".

5.3 Type Classes

Type-class Specification. Logically a type class is nothing more than an interpretation of a locale with *exactly one* type variable α, see [11]. Given such a locale $c[\alpha]$ with fixed type α, *class parameters* $\overline{g[\alpha]}$ and predicate $thy.c$, the corresponding type class c is established by the following interpretation, where the right column resembles the traditional **axclass** scheme [17]:

locale specification	class interpretation
locale $c =$	**classdecl** $c =$
fixes $\overline{g :: \tau[\alpha]}$	**const** $\overline{c.g :: \tau[\gamma{::}c]}$
assumes $thy.c\ \overline{g :: \tau[\alpha]}$	**axiom** $thy.c\ \overline{c.g[\gamma{::}c]}$

The class target augments the locale target (§5.2) by a second interpretation within the background theory where conclusions are relative to global constants $\overline{c.g[\gamma{::}c]}$ and class axiom $thy.c\ \overline{c.g[\gamma{::}c]}$, for arbitrary types γ of class c.

specification	**define** $f \equiv t[\alpha, \beta][\overline{g}]$
1. background theory	**constdef** $\forall \alpha\ \beta.\ thy.f \equiv \lambda\overline{x}.\ t[\alpha, \beta][\overline{x}]$
2a. locale target	**abbrev** $\forall \beta.\ loc.f \equiv thy.f[\alpha, \beta]\ \overline{g}$
2b. class target	**constdef** $\forall \gamma{::}c\ \beta.\ c.f \equiv thy.f[\gamma, \beta]\ \overline{c.g}$
3. auxiliary context	**vardef** $f \equiv thy.f[\alpha, \beta]\ \overline{g}$

specification	**note** $a = \langle B[\alpha, \beta][\overline{g}]\rangle$
1. background theory	**thmdef** $\forall \alpha\ \beta.\ thy.a = \langle\bigwedge\overline{x}.\ A[\alpha][\overline{x}] \Longrightarrow B[\alpha, \beta][\overline{x}]\rangle$
2a. locale target	**thm** $\forall \beta.\ loc.a = \langle B[\alpha, \beta][\overline{g}]\rangle$
2b. class target	**thmdef** $\forall \gamma{::}c\ \beta.\ c.a = \langle B[\gamma, \beta][\overline{c.g}]\rangle$
3. auxiliary context	**thm** $a = \langle B[\alpha, \beta][\overline{g}]\rangle$

The interpretation (2b) of **fixes** and **define** $f\ [\alpha,\ \beta]$ establishes a one-to-one correspondence with *class constants* $c.f\ [\gamma{::}c,\ \beta]$, such that $f\ [\alpha,\ \beta]$ becomes $c.f\ [\gamma{::}c,\ \beta]$. When interleaving **define** and **note** elements, the same situation occurs as described in §4 — hypothetical definitions need to be folded:

	note $a = \langle B[\alpha,\ \beta][f,\ \overline{g}]\rangle$
1.	**thmdef** $\forall\alpha\ \beta.\ thy.a = \langle\bigwedge\overline{x}.\ A[\alpha][\overline{x}] \Longrightarrow B[\alpha,\ \beta][thy.f[\alpha,\ \beta]\ \overline{g},\ \overline{g}]\rangle$
2a.	**thm** $\forall\beta.\ loc.a = \langle B[\alpha,\ \beta][thy.f[\alpha,\ \beta]\ \overline{g},\ \overline{g}]\rangle$
2b.	**thmdef** $\forall\gamma{::}c\ \beta.\ thy.a = \langle B[\gamma,\ \beta][c.f[\gamma,\ \beta],\ \overline{c.g}]\rangle$
3.	**thm** $a = \langle B[\alpha,\ \beta][f,\ \overline{g}]\rangle$

Type-class Instantiation. The **instantiation** target integrates type classes and overloading by providing a Haskell-like policy for class instantiation: each arity $\kappa :: (\overline{s})\ c$ is associated with a set of class parameters $c.g\ [(\overline{\delta{::}s})\ \kappa]$ for which specifications are given which respect the **assumes** of c. As auxiliary means, at the **begin** of an **instantiation**, each of these $c.g\ [(\overline{\delta{::}s})\ \kappa]$ is associated with a corresponding shadow variable $\kappa.g\ [\boxed{\delta{::}s}]$. These are treated specifically in subsequent **define** elements:

	define $\kappa.g\ \boxed{\delta{::}s} \equiv t$
1.	**constdef** $\forall\overline{\delta{::}s}.\ c.g\ [(\overline{\delta{::}s})\ \kappa] \equiv t$
2.	(omitted)
3.	**vardef** $\kappa.g\ \boxed{\delta{::}s} \equiv c.g\ [(\overline{\delta{::}s})\ \kappa]$

In other words, **define** is interpreted by overloaded **constdef**. Further occurrences of **define** or **note**, unrelated to the class instantiation, are interpreted as in the global theory target (§5.1). As an additional policy the **instantiation** target requires all class parameters to be specified before admitting the obligatory **instance** proof before the **end**.

Target Syntax. As seen in §2 both the **class** and **instantiation** context allows us to refer to whole families of corresponding constants uniformly. The idea is to let the user write $c.f$ unambiguously, and substitute this internally according to the actual instantiation of the class type parameter $[\gamma{::}c]$. This works by splitting conventional order-sorted type inference into three phases:

phase	**class** target	**instantiation** target
1. type inference	with class constraint $\gamma{::}c$ on $c.f$ disregarded	
2a. type improvement	$c.f\ [?\xi] \rightsquigarrow c.f\ [\alpha]$	$c.f\ [?\xi] \rightsquigarrow c.f\ [(\overline{\delta{::}s})\ \kappa]$
2b. substitution	$c.f\ [\alpha] \rightsquigarrow f\ [\alpha]$	$c.f\ [(\overline{\delta{::}s})\ \kappa] \rightsquigarrow \kappa.f\ [\overline{\delta}]$
3. type inference	with all constraints, fixing remaining inference parameters	

To permit writing terms $c.f\ [\alpha]$ for α even without a class constraint, first the class constraint $\gamma{::}c$ on $c.f$ is disregarded in phase (1) and is re-considered in phase (3); in between, types left open by type inference are still improvable type inference parameters $?\xi$.

Whenever $c.f$ is used with a characteristic type parameter (α in **class** case, $(\overline{\delta::s})\ \kappa$ in **instantiation** case), it is substituted by the appropriate parameter (f for **class** or $\kappa.f$ for **instantiation**) in phase (2b); more general occurrences $c.f\ [\gamma]$ are left unchanged. This allows to write $c.f$ uniformly for both local and global versions.

To relieve the user from cumbersome type annotations, a *type improvement* step is carried out (2a): if $c.f$ carries a type inference parameter $?\xi$, this is specialized to the characteristic type parameter. This step will hand over $c.f$ with completely determined type information.

As a particularity of the **instantiation** target, the substitution $c.f\ [(\overline{\delta::s})\ \kappa] \rightsquigarrow \kappa.f\ [\overline{\delta}]$ is only carried out while no **define** $\kappa.f\ [\overline{\delta::s}] \equiv t$ has occurred yet; afterwards, occurrences of $c.f\ [(\overline{\delta::s})\ \kappa]$ are taken literally.

When printing terms, substitutions are reverted: $f\ [\alpha] \rightsquigarrow c.f\ [\alpha]$ for **class**, and $\kappa.f\ [\overline{\delta}] \rightsquigarrow c.f\ [(\overline{\delta::s})\ \kappa]$ for **instantiation.** Thus the surface syntax expected by the end-user is recovered.

6 Conclusion

Related Work. Structured specifications depending on parameters and assumptions are closely related to any variety of "*modular logic*", which may appear in the guise of *algebraic specification*, *little theories* etc. Many module systems for proof assistants have been developed in the past, and this is still a matter of active research. Taking only Coq [4] as example, there are "structures" (a variety of record types), "sections" (groups of definitions and proofs depending on parameters and assumptions), and "modules" that resemble ML functors.

Our approach of building up specification contexts and the canonical interpretation of body elements by λ-lifting is closely akin to "sections" in Coq. Here we continue the original locale idea [12], which was presented as a "sectioning concept for Isabelle" in the first place. There are two main differences to Coq sections: our axiomatic target needs to be fixed once and for all, but the definitional body may be extended consecutively. In Coq both parts are intermingled, and cannot be changed later. Note that Coq sections vanish when the scope is closed, but a local theory may be recommenced.

Beyond similarities to particular module systems our approach is different in providing a broader scope. Acknowledging the existence of different module concepts, we offer a general architecture for integrating them into a common framework. After implementing a suitable target mechanism, a particular module concept will immediately benefit from body specification elements, as produced by existing definitional packages (**inductive**, **primrec**, **function**, etc.).

Fitting a module system into our framework requires a representation of its logical aspects within Isabelle/Pure, and any auxiliary infrastructure as Isabelle/Isar context. The latter is very flexible thanks to generic context data (covering arbitrary ML values in a type-safe fashion), and generic "declarations" for maintaining such data depending on a logical morphism. The Pure framework [15] supports higher-order logic, but only simple types without type quantification [8]. The particular targets presented here demonstrate that non-trivial modular concepts can indeed be mapped to the Pure logic, including an illusion of local type-quantification (for definitions) according to Hindley-Milner [13].

In fact, there is no need to stay within our canonical interpretation of λ-lifting at all. This template may be transcended by using explicit proof terms in Pure, to enable more general "admissible" principles in the interpretation. For example, the AWE tool [6] applies theory interpretation techniques directly to global type constructors, constants and axioms. This allows one to operate on polymorphic entities, as required for an abstract theory of monads, for example. In AWE, definitions and theorems depending on such global axiomatizations are transformed extra-logically by mapping the corresponding proof objects, and replaying them in the target context. The present implementation needs to redefine some common specification elements. Alternatively, one could present this mechanism as another local theory target, enabling it to work with any local definitional package, without requiring special patches just for AWE.

Implementation and Applications. Since Isabelle2008, local theories are the official interface for implementing derived specification mechanisms within the Isabelle framework. The distribution includes the general framework, with a couple of targets and definitional packages for the new programming interface. We already provide target mechanisms for global theories, locales, type classes and class instantiations as described above. There is another target for raw overloading without the type-class discipline. Moreover, the following body specifications are available:

- **definition** and **theorem** as wrappers for the **define** and **note** primitives
- **primrec** for structural recursion over datatypes
- **inductive** and **coinductive** for recursive predicates and sets (by Stefan Berghofer)
- **function** for general recursive functions (by Alexander Krauss)
- **nominal-inductive** and **nominal-primrec** for specifications in nominal logic (by Christian Urban and Stefan Berghofer)

Experience with such "localization" efforts of existing packages indicates that conversion of old code is reasonably easy; package implementations can usually be simplified by replacing primitive specifications by our streamlined local theory elements. Some extra care is required since packages may no longer maintain "global handles" on results, e.g. the global constant name of an inductively defined predicate. Such references to logical entities need to be generalized to arbitrary terms. Due to interpretation of the original specification in a variety of targets, one cannot count on particular global results, but needs to work with explicit Isar contexts and morphisms on associated data.

Acknowledgments. Tobias Nipkow and Alexander Krauss greatly influenced the initial "local theory" design (more than 2 years ago), by asking critical questions about definitions within locales. Early experiments with inductive definitions by Stefan Berghofer showed that the concept of "auxiliary context" is really required, apart from the "target context". Amine Chaieb convinced the authors that serious integration of locales and type-classes is really needed for advanced algebraic proof tools. Clemens Ballarin helped to separate general local theory principles from genuine features of locales. Norbert Schirmer and other early adopters helped to polish the interfaces.

References

1. Ballarin, C.: Locales and locale expressions in Isabelle/Isar. In: Berardi, S., et al. (eds.) TYPES 2003. LNCS, vol. 3085, pp. 34–50. Springer, Heidelberg (2004)
2. Ballarin, C.: Interpretation of locales in Isabelle: Theories and proof contexts. In: Borwein, J.M., Farmer, W.M. (eds.) MKM 2006. LNCS(LNAI), vol. 4108, pp. 31–43. Springer, Heidelberg (2006)
3. Barendregt, H., Geuvers, H.: Proof assistants using dependent type systems. In: Robinson, A., Voronkov, A. (eds.) Handbook of Automated Reasoning. Elsevier, Amsterdam (2001)
4. Barras, B., et al.: The Coq Proof Assistant Reference Manual, v. 8.1. INRIA (2006)
5. Bertot, Y., Dowek, G., Hirschowitz, A., Paulin, C., Théry, L.: TPHOLs 1999. LNCS, vol. 1690. Springer, Heidelberg (1999)
6. Bortin, M., Broch Johnsen, E., Lüth, C.: Structured formal development in Isabelle. Nordic Journal of Computing 13 (2006)
7. Chaieb, A., Wenzel, M.: Context aware calculation and deduction — ring equalities via Gröbner Bases in Isabelle. In: Kauers, M., Kerber, M., Miner, R., Windsteiger, W. (eds.) MKM/CALCULEMUS 2007. LNCS(LNAI), vol. 4573, pp. 27–39. Springer, Heidelberg (2007)
8. Church, A.: A formulation of the simple theory of types. J. Symbolic Logic (1940)
9. Gordon, M.J.C., Melham, T.F. (eds.): Introduction to HOL: A theorem proving environment for higher order logic. Cambridge University Press, Cambridge (1993)
10. Gunter, E.L., Felty, A. (eds.): Theorem Proving in Higher Order Logics (TPHOLs 1997). LNCS, vol. 1275. Springer, Heidelberg (1997)
11. Haftmann, F., Wenzel, M.: Constructive type classes in Isabelle. In: Altenkirch, T., McBride, C. (eds.) TYPES 2006. LNCS, vol. 4502, pp. 160–174. Springer, Heidelberg (2007)
12. Kammüller, F., Wenzel, M., Paulson, L.C.: Locales: A sectioning concept for Isabelle. In: Bertot, et al. (eds.) [5]
13. Milner, R.: A theory of type polymorphism in programming. J. Computer and System Sciences 17(3) (1978)
14. Nipkow, T., Paulson, L.C., Wenzel, M. (eds.): Isabelle/HOL — A Proof Assistant for Higher-Order Logic. LNCS, vol. 2283. Springer, Heidelberg (2002)
15. Paulson, L.C.: Isabelle: the next 700 theorem provers. In: Odifreddi, P. (ed.) Logic and Computer Science. Academic Press, London (1990)
16. Slind, K.: Function definition in higher-order logic. In: Gunter, Felty (eds.) [10]
17. Wenzel, M.: Type classes and overloading in higher-order logic. In: Gunter, Felty (eds.) [10]
18. Wenzel, M.: Isar — a generic interpretative approach to readable formal proof documents. In: Bertot, et al. (eds.) [5]
19. Wenzel, M., Wolff, B.: Building formal method tools in the Isabelle/Isar framework. In: Schneider, K., Brandt, J. (eds.) TPHOLs 2007. LNCS, vol. 4732, pp. 352–367. Springer, Heidelberg (2007)

Axiom Directed Focusing

Clément Houtmann

Université Henri Poincaré Nancy 1 & LORIA*,
Campus Scientifique, BP 239
54506 Vandoeuvre-lès-Nancy Cedex, France
Clement.Houtmann@loria.fr

Abstract. Superdeduction and deduction modulo are two methods designed to ease the use of first-order theories in predicate logic. Superdeduction modulo combines both in a single framework. Although soundness is ensured, using superdeduction modulo to extend deduction with awkward theories can jeopardize cut-elimination or completeness *w.r.t.* predicate logic. In this paper our aim is to design criteria for theories which will ensure these properties. We revisit the superdeduction paradigm by comparing it with the focusing approach. In particular we prove a focalization theorem for cut-free superdeduction modulo: we show that permutations of inference rules can transform any cut-free proof in deduction modulo into a cut-free proof in superdeduction modulo and conversely, provided that some hypotheses on the synchrony of reasoning axioms are verified. It implies that cut-elimination for deduction modulo and for superdeduction modulo are equivalent. Since several criteria have already been proposed for theories that do not break cut-elimination of the corresponding deduction modulo system, these criteria also imply cut-elimination of the superdeduction modulo system, provided our synchrony hypotheses hold.

Keywords: Proof theory, superdeduction, focusing, deduction modulo.

1 Introduction

The construction of formal proofs usually relies on a proof system containing the discipline that the user has to follow. Deduction is also normally conducted with respect to a theory (a set of axioms), which brings deductive as well as computing abilities to the system. In frameworks such as first-order natural deduction or sequent calculus, the use of a theory is always uniform since the proofs only express atomic steps, which correspond to decompositions of logical connectives. Higher-level notions such as sets, induction or arithmetic have to be encoded in the first-order language and handled through these atomic inference steps, leading often to long, hardly-readable "assembly like" proofs. Deduction modulo [1] and superdeduction [2] are two approaches that propose to ease the use of theories in predicate logic. The first has been designed to remove irrelevant

* UMR 7503 CNRS-INPL-INRIA-Nancy2-UHP.

S. Berardi, F. Damiani, and U. de'Liguoro (Eds.): TYPES 2008, LNCS 5497, pp. 169–185, 2009.

computational arguments (such as $2+2=4$) from proofs. These omitted computations are meant to be redone during the proofchecking process. The second proposes custom-made inference rules, similarly to Girard's synthetic connectives approach [3], devoted to specific axioms of the theory.

We believe that the superdeduction approach is closely related to Focusing, introduced by Andreoli [4] and meant to remove irrelevant choices in backward reasoning for sequent calculus: Syntactically different proofs can still be identical up to permutations or simplifications of the applications of the inference rules. The superdeduction point of view is that the use of an axiom is usually done through a destruction of its connectives, and that these steps are always *identical up to permutations or simplifications of the applications of the inference rules*. In this paper we consequently propose to fill the gap between the two approaches. We show that superdeduction systems as well as a specific focusing system are instances of a more general paradigm, therefore obtaining an insight of the similarities and differences between these systems.

Since awkward theories can break completeness *w.r.t.* predicate logic, cut-elimination (admissibility of cuts) or normalisation (of a cut-elimination procedure), one has to propose criteria for safe theories to be used in superdeduction and deduction modulo. On the one hand, strong normalisation for deduction modulo has been studied, using reducibility candidates [5] leading to the notion of truth values algebras and superconsistency [6,7], and cut-elimination results have been obtained using semantic methods [8,9] or abstract completion [10]. On the other hand superdeduction is provided with a proofterm language in [2] together with criteria for its normalisation. In this context, our comparison between superdeduction and focusing will allow us to prove a focalization theorem: cut-free deduction modulo proofs can be transformed into cut-free superdeduction modulo proofs, provided some synchrony hypotheses are verified. Then it implies that criteria for cut-elimination in deduction modulo also hold for superdeduction modulo.

Several other paradigms propose *ad hoc* systems for specific theories. Let us cite Huang's Assertion level [11] motivated by the presentation of proofs in natural language. Another approach proposed by Negri and von Plato [12] expresses first-order axioms through inference rules, which however only act on the left-hand side of sequents and consequently poorly interact with an elimination of cuts. Our approach is much closer to Definitional Reflection [13], extended with induction [14], since it adds left and right introduction rules to an intuitionistic sequent calculus in order to reflect some *definitional clause*, corresponding to what we will call *proposition rewrite rules*. In addition, cut-elimination results are proved for these logics with definitions and induction. However, working in the classical sequent calculus allows us to add more general inference rules.

Superdeduction modulo can be an innovative foundation for proof assistants, as argued in [2,15]. First it allows to naturally encode custom reasoning and computational schemes, such as induction and arithmetic. Furthermore it is especially adapted to human interaction: Proofs constructed with existing proof-assistants such as Coq or Isabelle usually consist in tactics or proofterms that may convince the user but not actually *explain*. Indeed the main steps are often flooded with straightforward logical and computational arguments, which are meant to be hidden by superdeduction modulo. In addition, the expressiveness of this latter framework is promising: Higher-order logics can be formalized using rewrite systems in deduction modulo [16,17,18]. PTS can be encoded in superdeduction modulo [19] and deduction modulo is proved to admit unbounded proof size speed-up [20]. Finally deduction modulo leads to interesting automated theorem proving procedures like ENAR [1] or TaMeD [21].

Superdeduction modulo is presented in Section 2. In this context, our contributions are the following: In Section 3 we introduce an extension of sequent calculus and show that superdeduction systems as well as a specific focusing system are instances of this extension. It consequently leads to a clear comparison of superdeduction with the focusing approach. Then in Section 4 we prove a focalization theorem for superdeduction modulo, which states that a cut-free proof in deduction modulo can be translated into a cut-free proof in raw superdeduction modulo (the converse being obvious), and which leads to criteria for theories that can extend deduction through the superdeduction modulo paradigm without endangering cut-elimination. Detailed proofs of theorems are available in a long version of this paper [22].

2 Superdeduction Modulo

In this section we define superdeduction modulo, which is the combination of deduction modulo and superdeduction. We consider the following classical sequent calculus LK_{core}, which contains the *deduction core* of our superdeduction modulo systems.

$$\textsc{Ax}\ \frac{}{\Gamma, \varphi \vdash \varphi, \Delta} \quad \bot_L\ \frac{}{\Gamma, \bot \vdash \Delta} \quad \bot_R\ \frac{\Gamma \vdash \Delta}{\Gamma \vdash \bot, \Delta} \quad \top_R\ \frac{}{\Gamma \vdash \top, \Delta} \quad \top_L\ \frac{\Gamma \vdash \Delta}{\Gamma, \top \vdash \Delta}$$

$$\wedge_L\ \frac{\Gamma, \varphi_1, \varphi_2 \vdash \Delta}{\Gamma, \varphi_1 \wedge \varphi_2 \vdash \Delta} \quad \wedge_R\ \frac{\Gamma \vdash \varphi_1, \Delta \qquad \Gamma \vdash \varphi_2, \Delta}{\Gamma \vdash \varphi_1 \wedge \varphi_2, \Delta} \quad \vee_L\ \frac{\Gamma, \varphi_1 \vdash \Delta \qquad \Gamma, \varphi_2 \vdash \Delta}{\Gamma, \varphi_1 \vee \varphi_2 \vdash \Delta}$$

$$\Rightarrow_R \frac{\Gamma, \varphi_1 \vdash \varphi_2, \Delta}{\Gamma \vdash \varphi_1 \Rightarrow \varphi_2, \Delta} \quad \Rightarrow_L \frac{\Gamma \vdash \varphi_1, \Delta \qquad \Gamma, \varphi_2 \vdash \Delta}{\Gamma, \varphi_1 \Rightarrow \varphi_2 \vdash \Delta} \quad \vee_R\ \frac{\Gamma \vdash \varphi_1, \varphi_2, \Delta}{\Gamma \vdash \varphi_1 \vee \varphi_2, \Delta}$$

$$\forall_R\ \frac{\Gamma \vdash \varphi, \Delta}{\Gamma \vdash \forall x.\varphi, \Delta}\ {\scriptstyle x \notin \mathcal{FV}(\Gamma,\Delta)} \quad \forall_L\ \frac{\Gamma, \varphi[t/x] \vdash \Delta}{\Gamma, \forall x.\varphi \vdash \Delta} \quad \exists_L\ \frac{\Gamma, \varphi \vdash \Delta}{\Gamma, \exists x.\varphi \vdash \Delta}\ {\scriptstyle x \notin \mathcal{FV}(\Gamma,\Delta)}$$

$$\mathrm{C}_R\ \frac{\Gamma \vdash \varphi, \varphi, \Delta}{\Gamma \vdash \varphi, \Delta} \quad \mathrm{C}_L\ \frac{\Gamma, \varphi, \varphi \vdash \Delta}{\Gamma, \varphi \vdash \Delta} \quad \textsc{Cut}\ \frac{\Gamma \vdash \varphi, \Delta \quad \Gamma, \varphi \vdash \Delta}{\Gamma \vdash \Delta} \quad \exists_R\ \frac{\Gamma \vdash \varphi[t/x], \Delta}{\Gamma \vdash \exists x.\varphi, \Delta}$$

A *term rewrite rule* rewrites first-order terms into first-order terms and a *proposition rewrite rule* rewrites an atomic proposition into an arbitrary formula. For instance $\mathrm{plus}(\mathrm{zero}, x) \to x$ is a term rewrite rule while $a \subseteq b \to \forall x. x \in a \Rightarrow x \in b$ is a proposition rewrite rule. We will consider two sets of rewrite rules: $\mathcal{T}h_1$ will contain computational axioms, and will be used to extend the deduction system through deduction modulo ; $\mathcal{T}h_2$ will contain deductive axioms, and will be used to enrich the deduction system through superdeduction. $\mathcal{T}h_1$ contains both term and proposition rewrite rules while $\mathcal{T}h_2$ is a set of proposition rewrite rules. We suppose that each rewrite rule $P \rightarrow \varphi$ of $\mathcal{T}h_2$ is associated with some name R then denoted $\mathsf{R} : P \rightarrow \varphi$. For $i \in \{1,2\}$, the one-step rewrite reduction associated with $\mathcal{T}h_i$ is denoted $\rightarrow_i$. The reflexive and transitive closure of $\rightarrow_i$ is denoted $\rightarrow_i^*$. The symmetric, reflexive and transitive closure of $\rightarrow_i$ is denoted $\equiv_i$. The notations $\rightarrow_{1,2}$, $\rightarrow_{1,2}^*$, and $\equiv_{1,2}$ are used for $\mathcal{T}h_1 \cup \mathcal{T}h_2$. The first-order axiom associated with a proposition rewrite rule $P \rightarrow \varphi$ is $\forall \bar{x}.(P \Leftrightarrow \varphi)$, where $\bar{x}$ represents the free variables of P and φ, denoted $\mathcal{FV}(P, \varphi)$. The first-order axiom associated with a term rewrite rule $l \rightarrow r$ is $\forall \bar{x}.(l = r)$, where $\bar{x}$ represents the free variables of l and r. This way of representing a term rewrite rule as a first-order axiom supposes that the logic contains an equality symbol. If it does not, one may add this predicate and the corresponding axioms in a conservative way as detailed in [23]. Another way to proceed with the term rewrite rules of $\mathcal{T}h_1$ is to use axioms of the form $\forall \bar{x}.\ P \Leftrightarrow Q$ for all $P \equiv_1 Q$. When writing $\vdash^{+\mathcal{T}h_2}_{\equiv_{\mathcal{T}h_1}}$, $\mathcal{T}h_1$ and $\mathcal{T}h_2$ will represent the rewrite rules, and when writing $\mathcal{T}h_1 \vdash$ or $\mathcal{T}h_2 \vdash$, they will represent the first-order axioms. For some deduction system $\vdash^\star_*$, we will just write $\Gamma \vdash^\star_* \Delta$ instead of the sentence *there is a proof of* $\Gamma \vdash^\star_* \Delta$. The sequents in the corresponding cut-free deduction system will be denoted $\Gamma \vdash^{cf\star}_* \Delta$.

Superdeduction stands for the addition to LK_{core} of new superdeduction inference rules that are computed from $\mathcal{T}h_2$ in the following way.

Definition 1 (Superdeduction rules computation [2]). *Let* Calc *be the set of inference rules formed of* Ax, $\bot_L$, $\top_R$, $\vee_L$, $\vee_R$, $\wedge_L$, $\wedge_R$, $\Rightarrow_L$, $\Rightarrow_R$, $\forall_L$, $\forall_R$, $\exists_L$, $\exists_R$, $\top_L$ *and* $\bot_R$. *Let us suppose* $\mathsf{R} : P \rightarrow \varphi \in \mathcal{T}h_2$. *To get the right (resp. left) rules associated with* R, *apply bottom-up the rules of* Calc *to the sequent* $\Gamma \vdash \varphi, \Delta$ *(resp.* $\Gamma, \varphi \vdash \Delta$*) until no connective of* φ *remain, collect the premises and side conditions, and finally replace* φ *by* P *in the conclusion.*

For instance, the rules associated with $\subseteq_{\mathsf{def}}$: $a \subseteq b \rightarrow \forall x. x \in a \Rightarrow x \in b$ are

$$\subseteq_{\mathsf{def}\,R}\ \frac{\Gamma, x \in a \vdash x \in b, \Delta}{\Gamma \vdash a \subseteq b, \Delta}\ x \notin \mathcal{FV}(\Gamma, \Delta) \qquad \subseteq_{\mathsf{def}\,L}\ \frac{\Gamma, t \in b \vdash \Delta \qquad \Gamma \vdash t \in a, \Delta}{\Gamma, a \subseteq b \vdash \Delta}$$

Since the propositional rules of Calc commute with any other rules, they may be applied in any order to reach axioms. However the application order of rules concerning quantifiers can be significant and the resulting side condition may differ: Decomposing $P \rightarrow (\exists x.A(x)) \Rightarrow (\exists x.B(x))$ on the right can lead to

$$\frac{\Gamma, A(x) \vdash B(t), \Delta}{\Gamma \vdash P, \Delta}\ x \notin \mathcal{FV}(\Gamma, \Delta) \quad \text{or} \quad \frac{\Gamma, A(x) \vdash B(t), \Delta}{\Gamma \vdash P, \Delta}\ x \notin \mathcal{FV}(\Gamma, \Delta, t)$$

depending on which existential quantifier is decomposed first. Although both rules are sound, only the first is complete. Therefore when computing superdeduction rules, one should give $\exists_L$ and $\forall_R$ a higher priority than the other rules, consequently ensuring that a weakest side condition is obtained.

Let us also remark that there may be several inference rules for introducing the same proposition rewrite rule on the left or on the right. An anonymous referee sagely pointed out the example of $P \to (\exists x_1.\forall x_2.A(x_1,x_2)) \vee (\exists y_1.\forall y_2.B(y_1,y_2))$ whose most general superdeduction rules are

$$\frac{\Gamma \vdash A(t,x_2), B(u,y_2), \Delta}{\Gamma \vdash P, \Delta} \begin{cases} x_2 \notin \mathcal{FV}(\Gamma,\Delta,u) \\ y_2 \notin \mathcal{FV}(\Gamma,\Delta) \end{cases}$$

and

$$\frac{\Gamma \vdash A(t,x_2), B(u,y_2), \Delta}{\Gamma \vdash P, \Delta} \begin{cases} x_2 \notin \mathcal{FV}(\Gamma,\Delta) \\ y_2 \notin \mathcal{FV}(\Gamma,\Delta,t) \end{cases}$$

This is not problematic since all the obtainable rules are available any time. In addition, let us remark that the hypotheses which imply our focalization theorem forbid this to occur (see Hypothesis 1).

Definition 2 (Superdeduction modulo). *The superdeduction modulo system associated with $(\mathcal{Th}_1, \mathcal{Th}_2)$ is formed of the rules of LK_{core} and the rules built upon $\mathcal{Th}_2$, where each first-order term or proposition is considered modulo rewriting through $\mathcal{Th}_1$. Sequents in this system are denoted $\Gamma \vdash^{+\mathcal{Th}_2}_{\equiv_{\mathcal{Th}_1}} \Delta$, or simply $\Gamma \vdash^{+_2}_{\equiv_1} \Delta$.*

If $\mathcal{Th}_2$ is empty, then we are in raw deduction modulo and sequents are denoted $\Gamma \vdash_{\equiv_1} \Delta$. If $\mathcal{Th}_1$ is empty, then we are in raw superdeduction and sequents are denoted $\Gamma \vdash^{+_2} \Delta$. We may use two equivalent ways to present inferences modulo $\equiv_1$: using explicit conversion rules, which rewrites part of the sequent or including rewriting in the inference rules of the system, as presented in [1]. Let us recall that we allow rewriting on *formulæ* in $\mathcal{Th}_1$. This can lead to odd situations such as a cut on a formula which is (rewrites to) both a conjunction and a disjunction. In order to avoid these kind of situation, **we will always suppose that $\mathcal{Th}_1$ is confluent besides of only rewriting atoms.**

Superdeduction modulo is sound *w.r.t.* predicate logic: Indeed it is proved in [1] that $\Gamma \vdash_{\equiv_1} \Delta$ if and only if $\mathcal{Th}_1, \Gamma \vdash \Delta$. In addition, it is proved in [2] that $\Gamma \vdash^{+_2} \Delta$ if and only if $\mathcal{Th}_2, \Gamma \vdash \Delta$. The soundness of superdeduction modulo is then a direct consequence of the following lemma.

Lemma 1. *If $\Gamma \vdash^{+_2}_{\equiv_1} \Delta$, then $\Gamma \vdash_{\equiv_{1,2}} \Delta$. If $\Gamma \vdash^{cf+_2}_{\equiv_1} \Delta$, then $\Gamma \vdash^{cf}_{\equiv_{1,2}} \Delta$.*

Proof. Superdeduction rules are replaced by rules of Calc and a step $\varphi \equiv_2 P$.

Theorem 1 (Soundness of $\vdash^{+_2}_{\equiv_1}$). *If $\Gamma \vdash^{+_2}_{\equiv_1} \Delta$, then $\mathcal{Th}_1, \mathcal{Th}_2, \Gamma \vdash \Delta$.*

Proof. Using Lemma 1, the proof of $\Gamma \vdash^{+_2}_{\equiv_1} \Delta$ is translated into a proof of $\Gamma \vdash_{\equiv_{1,2}} \Delta$, then by soundness of deduction modulo into a proof of $\mathcal{Th}_1, \mathcal{Th}_2, \Gamma \vdash \Delta$.

Let us now demonstrate the use of superdeduction modulo through an example. Deduction modulo is convenient for representing the computational part of some theory. For instance one can define addition using $0 + y \to y$ and $S(x) + y \to S(x + y)$, and multiplication using $0 * x \to 0$ and $S(x) * y \to y + (y * x)$.

Let us also define $\mathrm{sum}(x) = \sum_{k=0}^{x-1}(2 * k + 1)$ with the rules $\mathrm{sum}(0) \to 0$ and $\mathrm{sum}(S(x)) \to S(x+(x+\mathrm{sum}(x)))$. We obtain a convergent rewrite system, which is suitable for representing computation in deduction modulo. Superdeduction is convenient for representing the deductive part of a theory. Let us define natural numbers (and induction) using the rules $\mathbb{N}(n) \to \forall P.\ 0 \in P \Rightarrow \mathfrak{H}(P) \Rightarrow n \in P$ and $\mathfrak{H}(P) \to \forall k.\ k \in P \Rightarrow S(k) \in P$ [1]. Let us also define Leibniz's equality with the rule $x = y \to \forall P.\ x \in P \Rightarrow y \in P$. In the three latter rules, we wrote $x \in P$ instead of $P(x)$, for some formula P. This is allowed if we use (in $\mathcal{T}h_1$) the axiom $x \in \tilde{P} \to P(x)$ coming from the theory of classes, where $\tilde{P}$ denotes fresh constants associated with each formula P. The new inference rules for natural numbers and equality are then

$$\mathbb{N}_R\ \frac{\Gamma, 0\in P, \mathfrak{H}(P) \vdash^{+2}_{\equiv_1} n\in P, \Delta}{\Gamma \vdash^{+2}_{\equiv_1} \mathbb{N}(n), \Delta}\ P \notin \mathcal{FV}(\Gamma,\Delta)$$

$$\mathbb{N}_L\ \frac{\Gamma \vdash^{+2}_{\equiv_1} 0\in P, \Delta \quad \Gamma \vdash^{+2}_{\equiv_1} \mathfrak{H}(P), \Delta \quad \Gamma, n\in P \vdash^{+2}_{\equiv_1} \Delta}{\Gamma, \mathbb{N}(n) \vdash^{+2}_{\equiv_1} \Delta}$$

$$\mathfrak{H}_L\ \frac{\Gamma \vdash^{+2}_{\equiv_1} m\in P, \Delta \quad \Gamma, S(m)\in P \vdash^{+2}_{\equiv_1} \Delta}{\Gamma, \mathfrak{H}(P) \vdash^{+2}_{\equiv_1} \Delta} \qquad \mathfrak{H}_R\ \frac{\Gamma, k\in P \vdash^{+2}_{\equiv_1} S(k)\in P, \Delta}{\Gamma \vdash^{+2}_{\equiv_1} \mathfrak{H}(P), \Delta}\ k \notin \mathcal{FV}(\Gamma,\Delta)$$

$$=_R\ \frac{\Gamma, x \in P \vdash^{+2}_{\equiv_1} y \in P, \Delta}{\Gamma \vdash^{+2}_{\equiv_1} x = y, \Delta}\ P \notin \mathcal{FV}(\Gamma,\Delta) \qquad =_L\ \frac{\Gamma, y \in P \vdash^{+2}_{\equiv_1} \Delta \quad \Gamma \vdash^{+2}_{\equiv_1} x \in P, \Delta}{\Gamma, x = y \vdash^{+2}_{\equiv_1} \Delta}$$

Besides, the system is considered modulo the rules for addition, multiplication and $\mathrm{sum}(x)$. Then one can easily prove in the system that $\mathbb{N}(n)$ implies $\sum_{k=0}^{n-1}(2 * k + 1) = n^2$. The proof is

$$\mathbb{N}_L\ \dfrac{=_R\dfrac{\mathsf{Ax}(0 \in A)}{\vdash^{+2}_{\equiv_1} 0 = 0} \qquad \mathfrak{H}_R\dfrac{=_L\dfrac{=_R\dfrac{\mathsf{Ax}(S(m+(m+\mathrm{sum}(m))) \in A)}{\vdash^{+2}_{\equiv_1} S(m+(m+\mathrm{sum}(m))) = S(m+(m+\mathrm{sum}(m)))} \quad \begin{array}{c}\mathsf{Ax}(S(m+(m+\mathrm{sum}(m))) = S(m+(m+(m*m))))\\ \vdots\end{array}}{\mathrm{sum}(m) = m*m \vdash^{+2}_{\equiv_1} S(m+(m+\mathrm{sum}(m))) = S(m+(m+(m*m)))}}{\begin{array}{c}\vdash^{+2}_{\equiv_1} \mathfrak{H}(\tilde{P})\\ \vdots\end{array}} \qquad \mathsf{Ax}(\mathrm{sum}(n) = n * n)}{\mathbb{N}(n) \vdash^{+2}_{\equiv_1} \mathrm{sum}(n) = n * n}$$

where $\tilde{P}$ is the first-order constant associated with the formula $\mathrm{sum}(x) = x * x$ and $\mathsf{Ax}(\varphi)$ stands for the proof $\mathrm{Ax}\ \dfrac{}{\varphi \vdash^{cf+2}_{\equiv_1} \varphi}$. The premises of the rule $=_L$ are $\vdash \mathrm{sum}(m) \in \tilde{Q}$ and $m{*}m \in \tilde{Q} \vdash S(m{+}(m{+}\mathrm{sum}(m))) = S(m{+}(m{+}(m{*}m)))$ where $\tilde{Q}$ is the first-order constant associated with the formula $S(m+(m+\mathrm{sum}(m))) = S(m+(m+x))$. Let us notice that the proof uses only superdeduction rules and Ax rules.

[1] The intermediate predicate $\mathfrak{H}(P)$ is introduced to ensure completeness, as we will see in Section 3. I also refer the reader to [2] for full explanations.

3 Superdeduction and Focusing Systems

The aim of this section is to explore how focusing and superdeduction are related. For example the distinction between the two kinds of sequents that focusing introduces, namely unfocused and focused sequents, is very similar to the distinction between sequents used in the superdeduction toplevel and (meta)sequents used during the superdeduction inference rules computation: In this latter case, a specific formula φ appearing in a proposition rewrite rule $P \rightarrow \varphi$ replaces P and is then focused until the complete decomposition of its connectives.

Let us propose an extension of classical sequent calculus based on this distinction between focused and unfocused sequents. We will see that superdeduction systems as well as a specific focusing system for LK_{core} are instances of this extension. Starting from LK_{core}, we add the following rules.

$$\frac{\Gamma(\vdash \varphi)\Delta}{\Gamma \vdash \psi, \Delta}\,\textsc{In}_R \qquad \frac{\Gamma(\varphi \vdash)\Delta}{\Gamma, \psi \vdash \Delta}\,\textsc{In}_L \qquad \frac{\Gamma, \Gamma' \vdash \Delta', \Delta}{\Gamma(\Gamma' \vdash \Delta')\Delta}\,\textsc{Out}$$

$$\frac{}{\Gamma(\Gamma', A \vdash A, \Delta')\Delta}\,\overline{\mathrm{Ax}} \qquad \frac{}{\Gamma(\Gamma', \bot \vdash \Delta')\Delta}\,\overline{\bot_L} \qquad \frac{}{\Gamma(\Gamma' \vdash \top, \Delta')\Delta}\,\overline{\top_R}$$

$$\frac{\Gamma(\Gamma' \vdash \Delta')\Delta}{\Gamma(\Gamma' \vdash \bot, \Delta')\Delta}\,\overline{\bot_R} \qquad \frac{\Gamma(\Gamma' \vdash \Delta')\Delta}{\Gamma(\Gamma', \top \vdash \Delta')\Delta}\,\overline{\top_L}$$

$$\frac{\Gamma(\Gamma' \vdash \varphi_1, \Delta')\Delta \quad \Gamma(\Gamma' \vdash \varphi_2, \Delta')\Delta}{\Gamma(\Gamma' \vdash \varphi_1 \wedge \varphi_2, \Delta')\Delta}\,\overline{\wedge_R} \qquad \frac{\Gamma(\Gamma', \varphi_1, \varphi_2 \vdash \Delta')\Delta}{\Gamma(\Gamma', \varphi_1 \wedge \varphi_2 \vdash \Delta')\Delta}\,\overline{\wedge_L}$$

$$\frac{\Gamma(\Gamma' \vdash \varphi_1, \varphi_2, \Delta')\Delta}{\Gamma(\Gamma' \vdash \varphi_1 \vee \varphi_2, \Delta')\Delta}\,\overline{\vee_R} \qquad \frac{\Gamma(\Gamma', \varphi_1 \vdash \Delta')\Delta \quad \Gamma(\Gamma', \varphi_2 \vdash \Delta')\Delta}{\Gamma(\Gamma', \varphi_1 \vee \varphi_2 \vdash \Delta')\Delta}\,\overline{\vee_L}$$

$$\frac{\Gamma(\Gamma', \varphi_2 \vdash \Delta')\Delta \quad \Gamma(\Gamma' \vdash \varphi_1, \Delta')\Delta}{\Gamma(\Gamma', \varphi_1 \Rightarrow \varphi_2 \vdash \Delta')\Delta}\,\overline{\Rightarrow_R} \qquad \frac{\Gamma(\Gamma', \varphi_1 \vdash \varphi_2, \Delta')\Delta}{\Gamma(\Gamma' \vdash \varphi_1 \Rightarrow \varphi_2, \Delta')\Delta}\,\overline{\Rightarrow_L}$$

$$\overline{\forall_R}\,\frac{\Gamma(\Gamma' \vdash \varphi, \Delta')\Delta}{\Gamma(\Gamma' \vdash \forall x.\varphi, \Delta')\Delta}\; x \notin \mathcal{FV}(\Gamma, \Gamma', \Delta, \Delta') \qquad \overline{\forall_L}\,\frac{\Gamma(\Gamma', \varphi[t/x] \vdash \Delta')\Delta}{\Gamma(\Gamma', \forall x.\varphi \vdash \Delta')\Delta}$$

$$\overline{\exists_R}\,\frac{\Gamma(\Gamma' \vdash \varphi[t/x], \Delta')\Delta}{\Gamma(\Gamma' \vdash \exists x.\varphi, \Delta')\Delta} \qquad \overline{\exists_L}\,\frac{\Gamma(\Gamma, \varphi \vdash \Delta')\Delta}{\Gamma(\Gamma, \exists x.\varphi \vdash \Delta')\Delta}\; x \notin \mathcal{FV}(\Gamma, \Gamma', \Delta, \Delta')$$

The system features then two kinds of sequents: Unfocused sequents denoted $\Gamma \vdash \Delta$ and focused sequents denoted $\Gamma(\Gamma' \vdash \Delta')\Delta$. Unfocused sequents are handled by the rules of classical sequent calculus; focused sequents are handled by the overlined rules; entering a focusing sequence is handled by rules $\textsc{In}_R$ and $\textsc{In}_L$; finally leaving a focusing sequence is handled by rule $\textsc{Out}$. In addition, the obtained sequent calculus has the three following parameters: A condition $\mathcal{C}_{\textsc{In}}$, which is enforced when applying the rules $\textsc{In}_R$ and $\textsc{In}_L$, a condition $\mathcal{C}_{\textsc{Out}}$, which is enforced when applying the rule $\textsc{Out}$, a condition $\mathcal{C}_{\textsc{Focus}}$, which is enforced when applying any of the overlined rules and a condition $\mathcal{C}_{\textsc{Unfocus}}$, which is enforced when applying any of the raw LK_{core} rules. Therefore the resulting deduction system is denoted $\mathrm{LK}(\mathcal{C}_{\textsc{In}}, \mathcal{C}_{\textsc{Out}}, \mathcal{C}_{\textsc{Focus}}, \mathcal{C}_{\textsc{Unfocus}})$. $\mathcal{C}_{\textsc{In}}$ and $\mathcal{C}_{\textsc{Out}}$ are meant to respectively control the bottom-up entrance and exit of focusing phases; $\mathcal{C}_{\textsc{Focus}}$ and $\mathcal{C}_{\textsc{Unfocus}}$ are meant to respectively control the deduction inside of focused and unfocused phases. For instance if $\mathcal{C}_{\mathcal{T}h_1}$ is the condition that forces $\varphi \equiv_1 \psi$ in the rules IN_R and IN_L, then $\text{LK}(\mathcal{C}_{Th_1}, \text{true}, \text{false}, \text{true})$ is equivalent to deduction modulo Th_1: The overlined rules are simply discarded, and the IN_R and IN_L rules are turned into conversion rules for Th_1. Both focusing proofs and superdeduction can also be formulated as instances of $\text{LK}(\mathcal{C}_{\text{IN}}, \mathcal{C}_{\text{OUT}}, \mathcal{C}_{\text{FOCUS}}, \mathcal{C}_{\text{UNFOCUS}})$.

Focusing. We will say that a (focused or unfocused) sequent is positive if the head connectives of its non-atomic formulæ are $\forall$ on the left and $\exists$ on the right. Then one can define the following focusing system for LK_{core}: $\mathcal{F}_+$ is the condition that restricts the conclusion of the applied inference rule to be a positive sequent. $\mathcal{F}_-$ is the condition that restricts the conclusion of the applied inference rule not to be a positive sequent. Then $\text{LK}(\mathcal{F}_+ \wedge \varphi = \psi, \mathcal{F}_-, \mathcal{F}_+, \mathcal{F}_-)$ is a focusing system for LK_{core}. It is quite different from the focusing system LKF presented in [24], in particular regarding our asynchronous treatment of propositional connectives. However both systems only allow to focus on a single formula. It is syntactically ensured in LKF, while ensured by $\mathcal{F}_+$ together with the shape of $\overline{\exists_R}$ and $\overline{\forall_L}$ (the only overlined rules allowed by $\mathcal{F}_+$) in $\text{LK}(\mathcal{F}_+ \wedge \varphi = \psi, \mathcal{F}_-, \mathcal{F}_+, \mathcal{F}_-)$. Focused sequents, which are denoted $\mapsto [\Theta], P$ in [24], are denoted $(\vdash P)\Theta$ here.

Superdeduction. The explicit superdeduction system associated with Th_2 is defined as $\text{LK}(\mathcal{C}_{Th_2}, \mathcal{C}_{\text{atoms}}, \text{true}, \text{true})$ where $\mathcal{C}_{Th_2}$ is the condition (for the application of rules IN_R and IN_L) that states that $\psi \rightarrow \varphi$ is a rule of Th_2 and $\mathcal{C}_{\text{atoms}}$ is the condition that restricts the focused part of the conclusion of the rule OUT to contain only atoms. $\text{LK}(\mathcal{C}_{Th_2}, \mathcal{C}_{\text{atoms}}, \text{true}, \text{true})$ is equivalent to the superdeduction system associated with Th_2, since each focused phase exactly corresponds with the computation of some superdeduction inference rule.

On the one hand, $\text{LK}(\mathcal{C}_{\text{IN}}, \mathcal{C}_{\text{OUT}}, \mathcal{C}_{\text{FOCUS}}, \mathcal{C}_{\text{UNFOCUS}})$, as the *supremum* of a specific focusing system for LK_{core} and superdeduction systems, makes clear the similarities between these two approaches:

- Focused and unfocused sequents are syntactically identified.
- Focused sequents focus on a special part of the sequent.
- The system features three kinds of inferences, which either handle focused sequents, handle unfocused sequents or controls the enter/exit of the focus.

On the other hand, it also underlines the dissimilarities:

- Focused sequents usually focus on a single formula in the focusing approach (there are exceptions, such as multifocusing [25]), but they may focus on several in the superdeduction approach.
- Focusing phases of superdeduction contain an unfolding step (IN rules) besides purely logical steps.
- In the focusing approach, the De Morgan dual of a positive connective is usually negative and conversely. On the contrary, superdeduction locally divides occurrences of connectives into toplevel occurrences (decomposed in unfocused phases) and connectives appearing in axioms of Th_2, but a conjunction, for instance, can appear as a toplevel connective somewhere and in the explicit decomposition of a superdeduction axiom.

– In the focusing approach, $\mathcal{F}_+$ and $\mathcal{F}_-$ state that the focusing phase is entered (bottom-up) only when no negative connective remains (the latest possible), and is left as soon as a negative connective is found (the soonest possible). This aspect plays an important role in the completeness of the focusing approach. In the superdeduction approach, $\mathcal{C}_{\mathcal{T}h_2}$ and $\mathcal{C}_{\text{atoms}}$ state that the focusing phase is entered anytime a superdeduction axiom is unfolded, and is left only when no connective of the axiom remains (the latest).

More generally, let us stress the fact in the superdeduction approach, the focused phases are *directed* by axioms of $\mathcal{T}h_2$ (therefore *axiom directed focusing*). In particular these axioms direct when the focusing phases can be entered and exited. They also direct which connectives will be decomposed during a single focusing phase. Consequently one can propose syntactical criteria on $\mathcal{T}h_2$ to ensure completeness or cut-elimination of the system (the later will be dealt with in the next section). Bipoles, which come from the focusing approach and whose definition is adapted from [26], may be seen as such a characterization of which formulæ can be used in $\mathcal{T}h_2$ in order to obtain a complete superdeduction system: If for some $\mathsf{R} : P \to \varphi \in \mathcal{T}h_2$, φ and $\neg\varphi$ are both bipoles, then the rules R_R and R_L are complete. The definition of bipoles strongly relies on the non-permutability cases: As analysed by Kleene in [27], there are four cases for which two steps of LK_{core} cannot be permuted:

$$\forall_R \dfrac{\exists_R \dfrac{\Gamma \vdash^{cf+2}_{\equiv_1} A(x), B(x), \Delta}{\Gamma \vdash^{cf+2}_{\equiv_1} A(x), \psi, \Delta}\ \psi \equiv_1 \exists x.B(x)}{\Gamma \vdash^{cf+2}_{\equiv_1} \varphi, \psi, \Delta}\ \varphi \equiv_1 \forall x.A(x) \qquad \begin{array}{l}\exists_L \text{ receives the variable } x\\ \forall_R \text{ emits the variable } x\end{array}$$

and similarly where $\forall_R$ is replaced by $\exists_L$ and/or $\exists_R$ is replaced by $\forall_L$. Kleene's analysis in [27] leads to strong results about the permutations of inferences in proofs. This analysis transfered to deduction modulo already allowed Hermant to prove cut-elimination results for deduction modulo in [8] and we will see in Section 4 that transfering this analysis to superdeduction modulo allows to prove cut-elimination results for superdeduction modulo. Before that, let us define positive and negative connectives as well as bipoles and monopoles.

Definition 3 (Polarity of a subformula). *The polarity $pol_\varphi(\psi)$ of ψ in φ where ψ is an occurrence of a subformula of φ is defined as*
true *if* $\varphi = \psi$;
$pol_{\varphi_i}(\psi)$ *if* $\varphi = \varphi_1 \wedge \varphi_2$ *or* $\varphi_1 \vee \varphi_2$ *or* $\forall x.\varphi_1$ *or* $\exists x.\varphi_1$*, and* ψ *occurs in* φ_i;
$\neg pol_{\varphi_1}(\psi)$ *if* $\varphi = \varphi_1 \Rightarrow \varphi_2$ *and if* ψ *occurs in* φ_1;
$pol_{\varphi_2}(\psi)$ *if* $\varphi = \varphi_1 \Rightarrow \varphi_2$ *and if* ψ *occurs in* φ_2.

Definition 4 (Neutral/Positive/Negative connectives). *In a formula φ, we will say that an occurrence of a connective is* neutral *if it is not a quantifier,* positive *(or synchronous) if it is a universal quantifier of polarity* true *or an existential quantifier of polarity* false, negative *(or asynchronous) if it is a universal quantifier of polarity* false *or an existential quantifier of polarity* true.

Definition 5 (Monopoles/Bipoles). *Monopoles are formulæ built from atoms with neutral and negative connectives. Bipoles are formulæ built from monopoles with neutral and positive connectives.*

These pictures represent respectively a bipole, a formula whose negation is a bipole and a bipole whose negation is also a bipole. Minuses represent negative connectives, pluses represent positive connectives and the grey background represents neutral connectives.

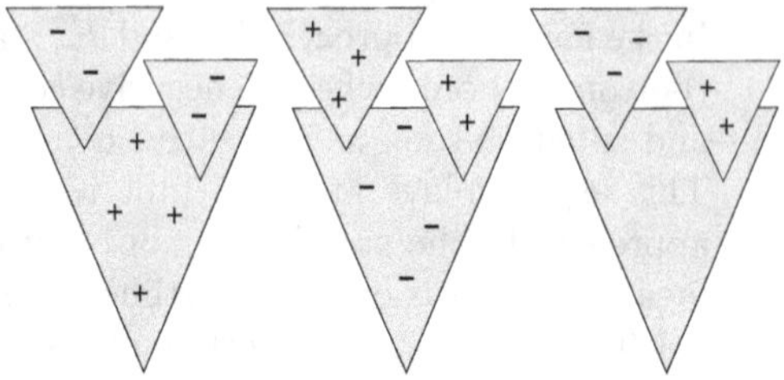

Theorem 2 (Completeness of superdeduction modulo) *If for all φ appearing in $P \rightarrow \varphi \in \mathit{Th}_2$, φ and $\neg\varphi$ are both bipoles, then $\mathit{Th}_1, \mathit{Th}_2, \Gamma \vdash \Delta$ implies $\Gamma \vdash^{+_2}_{\equiv_1} \Delta$.*

Proof. From the completeness of superdeduction, which is proved in [15], for all $\varphi \in \mathit{Th}_2$, there exists a proof of $\vdash^{+_2} \varphi$. From the completeness of deduction modulo, for all $\varphi \in \mathit{Th}_1$, there exists a proof of $\vdash_{\equiv_1} \varphi$. Starting from a proof of $\mathit{Th}_1, \mathit{Th}_2, \Gamma \vdash \Delta$, using cuts with the proofs of $\vdash^{+_2}_{\equiv_1} \varphi$ for all $\varphi \in \mathit{Th}_1 \cup \mathit{Th}_2$, we get a proof of $\Gamma \vdash^{+_2}_{\equiv_1} \Delta$.

Then for the rest of the paper we make the same hypothesis as in [2,15]:

Hypothesis 1. *If $P \rightarrow \varphi \in \mathit{Th}_2$, then φ and $\neg\varphi$ are both bipoles.*

Nevertheless, this hypothesis does not restrict the set of theories that superdeduction can handle. Indeed a procedure, greatly inspired by Andreoli's Bipolarisation [26], has been proposed in [2] to transform any set of proposition rewrite rules into an equivalent one verifying this hypothesis, namely, for all $P \rightarrow \varphi \in \mathit{Th}_2$, replacing φ by its greatest prefix that is a bipole and whose negation is also a bipole; any subformula ψ of φ that is consequently separated from φ is then replaced in the prefix by a fresh predicate symbol Q parametrised by the free variables of ψ; finally the rule $Q \rightarrow \psi$ is recursively processed by the procedure and then added to Th_2. For instance the proposition rewrite rule $P \rightarrow (\forall x A(x) \Rightarrow (\forall x.\exists y.B(x,y))$ is transformed into

$$P \rightarrow (\forall x.A(x)) \Rightarrow (\forall x.Q(x)) \quad \text{and} \quad Q(x) \rightarrow \exists y.B(x,y) \ .$$

Before going to the next section and proving cut-elimination results for superdeduction modulo, we make another syntactic hypothesis on axioms of Th_2.

Hypothesis 2. *If $P \rightarrow \varphi \in \mathit{Th}_2$, then φ is in prenex normal form.*

Let us notice that together with Hypothesis 1, it implies the following lemma.

Lemma 2. *If $P \rightarrow \varphi \in \mathit{Th}_2$, then φ is either $\forall x_1 \ldots x_n.\psi$ or $\exists x_1 \ldots x_n.\psi$ for some prop. formula ψ. We will resp. say that φ is* right-handed *and* left-handed.

Hypothesis 2 does not restrict the set of theories that superdeduction can handle, since the procedure proposed in [2] that turns any set of proposition rewrite rules into an equivalent one verifying Hypothesis 1 can easily be strengthened into a procedure that turns any set of proposition rewrite rules into an equivalent one verifying both Hypotheses 1 and 2.

Hypotheses 2 and 1 can be replaced by the following single hypothesis (which we will call Hypothesis 3): For any φ appearing in $P \to \varphi \in \mathcal{Th}_2$, either φ or its negation is a monopole. Then the only quantifiers of φ are either positive $\forall$ and negative $\exists$, or negative $\forall$ and positive $\exists$. Indeed in this case, one can always find a prenex normal form ψ classically equivalent to φ that satisfies Hypotheses 2 and 1 and such as the inference rules associated with $P \to \varphi$ are the same as the rules associated with $P \to \psi$. The deduction system is consequently unchanged and then our final result also holds if we just replace Hypotheses 2 and 1 by Hypothesis 3. Let us illustrate this with an example: The proposition rewrite rule $\mathsf{R} : P \to (\forall x.A(x)) \Rightarrow (\exists y.B(y))$ verifies Hypothesis 3 (which in turn implies Hypothesis 1), but does not verify Hypothesis 2. However $(\forall x.A(x)) \Rightarrow (\exists y.B(y))$ is classically equivalent to $\exists x.\exists y.(A(x) \Rightarrow B(y))$. Furthermore we obtain the same superdeduction inference rules if we replace the proposition rewrite rule R by $\mathsf{R}' : P \to \exists x.\exists y.(A(x) \Rightarrow B(y))$, namely

$$\frac{\Gamma, A(t_1) \vdash B(t_2), \Delta}{\Gamma \vdash P, \Delta} \qquad \frac{\Gamma, B(y) \vdash \Delta \quad \Gamma \vdash A(x), \Delta}{\Gamma, P \vdash \Delta}\ x, y \notin \mathcal{FV}(\Gamma, \Delta)$$

Finally R' satisfies both Hypotheses 1 and 2.

4 Focalization in Cut-Free Superdeduction Modulo

Using awkard theories to extend deduction through deduction modulo or superdeduction is known to jeopardize cut-elimination (completeness of the cut-free deduction system) and normalisation (termination of a cut-elimination procedure). For example if $A \equiv_{1,2} A \Rightarrow A$, one can easily build a proof of $\vdash^{+_2}_{\equiv_1} A$ [2] that does not normalize. However cut-elimination and normalisation are well-studied for deduction modulo. Several proofs and criteria have been presented for cut-elimination and normalisation. Let us cite first the early work of Dowek and Werner [5]. Using reducibility candidates they prove that the existence of a *pre-model*[3] for some rewrite system implies the normalisation of intuitionistic natural deduction modulo this rewrite system. Then they transfer this result to classical logic using *light double negation*[3].

Theorem 3 ([5]). *If the light double negation of a rewrite system $\mathcal{R}$ has a premodel, cut-elimination holds for the classical sequent calculus modulo $\mathcal{R}$.*

Hermant used semantic methods to study cut-elimination for the intuitionistic and classical sequent calculus in [9,28,8]. Among others, he proved the following cut-elimination theorem for classical sequent calculus modulo.

[2] The proof may be written $(\lambda x.\ x\ x)\ (\lambda x.\ x\ x)$ in a λ-calculus style.

[3] Both definitions of light double negation and pre-model can be found in [5].

Theorem 4 ([8]). *If $\mathcal{R}$ is a rewrite system compatible with a well-founded order, if $\mathcal{R}^+$ is a positive rewrite system*[4] *whose right-hand sides are $\mathcal{R}$-normal forms and if $\mathcal{R} \cup \mathcal{R}^+$ is confluent, then cut-elimination holds for the classical sequent calculus modulo $\mathcal{R} \cup \mathcal{R}^+$.*

Finally Burel and Kirchner proposed another approach in [10] where they use abstract canonical systems and abstract completion in order to mechanically transform a classical sequent calculus system into an equivalent one having the cut-elimination property.

Theorem 5 ([10]). *A sequent has a proof in classical sequent calculus modulo some theory $\mathcal{R}$ if and only if it has a cut-free proof in the* saturated theory *corresponding to $\mathcal{R}$.*

Our aim is now to relate superdeduction modulo to deduction modulo in order to prove that all these criteria are extendible to superdeduction modulo. We will only consider cut-free deduction systems and show that they allow to prove exactly the same sequents: A sequent $\Gamma \vdash^{cf+2}_{\equiv_1} \Delta$ is provable if and only if $\Gamma \vdash^{cf}_{\equiv_{1,2}} \Delta$ is. The translation of a proof in superdeduction modulo into sheer deduction modulo is already shown in Lemma 1. Now let us prove the converse: If $\Gamma \vdash^{cf}_{\equiv_{1,2}} \Delta$, then $\Gamma \vdash^{cf+2}_{\equiv_1} \Delta$. To achieve this goal, we will first consider some proposition rewrite rule $\mathsf{R} : P \to \varphi$ of $\mathcal{T}h_2$ and prove that if $\Gamma \vdash^{cf+2}_{\equiv_1} \varphi, \Delta$, then $\Gamma \vdash^{cf+2}_{\equiv_1} P, \Delta$; if $\Gamma, \varphi \vdash^{cf+2}_{\equiv_1} \Delta$, then $\Gamma, P \vdash^{cf+2}_{\equiv_1} \Delta$. This property then implies the admissibility of rules $\equiv_{2R} \dfrac{\Gamma \vdash \varphi, \Delta}{\Gamma \vdash \psi, \Delta} \varphi \equiv_2 \psi$ and $\equiv_{2L} \dfrac{\Gamma, \varphi \vdash \Delta}{\Gamma, \psi \vdash \Delta} \varphi \equiv_2 \psi$ in *cut-free* superdeduction modulo, which in turn implies that if $\Gamma \vdash^{cf}_{\equiv_{1,2}} \Delta$, then $\Gamma \vdash^{cf+2}_{\equiv_1} \Delta$. The proof of the property deals with permutability problems in classical sequent calculus: Indeed the intuition is to unite the steps of the proof that decompose φ into a one-step decomposition of P. Considering the permutability problems that one can have when dealing with quantifiers (see Section 3), we can easily build a proof of some sequent $\Gamma \vdash^{+2}_{\equiv_1} \varphi, \Delta$ where the three steps decomposing φ, marked with the symbol $*$, cannot be united using sheer permutations:

$$
\Rightarrow_R \dfrac{\forall_R \dfrac{\forall_L \dfrac{\exists_L \dfrac{\forall_L \dfrac{\dfrac{\dots}{A(y_0) \Rightarrow B(x_0), A(y_0) \vdash B(x_0)}}{A(y_0) \Rightarrow B(x_0), (\forall x.A(x)) \vdash B(x_0)} *}{\exists y.(A(y) \Rightarrow B(x_0)), (\forall x.A(x)) \vdash B(x_0)}}{\forall x.\exists y.(A(y) \Rightarrow B(x)), (\forall x.A(x)) \vdash B(x_0)}}{\forall x.\exists y.(A(y) \Rightarrow B(x)), (\forall x.A(x)) \vdash (\forall x.B(x))} *}{\forall x.\exists y.(A(y) \Rightarrow B(x)) \vdash (\forall x.A(x)) \Rightarrow (\forall x.B(x))} *
$$

where φ is $(\forall x.A(x)) \Rightarrow (\forall x.B(x))$. Let us notice both φ and $\neg\varphi$ are bipoles (Hypothesis 1 is verified), but Hypothesis 2 is not verified. A solution is to combine the permutations with contractions as follows.

[4] The definition of a *positive* rewrite system can be found in [8].

$$
\begin{array}{lcl}
 & \dots & \\ \hline
 & A(y_0)\Rightarrow B(x_0), A(x_0), A(y_0)\vdash B(x_1), B(x_0) & \\ \hline
\forall_R & A(y_0)\Rightarrow B(x_0), A(x_0), A(y_0)\vdash \forall x.B(x), B(x_0) & * \\ \hline
\forall_L & A(y_0)\Rightarrow B(x_0), A(x_0), \forall x.A(x)\vdash \forall x.B(x), B(x_0) & * \\ \hline
\Rightarrow_R & A(y_0)\Rightarrow B(x_0), A(x_0)\vdash \varphi, B(x_0) & * \\ \hline
\exists_L & \exists y.A(y)\Rightarrow B(x_0), A(x_0)\vdash \varphi, B(x_0) & \\ \hline
\forall_L & \forall x.\exists y.A(y)\Rightarrow B(x), A(x_0)\vdash \varphi, B(x_0) & \\ \hline
\forall_L & \forall x.\exists y.A(y)\Rightarrow B(x), \forall x.A(x)\vdash \varphi, B(x_0) & * \\ \hline
\forall_R & \forall x.\exists y.A(y)\Rightarrow B(x), \forall x.A(x)\vdash \varphi, \forall x.B(x) & * \\ \hline
\Rightarrow_R & \forall x.\exists y.A(y)\Rightarrow B(x)\vdash \varphi, (\forall x.A(x))\Rightarrow(\forall x.B(x)) & * \\ \hline
\textsc{Contr}_R & \forall x.\exists y.(A(y)\Rightarrow B(x))\vdash (\forall x.A(x))\Rightarrow(\forall x.B(x)) &
\end{array}
$$

In this proof, all the steps decomposing $(\forall x.A(x)) \Rightarrow (\forall x.B(x))$ have been duplicated and then united. It seems that this manipulation allows to transform any proof using several LK inference rules to decompose a formula φ into a proof using several occurrences of a single superdeduction inference rule decomposing the corresponding predicate P ($P \to \varphi \in \mathit{Th}_2$). However proving that this manipulation always solves the problem remains an open question to our knowledge. In addition, the cunjonction of Hypotheses 1 and 2 prevents these permutability problems to appear.

Now let us prove that $\vdash^{cf+_2}_{\equiv_1}$ is equivalent to $\vdash^{cf}_{\equiv_{1,2}}$. Dealing with right-handed formulæ on the right and left-handed formulæ on the left of a sequent needs the following lemma, adapted from a central lemma in Hermant's semantic proofs of cut-elimination for deduction modulo [8].

Lemma 3 (Kleene's lemma adapted to $\vdash^{+_2}_{\equiv_1}$)
Let $A_1 \equiv_1 A_2 \equiv_1 \dots A_n \equiv_1 \varphi$ be some propositions. Let $\Theta = A_1, A_2 \dots A_n$.

- *If $\varphi = \neg A$ and $\Gamma, \Theta \vdash^{cf+_2}_{\equiv_1} \Delta$ then $\Gamma \vdash^{cf+_2}_{\equiv_1} A, \Delta$.*
- *If $\varphi = A \wedge B$ and $\Gamma, \Theta \vdash^{cf+_2}_{\equiv_1} \Delta$ then $\Gamma, A, B \vdash^{cf+_2}_{\equiv_1} \Delta$.*
- *If $\varphi = A \vee B$ and $\Gamma, \Theta \vdash^{cf+_2}_{\equiv_1} \Delta$ then $\Gamma, A \vdash^{cf+_2}_{\equiv_1} \Delta$ and $\Gamma, B \vdash^{cf+_2}_{\equiv_1} \Delta$.*
- *If $\varphi = A \Rightarrow B$ and $\Gamma, \Theta \vdash^{cf+_2}_{\equiv_1} \Delta$ then $\Gamma, B \vdash^{cf+_2}_{\equiv_1} \Delta$ and $\Gamma \vdash^{cf+_2}_{\equiv_1} A, \Delta$.*
- *If $\varphi = \exists x.Q$ and $\Gamma, \Theta \vdash^{cf+_2}_{\equiv_1} \Delta$ then $\Gamma, Q[c/x] \vdash^{cf+_2}_{\equiv_1} \Delta$ for some fresh variable c.*

- *If $\varphi = \neg A$ and $\Gamma \vdash^{cf+_2}_{\equiv_1} \Theta, \Delta$ then $\Gamma, A \vdash^{cf+_2}_{\equiv_1} \Delta$.*
- *If $\varphi = A \wedge B$ and $\Gamma \vdash^{cf+_2}_{\equiv_1} \Theta, \Delta$ then $\Gamma \vdash^{cf+_2}_{\equiv_1} A, \Delta$ and $\Gamma \vdash^{cf+_2}_{\equiv_1} B, \Delta$.*
- *If $\varphi = A \vee B$ and $\Gamma \vdash^{cf+_2}_{\equiv_1} \Theta, \Delta$ then $\Gamma \vdash^{cf+_2}_{\equiv_1} A, B, \Delta$.*
- *If $\varphi = A \Rightarrow B$ and $\Gamma \vdash^{cf+_2}_{\equiv_1} \Theta, \Delta$, then $\Gamma, A \vdash^{cf+_2}_{\equiv_1} B, \Delta$.*
- *If $\varphi = \forall x.Q$ and $\Gamma \vdash^{cf+_2}_{\equiv_1} \Theta, \Delta$ then $\Gamma \vdash^{cf+_2}_{\equiv_1} Q[c/x], \Delta$ for some fresh variable c.*

Proof. For each assertion, by induction. The proof is detailed in [22].

It implies the following lemma.

Lemma 4. *Let us consider some $\mathsf{R} : P \to \varphi \in \mathit{Th}_2$.*

- *If φ is right-handed and $\Gamma \vdash^{cf+_2}_{\equiv_1} \varphi, \Delta$, then there exist (cut-free) proofs of each premise of R_R. Therefore $\Gamma \vdash^{cf+_2}_{\equiv_1} P, \Delta$.*

– *If φ is left-handed and $\Gamma, \varphi \vdash^{cf+_2}_{\equiv_1} \Delta$, then there exist (cut-free) proofs of each premise of* R_L. *Therefore* $\Gamma, P \vdash^{cf+_2}_{\equiv_1} \Delta$.

Proof. By iteration of Lemma 3. The proof is detailed in [22]. The idea is to unite the decomposition of φ at the root of the proof: All these steps can move downward since φ is right-handed and decomposed on the right.

Lemma 5 deals with right-handed formulæ on the left and vice versa.

Lemma 5. *Let us consider* $\mathsf{R} : P \rightarrow \varphi \in \mathit{Th}_2$.

– *If φ is right-handed and $\Gamma, \varphi \vdash^{cf+_2}_{\equiv_1} \Delta$, then $\Gamma, P \vdash^{cf+_2}_{\equiv_1} \Delta$.*
– *If φ is left-handed and $\Gamma \vdash^{cf+_2}_{\equiv_1} \varphi, \Delta$, then $\Gamma \vdash^{cf+_2}_{\equiv_1} P, \Delta$.*

Proof. The proof is detailed in [22]. The idea is to fully decompose φ, partially eliminate contractions, and then to pull up the decomposition of φ.

Lemmas 4 and 5 are concentrated in the following theorem, which proves that rewriting through $\mathit{Th}_1 \cup \mathit{Th}_2$ preserves provability in superdeduction modulo.

Theorem 6. *If* $\varphi \equiv_{1,2} \psi$, *then* $\begin{cases} \Gamma \vdash^{cf+_2}_{\equiv_1} \varphi, \Delta & \textit{if and only if} \quad \Gamma \vdash^{cf+_2}_{\equiv_1} \psi, \Delta \\ \Gamma, \varphi \vdash^{cf+_2}_{\equiv_1} \Delta & \textit{if and only if} \quad \Gamma, \psi \vdash^{cf+_2}_{\equiv_1} \Delta \end{cases}$

Proof. First by Lemmas 4 and 5, provability is preserved by one-step head reduction through Th_2 (*i.e.* if $\varphi = P\sigma$ for some substitution σ and $\psi = \phi\sigma$ and $P \rightarrow \phi \in \mathit{Th}_2$). By induction on φ, we prove that it extends to any one-step reduction ($\varphi \rightarrow_2 \psi$). Then by induction on $\varphi \equiv_{1,2} \psi$, we obtain the final result.

Lemma 6 (From $\vdash^{cf}_{\equiv_{1,2}}$ to $\vdash^{cf+_2}_{\equiv_1}$). *If $\Gamma \vdash^{cf}_{\equiv_{1,2}} \Delta$, then $\Gamma \vdash^{cf+_2}_{\equiv_1} \Delta$.*

Proof. By induction on the proof of $\Gamma \vdash^{cf}_{\equiv_{\mathit{Th}_1 \cup \mathit{Th}_2}} \Delta$: The raw deductive steps and the steps modulo Th_1 are unchanged. A step $\dfrac{\Gamma \vdash^{cf}_{\equiv_{1,2}} \psi, \Delta}{\Gamma \vdash^{cf}_{\equiv_{1,2}} \varphi, \Delta}\ \psi \equiv_2 \varphi$ is translated (by induction hypothesis) into a proof of $\Gamma \vdash^{cf+_2}_{\equiv_1} \psi, \Delta$, which is translated using Theorem 6 into a proof of $\Gamma \vdash^{cf+_2}_{\equiv_1} \varphi, \Delta$.

The transformation from $\vdash^{cf}_{\equiv_{1,2}}$ to $\vdash^{cf+_2}_{\equiv_1}$ is called a focalization procedure in the focusing terminology. It transforms (cut-free) unstructured proofs into (cut-free) structured focusing proofs (superdeduction proofs here). In addition, using superdeduction rules instead of atomic deduction steps gives intuitive informations about the structure of proofs. We can therefore imagine building proofs only through superdeduction rules, as in the proof of the example of Section 2. An important difference with usual focalization is that in superdeduction, whenever this transformation has to focalize a sequence of connectives on one side of sequents, then it also has to focalize it on the other side of sequents, potentially in the same proof. Difficulties also arise from the fact that superdeduction *modulo* allows inferences to spawn new connectives from atoms. Finally we get that cut-free deduction modulo and cut-free superdeduction modulo are equivalent.

Theorem 7. *$\Gamma \vdash^{cf}_{\equiv_{1,2}} \Delta$ if and only if $\Gamma \vdash^{cf+_2}_{\equiv_1} \Delta$.*

Proof. By Lemmas 1 and 6.

As a corollary we directly obtain that cut-elimination holds for deduction modulo $\mathcal{T}h_1 \cup \mathcal{T}h_2$ if and only if it holds for the superdeduction modulo system associated with $(\mathcal{T}h_1, \mathcal{T}h_2)$. In particular it holds if $(\mathcal{T}h_1, \mathcal{T}h_2)$ verifies the criterion of Theorem 3, 4 or 5.

5 Conclusion

Superdeduction modulo is the combination of superdeduction and deduction modulo with both inference rules systematically derived from an axiomatic theory and the ability to conduct deduction modulo computation. In this paper, we have filled the gap with the focusing approach by proposing an extension of classical sequent calculus and by showing that superdeduction systems as well as a specific focusing system are instances of this extension. Our analysis then indicates that superdeduction stands in fact for introducing focusing phases that are directed by the unfolding of axioms, and that are made explicit in the explicit superdeduction system we have presented. Then we proved a focalization result for superdeduction, namely that cut-free deduction modulo can be translated into cut-free superdeduction modulo using permutations of the applications of the inference rules, provided that some hypotheses on the synchrony of the superdeduction axioms are verified. The inverse translation being trivial, we acquired as a corollary that cut-elimination for superdeduction modulo is equivalent to cut-elimination for deduction modulo, consequently obtaining that the numerous criteria for cut-elimination in deduction modulo also hold for superdeduction modulo.

Our comparison of superdeduction and focusing is meant to be carried forward. In particular we believe that our focalization proof is rather complicated. This is greatly due to the fact that our proof handles superdeduction *modulo*, consequently allowing inference rules to spawn new connectives from atoms. However we wish to compare our focalization proof with Andreoli's original one and with the elegant focalization graph technique introduced by Miller and Saurin in [25]. This analysis could lead us to simpler focalization proofs and weaker conditions for deductive theories that can be safely used in superdeduction modulo.

Superdeduction modulo is a promising framework for proof engineering. In particular a tableau method for superdeduction modulo, inspired from TaMeD [21], is presented in [22]. Its completeness is a consequence of our cut-elimination result. Besides let us notice that superdeduction modulo is already the core of a small proof assistant named Lemuridæ, which can be downloaded at `http://rho.loria.fr/lemuridae.html`. Superdeduction modulo is also used in [19] in a restricted manner since modulo is only used on first-order terms (and *not* first-order propositions). Its expressiveness is nevertheless demonstrated by an encoding of functional PTS.

Building bridges between superdeduction and deduction modulo is also done in a related approach by Brauner and Dowek [29]. Whereas we proved the equivalence of cut-elimination for superdeduction modulo and deduction modulo, they proved the equivalence of normalisation for deduction modulo and supernatural deduction. Superdeduction and supernatural deduction stand for the same paradigm applied to classical sequent calculus for the first and to intuitionistic natural deduction for the second. However in this latter system, permutability problems forbid the paradigm to handle disjunctions or existential quantifications. Therefore an interesting extension would be to apply their approach to (classical sequent calculus) superdeduction.

Acknowledgements. The author thanks Claude Kirchner for his useful comments and advices. Many thanks also to Richard Bonichon and Cody Roux for fertile discussions, to anonymous referees for their comments on a previous version of this paper, to the Modulo meetings and to the Pareo team for many interactions.

References

1. Dowek, G., Hardin, T., Kirchner, C.: Theorem proving modulo. Journal of Automated Reasoning 31(1), 33–72 (2003)
2. Brauner, P., Houtmann, C., Kirchner, C.: Principles of superdeduction. In: Ong, L. (ed.) Proceedings of LICS, July 2007, pp. 41–50 (2007)
3. Girard, J.Y.: On the meaning of logical rules II : multiplicatives and additives. In: Bauer, Steinbrüggen (eds.) Foundation of Secure Computation, pp. 183–212. IOS Press, Amsterdam (2000)
4. Andreoli, J.M.: Logic programming with focusing proofs in linear logic. Journal of Logic and Computation 2(3), 297–347 (1992)
5. Dowek, G., Werner, B.: Proof normalization modulo. Journal of Symbolic Logic 68(4), 1289–1316 (2003)
6. Dowek, G.: Truth values algebras and proof normalization. In: Altenkirch, T., McBride, C. (eds.) TYPES 2006. LNCS, vol. 4502, pp. 110–124. Springer, Heidelberg (2007)
7. Dowek, G., Hermant, O.: A simple proof that super-consistency implies cut elimination. In: Baader, F. (ed.) RTA 2007. LNCS, vol. 4533, pp. 93–106. Springer, Heidelberg (2007)
8. Hermant, O.: Méthodes Sémantiques en Déduction Modulo. PhD thesis, École Polytechnique (2005)
9. Hermant, O.: Semantic cut elimination in the intuitionistic sequent calculus. In: Urzyczyn, P. (ed.) TLCA 2005. LNCS, vol. 3461, pp. 221–233. Springer, Heidelberg (2005)
10. Burel, G., Kirchner, C.: Cut elimination in deduction modulo by abstract completion. In: Artemov, S.N., Nerode, A. (eds.) LFCS 2007. LNCS, vol. 4514, pp. 115–131. Springer, Heidelberg (2007)
11. Huang, X.: Reconstruction proofs at the assertion level. In: Bundy, A. (ed.) CADE 1994. LNCS, vol. 814, pp. 738–752. Springer, Heidelberg (1994)
12. Negri, S., von Plato, J.: Cut elimination in the presence of axioms. Bulletin of Symbolic Logic 4(4), 418–435 (1998)

13. Hallnäs, L., Schroeder-Heister, P.: A proof-theoretic approach to logic programming. II. programs as definitions. Journal of Logic and Computation 1(5), 635–660 (1991)
14. McDowell, R., Miller, D.: Cut-elimination for a logic with definitions and induction. Theoretical Computer Science 232(1-2), 91–119 (2000)
15. Brauner, P., Houtmann, C., Kirchner, C.: Superdeduction at work. In: Comon-Lundh, H., Kirchner, C., Kirchner, H. (eds.) Jouannaud Festschrift. LNCS, vol. 4600, pp. 132–166. Springer, Heidelberg (2007)
16. Dowek, G., Hardin, T., Kirchner, C.: HOL-$\lambda\sigma$: an intentional first-order expression of higher-order logic. Math. Struct. in Comp. Science 11(1), 21–45 (2001)
17. Dowek, G.: Proof normalization for a first-order formulation of higher-order logic. In: Gunter, E.L., Felty, A.P. (eds.) TPHOLs 1997. LNCS, vol. 1275, pp. 105–119. Springer, Heidelberg (1997)
18. Cousineau, D., Dowek, G.: Embedding pure type systems in the lambda-pi-calculus modulo. In: Ronchi Della Rocca, S. (ed.) TLCA 2007. LNCS, vol. 4583, pp. 102–117. Springer, Heidelberg (2007)
19. Burel, G.: Superdeduction as a logical framework. In: LICS 2008 (accepted, 2008)
20. Burel, G.: Unbounded proof-length speed-up in deduction modulo. In: Duparc, J., Henzinger, T.A. (eds.) CSL 2007. LNCS, vol. 4646, pp. 496–511. Springer, Heidelberg (2007)
21. Bonichon, R.: TaMeD: A tableau method for deduction modulo. In: Basin, D., Rusinowitch, M. (eds.) IJCAR 2004. LNCS, vol. 3097, pp. 445–459. Springer, Heidelberg (2004)
22. Houtmann, C.: Axiom directed focusing. Long version (2008), `http://hal.inria.fr/inria-00212059/en/`
23. Dowek, G.: La part du calcul. Mémoire d'habilitation, Université de Paris 7 (1999)
24. Liang, C., Miller, D.: Focusing and polarization in linear, intuitionistic, and classical logic. In: TCS (March 2008) (accepted)
25. Miller, D., Saurin, A.: From proofs to focused proofs: A modular proof of focalization in linear logic. In: Duparc, J., Henzinger, T.A. (eds.) CSL 2007. LNCS, vol. 4646, pp. 405–419. Springer, Heidelberg (2007)
26. Andreoli, J.M.: Focussing and proof construction. Annals Pure Applied Logic 107(1-3), 131–163 (2001)
27. Kleene, S.C.: Permutability of inferences in Gentzen's calculi LK and LJ. In: Two Papers on the Predicate Calculus, pp. 1–26. American Mathematical Society (1952)
28. Hermant, O.: A model-based cut elimination proof. In: 2nd St-Petersburg Days of Logic and Computability (2003)
29. Brauner, P., Dowek, G., Wack, B.: Normalization in supernatural deduction and in deduction modulo (2007) (available on the author's webpage)

A Type System for Usage of Software Components

Dag Hovland

Department of Informatics, The University of Bergen,
PB 7803, N-5020 Bergen, Norway
dagh@ii.uib.no

Abstract. The aim of this article is to support component-based software engineering by modelling exclusive and inclusive usage of software components. Truong and Bezem describe in several papers abstract languages for component software with the aim to find bounds of the number of instances of components. Their language includes primitives for instantiating and deleting instances of components and operators for sequential, alternative and parallel composition and a scope mechanism. The language is here supplemented with the primitives `use`, `lock` and `free`. The main contribution is a type system which guarantees the safety of usage, in the following way: When a well-typed program executes a subexpression `use`$[x]$ or `lock`$[x]$, it is guaranteed that an instance of x is available.

Keywords: Component Software, Type System, Parallel Execution, Component Usage, Process Model.

1 Introduction

The idea of "Mass produced software components" was first formulated by McIlroy [1] in an attempt to encourage the production of software routines in much the same way industry manufactures ordinary, tangible products. The last two decades "component" has got the more general meaning of a highly reusable piece of software. According to Szyperski [2] (p. 3), "(...) software components are executable units of independent production, acquisition, and deployment that can be composed into a functioning system". We will model software that is constructed of such components, and assume that during the execution of such a program, instances of the components can be created, used and deleted.

Efficient component software engineering is not compatible with programmers having to acquire detailed knowledge of the internal structure of components that are being used. Components can also be constructed to use other components, such that instantiating one component, could lead to several instances of other components. This lack of knowledge in combination with nested dependencies weakens the control over resource usage in the composed software.

The goal of this article is to guarantee the safe usage of components, such that one can specify that some instances must be available, possibly exclusively to

S. Berardi, F. Damiani, and U. de'Liguoro (Eds.): TYPES 2008, LNCS 5497, pp. 186–202, 2009.

the current thread of execution. In [3,4,5], Truong and Bezem describe abstract languages for component software with the aim of finding bounds of the number of instances of components existing during and remaining after execution of a component program. Their languages include primitives for instantiating and deleting instances of components and have operators for sequential, alternative and parallel composition and a scope mechanism. The first three operators are well-known, and have been treated by for example Milner [6] (where alternative composition is called *summation*). The scope mechanism works like this: Any component instantiated in a scope has a lifetime limited to the scope. Furthermore, inside a scope, only instances in the local store of the same scope can be deleted. The types count the maximum number of active component instances during execution and remaining after execution of a component program.

The languages described by Truong and Bezem lack a direct way of specifying that one or more instances of a component must exist at some point in the execution. In this paper we have added the primitives `use`, `lock` and `free` in order to study the usage of components. The first (`use`) is used for "inclusive usage", that is, when a set of instances must be available, but these instances may be shared between threads. The other form (`lock` and `free`) is used when the instances must exclusively be available for this execution thread. The difference between exclusive and inclusive usage can be seen by comparing the expressions $\texttt{new}x(\texttt{use}[x] \parallel \texttt{use}[x])$ and $\texttt{new}x(\texttt{lock}[x]\texttt{free}[x] \parallel \texttt{use}[x])$. The first expression is safe to execute, while executing the latter expression can lead to an error if x is locked, but not freed, by the left thread before it is used by the right thread. Instances of the same component cannot be distinguished, such that locking and freeing is not applied to specific instances, but to the number of instances of each component.

The type system must guarantee that the instances that are to be used are available. The system will not test whether the deletion of instances in local stores is safe, as this can be tested using the type systems in [7,3,4,5] together with an easy translation described in Section 7. Only non-recursive programs are treated, but an extension with loops and simple recursion, described in [7], can also be applied to this system.

Section 2 introduces an example using C++, which is applied to the type system in Section 6. The language of component programs is defined in Section 3, and the operational semantics is defined in Section 4. The types and the type system are explained in Section 5. Important properties of the type system are formulated in Section 7, while the main results concerning correctness are collected in Section 8. The article ends with a section on related work and a conclusion.

2 Example: Objects on the Free Store in C++

We will introduce an example with dynamically allocated memory in C++ [8]. In Section 6 we will apply the type system to the example. The example is inspired by a similar one in [7].

In the program fragment in Figure 1, so-called POSIX threads [9] are used for parallelism. After creating an instance of the class `C`, the function `pthread_create` launches a new thread calling the function which is third in the parameter list with the argument which is fourth. This function call, either `P1(C_instance)` or `P2(C_instance)`, is executed in parallel to `P3(C_instance)`, and the two threads are joined in `pthread_join` before the instance of `C` is deleted.

The dynamic data type `C` and the functions `P1, P2, P3` are left abstract. We will assume the latter three functions use the instance of `C` in some way, and that `P2` needs exclusive access to the instance.

```
void EX(int choice) {
  pthread_t pth;
  C* C_instance = new C();
  pthread_create(&pth, NULL, choice ? P1 : P2 , C_instance);
  P3(C_instance);
  pthread_join(pth, NULL);
  delete C_instance;
}
```

Fig. 1. C++ code using threads and objects on the free store

The question in this example is whether we can guarantee that `P2` gets exclusive access to the instance of `C`. In this small example it is possible to see that this is not the case. After the grammar is explained in the next section we will model the program in the language as shown in Figure 2, and use the type system to answer the question and correct the program.

3 Syntax

The language for components is parametrized by an arbitrary set $\mathbb{C}=\{a,b,c,\ldots\}$ of *component names*. We let variables x, y, z range over $\mathbb{C}$. Bags and multisets are used frequently in this paper, and will therefore be explained here.

3.1 Bags and Multisets

Bags are like sets but allow multiple occurrences of elements. Formally, a *bag* with underlying set of elements $\mathbb{C}$ is a mapping $M : \mathbb{C}{\to}\mathbb{N}$. Bags are often also called multisets, but we reserve the term multiset for a concept which allows one to express a deficit of certain elements as well. Formally, a *multiset* with underlying set of elements $\mathbb{C}$ is a mapping $M : \mathbb{C}{\to}\mathbb{Z}$. We shall use the operations $\cup, \cap, +, -$ defined on multisets, as well as relations $\subseteq$ and $\in$ between multisets and between an element and a multiset, respectively. We recall briefly their definitions: $(M \cup M')(x) = \mathsf{max}(M(x), M'(x))$, $(M \cap M')(x) = \mathsf{min}(M(x), M'(x))$, $(M+M')(x) = M(x) + M'(x)$, $(M - M')(x) = M(x) - M'(x)$, $M \subseteq M'$ iff $M(x) \leq M'(x)$ for all $x \in \mathbb{C}$. The operation $+$ is sometimes called *additive union*. Bags are closed under all operations above with the exception of $-$. Note that the operation $\cup$

returns a bag if at least one of its operands is a bag. For convenience, multisets with a limited number of elements are sometimes denoted as, for example, $M = [2x, -y]$, instead of $M(x) = 2,\ M(y) = -1,\ M(z) = 0$ for all $z \neq x, y$. In this notation, $[\,]$ stands for the *empty* multiset, i.e., $[\,](x) = 0$ for all $x \in \mathbb{C}$. We further abbreviate $M + [x]$ by $M + x$ and $M - [x]$ by $M - x$. Both multisets and bags will be denoted by M or N (with primes and subscripts), it will always be clear from the context when a bag is meant. For any bag, let $\mathsf{set}(M)$ denote its set of elements, that is, $M = \{x \in \mathbb{C} \mid M(x) > 0\}$. Note that a bag is also a multiset, while a multiset is also a bag only if it maps all elements to non-negative numbers.

3.2 Grammar

Component expressions are given by the syntax in Table 1. We let capital letters $A, \ldots, E$ (with primes and subscripts) range over *Expr*. A *component program* P is a comma-separated list starting with `nil` and followed by zero or more *component declarations*, which are of the form $x \multimap Expr$, with $x \in \mathbb{C}$ (`nil` will usually be omitted, except in the case of a program containing no declarations). $\mathsf{dom}(P)$ denotes the set of component names declared in P (so $\mathsf{dom}(\mathtt{nil}) = \varnothing$). Declarations of the form $x \multimap \mathtt{nop}$ are used for *primitive* components, i.e., components that do not use *subcomponents*.

Table 1. Syntax

$$
\begin{array}{lcl}
\mathit{Expr} & ::= & \mathit{Factor} \mid \mathit{Expr} \cdot \mathit{Expr} \\
\mathit{Factor} & ::= & \mathtt{new}\,x \mid \mathtt{del}\,x \mid \mathtt{lock}\,M \mid \mathtt{free}\,M \mid \mathtt{use}\,M \mid \mathtt{nop} \\
 & & \mid (\mathit{Expr} + \mathit{Expr}) \mid (\mathit{Expr} \parallel \mathit{Expr}) \mid \mathit{ScExp} \\
\mathit{ScExp} & ::= & \{M, \mathit{Expr}\} \\
M & ::= & \text{bag of elements from } \mathbb{C} \\
\mathit{Prog} & ::= & \mathtt{nil} \mid \mathit{Prog}, x \multimap \mathit{Expr}
\end{array}
$$

We have two primitives `new` and `del` for creating and deleting instances of a component, three primitives `free`, `lock` and `use` for specifying usage of instances of components and four primitives for composition: sequential composition denoted by juxtaposition, $+$ for choice (also called sum), $\parallel$ for parallel and $\{\ldots\}$ for scope. Note that instances of the same component cannot be distinguished. The effect of `lock` is therefore to decrease the number of instances available for usage, while `free` increases this number.

Executing the sum $E_1 + E_2$ means choosing either one of the expressions E_1 or E_2 and executing that one. Executing E_1 and E_2 in parallel, that is, executing $(E_1 \parallel E_2)$, means executing both expressions in some arbitrary interleaved order. Executing an expression inside a scope, $\{[\,], E\}$ means executing E, while only allowing deletion of instances inside the same scope, and after the execution of E, deleting all instances inside the scope.

The grammatical ambiguity in the rule for *Expr* is unproblematic. Like in process algebra, sequential composition can be viewed as an associative multiplication operation and products may be denoted as $E\,E'$ instead of $E \cdot E'$. The operations $+$ and $\|$ are also associative and we only parenthesize if necessary to prevent ambiguity. Sequential composition has the highest precedence, followed by $\|$ and then $+$. The primitive `nop` models zero or more operations that do not involve component instantiation or deallocation.

In the third clause of the grammar we define *scope expressions*, used to limit the lifetime of instances and the scope of deletion. A scope expression is a pair of a bag, called the local store, and an expression. Scope expressions appearing in a *component declaration* in a program are required to have an empty local store. Non-empty local stores only appear *during execution* of a program.

Definition 1. *By* $\mathsf{var}(E)$ *we denote the set of component names occurring in* E*, formally defined by* $\mathsf{var}(\mathtt{nop}) = \varnothing$, $\mathsf{var}(\mathtt{new}\,x) = \mathsf{var}(\mathtt{del}\,x) = \{x\}$,$\mathsf{var}(\mathtt{use}\,M) = \mathsf{var}(\mathtt{free}\,M) = \mathsf{var}(\mathtt{lock}\,M) = \mathsf{set}(M)$, $\mathsf{var}(E_1 + E_2) = \mathsf{var}(E_1 \parallel E_2) = \mathsf{var}(E_1\,E_2) = \mathsf{var}(E_1) \cup \mathsf{var}(E_2)$ *and* $\mathsf{var}(\{M, E\}) = \mathsf{set}(M) \cup \mathsf{var}(E)$.

Definition 2. *The* size *of an expression* E*, denoted* $\sigma(E)$*, is defined by* $\sigma(\mathtt{new}\,x) = \sigma(\mathtt{del}\,x) = \sigma(\mathtt{use}\,N) = \sigma(\mathtt{lock}\,N) = \sigma(\mathtt{free}\,N) = \sigma(\mathtt{nop}) = 1$, $\sigma(\{M, E\}) = \sigma(E) + 1$ *and* $\sigma(A + B) = \sigma(AB) = \sigma(A \| B) = \sigma(A) + \sigma(B) + 1$*. The size of a program* P*, denoted* $\sigma(P)$*, is defined by* $\sigma(P, x \prec A) = \sigma(P) + 1 + \sigma(A)$ *and* $\sigma(\mathtt{nil}) = 1$.

3.3 Examples

We assume that a program is executed by executing $\mathtt{new}\,x$, where x is the last component declared in the program, starting with empty stores of component instances. Examples of programs that will execute properly and will be well-typed are

Example 1

$$x \prec \mathtt{nop}, y \prec \mathtt{new}\,x\,\mathtt{use}\,[x]\,\mathtt{lock}\,[x]\,\mathtt{free}\,[x]$$
$$x \prec \mathtt{nop}, y \prec \mathtt{new}\,x\,\mathtt{new}\,x\,\{[\,], (\mathtt{use}\,[x] \parallel \mathtt{lock}\,[x])\}\,\mathtt{free}\,[x]$$

Examples of programs that can, for some reason, produce an error are:

Example 2

$$x \prec \mathtt{nop}, y \prec \mathtt{new}\,x\,\mathtt{new}\,x\,\{[\,], (\mathtt{use}\,[x] \parallel \mathtt{lock}\,[x])\}$$
$$x \prec \mathtt{nop}, y \prec \mathtt{new}\,x\,\mathtt{lock}\,[x]\,\mathtt{use}\,[x]\,\mathtt{free}\,[x]$$
$$x \prec \mathtt{nop}, y \prec \mathtt{new}\,x\,\{[\,], (\mathtt{use}\,[x] \parallel \mathtt{lock}\,[x])\}\,\mathtt{free}\,[x]$$
$$x \prec \mathtt{nop}, y \prec \mathtt{new}\,x\,\mathtt{free}\,[x]\,\mathtt{lock}\,[x]$$
$$x \prec \mathtt{nop}, y \prec \mathtt{new}\,x\,\{[\,], (\mathtt{use}\,[x] + \mathtt{lock}\,[x])\}\,\mathtt{free}\,[x]$$

The first program leaves one instance of x locked after execution. The second will get stuck as no instance of x will be available for use by the `use`-statement.

The third might also get stuck. Note that there exists an error-free execution of the third program, where the left branch of $(\texttt{use}[x] \parallel \texttt{lock}[x])$ is executed before the right one. But as we do not wish to make any assumptions about the scheduling of the parallel execution, we consider this an error. The fourth program tries to free a component instance that is not locked. The fifth program has a run in which $\texttt{free}[x]$ is executed, but no instance of x has been locked.

C++ Example. We now describe the model of the example program in Figure 1. Functions (such as `EX`) as well as objects on the free store (such as `C_instance`) are modelled as components. We let $\texttt{call}f$ abbreviate $\texttt{new}f\,\texttt{del}f$ and use this expression to model a function call. Note that f is deleted automatically by $\texttt{call}f$, which models the (automatic) deallocation of stack objects created by f. However, the subcomponents of f are not deleted by $\texttt{del}f$. We use small letters for the component names and model functions as components, where the function body is given by the right hand side of the declaration. Since `P2` needs exclusive access to an instance of `C` we add $\texttt{lock}[c]\,\texttt{free}[c]$ to the declaration of p_2. For p_1 and p_3 we indicate the non-exclusive usage by $\texttt{use}[c]$. Collecting all declarations we get the program in Figure 2.

$$\begin{aligned}
&c \multimap \texttt{nop},\\
&p_1 \multimap \texttt{use}[c],\\
&p_2 \multimap \texttt{lock}[c]\,\texttt{free}[c],\\
&p_3 \multimap \texttt{use}[c],\\
&ex \multimap \texttt{new}c\,((\texttt{call}p_1 + \texttt{call}p_2) \parallel \texttt{call}p_3)\,\texttt{del}c
\end{aligned}$$

Fig. 2. Program P, a model of the C++ program in Figure 1

4 Operational Semantics

A *state*, or state expression, is a pair $(M_u, \{M, E\})$ consisting of a bag M_u (called the global store) with underlying set of elements $\mathbb{C}$, and a scope expression $\{M, E\}$. The store M in this scope expression is called the local store of the expression. An *initial state* is of the form $([\,], \{[\,], \texttt{new}x\})$, and a *terminal state* is of the form $(M_u, \{M, \texttt{nop}\})$.

A state $(M_u, \{M, E\})$ expresses that we execute E with a local bag M and a global bag M_u of instances of components. The operational semantics is given in Table 2 as a state transition system in the style of structural operational semantics [10]. The inductive rules are osPar1, osPar2, osScp and osSeq. The other rules are not inductive, but osNew, osDel, osLock, osUse and osPop are conditional with the condition specified as a premiss of the rule. The transition relation with respect to a program P is denoted by $\leadsto_P$, with transitive and reflexive closure by $\leadsto_P^*$.

Table 2. Transition rules for a component program P

(osNop)
$$\frac{}{(M_u, \{M, \mathtt{nop}\,E\}) \leadsto_P (M_u, \{M, E\})}$$

(osNew)
$$\frac{x \prec A \in P}{(M_u, \{M, \mathtt{new}\,x\}) \leadsto_P (M_u + x, \{M + x, A\})}$$

(osDel)
$$\frac{x \in (M \cap M_u)}{(M_u, \{M, \mathtt{del}\,x\}) \leadsto_P (M_u - x, \{M - x, \mathtt{nop}\})}$$

(osLock)
$$\frac{N \subseteq M_u}{(M_u, \{M, \mathtt{lock}\,N\}) \leadsto_P (M_u - N, \{M, \mathtt{nop}\})}$$

(osFree)
$$\frac{}{(M_u, \{M, \mathtt{free}\,N\}) \leadsto_P (M_u + N, \{M, \mathtt{nop}\})}$$

(osUse)
$$\frac{N \subseteq M_u}{(M_u, \{M, \mathtt{use}\,N\}) \leadsto_P (M_u, \{M, \mathtt{nop}\})}$$

(osScp)
$$\frac{(M_u, \{N, A\}) \leadsto_P (M'_u, \{N', A'\})}{(M_u, \{M, \{N, A\}\}) \leadsto_P (M'_u, \{M, \{N', A'\}\})}$$

(osPop)
$$\frac{N \subseteq M_u}{(M_u, \{M, \{N, \mathtt{nop}\}\}) \leadsto_P (M_u - N, \{M, \mathtt{nop}\})}$$

(osAlt$i$)
$$\frac{i \in \{1, 2\}}{(M_u, \{M, (E_1 + E_2)\}) \leadsto_P (M_u, \{M, E_i\})}$$

(osSeq)
$$\frac{(M_u, \{M, A\}) \leadsto_P (M'_u, \{M', A'\})}{(M_u, \{M, A\,E\}) \leadsto_P (M'_u, \{M', A'\,E\})}$$

(osParEnd)
$$\frac{}{(M_u, \{M, (\mathtt{nop} \parallel \mathtt{nop})\}) \leadsto_P (M_u, \{M, \mathtt{nop}\})}$$

(osPar1)
$$\frac{(M_u, \{M, E_1\}) \leadsto_P (M'_u, \{M', E'_1\})}{(M_u, \{M, (E_1 \parallel E_2)\}) \leadsto_P (M'_u, \{M', (E'_1 \parallel E_2)\})}$$

(osPar2)
$$\frac{(M_u, \{M, E_2\}) \leadsto_P (M'_u, \{M', E'_2\})}{(M_u, \{M, (E_1 \parallel E_2)\}) \leadsto_P (M'_u, \{M', (E_1 \parallel E'_2)\})}$$

4.1 Unsafe States

A *stuck state* is usually defined as a state which is not terminal, and where there is no possible next transition. We wish to use a different condition, because we want to assure that all possible runs are error-free. This means that we do not assume anything about the interleaving used in parallel executions. This is more in line with how parallelism works by default in many environments, for example with `pthreads` and C++ without mutex locking. Informally, we call a state *unsafe* if there is at least one transition which cannot be used in this state, but which would be possible with a larger global store. For example, $([], \{[x], \mathtt{lock}\,[x] \parallel \mathtt{free}\,[x]\})$ is an unsafe state, because using osPar1 is only possible with a larger global store.

Definition 3 (Unsafe states). *Given a component program P, a state $(M_u, \{M, E\})$ is called unsafe if and only if there exist bags M'_u, M and N and an expression E' such that $(M_u + N, \{M, E\}) \rightsquigarrow_P (M'_u + N, \{M', E'\})$, but not $(M_u, \{M, E\}) \rightsquigarrow_P (M'_u, \{M', E'\})$*

It is also possible to characterize the unsafe states with the following inductive rules parametrized by a program P and bags M_u and M: for all x and N, where $x \notin M_u$ and $N \not\subseteq M_u$, $(M_u, \{M, \mathtt{lock}N\})$, $(M_u, \{M, \mathtt{use}N\})$, $(M_u, \{M, \mathtt{del}x\})$ and $(M_u, \{M, \{N, \mathtt{nop}\}\})$ are unsafe, and for all expressions E and F, if $(M_u, \{M, E\})$ is unsafe then for all bags N, also $(M_u, \{N, \{M, E\}\})$, $(M_u, \{M, EF\})$, $(M_u, \{M, E \parallel F\})$ and $(M_u, \{M, F \parallel E\})$ are unsafe. Recall that deletion of component instances in the local store is assumed to always be safe, as this can be assured by the system in [7]. A state which is not unsafe is called *safe*.

4.2 Valid States

For some state $(M_u, \{M, E\})$ in a run, M_u models all component instances available for usage. We must therefore have M_u no larger than the sum of N in all subexpressions $\{N, A\}$ of E. For example $([x], \{[\,], \mathtt{nop}\})$ should not appear in a run because $M_u \supset [\,]$. Conditions for this to be true will be stated later. However, there are transitions where the states in the transition fulfil this condition, while the derivation of the transition contains states which do not fulfil the condition. An example is the transition $([x], \{[x], \{[], \mathtt{use}\,[x]\}\}) \rightsquigarrow_P ([x], \{[x], \{[], \mathtt{nop}\}\})$, in which both states fulfil this condition, while it is the result of applying osScp to the premiss $([x], \{[], \mathtt{use}\,[x]\}) \rightsquigarrow_P ([x], \{[], \mathtt{nop}\})$, where none of the two states fulfil the condition.

To express this property more formally we need a way to sum all the local stores in an expression. In doing so, however, one counts in instances that will never coexist, such as in $\{M_1, E_1\} + \{M_2, E_2\}$ and $\{M_1, E_1\}\,\{M_2, E_2\}$. Therefore we also define the notion of a valid expression, in which irrelevant bags are empty.

Definition 4 (Sum of local stores). *For any expression E, let ΣE be the sum of all N in subexpressions $\{N, A\}$ of E. More formally: $\Sigma\{M, E\} = M + \Sigma E$ and $\Sigma(E_1 \parallel E_2) = \Sigma(E_1\, E_2) = \Sigma(E_1 + E_2) = \Sigma E_1 + \Sigma E_2$ and $\Sigma\mathtt{del}x = \Sigma\mathtt{new}x = \Sigma\mathtt{use}N = \Sigma\mathtt{lock}N = \Sigma\mathtt{free}N = \Sigma\mathtt{nop} = [\,]$. An expression E is valid if for all subexpressions of the form $(E_1 + E_2)$ we have $\Sigma(E_1 + E_2) = [\,]$, and for all subexpressions of the form $F\,E'$, F a factor, we have $\Sigma E' = [\,]$.*

Note that an expression is valid if and only if all its subexpressions are valid. We will say that a state $(M_u, \{M, E\})$ is valid if and only if E is valid. The initial state is valid by definition. In any declaration $x \prec E$, since only empty bags are allowed to occur in E, E is obviously valid and $\Sigma E = [\,]$.

5 Type System

5.1 Types

A *type* of a component expression is a tuple $X = \langle X^u, X^n, X^l, X^d, X^p, X^h \rangle$, where X^n, X^u and X^p are bags and X^l, X^d and X^h are multisets. We use

Table 3. The parts of the types

X^u:	Minimum safe size of the global store.
X^n:	Largest decrease of the global store during execution.
X^l:	Lower bound of the net effect on the global store.
X^d:	Net change in the difference between the local and the global store.
X^p:	Maximum increase, during execution, of the difference between the global store and the sum of all local stores.
X^h:	Maximum net effect on the difference between the global store and the sum of all the local stores.

$U, \ldots, Z$ to denote types. The properties of the different parts of the types are summarized in Table 3, and will be explained below. The bag X^u (u for "usage") contains the minimum size the global store must have for an expression to be safely executed.

Because of sequential composition, we also need a multiset X^l. To run the expression $E_1\, E_2$, we must not only know the minimum safe sizes for executing E_1 and E_2 separately, but also how much E_1 decreases or increases the global store. The multiset X^l therefore contains, for each $x \in \mathbb{C}$, the *lowest* net increase in the number of instances in the global store after the execution of the expression. (Where a decrease is negative increase.) This implies that, if the type of E is X and if $(M_u, \{M, E\}) \rightsquigarrow_P^* (M'_u, \{M', \texttt{nop}\})$, then $X^l \subseteq M'_u - M_u$.

The scope operator makes necessary the component X^d. When a scope is *popped* with the rule osPop, the remaining bag in the scope is subtracted from the global store. The difference between these two bags must therefore be controlled by X^d. In addition, concerning the two alternatives joined in a choice expression, the net effect on the difference between the global store and the local store must be equal. An example of an invalid expression excluded by this rule is $(\texttt{lock}x + \texttt{use}x)$. If the latter expression was allowed in a program, it would not be possible to give the guarantees needed for osPop to the number of instances of x locked after execution. The multiset X^d therefore contains the exact change in the difference between the local store and the global store made by execution of the expression. This difference is independent of how the expression is executed. This implies that, if the type of E is X and if $(M_u, \{M, E\}) \rightsquigarrow_P^* (M'_u, \{M', \texttt{nop}\})$, then $X^d = (M'_u - M') - (M_u - M)$.

Parallel composition necessitates the bag X^n. The minimum safe size for executing $(E_1 \parallel E_2)$ depends not only on the minimum safe size for executing each of E_1 and E_2, but also on how much each of them decreases the global store. For example, both $\texttt{use}x$ and $\texttt{lock}x\,\texttt{free}x$ need one instance of x, but $\texttt{use}x \parallel \texttt{use}x$ also needs only one, whereas $\texttt{lock}x\,\texttt{free}x \parallel \texttt{lock}x\,\texttt{free}x$ needs two instances of x. X^n contains, for each $x \in \mathbb{C}$, the highest *negative* net change in the number of instances in the global store during the execution of the expression. This implies that, if the type of E is X and if $(M_u, \{M, E\}) \rightsquigarrow_P^* (M'_u, \{M', E'\})$, then $-X^n \subseteq M'_u - M_u$.

As seen in Example 2 in Section 3.3, there are grammatically correct programs that "free" instances that are not locked. So far, we have not distinguished between $\texttt{free}[x]\,\texttt{lock}[x]$ and $\texttt{lock}[x]\,\texttt{free}[x]$. Obviously, these expressions cannot be assigned the same type. For example, the program $x \prec \texttt{nop}, y \prec \texttt{new}\,x\,\texttt{free}[x]\,\texttt{lock}[x]$ is wrong, and should not be well-typed, while the program $x \prec \texttt{nop}, y \prec \texttt{new}\,x\,\texttt{lock}[x]\,\texttt{free}[x]$ is correct and should be well-typed. There is a need for types concerned with the difference between the number of instances in the sum of all local stores and the number of instances in the global store. If $(M_u, \{M, E\})$ is a state during the execution of a component program, then the value of $(M_u - \Sigma\{M, E\})(x)$ for a component x is negative if an instance of x is locked, but not yet freed, and positive if it has been freed without being locked. The latter is seen as an error and should not occur in the run of a well-typed program. The bag X^p and multiset X^h are used for keeping track of the set $M_u - \Sigma\{M, E\}$, and contain, the highest *positive* net change during execution and the *highest* net increase of this bag after execution. This implies that if the type of E is X, then if $(M_u, \{M, E\}) \rightsquigarrow_P^* (M'_u, \{M', E'\})$ then $X^p \supseteq (M'_u - \Sigma\{M', E'\}) - (M_u - \Sigma\{M, E\})$, and if $(M_u, \{M, E\}) \rightsquigarrow_P^* (M'_u, \{M', \texttt{nop}\})$, we get $X^h \supseteq (M'_u - M') - (M_u - \Sigma\{M, E\})$. In the type of a well-typed program these parts must be empty bags.

5.2 Typing Rules

The typing rules in Table 4 and Table 5 must be understood with the above interpretation in mind. They define a ternary typing relation $\Gamma \vdash E : X$ and a binary typing relation $\vdash P : \Gamma$ in the usual inductive way. Here Γ is usually called a *basis*, mapping component names to the type of the expression in its declaration. In the relation $\vdash P : \Gamma$, Γ can be viewed as a type of P. An expression of the form $\Gamma \vdash E : X$ or $\vdash P : \Gamma$ will be called a *typing* and will also be phrased as 'expression E has type X in Γ' or 'program P has type Γ', respectively.

A basis Γ is a partial mapping of components $x \in \mathbb{C}$ to types. By $\mathsf{dom}(\Gamma)$ we denote the domain of Γ, and for any $x \in \mathsf{dom}(\Gamma)$, $\Gamma(x)$ denotes its type in Γ. For a set $S \subseteq \mathsf{dom}(\Gamma)$, $\Gamma|_S$ is Γ restricted to the domain S. For any $x \in \mathbb{C}$ and type X, $\{x \mapsto X\}$ denotes a basis with domain $\{x\}$ and which maps x to X. An expression E is called *typable* in Γ if $\Gamma \vdash E : X$ for some type X. The latter type X will be proved to be unique and will sometimes be denoted by $\Gamma(E)$.

Definition 5 (Well-typed program). *A program P with at least one declaration is* well-typed *if there are Γ and X such that $\vdash P : \Gamma$, $\Gamma \vdash \texttt{new}\,x : X$ and $X^d = X^u = X^p = []$, where x is the last component declared in P.*

The condition in Definition 5 that parts X^d, X^u and X^p be empty deserves an explanation. X^d must be empty, because the global and local store must be equal in the final state, that is, no instances are still locked when the program ends. X^u is the minimum safe size of the global store, and we assume the program is executed starting with an empty global store, so X^u must be empty. X^p must

Table 4. Typing Rules

(AxmP)
$$\frac{}{\vdash \texttt{nil} : \varnothing}$$

(New)
$$\frac{\Gamma(x) = X}{\Gamma \vdash \texttt{new}\,x : \langle X^u, X^n, X^l + x, X^d, X^p, X^h \rangle}$$

(Axm)
$$\frac{}{\Gamma \vdash \texttt{nop} : \langle [\,], [\,], [\,], [\,], [\,], [\,] \rangle}$$

(Del)
$$\frac{\Gamma(x) = X}{\Gamma \vdash \texttt{del}\,x : \langle [x], [x], [-x], [\,], [\,], [\,] \rangle}$$

(Lock)
$$\frac{\mathsf{set}(N) \subseteq \mathsf{dom}(\Gamma)}{\Gamma \vdash \texttt{lock}\,N : \langle N, N, -N, -N, [\,], -N \rangle}$$

(Use)
$$\frac{\mathsf{set}(N) \subseteq \mathsf{dom}(\Gamma)}{\Gamma \vdash \texttt{use}\,N : \langle N, [\,], [\,], [\,], [\,], [\,] \rangle}$$

(Free)
$$\frac{\mathsf{set}(N) \subseteq \mathsf{dom}(\Gamma)}{\Gamma \vdash \texttt{free}\,N : \langle [\,], [\,], N, N, N, N \rangle}$$

(Prog)
$$\frac{\Gamma \vdash E : X, \ \ \vdash P : \Gamma, \ x \notin \mathsf{dom}(\Gamma)}{\vdash P, x \prec E : \Gamma \cup \{x \mapsto X\}}$$

be empty, because this is the only way to guarantee that, during execution, no instance is freed, unless there already is a locked instance of the same component.

Type inference in this system is similar to [7,3,4,5]. In particular, the type inference algorithm has quadratic runtime. An implementation of the type system can be downloaded from the author's website.

6 C++ Example Continued

Recall the C++ program in Figure 1 and the component program in Figure 2. Type inference gives the following results:

$$\begin{aligned}
&\texttt{call}\,p_1 : \langle [c], [], [], [], [], [] \rangle, \\
&\texttt{call}\,p_2 : \langle [c], [c], [], [], [], [] \rangle, \\
&\texttt{call}\,p_3 : \langle [c], [], [], [], [], [] \rangle, \\
&\texttt{call}\,ex : \langle [c], [], [], [], [], [] \rangle
\end{aligned}$$

This signals in the first multiset $(\cdot^u)$ of the type of $\texttt{call}\,ex$ that one instance of c is needed before execution of $\texttt{call}\,ex$. This is caused by the possible choice of $\texttt{call}\,p_2$ instead of $\texttt{call}\,p_1$ by ex, whereby there could be parallel calls to p_4 and p_5. One way to fix this is to instantiate two instances of `C` instead of just one. Then one instance could be passed to `P1` or `P2` and the second to `P3`. This means that P is changed by changing ex into $ex' \prec \texttt{new}\,c\,\texttt{new}\,c\,((\texttt{call}\,p_1 + \texttt{call}\,p_2) \parallel \texttt{call}\,p_3)\,\texttt{del}\,c$. The type of $\texttt{call}\,ex'$ is $\langle [], [], [c], [], [], [] \rangle$ which signals that the expression now can be executed starting with an empty store. But the third multiset $(\cdot^l)$ signals that there is one instance of c left after execution. This can be fixed by deleting one more instance, that is, changing ex' to

Table 5. Typing Rules

(Par)

$$\frac{\Gamma \vdash E_1 : X_1,\ \Gamma \vdash E_2 : X_2}{\Gamma \vdash E_1 \parallel E_2 : \left\langle \begin{array}{l} (X_1^u + X_2^n) \cup (X_2^u + X_1^n), X_1^n + X_2^n, \\ X_1^l + X_2^l, X_1^d + X_2^d, X_1^p + X_2^p, X_1^h + X_2^h \end{array} \right\rangle}$$

(Alt)

$$\frac{\Gamma \vdash E_1 : X_1,\ \Gamma \vdash E_2 : X_2,\ X_1^d = X_2^d}{\Gamma \vdash E_1 + E_2 : \langle X_1^u \cup X_2^u, X_1^n \cup X_2^n, X_1^l \cap X_2^l, X_1^d, X_1^p \cup X_2^p, X_1^h \cup X_2^h \rangle}$$

(Seq)

$$\frac{\Gamma \vdash E_1 : X_1,\ \Gamma \vdash E_2 : X_2}{\Gamma \vdash E_1\, E_2 : \left\langle \begin{array}{l} X_1^u \cup (X_2^u - X_1^l), X_1^n \cup (X_2^n - X_1^l), \\ X_1^l + X_2^l, X_1^d + X_2^d, X_1^p \cup (X_2^p + X_1^h), X_1^h + X_2^h \end{array} \right\rangle}$$

(Scope)

$$\frac{\Gamma \vdash E : X,\ \mathsf{set}(M) \subseteq \mathsf{dom}(\Gamma)}{\Gamma \vdash \{M, E\} : \langle X^u \cup (M - X^d), X^n \cup (M - X^d), X^d - M, X^d - M, X^p, X^h \rangle}$$

$ex'' \prec \texttt{new}\,c\,\texttt{new}\,c\,((\texttt{call}\,p_1 + \texttt{call}\,p_2) \parallel \texttt{call}\,p_3\,\texttt{del}\,c)\,\texttt{del}\,c$. The type of $\texttt{call}\,ex''$ is $\langle [], [], [], [], [], [] \rangle$.

Another way of solving the original problem is to remove the parallelism from the program, such that ex is changed to $ex''' \prec \texttt{new}\,c\,(\texttt{call}\,p_1 + \texttt{call}\,p_2)\,\texttt{call}\,p_3\,\texttt{del}\,c$. The type of $\texttt{call}\,ex'''$ is also $\langle [], [], [], [], [], [] \rangle$.

7 Properties of the Type System

This section contains several basic lemmas about the type system. Proofs in this and the next section are omitted for space considerations. Contact the author for a full version including proofs.

It should be noted again that the type systems in [7,3,4,5] can be used to test whether the deletion of instances is safe, by first translating `use`, `lock` and `free` to `nop`. We can therefore regard only the programs where deletion of instances from the local store is safe.

Lemma 1 (Basics)

1. *If* $\Gamma \vdash E : X$, *then* $\mathsf{var}(E) \subseteq \mathsf{dom}(\Gamma)$.
2. *If* $\vdash P : \Gamma$ *and* $\Gamma \vdash E : X$, *then* $\mathsf{dom}(P) = \mathsf{dom}(\Gamma)$ *and* $-X^u \subseteq -X^n \subseteq X^l$ *and* $X^h \subseteq X^p$.

Lemma 2 (Associativity). *If* $\Gamma \vdash A : X$, $\Gamma \vdash B : Y$ *and* $\Gamma \vdash C : Z$, *then the two ways of typing the expression* $A\,B\,C$ *by the rule* Seq, *corresponding to the different parses* $(A\,B)\,C$ *and* $A\,(B\,C)$, *lead to the same type.*

The following lemma is necessary since the typing rules are not fully syntax-directed. If, e.g., $E_1 = A \cdot B$, then the type of $E_1 \cdot E_2$ could have been inferred by an application of the rule Seq to A and $B\,E_2$. In that case we apply the previous lemma.

Lemma 3 (Inversion)

1. *If* $\vdash P{:}\Gamma$ *and* $\Gamma(x) = X$, *then there exists a program* P' *and an expression* A *such that* $P', x \prec A$ *is the initial segment of* P *and* $\vdash P' : \Gamma|_{\mathsf{dom}(P')}$ *and* $\Gamma|_{\mathsf{dom}(P')} \vdash A{:}X$.
2. *If* $\Gamma \vdash \mathtt{new}x{:}X$, *then* $X = \langle \Gamma(x)^u, \Gamma(x)^n, \Gamma(x)^l + x, \Gamma(x)^d, \Gamma(x)^p, \Gamma(x)^h \rangle$.
3. *If* $\Gamma \vdash \mathtt{del}x{:}X$, *then* $X = \langle [x], [x], [-x], [\,], [\,], [\,] \rangle$.
4. *If* $\Gamma \vdash \mathtt{lock}N{:}X$, *then* $X = \langle N, N, -N, -N, [\,], -N \rangle$.
5. *If* $\Gamma \vdash \mathtt{free}N{:}X$, *then* $X = \langle [\,], [\,], N, N, N, N \rangle$.
6. *If* $\Gamma \vdash \mathtt{use}N{:}X$, *then* $X = \langle N, [\,], [\,], [\,], [\,], [\,] \rangle$.
7. *If* $\Gamma \vdash \mathtt{nop}{:}X$, *then* $X = \langle [\,], [\,], [\,], [\,], [\,], [\,] \rangle$.
8. *For* $\circ \in \{+, \|, \cdot\}$, *if* $\Gamma \vdash (E_1 \circ E_2){:}X$, *then there exists* X_i *such that* $\Gamma \vdash E_i{:}X_i$ *for* $i = 1, 2$. *Moreover,*
   $X = \langle X_1^u \cup X_2^u, X_1^n \cup X_2^n, X_1^l \cap X_2^l, X_1^d, X_1^p \cup X_2^p, X_1^h \cup X_2^h \rangle$ *and* $X_1^d = X_2^d$ *if* $\circ = +$,
   $$X = \left\langle \begin{matrix} (X_1^u + X_2^n) \cup (X_2^u + X_1^n), X_1^n + X_2^n, \\ X_1^l + X_2^l, X_1^d + X_2^d, X_1^p + X_2^p, X_1^h + X_2^h \end{matrix} \right\rangle \text{ if } \circ = \|, \text{ and}$$
   $$X = \left\langle \begin{matrix} X_1^u \cup (X_2^u - X_1^l), X_1^n \cup (X_2^n - X_1^l), \\ X_1^l + X_2^l, X_1^d + X_2^d, X_1^p \cup (X_2^p + X_1^h), X_1^h + X_2^h \end{matrix} \right\rangle \text{ if } \circ = \cdot.$$
9. *If* $\Gamma \vdash \{M, A\} : X$, *then there exists a type* Y, *such that* $\Gamma \vdash A : Y$ *and* $X = \langle Y^u \cup (M - Y^d), Y^n \cup (M - Y^d), Y^d - M, Y^d - M, Y^p, Y^h \rangle$.

The last lemma in this section is concerned with three forms of uniqueness of the types inferred in the type system. This is necessary in some of the proofs, and for an algorithm for type inference.

Lemma 4 (Uniqueness of types)

1. *If* $\Gamma_1 \vdash E{:}X$, $\Gamma_2 \vdash E{:}Y$ *and* $\Gamma_1|_{\mathsf{var}(E)} = \Gamma_2|_{\mathsf{var}(E)}$, *then* $X = Y$.
2. *If* $\vdash P{:}\Gamma$ *and* $\vdash P{:}\Gamma'$, *then* $\Gamma = \Gamma'$.
3. *If* $\vdash P_1 : \Gamma_1$ *and* $\vdash P_2 : \Gamma_2$ *and* P_2 *is a reordering of* a subset of P_1, *then* $\Gamma_1|_{\mathsf{dom}(P_2)} = \Gamma_2$.

8 Correctness

This section contains lemmas and theorems connecting the type system and the operational semantics. Included are theorems comparable to what is often called preservation and progress, for example in [11]. The following lemma implies that all states in sequences representing the execution of a well-typed program are valid, as defined in Definition 4.

Lemma 5. *If* $\vdash P{:}\Gamma$, $\Gamma \vdash E{:}X$, E *is valid and* $(M_u, \{M, E\}) \rightsquigarrow_P (M'_u, \{M', E'\})$ *is a step in the operational semantics, then also* E' *is valid.*

The next lemma fixes several properties of two states connected by a single step in the operational semantics. This is used heavily in the main theorems below. The first part is known under the names *subject reduction* and *type preservation*. The remaining parts reflect the fact that every step reduces the set of reachable states. Hence maxima do not increase and minima do not decrease.

Lemma 6 (Invariants). *Let P be a component program, E a valid expression, Γ a basis and U a type such that $\vdash P:\Gamma$, $\Gamma \vdash E:U$, and $(M_u, \{M, E\}) \rightsquigarrow_P (M'_u, \{M', E'\})$ is a step in the operational semantics. Then we have for some type V:*

1. $\Gamma \vdash E':V$.
2. $M'_u - V^u \supseteq M_u - U^u$, *i.e., the safety margin of the global store does not decrease.*
3. $M'_u - V^n \supseteq M_u - U^n$, *i.e., the lower bound on the global store in all reachable states does not decrease.*
4. $M'_u + V^l \supseteq M_u + U^l$, *i.e., the lower bound on the global store in the terminal state does not decrease.*
5. $M'_u - M' + V^d = M_u - M + U^d$, *i.e., the difference between the local and the global store in the terminal state does not change.*
6. $M'_u - \Sigma\{M', E'\} + V^p \subseteq M_u - \Sigma\{M, E\} + U^p$, *i.e., the upper bound on the difference, in any reachable state, between the global store and the sum of the local stores, does not increase.*
7. $M'_u - \Sigma\{M', E'\} + V^h \subseteq M_u - \Sigma\{M, E\} + U^h$, *i.e., the upper bound on the net effect on the difference between the global store and the sum of the local stores does not increase.*

The following Theorem 1 is a combination of several statements which in combination are often called *soundness* or *safety*. Items 1, 2 and 3 are similar to the properties often called *preservation*, *progress* and *termination*, respectively. (See for example [11]). Items 1, 4 and 5 assert that the parts of the types have the meanings given in 5.1.

Theorem 1 (Soundness). *If $\vdash P:\Gamma$, $\Gamma \vdash E:X$, E is valid and $X^u \subseteq M_u$, then the following holds:*

1. *If $(M_u, \{M, E\}) \rightsquigarrow_P (M'_u, \{M', E'\})$ and $\Sigma\{M, E\} - M_u \supseteq X^p$, then there is Y such that $\Gamma \vdash E' : Y$, $M'_u \supseteq Y^u$ and $\Sigma\{M', E'\} - M'_u \supseteq Y^p$.*
2. *If E is not* **nop**, *we have $(M_u, \{M, E\}) \rightsquigarrow_P (M'_u, \{M', E'\})$ for some $(M'_u, \{M', E'\})$.*
3. *All $\rightsquigarrow_P$-sequences starting in state $(M_u, \{M, E\})$ are finite.*
4. *If $(M_u, \{M, E\}) \rightsquigarrow^*_P (M'_u, \{M', \mathbf{nop}\})$, then $X^l \subseteq M'_u - M_u$, $X^d = (M'_u - M') - (M_u - M)$ and $X^h \supseteq (M'_u - M') - (M_u - \Sigma\{M, E\})$.*
5. *If $(M_u, \{M, E\}) \rightsquigarrow^*_P (M'_u, \{M', E'\})$ then $-X^n \subseteq M'_u - M_u$ and $X^p \supseteq (M'_u - \Sigma\{M', E'\}) - (M_u - \Sigma\{M, E\})$.*
6. *All states reachable from $(M_u, \{M, E\})$ are safe.*

Finally, we summarize the properties of the type system for well-typed programs, as defined in Definition 5 on page 195. The reader is referred to the paragraph following Definition 5 for an explanation of the three bags required to be empty, to Section 4.1 and Definition 3 for an explanation of safe states, and to Section 4.2 for an explanation of why it is important that $M'_u \subseteq \Sigma\{M', E'\}$.

Corollary 1. *If* $\vdash P : \Gamma$ *and* $\Gamma \vdash \mathtt{new}x : X$*, where* x *is the last component declared in* P *and* $X^d = X^u = X^p = []$*, then*

- *All maximal transition sequences starting with* $([], \{[], \mathtt{new}x\})$ *end with* $(M, \{M, \mathtt{nop}\})$ *for some bag* M*.*
- *All states* $(M'_u, \{M', E'\})$ *reachable from* $([], \{[], \mathtt{new}x\})$ *are safe, and such that* $M'_u \subseteq \Sigma\{M', E'\}$*.*

The following theorem states that the types are *sharp*. Informally, this means, they are as small as they can be, while still guaranteeing safety of execution. The part X^d is not included as it is already stated in Theorem 1 to be exact. The property is formulated differently for the part X^u because of its nature — the other parts contain information about how some of the bags or the difference between them change, while X^u only states the minimum safe size of the bag M_u.

Theorem 2 (Sharpness). *Assume some program* P*, bags* M *and* M_u *and valid expression* E *such that* $\vdash P{:}\Gamma$ *and* $\Gamma \vdash E{:}X$ *and* $M_u \subseteq \Sigma\{M, E\}$

1. *If* $M_u \not\supseteq X^u$*, then an unsafe state is reachable from* $(M_u, \{M, E\})$*.*
2. *If* $M_u \supseteq X^u$*:*
   - n *For every* $y \in \mathbb{C}$ *there exists a state* $(M'_u, \{M', E'\})$ *such that* $(M_u, \{M, E\}) \rightsquigarrow^*_P (M'_u, \{M', E'\})$ *and* $(M'_u - M_u)(y) = -X^n(y)$*.*
   - l *For every* $y \in \mathbb{C}$ *there exists a terminal state* $(M'_u, \{M', \mathtt{nop}\})$ *such that* $(M_u, \{M, E\}) \rightsquigarrow^*_P (M'_u, \{M', \mathtt{nop}\})$ *and* $(M'_u - M_u)(y) = X^l(y)$*.*
   - p *For every* $y \in \mathbb{C}$ *there exists a state* $(M'_u, \{M', E'\})$ *such that* $(M_u, \{M, E\}) \rightsquigarrow^*_P (M'_u, \{M', E'\})$ *and* $(M'_u - \Sigma\{M', E'\}) - (M_u - \Sigma\{M, E\})(y) = X^p(y)$*.*
   - h *For every* $y \in \mathbb{C}$ *there exists a terminal state* $(M'_u, \{M', \mathtt{nop}\})$ *such that* $(M_u, \{M, E\}) \rightsquigarrow^*_P (M'_u, \{M', \mathtt{nop}\})$ *and* $(M'_u - M') - (M_u - \Sigma\{M, E\})(y) = X^h(y)$*.*

9 Related Work and Conclusion

There is a large amount of work related to similar problems. Most approaches differ from this article by using super-polynomial algorithms, by assuming more on the runtime scheduling of parallel executions, or by treating only memory consumption. For the functional languages, see e.g. [12,13,14,15]. Popea and Chin in [16] also discuss usage in a related way. Their algorithm depends on solving constraints in Presburger arithmetic, which in the worst case uses doubly exponential time. Igarashi and Kobayashi in [17], analyse the *resource usage problem* for an extension of simply typed lambda calculus including resource usage. The algorithm extracts the set of possible traces of usage from the program, and then decides whether all these traces are allowed by the specification. This latter problem is still computationally hard to solve and undecidable in the worst case. Parallel composition is not considered. For the imperative paradigm, which is closer to the system described here, e.g. [18,19,20] treat memory usage. The problem of component usage in a parallel setting is related to prevention of deadlocks and race conditions. Boyapati et al. describe in [21] an explicitly typed system for verifying there are no deadlocks or race conditions in Java programs. In addition to the higher level of detail, the main difference from the system described in this article is the assumptions on the scheduling of parallel executions, namely the ability of a thread to wait until another thread frees/releases a lock. This scheduling has of course a cost in terms of added runtime and of complexity of the implementation.

We have defined a component language with a small-step operational semantics and a type system. The type system combined with the system in [7] or the system in [4] guarantees that the execution of a well-typed program will terminate and cannot reach an unsafe state. The language described in this article is an extension of the language first described in [5], and uses the results from [5,7]. The properties proved in the current article are new, though, and in some ways orthogonal to those shown in [5,7]. The language we introduced is inspired by CCS [6], with the atomic actions interpreted as component instantiation, deallocation and usage. The basic operators are sequential, alternative and parallel composition and a scope operator. The operational semantics is SOS-style [10], with the approach to soundness similar in spirit to [22]. We have presented a type system for this language which predicts sharp bounds of the number of instances of components necessary for safe execution. The type inference algorithm has quadratic runtime.

References

1. McIlroy, M.D.: Mass produced software components. In: Naur, P., Randell, B. (eds.) Software Engineering: Report of a conference sponsored by the NATO Science Committee, pp. 79–87. Scientific Affairs Division, NATO (1968)
2. Szyperski, C.: Component Software—Beyond Object–Oriented Programming, 2nd edn. Addison–Wesley / ACM Press (2002)
3. Bezem, M., Truong, H.: A type system for the safe instantiation of components. Electronic Notes in Theoretical Computer Science 97, 197–217 (2004)
4. Truong, H.: Guaranteeing resource bounds for component software. In: Steffen, M., Zavattaro, G. (eds.) FMOODS 2005. LNCS, vol. 3535, pp. 179–194. Springer, Heidelberg (2005)
5. Truong, H., Bezem, M.: Finding resource bounds in the presence of explicit deallocation. In: Van Hung, D., Wirsing, M. (eds.) ICTAC 2005. LNCS, vol. 3722, pp. 227–241. Springer, Heidelberg (2005)
6. Milner, R.: A Calculus of Communication Systems. LNCS, vol. 92. Springer, Heidelberg (1980)

7. Bezem, M., Hovland, D., Truong, H.: A type system for counting instances of software components. Technical Report 363, Department of Informatics, The University of Bergen, P.O. Box 7800, N-5020 Bergen, Norway (2007)
8. Stroustrup, B.: The C++ Programming Language, Third Edition. Addison-Wesley, Reading (2000)
9. IEEE: The open group base specifications issue 6 (2004)
10. Plotkin, G.D.: A structural approach to operational semantics. Journal of Logic and Algebraic Programming 60-61, 17–139 (2004)
11. Pierce, B.C.: Types and Programming Languages. MIT Press, Cambridge (2002)
12. Crary, K., Weirich, S.: Resource bound certification. In: POPL 2000: Proceedings of the 27th ACM SIGPLAN–SIGACT symposium on Principles of programming languages, pp. 184–198. ACM Press, New York (2000)
13. Hofmann, M., Jost, S.: Static prediction of heap space usage for first-order functional programs. In: POPL 2003: Proceedings of the 30th ACM SIGPLAN-SIGACT symposium on Principles of programming languages, pp. 185–197. ACM Press, New York (2003)
14. Kobayashi, N., Suenaga, K., Wischik, L.: Resource usage analysis for the π-calculus. Logical Methods in Computer Science 2(3) (2006)
15. Unnikrishnan, L., Stoller, S.D., Liu, Y.A.: Optimized live heap bound analysis. In: Zuck, L.D., Attie, P.C., Cortesi, A., Mukhopadhyay, S. (eds.) VMCAI 2003. LNCS, vol. 2575, pp. 70–85. Springer, Heidelberg (2002)
16. Popeea, C., Chin, W.N.: A type system for resource protocol verification and its correctness proof. In: Heintze, N., Sestoft, P. (eds.) PEPM, pp. 135–146. ACM, New York (2004)
17. Igarashi, A., Kobayashi, N.: Resource usage analysis. ACM Trans. Program. Lang. Syst. 27(2), 264–313 (2005)
18. Braberman, V., Garbervetsky, D., Yovine, S.: A static analysis for synthesizing parametric specifications of dynamic memory consumption. Journal of Object Technology 5(5), 31–58 (2006)
19. Chin, W.N., Nguyen, H.H., Qin, S., Rinard, M.C.: Memory usage verification for OO programs. In: Hankin, C., Siveroni, I. (eds.) SAS 2005. LNCS, vol. 3672, pp. 70–86. Springer, Heidelberg (2005)
20. Hofmann, M., Jost, S.: Type-based amortised heap-space analysis (for an object-oriented language). In: Sestoft, P. (ed.) ESOP 2006. LNCS, vol. 3924, pp. 22–37. Springer, Heidelberg (2006)
21. Boyapati, C., Lee, R., Rinard, M.C.: Ownership types for safe programming: preventing data races and deadlocks. In: OOPSLA, pp. 211–230 (2002)
22. Wright, A.K., Felleisen, M.: A syntactic approach to type soundness. Information and Computation 115(1), 38–94 (1994)

Merging Procedural and Declarative Proof

Cezary Kaliszyk and Freek Wiedijk

Institute for Computing and Information Sciences,
Radboud University Nijmegen, The Netherlands
{cek,freek@cs.ru.nl}

Abstract. There are two different styles for writing natural deduction proofs: the 'Gentzen' style in which a proof is a tree with the conclusion at the root and the assumptions at the leaves, and the 'Fitch' style (also called 'flag' style) in which a proof consists of lines that are grouped together in nested boxes.

In the world of proof assistants these two kinds of natural deduction correspond to procedural proofs (tactic scripts that work on one or more subgoals, like those of the Coq, HOL and PVS systems), and declarative proofs (like those of the Mizar and Isabelle/Isar languages).

In this paper we give an algorithm for converting tree style proofs to flag style proofs. We then present a rewrite system that simplifies the results.

This algorithm can be used to convert arbitrary procedural proofs to declarative proofs. It does not work on the level of the proof terms (the basic inferences of the system), but on the level of the statements that the user sees in the goals when constructing the proof.

The algorithm from this paper has been implemented in the ProofWeb interface to Coq. In ProofWeb a proof that is given as a Coq proof script (even with arbitrary Coq tactics) can be displayed both as a tree style and as a flag style proof.

1 Introduction

Proof assistants are computer programs for constructing and checking proofs. In these systems one can distinguish between two quite different kind of entities that both might be considered the 'proofs' that are being checked:

- First there are the low level proofs of the logic of the system. In type theoretical systems these are the *proof terms*. In other systems they are built from tiny proof steps called *basic inferences*. Generally such proof objects are huge and constructed from a small number of basic elements.
- Then there also are the high level proof texts that the user of the system works with. Often these texts are *scripts* of commands from the user to the proof assistant. These texts are of a size comparable to traditional mathematical texts, and contain a much larger variety of proof steps. For instance both the Coq and HOL systems have dozens of *tactics* that can occur in this kind of proof.

S. Berardi, F. Damiani, and U. de'Liguoro (Eds.): TYPES 2008, LNCS 5497, pp. 203–219, 2009.

The proof assistant does two things for the user. First it translates high level proofs into low level proofs, and secondly it checks the low level proofs obtained in this way with respect to the rules of the logic of the system.

As an example, the following 'high level' Coq proof script:

```
Lemma example : forall n : nat, n <= n.
intros.
omega.
Qed.
```

is translated to the following 'low level' proof term:

```
fun n : nat => Decidable.dec_not_not (n <= n) (dec_le n n) (fun H : ~ n <= n =>
  ex_ind (fun (Zvar1 : Z) (Omega5 : Z_of_nat n = Zvar1 / (0 <= Zvar1 * 1 + 0)%Z) =>
    and_ind (fun (Omega3 : Z_of_nat n = Zvar1) (_ : (0 <= Zvar1 * 1 + 0)%Z) =>
     let H0 := eq_ind_r (fun x : Z => (0 <= x + -1 + - Z_of_nat n)%Z -> False)
      (eq_ind_r (fun x : Z => (0 <= Zvar1 + -1 + - x)%Z -> False)
       (fast_Zopp_eq_mult_neg_1 Zvar1 (fun x : Z => (0 <= Zvar1 + -1 + x)%Z -> False)
        (fast_Zplus_comm (Zvar1 + -1) (Zvar1 * -1) (fun x : Z => (0 <= x)%Z -> False)
         (fast_Zplus_assoc (Zvar1 * -1) Zvar1 (-1) (fun x : Z => (0 <= x)%Z -> False)
          (fast_Zred_factor3 Zvar1 (-1) (fun x : Z => (0 <= x + -1)%Z -> False)
           (fast_Zred_factor5 Zvar1 (-1) (fun x : Z => (0 <= x)%Z -> False)
            (fun Omega4 : (0 <= -1)%Z => Omega4 (refl_equal Gt))))))) Omega3) Omega3 in
     H0 (Zgt_left (Z_of_nat n) (Z_of_nat n) (inj_gt n n (not_le n n H))))
    Omega5) (intro_Z n))
```

which then is type checked and found to be correct.

A good proof assistant should hide low level proofs from the user of the system as much as possible. Just like a user of a high level programming language should not need to be aware that the program internally is translated into machine code or bytecode, the user of a proof assistant should not have to be aware that internally a low level proof is being constructed.

It depends much on the specific proof assistant what the high level proofs look like. There are two basic groups of systems, as first introduced in [7]:

The procedural systems such as Coq, HOL and PVS. These systems generally are descendants of the LCF system. The proofs of a procedural system consist of tactics operating on goals. This leads to proofs that can naturally be represented as *tree shaped* derivations in the style of Gentzen. For instance, the example Coq proof then looks like:

```
    ―――――――――――――――― omega
    n:nat ⊢ n <= n
   ―――――――――――――――――― intros
   ⊢ ∀n:nat, n <= n
```

The above is a screenshot from the display of our ProofWeb system. In practice it is more useful to have ProofWeb display the tree without contexts:

```
      ―――――― omega
      n <= n
    ―――――――――――――― intros
    ∀n:nat, n <= n
```

The declarative systems. The main two systems of this kind are Mizar and Isabelle (when used with its declarative proof language Isar), but also automated theorem provers like ACL2 and Theorema can be considered to be declarative. There are experimental declarative proof languages, like the ones by Pierre Corbineau for Coq and by John Harrison for HOL Light.

The proofs of a declarative system are *block structured.* They basically consist of a *list* of statements, where each statement follows from the previous ones, with the system being responsible for automatically constructing the low level proof that shows this to be the case. Apart from these basic steps declarative proofs have other steps, like the `assume` step which introduces an assumption.

In declarative systems these proof steps are grouped into a hierarchical structure of *blocks*, just like in block structured programming languages. In declarative proofs these blocks are delimited by keywords like `proof` and `qed`.

Some systems might be considered not to be *fully* declarative in the sense that they still require the user to indicate *how* a statement follows from earlier statements. For instance this holds for Isabelle, where the user can (and sometimes must) give explicit inference rules. Indeed, it is common among the users of Isabelle to refer to the Isar proofs not as 'declarative' but 'structured'. However, for the purposes of this paper this distinction does not matter. In fact, the declarative proofs that we generate with our ProofWeb system also have the property that they contain an explicit tactic at each step in the proof.

Mathematicians generally think of their proofs in a declarative way. Declarative proofs are similar (although more precise and, with current technology, much more fine-grained) to the language that one finds in mathematical articles and textbooks.

The contribution of this paper is a generic method for converting a procedural proof to a declarative proof. For Coq this method has been implemented in the ProofWeb system. ProofWeb can display a high level Coq proof as a block structured list of statements. Here is how it will display the example proof:

1	n: nat	assumption
2	n <= n	omega
3	∀n:nat. n <= n	intros 1-2

The rest of the paper details the algorithms used for this.

In 2006–2008 we ran a project called *Web deduction for education in formal thinking*, in which we built a system for logic education. Our system allows students to practice natural deduction proofs. It has the following design choices:

- Our system runs on a web server. This means that students do not need to install anything, can access their work from anywhere (as long as they have Internet access), and that teachers can easily keep track of the progress of

their students. Our ProofWeb server is at `http://proofweb.cs.ru.nl/`. It can be tried using the guest login, with no registration.
- The system uses the Coq proof assistant, and the Coq proof language is not hidden from the user. Students are typing actual Coq proof scripts that use a restricted set of custom tactics to make their proofs correspond exactly to the proofs from their textbook. The use of Coq makes ProofWeb especially attractive for teachers who want their students to work on non-trivial examples.
- The system allows the students to both work in Gentzen style as well as in Fitch style. Proofs are displayed in (almost) exactly the same way that they are shown in the textbook. We decided to have our system be compatible with a popular logic textbook by Michael Huth and Mark Ryan [8].

The ProofWeb system can present the tree shaped proof that corresponds to the Coq proof script as a Fitch style proof. This means that it converts a procedural proof (the Coq script) to a declarative proof (the Fitch display). The method that it uses to do this is *generic*. It will work for converting *any* procedural proof to *any* declarative proof text, independent of the specific proof assistants involved or their logical foundations.[1]

We decided against presenting the conversion method that we used generically. In this paper we present just the method for the very specific situation of natural deduction proofs for first order predicate logic with equality. However, the method *is* perfectly generic. Also, our implementation already is not restricted to the small set of tactics that the users of ProofWeb are supposed to use. It will work with *any* Coq proof, providing a block structured Fitch style display of that proof.

The specifics of the first order logics that ProofWeb uses can be found in the ProofWeb manual [9]. We here just show an example for both logics in Figure 1. In ProofWeb flags are rendered as boxes (like in Huth and Ryan), with the right hand border of the boxes omitted to conserve space.

Declarative proofs are much more robust than procedural proofs, and for this reason can be expected to have a longer useful lifetime than procedural proofs. For this reason, development of the technology presented here might mean current formalizations get a longer useful lifetime. A current version of the procedural system can be used to export a formalisation declaratively. Keeping the declarative proof instead of the procedural one gives a much higher chance of the proof being accepted by future versions of the proof assistant.

The conversion algorithm presented here also works on proofs that have not been completed yet. In that case one gets a declarative proof with *gaps*. For instance in ProofWeb, the Fitch style display of the proof *before* the `omega` tactic is executed will be:

---

[1] The proof might contain some statements that have no good equivalent in the target system, and the automation of the target system might not always be able to bridge the gaps between the steps, but apart from those issues, a good starting point for a formalization in the target system can always be generated.

```
1 | n: nat                   assumption
  |   ...
2 |   n <= n
3   ∀n:nat. n <= n         intros 1-2
```

ProofWeb users often use the system through this feature. They do not look at the Coq proof state (which is also available to them), but just think in terms of the incomplete Fitch style proof.

This leads us to propose a new kind of prover interface. We call it a *luxury* declarative proof assistant (after a suggestion by Henk Barendregt). In a luxury system, the user does not see goals, but works on an incomplete declarative proof. This proof then can be modified in two ways:

- Either the user just edits the text, the common way to work in a declarative proof assistant. This is flexible but gives the user no help in writing the proof.
- Alternatively the user executes a *tactic* at a step in the proof that has not been sufficiently justified yet, i.e., for which the system has not yet generated

$$\dfrac{\dfrac{\dfrac{[\exists x(P(x)\lor\neg Q(a))]^{\mathtt{H1}} \qquad \dfrac{[P(b)\lor\neg Q(a)]^{\mathtt{H3}} \qquad \dfrac{[P(b)]^{\mathtt{H4}}}{\exists x\,P(x)}\,\exists\mathrm{i} \qquad \dfrac{\dfrac{[\neg Q(a)]^{\mathtt{H5}} \quad [Q(a)]^{\mathtt{H2}}}{\bot}\,\neg\mathrm{e}}{\exists x\,P(x)}\,\bot\mathrm{e}}{\exists x\,P(x)}\,\lor\mathrm{e}\,[\mathtt{H4},\mathtt{H5}]}{\exists x\,P(x)}\,\exists\mathrm{e}\,[\mathtt{H3}]}{Q(a)\to\exists x\,P(x)}\,\to\mathrm{i}\,[\mathtt{H2}]}{\exists x(P(x)\lor\neg Q(a))\to Q(a)\to\exists x\,P(x)}\,\to\mathrm{i}\,[\mathtt{H1}]$$

1	$\exists x(P(x)\lor\neg Q(a))$	
2	$Q(a)$	
3	b $P(b)\lor\neg Q(a)$	
4	$P(b)$	
5	$\exists x\,P(x)$	$\exists$i 4
6	$\neg Q(a)$	
7	$\bot$	$\neg$e 6,2
8	$\exists x\,P(x)$	$\bot$e 7
9	$\exists x\,P(x)$	$\lor$e 3,4–5,6–8
10	$\exists x\,P(x)$	$\exists$e 1,3–9
11	$Q(a)\to\exists x\,P(x)$	$\to$i 2–10
12	$\exists x(P(x)\lor\neg Q(a))\to Q(a)\to\exists x\,P(x)$	$\to$i 1–11

Fig. 1. Example derivation in Gentzen's and Fitch's systems

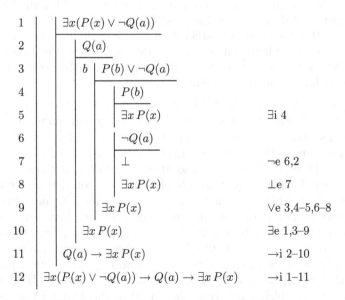

a low level proof. The 'goal' that this tactic sees has the statement of this step as the conclusion, and all the statements before it that are in scope as the assumptions. The tactic then will generate subgoals, which will be added to the proof text as new steps in, if needed, new sub-blocks.

Modifying a proof in this style (by executing a tactic at a not yet justified step), needs exactly the same algorithms that the conversion from a procedural proof to a declarative proof needs.

If one 'grows' a declarative proof in such a way, it basically will consist of a merged version of all the subgoals that the proof would have gone through in the procedural system.

It is important that in a luxury system *both* ways of working are available simultaneously. It should not be *required* to use tactics to modify a proof.

A simple version of this luxury concept is the following. In a declarative prover the user has to formulate the appropriate `assume` steps himself, while in a procedural prover he just can type `intros`. However in a luxury prover, the `intros` command will be available, which then will generate all the needed `assume` steps automatically. Similarly, appropriate statements in the case of an induction or application of a lemma can be generated automatically by the system.

The conversion from a tree style proof to a block structured proof is straightforward. It consists of two phases:

- First the tree is converted to a series of nested blocks in a naive way. This is trivial. However, it does not lead to a proof that a user will want to see, as there are many duplicate lines and boxes that are not necessary.
- The second phase is to *reduce* the proof. We use a rewrite system for this that eliminates various unwanted structures from the proof:
  - If a subproof has no new assumptions nor new variables, the block for it is not needed and can be flattened into the main proof.
  - Lines that are copies of earlier lines can generally be removed, as references to those lines can be replaced by references to the earlier lines.
  - 'Cuts' also can be removed from the proofs, as the declarative proofs really have a cut (in the Gentzen sense) at every line.

Below we will give the details of this rewrite system for proofs for the specific case of first order logic. We prove it to be terminating and confluent.

Our method is designed to convert proofs preserving the level of detail present in the original proof. When building a proof using automated tactics (decision procedures), the user might be curious after the proof that those tactics constructed internally. This is analogous to the rare occasion that a compiler user wants to see the machine code that was generated by the compiler. Our method does not work well for obtaining information on this level. However, Coq allows decomposition of tactics into smaller tactics using the `info` prefix, which means that getting such information is possible even when using our approach.

There have been various projects for translating proofs from a procedural proof assistant into a declarative presentation, most notably the HELM system by Asperti et al., which was further developed in the MoWGLI project [1,2]. However, those systems almost always work on the level of proof terms and not on the level of tactics. For this reason the declarative proofs that these systems produced tend to be too convoluted for human consumption.

An exception is the system by Guilhot, Naciri and Pottier where Coq proofs are considered on the level of the tactics, by converting Coq proof trees just like we do [6]. However in this work the generated text is only considered a *presentation* – they call it an *explanation* – and not a proof in a formal system like Fitch-style natural deduction in its own right.

Geuvers and Nederpelt [4] define a translation of natural deductions in Fitch style to simply typed λ-terms (i.e., their translation goes the opposite way from ours). They present reduction relations for Fitch-style deductions that allow simpler λ-terms to be obtained. These reductions remove unnecessary subproofs, remove repeats and unshare shared subproofs. They prove that Fitch deductions are mapped to the same λ-term if and only if they are equal under these relations; which shows that there is an isomorphism between these classes.

Proof nets [5] allow representing proofs in a geometrical way where the order of the application of rules as well as irrelevant features of regular natural deduction proofs can be eliminated. Geuvers and Loeb [3] show the correspondence between deduction graphs and proof nets and give translations from minimal propositional logic to proof nets via context nets. They also shows how an operation of cut elimination in deduction graphs can be performed after the translation to a context net.

2 Translating Minimal Logic Tree Style Proofs to Flag Style Proofs

We first will restrict ourselves to minimal propositional logic. We introduce a translation operation ($\longmapsto$) that translates a tree style proof **G** of a proposition A to a flag style proof **F**. An example of such a translation is:

$$\emptyset : \dfrac{\dfrac{[A]^x}{B \to A}\,{\to}\mathrm{i}[y]}{A \to B \to A}\,{\to}\mathrm{i}[x] \quad \longmapsto$$

1	$\quad$ | A	
2	$\quad$ | $\quad$ | B	
3	$\quad$ | $\quad$ | A	copy 1
4	$\quad$ | $B \to A$	$\to$i 2–3
5	$A \to B \to A$	$\to$i 1–4

This operation always preserves the conclusion, and the conclusion will be most often the part of the proof that we match, so we write it explicitly:

$$\Gamma : \left(\begin{array}{c} \vdots\,\mathbf{G} \\ A \end{array} \;\longmapsto\; \begin{array}{|c} \vdots\,\mathbf{F} \\ A \end{array} \right)$$

The translation operates in a context Γ. This context is a list of assumptions accompanied by labels that can be used in the proofs **G** and **F**. The assumptions that are discharged in the proof are no longer in the context. Sometimes

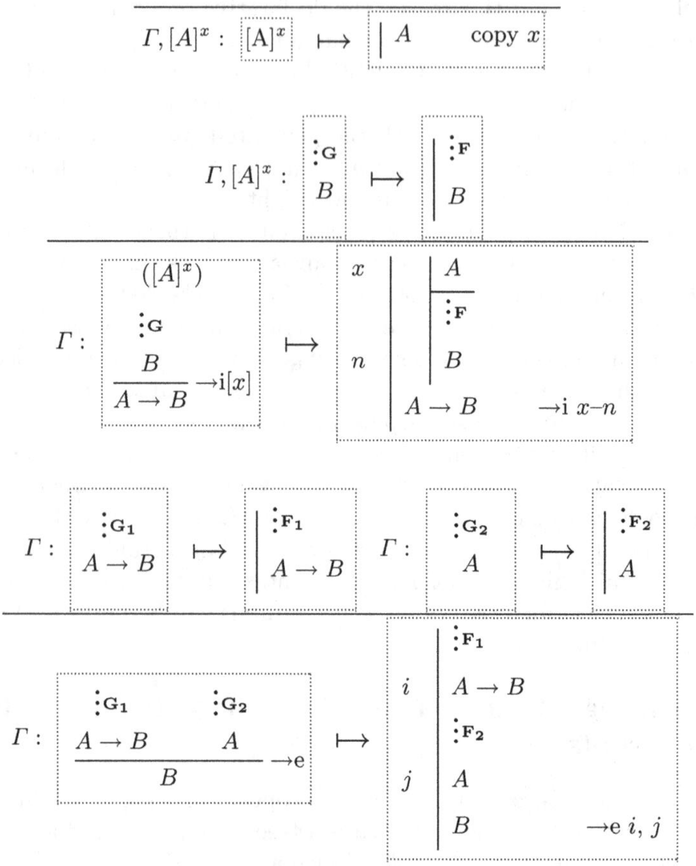

Fig. 2. The translation rules for minimal propositional logic

for clarity we will mark assumptions available in particular branches of proofs and discharged after by additional brackets. Below we give an example of a translation of proof styles in a non-empty context:

$$[A]^x, [B]^y : \quad \frac{[A]^x \quad [B]^y}{A \wedge B} \wedge i \quad \mapsto \quad \left| A \wedge B \quad \wedge i \ x, y \right.$$

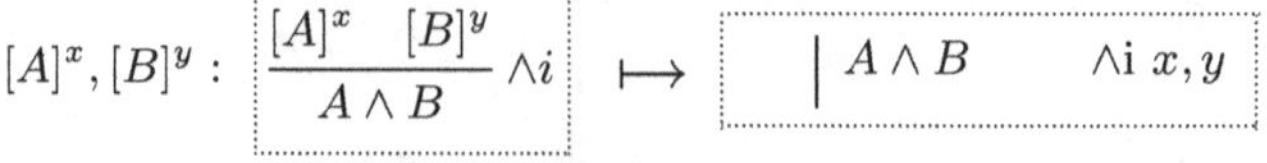

We define the translation operation inductively via the translation rules in Figure 2. The translation rules match the conclusion and the rule used and give a rule to build the flag style proof. All new labels introduced by the translation operation are fresh identifiers. The usual presentation of flag style proofs is with line numbers and rules that reference those numbers, but in our translation we will use identifiers. An implementation may render such proofs with lines numbered in the customary way, and we do indeed provide this in our ProofWeb implementation as described in Section 6.

The first rule translates the use of an assumption. We replace the use of an assumption with a copy line, and label this line with the name of the assumption variable.

If the derivation ends with implication introduction we translate it to the implication introduction rule in the flag style. We use the name of the introduced assumption in the tree style as the label of assumption line in the flag style. The assumption $[A]$ is not in the context since it is discharged, but for readability we mark it in brackets in the tree style proof. This means that the proof **G** can use this assumption. We provide fresh identifiers for new lines.

Implication elimination is analogous. We do not need to introduce a flag for the subtree of the tree style proof. This is what makes the depth of flag proofs much lower then the depth of tree style proofs.

3 Translating Proofs in More Complicated Logical Systems

To translate a proof in tree style of an arbitrary deduction system we will first translate it to a non-optimized proof.

We often need to open a number of flags depending on a list of assumptions. This is why we introduce a shorthand notation. We will write flags with a *list* above the assumption line to denote opening a number of flags. The last flag is opened with the rule provided in the shorthand notation, while all other flags are introduced one by one using implication introduction:

$$
\begin{array}{ll}
i & \;\big|\; A_1, A_2, \ldots, A_n \\[2pt]
  & \;\big|\; \vdots\, \mathbf{F} \\[2pt]
j & \;\big|\; B \\[2pt]
  & C \qquad\qquad\qquad\quad \text{R } i\text{--}j
\end{array}
$$

This stands for:

$$
\begin{array}{ll}
i & \big|\; A_1 \\
j & \quad\big|\; A_2 \\
  & \quad\quad \cdots \\
k & \quad\quad\big|\; A_n \\
  & \quad\quad\big|\; \vdots\, \mathbf{F} \\
l & \quad\quad\big|\; B \\
  & \quad\quad A_n \to B \qquad\qquad \to\!\text{i } k\text{--}l \\
m & \quad \cdots \\
n & \big|\; A_2 \to A_3 \to \ldots \to A_n \to B \qquad \to\!\text{i } j\text{--}m \\
  & C \qquad\qquad\qquad\qquad\qquad\qquad \text{R } i\text{--}n
\end{array}
$$

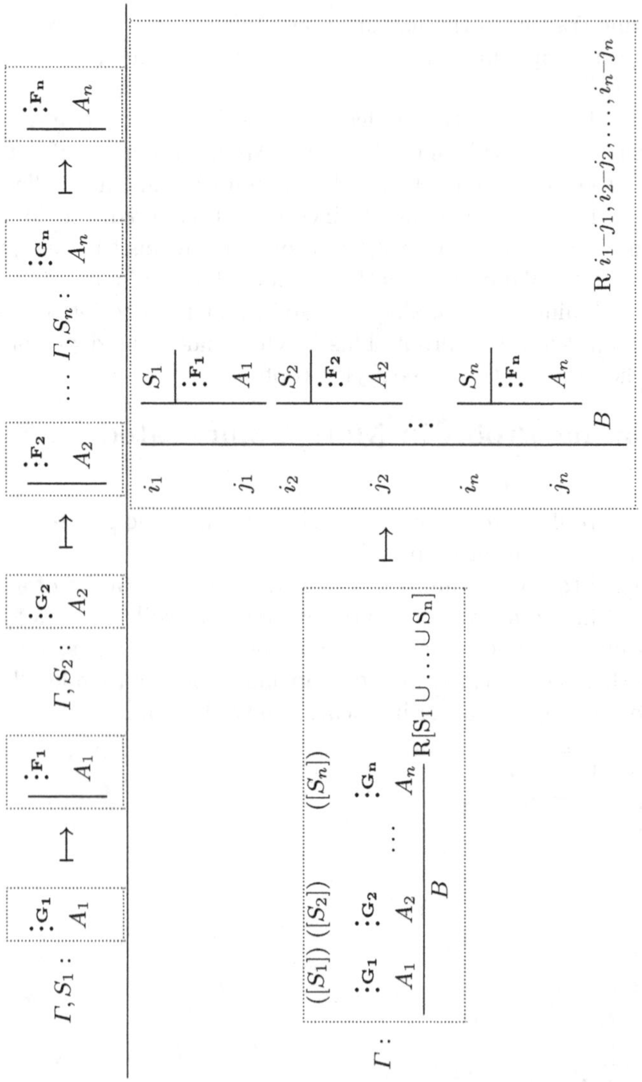

Fig. 3. The general schema for translating a rule of the logic

For a list with just one assumption this is equivalent to opening one flag with just the given rule. For a flag with an empty list of assumptions no flags need to be opened:

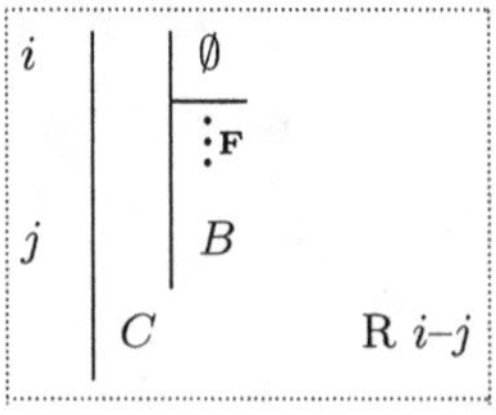

stands for:

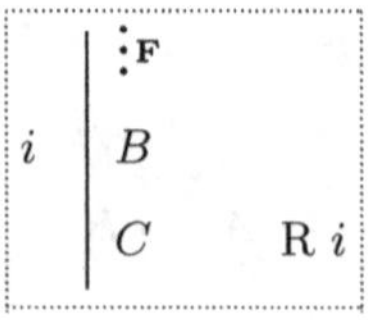

We show the translation of a given tree style proof in terms of a general schema. This schema will be instantiated for every rule of the logic. Given a rule R that proves the formula B from the tree style proofs $G_1, G_2, \ldots, G_n$ that have conclusions $A_1, A_2, \ldots, A_n$, which discharge assumption lists (possibly empty) $S_1, S_2, \ldots, S_n$ we recursively translate all subproofs to generate the final flag style proof (Figure 3). The subproofs $A_1, \ldots, A_n$ can use the assumptions from their appropriate lists and this is marked in the schema by brackets. An example of instantiation of the schema for a rule for is given in Figure 4.

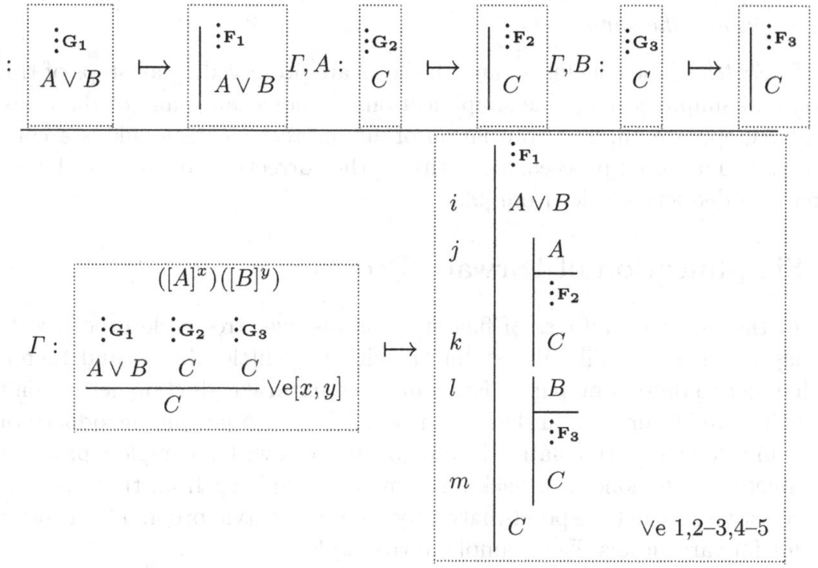

Fig. 4. Example of the general schema instantiated for ∨-elimination

4 Simplification of Obtained Proofs

We can remove many of the copy lines by 'path compression', i.e., if a copy line is not the last line under a flag, the copy line can be removed and all further references should be renumbered to refer to the line that was copied:

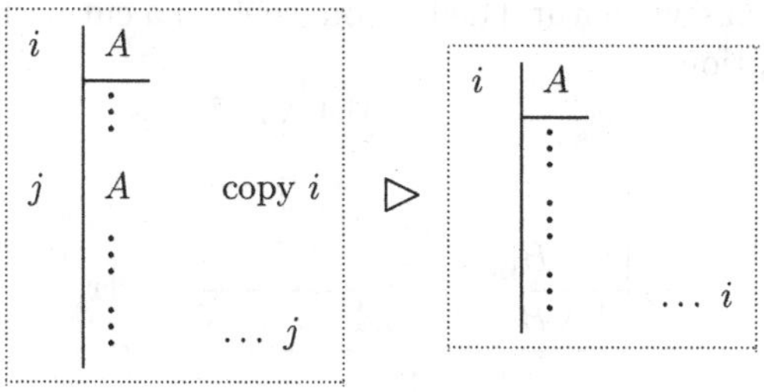

If the copy line is the only line under a flag and is the copy of the assumption introduced under this flag, the copy line can be removed. This creates proofs that resemble customary Fitch deduction drawing style:

$$
\begin{array}{l|l}
i & A \\ \hline
 & A \qquad \text{copy } i
\end{array}
\quad \triangleright \quad
\boxed{\underline{A}}
$$

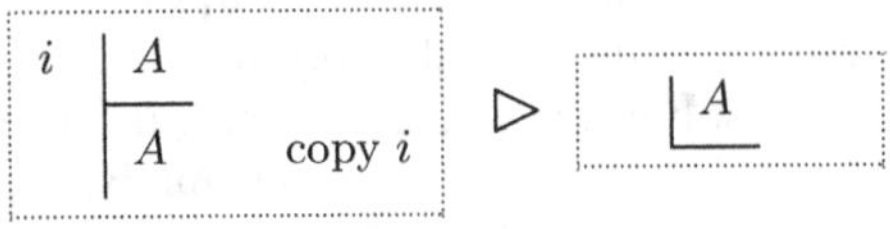

Theorem 1. *The use of the translation followed by performing the above simplifications on a correct Gentzen style natural deduction proof results in a flag style proof that is a correct Fitch style natural deduction proof with the same conclusion and the same rules.*

Proof (Sketch). The conclusion and the rules are preserved by all steps of translation and simplification. The simplifications do not change any of the rules or lines they operate on. The translation of any correct Gentzen rule is a correct Fitch rule. The proof proceeds by verifying the correctness of the translation of all natural deduction rules from [9]. □

5   Simplification of Forward Proofs

One of the main advantages of flag style proofs over tree style proofs, is that the flag proof is typically almost linear, with very little nesting and therefore much easier to present on paper. For completed natural deduction derivation the proof that we obtain by translation is mostly flat, with nesting introduced only for assumptions. Our translation is also able to work with incomplete proofs. For incomplete proofs done in a backwards manner (starting from the conclusion) the tree style proof corresponds naturally to the flag style proof. This is not the case for forward proofs. For example in tree style:

$$
\begin{array}{l|l}
i & A \\
j & \quad B \\
k & \quad A \wedge B \qquad \wedge i\ i,\ j \\
 & \quad \vdots \\
 & \quad C
\end{array}
$$

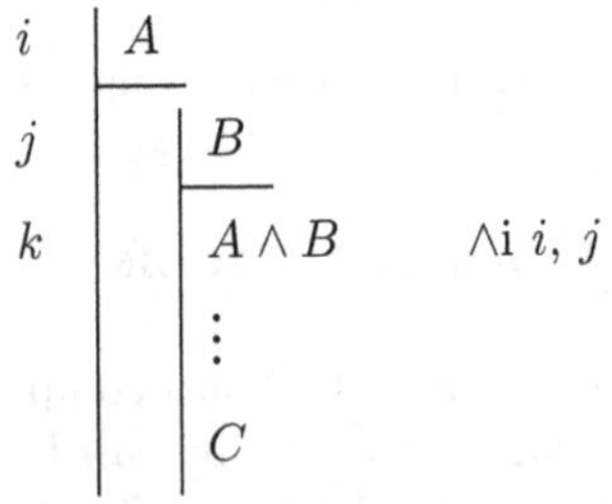

The line labeled k is obtained by $\wedge$-introduction from lines i and j. To represent this proof in Gentzen style natural deduction we need a cut with a branch where $A \wedge B$ is an assumption:

$$([A \wedge B]^x)$$

$$\vdots$$

$$
\dfrac{\dfrac{[A]\ \ [B]}{A \wedge B}\wedge i \qquad \dfrac{C}{A \wedge B \to C}\to i[x]}{C}\to e
$$

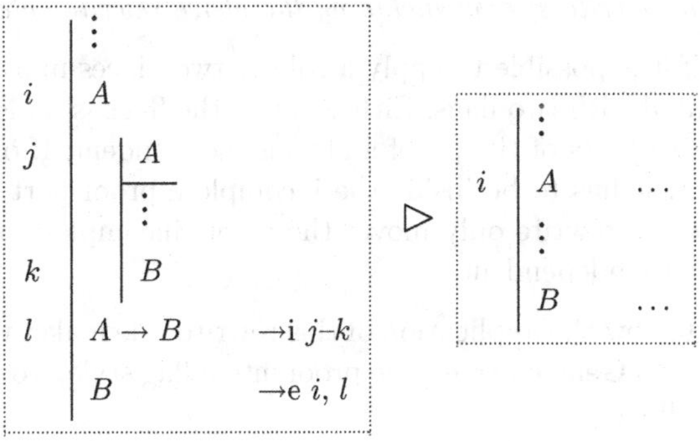

Fig. 5. Rewrite rule for eliminating explicit cuts from a Fitch deduction

$$\cfrac{\cfrac{\cdots}{\exists x, P\ x}}{(\exists y,\ (P\ x\theta \vee P\ y)) \to \exists x, P\ x} \to i\,[F1] \qquad \cfrac{[\forall x,\ \exists y,\ (P\ x \vee P\ y)]^{F}}{\exists y,\ (P\ x\theta \vee P\ y)}\,\forall e$$

$$\cfrac{\exists x, P\ x}{\forall x,\ \exists y,\ (P\ x \vee P\ y) \to \exists x, P\ x}\to i\,[F]$$

with the intermediate step marked $\to e$ between the two upper derivations and the lower $\exists x, P\ x$.

1	F:	∀x, ∃y, (P x ∨ P y)	assumption
2	F1:	∃y, (P x0 ∨ P y)	∀e 1
		. . .	
3		∃x, P x	
4		∀x, ∃y, (P x ∨ P y) → ∃x, P x	→i 1-3

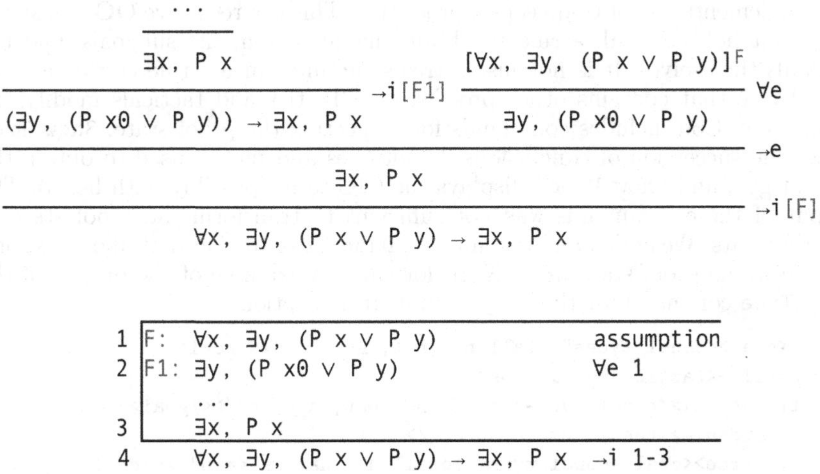

Fig. 6. An incomplete proof in Gentzen natural deduction and its translation to a Fitch deduction, as rendered by the implementation

The cut in the above proof cannot be eliminated until the proof is completed. However, this is not the case for flag style proofs, where this kind of cut can be eliminated without influence on the rest of the proof (assuming the rest of the proof is translated as well).

We want to give a mechanism that allows translating the above tree style proof with a cut to a flag style proof without a cut. The use of cut is a general technique; it is often used for inserting a subgoal that can be used further in the proof. This is why we will eliminate all the implication cuts that could have been obtained in this way. To do this we present the rewrite rule in Figure 5, which can be applied only if line l is not used further in the proof.

Theorem 2. *The rewrite system including the above rewrite rule terminates.*

Proof (Sketch). By induction on the number of flags.   □

Theorem 3. *The rewrite system including the above rewrite rule is confluent.*

Proof (Sketch). If it is possible to apply a rule at two places in a proof, the two places are associated with two flags. Either one of the flags is under the other or they are in separate parts of the proof and thus independent. If one of the flags is under the other, it has to be inside the incomplete proof part of the rewrite rule. In that case the rewrite only moves the whole incomplete proof and thus the rewrites also are independent. □

We see in Figure 6, how the application of this rewrite rule makes the translation of a Coq proof from a Gentzen tree style proof into a flag style proof with a small number of nested flags.

6 Implementation for Coq Proofs

The implementation of Coq keeps a proof tree. This is a recursive OCAML structure, that holds a goal, a rule to obtain this goal from the subgoals, and the subgoals themselves. It is not just a tree structure, since a rule can be a compound rule that contains other proof states. Tactics and tacticals modify the proof state. Coq includes commands for inspecting the proof state. `Show Tree` shows the succession of conclusions, hypotheses and tactics used to obtain the current goal and `Show Proof` displays the CIC term (possibly with holes). The output of these commands was not sufficient to transform the proof state in other formats. We added a new command `Dump Tree` to Coq that allows exporting the whole proof state in an XML format. An example of the output of the `Dump Tree` command for the Coq example from Section 1 is:

```
<tree><goal><concl type="forall n : nat, n <= n"/></goal>
  <cmpdrule><tactic cmd="intros"/>
    <tree><goal><concl type="forall n : nat, n <= n"/></goal>
      <cmpdrule><tactic cmd="intros"/>
        <tree><goal><concl type="forall n : nat, n <= n"/></goal>
          <rule text="intro n"/>
          <tree><goal><concl type="n <= n"/><hyp id="n" type="nat"/>
        </goal></tree></tree></cmpdrule>
      <tree><goal><concl type="n <= n"/><hyp id="n" type="nat"/>
    </goal></tree></tree>
  </cmpdrule><tree><goal><concl type="n <= n"/><hyp id="n" type="nat"/>
</goal></tree></tree>
```

This is the proof tree that corresponds to the incomplete Fitch proof on page 207.

The communication between ProofWeb and Coq is very narrow. The `Dump Tree` command is all that had to be added to Coq to allow our system to convert proofs, and its implementation only took a small amount of OCAML code. This code has now been integrated into the Coq code base, which means that the `Dump Tree` command will be standardly available in Coq from version 8.2.

Our system is intended to be used with simple tactics that correspond to the inference rules of first order logic, so currently we forget the information generated by automated tactics (the content of compound rules). We first transform the tree to a non-optimized flag proof. For every node of the Coq tree we create a new flag. This flag first contains all the assumptions. The notation presented in the previous sections where a flag is allowed to have an arbitrary number of assumptions is also used in the implementation; at a later step this gets translated according to the meaning of the notation. Then if a tree has subgoals, the transformed subgoals are attached. Otherwise, if the goal is not proved ellipses are attached. Finally the flag contains a line for the conclusion of the Coq node.

When rendering a flag style proof that was translated from a tree style proof done with the Coq tactics, the tactic names are printed in a special way. For tactics that match natural deduction rules, the names are changed to their natural deduction names. Furthermore we add the consecutive line numbers on the left of assumption lines and conclusion lines. We then replace references to labels with the appropriate numbers.

As an example of a flag style version of a serious Coq proof, consider the following proof from the Coq standard library:

```
Lemma leb_complete : forall m n:nat, leb m n = true -> m <= n.
Proof.
  induction m. trivial with arith.
  destruct n. intro H. discriminate H.
  auto with arith.
Qed.
```

This proof is rendered by ProofWeb as:

```
1          ∀n:nat, leb 0 n = true → 0 <= n              trivial[with,arith]
2   | m:    nat                                          assumption
3   | IHm:  ∀n:nat, leb m n = true → m <= n              assumption
4   | |H:   leb (S m) 0 = true                           assumption
5   | |     S m <= 0                                     discriminate[H]
6   |       leb (S m) 0 = true → S m <= 0                intro[H] 4-5
7   | |n:   nat                                          assumption
8   | |     leb (S m) (S n) = true → S m <= S n          auto[with,arith]
9   |       ∀n:nat, leb (S m) n = true → S m <= n        destruct[n] 6,7-8
10         ∀m:nat, ∀n:nat, leb m n = true → m <= n      induction[m] 1,2-9
```

7 Conclusion

The future work of this paper is to develop a *luxury* proof interface, as described in Section 1, for a serious proof assistant. The main difference with the ProofWeb system will then be that the system can also *input* a declarative proof. The declarative proofs then becomes the text that the user works on.

We implemented a rough prototype of a luxury proof language for the HOL Light system, and the approach seems to work quite well there. Currently we are redoing this system in a more systematic and structured manner. This experiment shows that our approach for converting procedural proofs into declarative proofs is not tied to any Coq specifics. It works just as well, and in exactly the same way, in a HOL environment.

A difference with ProofWeb will be to have one further rewrite rule for proofs. In the declarative language of the Mizar system there exists the `consider` statement that is used for existential elimination. If one knows that there exists an x that satisfies $P[x]$, one can write:

```
proof
  ...1
  consider x being A such that P[x] by ...2;
  ...3
  thus Q by ...4;
end;
```

This can be seen as a condensed version of

```
proof
  ...1
  proof
    let x be A;
    assume P[x];
    ...3
    thus Q by ...4;
  end;
  thus Q by ...2;
end;
```

In the case of the ProofWeb system we did *not* want the system to rewrite the latter to get the structure the former, as it would not leave Fitch-style proofs the way that student users would expect them to be. However, in a system for significant formalizations, an optimization like this will be essential.

We claim that a *luxury* proof interface – that is, an interface in which the user edits a declarative proof, but also can ask the system to extend that proof by executing tactics – combines the best of the procedural and declarative worlds. We expect that it will be straight-forward to implement such an interface using the methodology presented in this paper.

References

1. Asperti, A., Padovani, L., Sacerdoti Coen, C., Guidi, F., Schena, I.: Mathematical Knowledge Management in HELM. Annals of Mathematics and Artificial Intelligence, Special Issue on Mathematical Knowledge Management 38, 1–3 (2003)
2. Asperti, A., Wegner, B.: MOWGLI – A New Approach for the Content Description in Digital Documents. In: Proceedings of the Nineth International Conference on Electronic Resources and the Social Role of Libraries in the Future, vol. 1 (2002)
3. Geuvers, H., Loeb, I.: From Deduction Graphs to Proof Nets: Boxes and Sharing in the Graphical Presentation of Deductions. In: Královič, R., Urzyczyn, P. (eds.) MFCS 2006. LNCS, vol. 4162, pp. 39–57. Springer, Heidelberg (2006)
4. Geuvers, H., Nederpelt, R.: Rewriting for Fitch Style Natural Deductions. In: van Oostrom, V. (ed.) RTA 2004. LNCS, vol. 3091, pp. 134–154. Springer, Heidelberg (2004)

5. Girard, J.Y.: Linear Logic. Theor. Comput. Sci. 50, 1–102 (1987)
6. Guilhot, F., Naciri, H., Pottier, L.: Proof explanations: using natural language and graph view, Slides for a talk at a MoWGLI presentation (2003)
7. Harrison, J.R.: Proof Style. In: Giménez, E., Paulin-Möhring, C. (eds.) TYPES 1996. LNCS, vol. 1512, pp. 154–172. Springer, Heidelberg (1996)
8. Huth, M., Ryan, M.: Logic in Computer Science: Modelling and Reasoning about Systems, 2nd edn. Cambridge University Press, Cambridge (2004)
9. Kaliszyk, C., van Raamsdonk, F., Wiedijk, F., Wupper, H., Hendriks, M., de Vrijer, R.: Deduction using the ProofWeb system. Technical Report ICIS–R08016, Radboud University Nijmegen (September 2008)

Using Structural Recursion for Corecursion*

Yves Bertot[1] and Ekaterina Komendantskaya[2]

[1] INRIA Sophia Antipolis, France
Yves.Bertot@inria.fr
[2] School of Computer Science, University of St Andrews, UK
ek@cs.st-andrews.ac.uk

Abstract. We propose a (limited) solution to the problem of constructing stream values defined by recursive equations that do not respect the guardedness condition. The guardedness condition is imposed on definitions of corecursive functions in Coq, AGDA, and other higher-order proof assistants. In this paper, we concentrate in particular on those non-guarded equations where recursive calls appear under functions. We use a correspondence between streams and functions over natural numbers to show that some classes of non-guarded definitions can be modelled through the encoding as structural recursive functions. In practice, this work extends the class of stream values that can be defined in a constructive type theory-based theorem prover with inductive and coinductive types, structural recursion and guarded corecursion.

Keywords: Constructive Type Theory, Structural Recursion, Coinductive types, Guarded Corecursion, Coq.

1 Introduction

Interactive theorem provers with inductive types [27,28,20,16] provide a restricted programming language together with a formal meta-theory for reasoning about the language. This language is very close to functional programming languages, so that the verification of a program in a conventional functional programming language can often be viewed as a simple matter of adapting the program's formulation to a theorem prover's syntax, thus obtaining a faithful prover-level model. Then one can reason about this model in the theorem prover. This approach has inspired studies of a large collection of algorithms, starting from simple examples like sorting algorithms to more complex algorithms, like the ones used in the computation of Gröbner bases, the verification of the four-colour theorem, or compilers.

However, the prover's programming language is restricted, especially concerning recursion. For instance, *structural* restriction ensures that all programs terminate, so that values are never undefined; we give details in Section 2. Approaches

* Work is partially supported by the INRIA CORDI post-doctoral program, the ANR project "A3Pat" ANR-05-BLAN-0146 and by EPSRC postdoctoral grant EP/F044046/1.

S. Berardi, F. Damiani, and U. de'Liguoro (Eds.): TYPES 2008, LNCS 5497, pp. 220–236, 2009.

to cope with potentially non-terminating programs are available, especially by encoding domain theory as in HOLCF [24], but these approaches tend to make the description of programs more cumbersome, because the exceptional case where a computation may not terminate needs to be covered at every stage. An alternative is to manage a larger class of terminating functions, mainly using well-founded recursion [22,26], and this approach is now widely spread among all interactive theorem provers.

A few theorem provers [27,28,20] also support coinduction. Coinductive datatypes provide a way to look at infinite data objects. In particular, streams of data can be viewed as infinite lists. Coinductive datatypes also provide room for a new class of recursive objects, known as corecursive objects.

Termination is not required anymore for these functions, but termination still plays a role, since every finite value should still be computable in finite time, even if the computation involves an interaction with a corecursive value. This constraint boils down to a concept of *productivity*. Roughly speaking, infinite sequences of recursive calls where no data is being produced must be avoided. For recursive programs, productivity is undecidable for the same reason that termination is. For this reason, a more restrictive criterion is used to describe corecursive functions that are legitimate in theorem provers.

A theorem prover like Coq provides two kinds of recursion: terminating recursion, initially based on structural recursion for inductive types, which can also handle well-founded recursion; and productive corecursion, based on "guarded" corecursion [10,15]. Efforts have been made to extend the basic guarded corecursion in the same spirit that well-founded recursion extends the basic structural recursion. We can mention [14] and [4,7], which basically incorporate well-founded recursion to make sure several non-productive recursive calls are allowed as long as they ultimately become productive. In particular, [14] introduces a generalization of the concept of well-founded relation that uses an extra dimension to cover at the same time recursive or co-recursive functions; since there is an extra dimension, two notions of limits can be used and recursive values can mix terminating recursive and productive co-recursive aspects in a seamless fashion.

One essential characteristic of well-founded induction and the complete ordered families of equivalences in [14] is that the well-founded relation or families of equivalences must be given as extra data to make it possible to start the definition process. In the alternative approach described in this paper, we want to avoid this extra burden imposed on the user, and we attempt to develop a methodology that remains syntactic in nature.

We will concentrate on a class of recursive definitions where *mapping* functions interfere in the recursive equation, thus preventing the recursive equation to be recognised as *guarded by constructors.* The infinite sequences of Fibonacci numbers (considered e.g., in [1]) and of natural numbers (see Example 5) are famous representatives of the class. Many of the corecursive values studied, for example, in [25,12,13] fail to satisfy the guardedness condition, precisely because functions like `map` interfere in the recursive definition. A very elegant method of lazy differentiation [19] also gives rise to a function of multiplication for infinite streams of derivatives in the same class of definitions.

A simple example is the following recursive equation (studied later as Example 5):

```
nats = 1::map S nats
```

A quick analysis shows that we can use this equation to infer the value of each element in the stream: the first value is given directly, the second element is obtained from the first one through the behaviour of the `map` function, and so on. This recursive equation is a legitimate specification of a stream, and it can actually be used as a definition in a conventional lazy functional programming language like Haskell.

Thus the question studied in this article is: given a recursive equation like the one concerning `nats`, can we build a corecursive value that satisfies this equation, using only structural recursion and guarded corecursion? We will describe a partial solution to this problem. We will also show that this solution can incorporate other interfering functions than `map`. In Section 3, we briefly overview the class of the functions we target.

Our proposed approach is to map every stream value to a function over natural numbers in a reversible way: a stream $s_0{::}s_1{::}\cdots$ is mapped to the function $[\![s]\!] : i \mapsto s_i$, and the reverse map is an easily defined guarded corecursive function. It appears that all legitimate guarded corecursive values are mapped to structurally recursive functions and that the question of productivity is transformed into a question of termination. We discuss it in Section 5.

Moreover, uses of the `map` function and similar operations are transformed into program fragments that still respect the constraints of structural recursion. Thus, there are stream values whose recursive definitions as streams are mapped to structural recursive definitions, even though the initial equations did not respect guardedness constraints. For these stream values, we propose to define the corresponding recursive function using structural recursion, and then to produce the stream value using the reverse map from functions over natural numbers to streams. We present this method in Section 6.

2 Structurally Recursive Functions

We start with defining the notions of inductive and coinductive types, and recursive/corecursive functions. We will use the syntax of Coq throughout. For a more detailed introduction to Coq, see [5]. One can also handle inductive and coinductive types within HOL (proof assistant Isabelle) [23], and within Martin-Löf type theory (proof assistant AGDA) [27].

Inductive data types are defined by introducing a few basic constructors that generate elements of the new type.

Definition 1. *The definition of the inductive type of natural numbers is built using two constructors* `O` *and* `S`*:*

```
Inductive nat : Set := O : nat | S : nat -> nat.
```

This definition also implies that the type supports both pattern-matching and recursion: on the one hand, all values in the type are either of the form `O` or of the form `(S x)`; on the other hand, all values are finite and a function is well defined when its value on `O` is given and the value for `S x` can be computed from the value for `x`.

After the inductive type is defined, one can define its inhabitants and functions on it. Most functions defined on the inductive type must be defined recursively, that is, by describing values for different patterns of the constructors and by allowing calls to the same function on variables taken from the patterns.

Example 1. The recursive function below computes the n-th Fibonacci number.

```
Fixpoint fib (n:nat) : nat :=
  match n with
  | O => 1
  | S O => 1
  | S (S p as q) => fib p + fib q
  end.
```

There is one important property we wish every function defined in Coq to possess: it is termination. To guarantee this, Coq uses a syntactic restriction on definitions of functions, called *structural recursion*. A *structurally recursive* definition is such that every recursive call is performed on a structurally smaller argument. The function `fib` is *structurally recursive*: all recursive calls are made on variables (here `p` and `q`) that were obtained through pattern-matching from the initial argument.

There are many useful functions and algorithms that are not structurally recursive, but general recursive. They are not accepted by Coq or similar proof assistants directly, but they can be defined using various forms of well-founded induction or induction with respect to a *predicate* [5,8].

It is perhaps worth mentioning that there exists an approach to termination called "type-based termination" [1,3,17]. The essence of different methods proposed under this name is rejection of the structural recursion as being a too restrictive and narrow method for guaranteeing termination. Instead, this job is delegated to sized higher-order types. The type-based termination promises to be a powerful tool, but it is not easy to implement it. As for today, the major proof assistants still rely on structural recursion. Some non-guarded functions we formalise in this paper, can also be handled by methods of type-based termination. However, yet it gives little from the point of view of practical programming and automated proving. Therefore the value of this paper, as well as (e.g.) [5,7,8] is in the technical elegance and practical implementation in the existing proof assistants.

3 Guardedness

We now consider corecursion.

The following is the definition of a coinductive type of infinite streams, built using one constructor `Cons`.

Definition 2. *The type of streams is given by*

```
CoInductive Stream (A:Set) : Set :=
   Cons: A -> Stream A -> Stream A.
```

In the rest of this paper, we will write `a::tl` for `Cons _ a tl`, leaving the argument `A` to be inferred from the context.

While a structurally recursive function is supposed to rely on an inductive type for its domain and is restricted in the way recursive calls are using this input, a corecursive function is supposed to rely on a co-inductive type for its co-domain and is restricted in the way recursive calls are used for producing the output.

Definition 3 (Guardedness). *A position in an expression is* pre-guarded *if it occurs as the root of the expression, or if it is a direct sub-term of a pattern-matching construct or a conditional statement, which is itself in a pre-guarded position.*

A position is guarded if it occurs as a direct sub-term of a constructor for the co-inductive type that is being defined and if this constructor occurs in a pre-guarded position or a guarded position. A corecursive function is guarded *if all its corecursive calls occur in guarded positions.*

Example 2. The coinductive function `map` applies a given function `f` to a given infinite stream.

```
CoFixpoint map (A B :Type)(f: A -> B)(s: Stream A): Stream B :=
match s with x::s' => f x::map A B f s' end.
```

In this definition's right-hand side the match construct and the expression `f x::...` are in pre-guarded positions, the expression `map A B f s'` is in guarded position, and the definition is guarded.

Example 3. The coinductive function `nums` takes as argument a natural number n and produces a stream of natural numbers starting from n.

```
CoFixpoint nums (n: nat): Stream nat := n::nums (S n).
```

In this definition's right-hand side, the expression `n::nums (S n)` is in a pre-guarded position, the expression `nums (S n)` is in a guarded position.

Example 4. The following function `zipWith` is guarded:

```
CoFixpoint zipWith (A B C: Set)(f: A -> B -> C)
     (s: Stream A)(t: Stream B) : Stream C :=
match (s, t) with  (x :: s',  y :: t')  =>
(f x y):: (zipWith A B C f s' t')
end.
```

Informally speaking, the guardedness condition insures that

* each corecursive call is made under at least one constructor;
** if the recursive call is under a constructor, it does not appear as an argument of any function.

Violation of any of these two conditions makes a function non-guarded. According to the two guardedness conditions above, we will be talking about the two classes of non-guarded functions - (*) and (**).

A more subtle analysis of the corecursive functions that fail to satisfy the guardedness condition * can be found in [14,4,21,7]. In particular, the mentioned papers offer a solution to the problem of formalising productive corecursive functions of this kind.

Till the rest of the paper, we shall restrict our attention to the second class of functions. To the extent of our knowledge, this paper is the first attempt to systematically formulate the functions of this class in the language of a higher-order proof assistant with guarded corecursion.

Example 5. Consider the following equation:

```
nats = 1::map S nats
```

This definition is not guarded, the expression `map S nats` occurs in a guarded position, but `nats` is not; see the guardedness condition **. Despite of this, the value `nats` is well-defined.

Example 6. The following definition describes the stream of Fibonacci numbers:

```
fib =  0 ::  1 :: (zipWith nat nat plus (tl fib) fib).
```

Again, this recursive equation fails to satisfy **.

Example 7. The next example shows the function `dTimes` that multiplies the sequences of derivatives in the elegant method of lazy differentiation of [19,9].

```
dTimes x y =  match x, y with
  | x0 :: x', y0 :: y' =>
  (x0 * y0) :: (zipWith Z Z plus (dTimes x'  y)  (dTimes x y'))
   end.
```

Again, this function fails to satisfy **.

In the next section, we will develop a method that makes it possible to express Examples 5 - 7 as guarded corecursive values.

Values in co-inductive types usually cannot be observed as a whole, because of their infiniteness. To prove some properties of infinite streams, we use a method of observation. For example, to prove that the two streams are *bisimilar*, we must observe that their first elements are the same, and continue the process with the rest.

Definition 4. *Bisimilarity is expressed in the definition of the following coinductive type:*

```
CoInductive EqSt:  Stream A -> Stream A -> Prop :=
| eqst : forall (a : A) (s s' : Stream A), EqSt s s' ->
                   EqSt (a::s)(a::s').
```

In the rest of this paper, we will write a==b for `EqSt` a b. The definition of a==b corresponds to the conventional notion of bisimilarity as given, e.g. in [18]. Lemmas and theorems analogous to the *coinductive proof principle* of [18] are proved in Coq and can be found in [5].

Bisimilarity expresses that two streams are observationally equal. Very often, we will only be able to prove this form of equality, but for most purposes this will be sufficient.

4 Soundness of Recursive Transformations for Streams

In this section, we show that streams can be replaced by functions. Because there is a wide variety of techniques to define functions, this will make it possible to increase the class of streams we can reason about. Our approach will be to start with a (possibly non-guarded) recursive equation known to describe a stream, to transform it systematically into a recursive equation for a structurally recursive function, and then to transform this function back into a stream using a guarded corecursive scheme.

As a first step, we observe how to construct a stream from a function over natural numbers:

Definition 5. *Given a function f over natural numbers, it can be transformed into a stream using the following function:*

```
Cofixpoint stroff (A:Type)(f:nat->Type) : Stream A :=
  f 0 :: stroff A (fun x => f (1+x)).
```

This definition is guarded by constructors. In the rest of this paper, we will write $\langle s \rangle$ for `stroff _` s leaving the argument `A` to be inferred from the context.

The function `stroff` has a natural inverse, the function `nth` which returns the element of a stream at a given rank:

Definition 6. *The function* `nth`[1] *is defined as follows:*

```
Fixpoint nth (A:Type) (n:nat) (s: Stream A) {struct n}: A :=
match s with  a :: s' =>
  match n with | O =>  a | S p => nth A p s' end
end.
```

In the rest of this paper, we will omit the first argument (the type argument) of `nth`, following Coq's approach to implicit arguments. We will use notation $[\![s]\!]$ when talking about `(fun n => nth n s)`.

It is easy to prove that $[\![\cdot]\!]$ and $\langle\cdot\rangle$ are inverse of each other. Composing these two functions is the essence of the method we develop here. The lemmas below are essential for guaranteeing the soundness of our method.

[1] In Coq's library, this function is defined under the name `Str_nth`.

Lemma 1. *For any function f over natural numbers,* $\forall$`n, nth n` $\langle f \rangle$ `=` f `n`.

Lemma 2. *For any stream s,* s `==` $\langle [\![s]\!] \rangle$.

Proof. Both proofs are done in Coq and available in [6].

We now want to describe a transformation for (non-guarded) recursive equations for streams. A recursive equation for a stream would normally have the form

$$a = e \tag{1}$$

where both a and e are streams, and a can also occur in the expression e; see Examples 5 - 7. We use this initial non-guarded equation to formulate a guarded equation for a of the form:

$$a = \langle e' \rangle \tag{2}$$

where e' is a function extensionally equivalent to $[\![e]\!]$. As we show later in this section, we often need to evaluate `nth` only partially or only at a certain depth, this is why the job cannot be fully delegated to `nth`.

The definition of e' will have the form

$$e'\ n = E \tag{3}$$

where e' can again occur in the expression E.

Example 8 (zeroes). For simple examples, we can go through steps (1)-(3) intuitively. Consider the corecursive guarded definition of a stream `zeroes` that contains an infinite repetition of `0`.

```
CoFixpoint zeroes := 0 :: zeroes.
```

We can model the body of this corecursive definition as follows:

```
Fixpoint nzeroes (n:nat) : nat :=
  match n with 0 => 0 | S p => nzeroes p end.
```

This is a legitimate structurally recursive definition for a function that maps any natural number to zero. Note that the obtained function is extensionally equal to $[\![$`zeroes`$]\!]$.

```
Lemma nth_zeroes: forall n, nth n zeroes = nzeroes n.
```

Thus, a stream that is bisimilar to `zeroes` can be obtained by the following commands:

```
Definition zeroes' := stroff _ nzeroes.
```

By Lemma `nth_zeroes` and Lemma 2, `zeroes` and `zeroes'` are bismilar, see [6] for a proof.

The main issue is to describe a systematic transformation from the expression e in the equation 1 to the expression E in the equation (3). This "recursive" part of the work will be the main focus of the next section.

5 Recursive Analysis of Corecursive Functions

We continue to systematise the steps (1)-(3) of the transformation for a recursive equation $a = e$.

The expression e can be seen as the application of a function F to a. In this sense, the recursive definition of a expresses that a is fixpoint of F. The type of F is `Stream` $A \rightarrow$ `Stream` A for some type A. We will derive a new function F' of type (`nat` $\rightarrow A$) $\rightarrow$ (`nat` $\rightarrow A$); the recursive function a' that we want to define is a fixed point of F'. We obtain F' from F in two stages:

Step 1. We compose F on the left with $\langle\cdot\rangle$ and on the right with $[\![\cdot]\!]$. This naturally yields a new function of the required type. In practice, we do not use an explicit composition function, but perform the syntactic replacement of the formal parameter with the $\langle\cdot\rangle$ expression everywhere.

Example 9. For instance, when considering the `zeroes` example, the initial function

```
Definition zeroes_F (zeroes:Stream nat) := 0::zeroes
```

is recursively transformed into the function:

```
Definition zeroes_F' (nzeroes : nat -> nat) :=
  nth n (0::stroff nzeroes).
```

The corecursive value we consider may be a function taking arguments in types $t_1, \ldots, t_n$, that is, the function F may actually be defined as a function of type $(t_1 \rightarrow \cdots \rightarrow t_n \rightarrow \texttt{Stream}A) \rightarrow (t_1 \rightarrow \cdots \rightarrow t_n \rightarrow \texttt{Stream}A)$. The reformulated function F' that is obtained after composition with $\langle\cdot\rangle$ and $[\![\cdot]\!]$ has the corresponding type where `Stream` A is replaced with `nat` $\rightarrow A$. Thus, it is the first argument that incurs a type modification. When one of the types t_i is itself a stream type, we can choose to leave it as a stream type, or we can choose to replace it also with a function type. When replacing t_i with a function type, we have to add compositions with $[\![\cdot]\!]$ and $\langle\cdot\rangle$ at all positions where the first argument f of F is used, to express that the argument of f at the rank i must be converted from a stream type to a function type and at all positions where the argument of the rank $i+1$ of F is used, to express that this argument must be converted from a function type to a stream type.

We choose to perform this transformation of a stream argument into a function argument only when the function being defined is later used for another recursive definition. In this paper, this happens only for the functions `map` and `zipWith`.

Example 10. Consider the function `map` from Example 2. The function F for this case has the following form:

```
Definition map_F
 (map : forall (A B:Type)(f: A -> B), Stream A -> Stream B) :=
 fun A B f s => match s with a::s' => f a::map A B f s' end.
```

The fourth argument to `map` and the fifth argument to `map_F` have type `Stream A` and we choose to replace this type with a function type. We obtain the following new function:

```
Definition map_F'
  (map : forall (A B:Type)(f:A -> B), (nat -> A) -> nat -> B :=
fun A B f s => ⟦match ⟨s⟩ with a::s' => f a::⟨map A B f ⟦s'⟧⟩ end⟧.
```

Step 2. We go on transforming the body of F' according to rewriting rules that express the interaction between $\langle\cdot\rangle$, $[\![\cdot]\!]$, and the usual functions and constructs that deal with streams.

The Table 1 gives a summary of the rewriting rules for the transformation.

Table 1. Transformation rules for function representations of streams

1. `nth` n $\langle f\rangle = f\ n$,
2. `nth` n $(a$`::`$s')$ = `match` n `with` 0 `=>` a `|` `S` p `=>` `nth` $p\ s'$ `end`,
3. `hd` $\langle f\rangle = f\ 0$,
4. `tl` $\langle f\rangle = \langle$`fun` n `=>` f `(S` n`)`$\rangle$,
5. `match` $\langle f\rangle$ `with` a`::`s `=>` $e\ a\ s$ `end` $= e\ (f\ 0)\ \langle$`fun` n `=>` f `(S` n`)`$\rangle$,
6. β-reduction.

All these rules can be proved as theorems in the theorem prover [6]: this guarantees soundness of our approach. However, this kind of rewriting cannot be done directly inside the theorem prover, since rewriting can only be done while proving statements, while we are in the process of defining a function. Moreover, the rewriting operations must be done thoroughly, even inside lambda-abstraction, even though an operation for that may not be supported by the theorem prover (for instance, in the calculus of constructions as it is implemented in Coq, rewriting does not occur inside abstractions).

The rewriting rules make the second argument of `nth` decrease. When the recursive stream definition is guarded, this process ends with a structural function definition.

Example 11. Let us continue with the definition for `map`.

```
map_F' map A B f s n =
  nth n match ⟨s⟩ with a:: s' => f a::⟨map A B f ⟦s'⟧⟩end
= nth n (f(s 0)::⟨map A B f ⟦s'⟧⟩) end
= match n with
    0 => f(s 0) | S p => nth p ⟨map A B f ⟦⟨fun n => s (S n)⟩⟧⟩
  end
= match n with 0=>f(s 0)|S p=>map A B f ⟦⟨fun n => s (S n)⟩⟧p end
= match n with 0=>f(s 0)|S p=>map A B f (fun n => s (S n)) p end
```

When considered as the body for a recursive definition of a function `map'`, the last right-hand side is a good structural recursive definition with respect to the initial parameter `n`. We can use this for a structural definition:

```
Fixpoint map' (A B : Type) (f : A -> B) (s: nat -> A)
     (n : nat) {struct n} :=
match n with 0 => f a|S p => map' A B f (fun n => s (S n)) p end.
```

This function models the map function on streams, as a function on functions. It enjoys a particular property, which plays a central role in this paper:

Lemma 3 (Form-shifting lemmas)

$\forall$ f s n, nth n (map f s) = f (nth n s)

$\forall$ f s n, $[\![$map$]\!]$ f s n = f (s n).

Proof. See [6].

Thanks to the second statement of the lemma, s can be moved from an argument position to an active function position, as will later be needed for verifying structural recursion of other values relying on map.

Now, we show the same formalisation for the function zipWith:

Example 12 (Zip). The function zipWith can also be transformed, with the choice that both stream arguments are transformed into functions over natural numbers.

```
Definition zipWith_F
  (zipWith : forall (A B C : Type), (A -> B -> C) ->
                                  Stream A -> Stream B -> Stream C)
  (A B C : Type)(f : A -> B -> C)(a : Stream A)(b : Stream B) :=
  match a, b with
    x :: a', y :: b' => f x y :: zipWith A B C f a' b'
  end.
```

Viewing arguments a and b as functions and applying the rules from Table 1 to this definition yields the following recursive equation:

```
zipWith_F' zipwith' A B C f a b n =
  match n with
    0 => f (a 0) (b 0)
  | S p => zipwith' (fun n => a (S n)) (fun n => b (S n)) p
  end
```

Here again, this leads to a legitimate structural recursive definition on the fourth argument of type nat. We also have form-shifting lemmas:

Lemma 4 (Form-shifting lemmas)

$\forall$ f s_1 s_2 n, nth n (zipWith f s_1 s_2) = f (nth n s_1) (nth n s_2)

$\forall$ f s_1 s_2 n, $[\![$zipWith$]\!]$ f s_1 s_2 n = f (s_1 n) (s_2 n).

Proof. See [6].

The second statement also moves s_1 and s_2 from argument position to function position.

Unfortunately, we do not know the way to automatically discover the Form-shifting lemmas; although the statements of these lemmas follow the same generic pattern and once stated, the proofs for them do not tend to be difficult. Instead, as we illustrate in the Conclusion, we sometimes can give a convincing argument showing that a form-shifting lemma for a particular function cannot be found; and this provides an evidence that our method is not applicable to this function. That is, existence or non/existence of the form-shifting lemmas can serve as a criterion for determining whether the function can be covered by the method.

6 Satisfying Non-guarded Recursive Equations

Form-shifting lemmas play a role when studying recursive equations that do not satisfy the guardedness condition **, that is, when the corecursive call is made under functions like `map` or `zipWith`. To handle these functions, we simply need to add one new rule, as in Table 2, which will handle occurrences of each function that has a form-shifting lemma.

Table 2. Rule for recursive transformation of non-guarded streams

7. Let `f` be a function and `C` be a context in which arguments of `F` appear. If a form shifting lemma has the following shape:

   $$\forall a_1 \cdots a_k\ s_1 \cdots s_l\ n, [\![\mathtt{f}]\!] a_1 \cdots a_k\ s_1 \cdots s_l\ n = C[a_1, \ldots, a_k, s_1\ n, \ldots, s_l\ n],$$

   then this equation should be used as an extra rewriting rule.

The extended set of transformation rules from Tables 1 and 2 can now be used to produce functional definitions of streams that were initially defined by non-guarded corecursive equations. The technique is as follows:

(a) Translate the equation's right-hand-side as prescribed by the rules in Tables 1 and 2,
(b) Use the equation as a recursive definition for a function,
(c) Use the function $\langle\cdot\rangle$ to obtain the corresponding stream value,
(d) Prove that this stream satisfies the initial recursive equation, using bisimilarity as the equality relation.

For the last step concerning the proof, we rely on two features provided in the Coq setting:

– For each recursive definition, the Coq system can generate a specialised induction principle, as described in [2],
– A proof that two streams are bisimilar can be transformed into a proof that their functional views are extensionally equal, using the theorem `ntheq_eqst`:

```
ntheq_eqst :
  ∀A (s1 s2:Stream A), (∀n, nth n s1 = nth n s2) -> s1 == s2
```

Using these two theorems and combining them with systematic rewriting with Lemma 1 and the form-shifting lemmas, we actually obtain a tactic we called `str_eqn_tac` in [6] which proves the recursive equations automatically.

We illustrate this method using our running examples.

Example 13. Consider the corecursive non-guarded definition of `nats` from Example 5. Here is the initial equation

```
nats = 1::map S nats
```

After applying all transformation rules we obtain the following equation between functions:

```
⟦nats⟧ = fun n => if n = 0 then 1 else S (⟦nats⟧ (n - 1)).
```

This is now a legitimate structurally recursive equation defining ⟦`nats`⟧, from which we define `nats` as `nats =` ⟨⟦`nats`⟧⟩. The next step is to show that `nats` satisfies the equation of Example 5.

```
nats == 1::map S nats
```

Using the theorem `ntheq_eqst` and Lemma 1 on the left-hand-side this reduces to the following statement:

∀ n, ⟦`nats`⟧ n `= nth` n `(1::map S` ⟨⟦`nats`⟧⟩`)`

We can now prove this statement by induction on the structure of the function ⟦`nats`⟧, as explained in [2]. This gives two cases:

```
1 = nth 0 (1::map S ⟦nats⟧)

S (⟦nats⟧ p) = nth (S p) (1 :: map S ⟨⟦nats⟧⟩)
```

The first goal is a direct consequence of the definition of `nth`. The second goal reduces as follows:

```
S (⟦nats⟧ p) = nth p (map S ⟨⟦nats⟧⟩)
```

Rewriting with the first form-shifting lemma for `map` yields the following goal:

```
S (⟦nats⟧ p) = S (nth p ⟨⟦nats⟧⟩)
```

Rewriting again with Lemma 1 yields the following trivial equality.

```
S (⟦nats⟧ p) = S (⟦nats⟧ p).
```

Example 14. The sequence of Fibonacci numbers can be defined by the following equation:

```
fib = 1::1::zipWith plus fib (tl fib)
```

When processing the left-hand side of this equation using the rules from Tables 1, 2 and the form-shifting lemma for `zipWith`, we obtain the following code:

```
[[fib]] = fun n =>
 match n with
 | 0 => 1
 | S p => match p with 0=>1 | S q=>[[fib]] q + [[fib]] (1+q) end
 end
```

This is still not accepted by the Coq system because `(1+q)` is not a variable term, however it is semantically equivalent to `p`, and the following text is accepted:

```
[[fib]] = fun n =>
 match n with
 | 0 => 1
 | S p => match p with 0=>1 | S q=>[[fib]] q + [[fib]] p end
 end
```

Again, by Definition 5, we can define a stream `fib` = ⟨⟦`fib`⟧⟩, and `fib` is proved to satisfy the initial recursive equation automatically.

It is satisfactory that we have a systematic method to produce a stream value for the defining recursive equation, but we should be aware that the implementation of `fib` through a structural recursive function does not respect the intended behaviour and has a much worse complexity —exponential— while the initial equation can be implemented using lazy data-structures and have linear complexity.

Finally, we illustrate the work of this method on the function `dTimes` from Example 7:

Example 15. For the function `dTimes`, we choose to leave the two stream arguments `x` and `y` as streams. We recover the structurally recursive function ⟦`dTimes`⟧ from Example 7:

```
[[dTimes]] (x y:Stream nat) (n:nat){struct n} =
  match x, y with
  | x0 :: x', y0 :: y' =>
    match n with
    | 0 => x0 * y0
    | S p => ([[dTimes]] x' y p) + ([[dTimes]] x y' p)
    end
  end.
```

It remains to define the stream ⟨⟦`dTimes`⟧⟩, which is a straightforward application of Definition 5, and to prove that it satisfies the initial recursive equation from Example 7. In [6], the proof is again handled automatically.

7 Conclusions

The practical outcome of this work is to provide an approach to model corecursive values that are not directly accepted by the "guarded-by-constructors" criterion, without relying on more advanced concepts like well-founded recursion of ordered families of equivalences. With this approach we can address formal verification for a wider class of functional programming languages. The work presented here is complementary to the work presented in [7], since the method in that paper only considers definitions where recursive calls occur outside of any constructor, while the method in this paper considers definitions where recursive calls are inside constructors, but also inside interfering functions.

The attractive features of this approach is that it is systematic and simple. It appears to be simpler than, e.g., related work done in [14,7,11] that involved introducing particular coinductive types and manipulating ad-hoc predicates. Although the current state of our experiments relies on manual operations, we believe the approach can be automated in the near future, yielding a command in the same spirit as the `Function` command of Coq recent versions.

The Coq system also provides a mechanism known as extraction which produces values in conventional functional programming languages. When it comes to producing code for the solution of one of our recursive equations on streams, we have the choice of using the recursive equation as a definition, or the extracted code corresponding to the structurally recursive model. We suggest that the initial recursive equation, which was used as our specification, should be used as the definition, because the structural recursive value may not respect the intended computational complexity. This was visible in the model we produced for the Fibonacci sequence, which does not take advantage of the value re-use as described in the recursive equation. We still need to investigate whether using the specification instead of the code will be sound with respect to the extracted code.

The method presented here is still very limited: it cannot cope with the example of the Hamming sequence, as proposed in [12]. A recursive definition of this stream is:

```
H = 1::merge (map (Zmult 2) H) (map (Zmult 3) H)
```

In this definition, `merge` is the function that takes two streams and produces a new stream by always taking the least element of the two streams: when the input streams are ordered, the output stream is an ordered enumeration of all values in both streams. Such a `merge` function is easily defined by guarded corecursion, but `merge` interferes in the definition of `H` in the same way that `map` interfered in our previous examples. This time, we do not have any good form-shifting lemma for this function. The hamming sequence can probably be defined using the techniques of [14] and we were also able to find another syntactic approach for this example, this new approach is a subject for another paper.

References

1. Abel, A.: Type-Based Termination. A Polymorphic Lambda-Calculus with Sized Higher-Order Types. PhD thesis, Fakultät für Mathematik, Informatik und Statistik der Ludwig-Maximilians-Universität München (2006)
2. Barthe, G., Courtieu, P.: Efficient Reasoning about Executable Specifications in Coq. In: Carreño, V.A., Muñoz, C.A., Tahar, S. (eds.) TPHOLs 2002. LNCS, vol. 2410, pp. 31–46. Springer, Heidelberg (2002)
3. Barthe, G., Frade, M.J., Giménez, E., Pinto, L., Uustalu, T.: Type-based termination of recursive definitions. Mathematical Structures in Computer Science 14, 97–141 (2004)
4. Bertot, Y.: Filters and co-inductive streams, an application to Eratosthenes' sieve. In: Urzyczyn, P. (ed.) TLCA 2005. LNCS, vol. 3461, pp. 102–115. Springer, Heidelberg (2005)
5. Bertot, Y., Castéran, P.: Interactive Theorem Proving and Program Development, Coq'Art: the Calculus of Constructions. Springer, Heidelberg (2004)
6. Bertot, Y., Komendantskaya, E.: Experiments on using structural recursion for corecursion: Coq code (2008), `http://hal.inria.fr/inria-00322331/`
7. Bertot, Y., Komendantskaya, E.: Inductive and coinductive components of corecursive functions in Coq. Electr. Notes Theor. Comput. Sci. 203(5), 25–47 (2008)
8. Bove, A.: General Recursion in Type Theory. PhD thesis, Department of Computing Science, Chalmers University of Technology (2002)
9. Bronson, S.: Posting to Coq club. Codata: problem with guardedness condition? (June 30, 2008), `http://pauillac.inria.fr/pipermail/coq-club/2008/003783.html`
10. Coquand, T.: Infinite objects in type theory. In: Barendregt, H., Nipkow, T. (eds.) TYPES 1993. LNCS, vol. 806, pp. 62–78. Springer, Heidelberg (1994)
11. Danielsson, N.A.: Posting to Coq club. Codata: problem with guardedness condition? An ad-hoc approach to productive definitions (1,4 August 2008), `http://pauillac.inria.fr/pipermail/coq-club/2008/003859.html` and `http://sneezy.cs.nott.ac.uk/fplunch/weblog/?p=109`
12. Dijkstra, E.W.: Hamming's exercise in SASL, circulated privately (June 1981)
13. Endrullis, J., Grabmayer, C., Hendriks, D., Isihara, A., Klop, J.W.: Productivity of stream definitions. In: Csuhaj-Varjú, E., Ésik, Z. (eds.) FCT 2007. LNCS, vol. 4639, pp. 274–287. Springer, Heidelberg (2007)
14. Gianantonio, P.D., Miculan, M.: A unifying approach to recursive and co-recursive definitions. In: Geuvers, H., Wiedijk, F. (eds.) TYPES 2002. LNCS, vol. 2646, pp. 148–161. Springer, Heidelberg (2003)
15. Giménez, E.: Un Calcul de Constructions Infinies et son Application à la Vérification des Systèmes Communicants. PhD thesis, Laboratoire de l'Informatique du Parallélisme, Ecole Normale Supérieure de Lyon (1996)
16. Gordon, M.J.C., Melham, T.F.: Introduction to HOL: a theorem proving environment for higher-order logic. Cambridge University Press, Cambridge (1993)
17. Hughes, J., Pareto, L., Sabry, A.: Proving the correctness of reactive systems using sized types. In: POPL, pp. 410–423 (1996)
18. Jacobs, B., Rutten, J.: A tutorial on (co)algebras and (co)induction. EATCS Bulletin 62, 222–259 (1997)
19. Karczmarczuk, J.: Functional differentiation of computer programs. Higher-Order and Symbolic Computation 14(1), 35–57 (2001)

20. Nipkow, T., Paulson, L.C., Wenzel, M. (eds.): Isabelle/HOL - A Proof Assistant for Higher-Order Logic. LNCS, vol. 2283. Springer, Heidelberg (2002)
21. Niqui, M.: Coinductive field of exact real numbers and general corecursion. In: Proc. of CMCS 2006. ENTCS, vol. 164, pp. 121–139. Elsevier, Amsterdam (2006)
22. Nordström, B.: Terminating general recursion. BIT 28, 605–619 (1988)
23. Paulson, L.: Mechanizing coinduction and corecursion in higher-order logic. Logic and Computation 2(7), 175–204 (1997)
24. Regensburger, F.: Holcf: Higher order logic of computable functions. In: Schubert, E.T., Alves-Foss, J., Windley, P. (eds.) HUG 1995. LNCS, vol. 971, pp. 293–307. Springer, Heidelberg (1995)
25. Sijtsma, B.: On the productivity of recursive list definitions. ACM Transactions on Programing Languages and Systems 11(4), 633–649 (1989)
26. Slind, K.: Function definition in higher order logic. In: von Wright, J., Harrison, J., Grundy, J. (eds.) TPHOLs 1996. LNCS, vol. 1125, Springer, Heidelberg (1996)
27. The Agda Development Team. The agda reference manual, `http://appserv.cs.chalmers.se/users/ulfn/wiki/agda.php`
28. The Coq Development Team. The Coq proof assistant reference manual, `http://coq.inria.fr`

Manifest Fields and Module Mechanisms in Intensional Type Theory

Zhaohui Luo*

Dept of Computer Science, Royal Holloway, Univ of London
Egham, Surrey TW20 0EX, UK
`zhaohui@cs.rhul.ac.uk`

Abstract. Manifest fields in a type of modules are shown to be expressible in intensional type theory without strong extensional equality rules. These *intensional manifest fields* are made available with the help of coercive subtyping. It is shown that, for both Σ-types and dependent record types, the `with`-clause for expressing manifest fields can be introduced by means of the intensional manifest fields. This provides not only a higher-order module mechanism with ML-style sharing, but a powerful modelling mechanism in formalisation and verification of OO-style program modules.

1 Introduction

A type of modules may be expressed in type theory as a Σ-type or a dependent record type. A field in such a type is usually *abstract* (of the form '$v : A$') in the sense that the data in that field can be any object of a given type. In contrast, a field is *manifest* (of the form '$v = a : A$') if the data in that field is not only of a given type but the 'same' as some specific object of that type. Intuitively, manifest fields allow internal expressions of definitional entries and are hence very useful in expressing various powerful constructions in a type-theoretic setting. For example, one can use Σ-types or dependent record types with manifest fields to express the powerful module mechanism with the so-called ML-style sharing (or sharing by equations) [31,20].

For a manifest field $v = a : A$, the 'sameness' of the data as object a may be interpreted as judgemental equality in type theory, as is done in all of the previous studies on manifest fields in type theory [16,37,13]. If so, this gives rise to an extensional notion of judgemental equality and such manifest fields may be called *extensional manifest fields*. In type theory, such extensional manifest fields may also be obtained by means of other extensional constructs such as the singleton type [5,17] and the extensional equality [32,11]. It is known, however, such an extensional notion of equality is meta-theoretically difficult (in the cases of the extensional manifest fields and the singleton types) or even lead to outright undecidability (in the case of the extensional equality).

* This work is partially supported by the research grant F/07-537/AA of the Leverhulme Trust in U.K. and the TYPES grant IST-510996 of EU.

S. Berardi, F. Damiani, and U. de'Liguoro (Eds.): TYPES 2008, LNCS 5497, pp. 237–255, 2009.

As shown in this paper, the 'sameness' in a manifest field does not have to be interpreted by means of an extensional equality. With the help of coercive subtyping [25], manifest fields are expressible in *intensional* type theories such as Martin-Löf's intensional type theory [35] and UTT [23]. The idea is very simple: for a of type A, a manifest field $v = a : A$ is simply expressed as the shorthand of an ordinary (abstract) field $v : \mathbf{1}(A, a)$, where $\mathbf{1}(A, a)$ is the inductive unit type parameterised by A and a. Then, with a coercion that maps the objects of $\mathbf{1}(A, a)$ to a, v stands for a in a context that requires an object of type A. This achieves exactly what we want with a manifest field. Such a manifest field may be called an *intensional manifest field* (IMF) and, to distinguish it from an extensional manifest field, we use the notation $v \sim a : A$ to stand for $v : \mathbf{1}(A, a)$.

Manifest fields may be introduced using the `with`-clause that intuitively expresses that a field is manifest rather than abstract. For both Σ-types and dependent record types, `with`-clauses can be introduced by means of IMFs and used as expected in the presence of the component-wise coercion that propagates subtyping relations. For Σ-types, it is shown that the employed coercions are coherent together and that the IMF-representation of `with`-clauses is adequate.

Our work on IMFs in record types is based on a novel formulation of dependent record types (without manifest fields), which is different from those in the previous studies [16,37,13] and has its own merits. Among other things, our formulation is independent of structural subtyping (as in [37]), allowing more flexible subtyping relations to be adopted in formalisation, and introduces kinds of record types, giving a satisfactory solution to the problem of how to ensure label distinctness in record types.

Intensional manifest fields can be used to express definitional entries and provide not only a higher-order module mechanism with ML-style sharing[1] but also a powerful modelling mechanism in formalisation and verification of OO-style programs. Using the record macro in Coq [12], we give examples to show, with IMFs, how ML-style sharing can be captured and how classes in OO-style programs can be modelled. Since intensional type theories are implemented in the current proof assistants, many of which support the use of coercions, the module mechanism supported by IMFs can also be used for modular development of proofs and dependently-typed programs.

The following subsection briefly describes the logical framework LF and coercive subtyping, establishing the notational conventions. In Section 2, we introduce manifest fields and explain how they may be expressed in extensional type theories. The IMFs in Σ-types are studied in Section 3. In Section 4, we formulate dependent record types, introduce the IMFs in record types, and illustrate their uses in expressing the module mechanism with ML-style sharing and in modelling OO-style classes in formalisation and verification. Some of the related and future work is discussed in the conclusion.

[1] Historically, expressing ML-style sharing is the main motivation behind the studies of manifest fields [16,20,37]. In fact, it has long been believed that, to express ML-style sharing in type theory, it is essential to have some construct with an extensional notion of equality. As shown in this paper, this is actually unnecessary.

1.1 The Logical Framework and Coercive Subtyping

The Logical Framework LF. LF [23] is the typed version of Martin-Löf's logical framework [35]. It is a dependent type system for specifying type theories such as Martin-Löf's intensional type theory [35] and the Unifying Theory of dependent Types (UTT) [23]. The types in LF are called *kinds*, including:

- $Type$ – the kind representing the universe of types (A is a type if $A : Type$);
- $El(A)$ – the kind of objects of type A (we often omit El); and
- $(x{:}K)K'$ (or simply $(K)K'$ when $x \notin FV(K')$) – the kind of dependent functional operations such as the abstraction $[x{:}K]k'$.

The rules of LF can be found in Chapter 9 of [23]. We sometimes use $M[x]$ to indicate that x may occur free in M and subsequently write $M[a]$ for the substitution $[a/x]M$.

When a type theory is specified in LF, its types are declared, together with their introduction/elimination operators and the associated computation rules. Examples include inductive types such as Nat of natural numbers, inductive families of types such as $Vect(n)$ of vectors of length n, and families of inductive types such as Π-types $\Pi(A, B)$ of functions $\lambda(x{:}A)b$ and Σ-types $\Sigma(A, B)$ of dependent pairs (a, b).[2] In a non-LF notation, $\Sigma(A, B)$, for example, will be written as $\Sigma x{:}A.B(x)$. A nested Σ-type can be seen as a type of tuples/modules.

Notation. We shall use $\sum[x_1 : A_1,\ x_2 : A_2,\ ...,\ x_n : A_n]$ to stand for $\Sigma x_1 : A_1 \Sigma x_2 : A_2\ ...\ \Sigma x_{n-1} : A_{n-1}.\ A_n$, where $n \geq 1$. (When $n = 1$, $\sum[x{:}A] =_{\text{df}} A$.) Similarly, $(a_1,\ a_2,\ ...,\ a_n)$ stands for $(a_1,\ (a_2,\ ...,\ (a_{n-1}, a_n)...))$. Furthermore, for any a of type $\sum[x_1 : A_1,\ x_2 : A_2,\ ...,\ x_n : A_n]$,

- $a.i =_{\text{df}} \pi_1(\pi_2(...\pi_2(\pi_2(a))...)$, where π_2 occurs $i - 1$ times $(1 \leq i < n)$, and
- $a.n =_{\text{df}} \pi_2(...\pi_2(\pi_2(a))...)$, where π_2 occurs $n - 1$ times.

For instance, when $n = 3$, $a.2 \equiv \pi_1(\pi_2(a))$ and $a.3 \equiv \pi_2(\pi_2(a))$. □

Types can be parameterised. For example, the unit type $\mathbf{1}(A, x)$ is parameterised by $A : Type$ and $x : A$ and can be formally introduced by declaring:

$$\mathbf{1} : (A{:}Type)(x{:}A)\ Type$$
$$* : (A{:}Type)(x{:}A)\ \mathbf{1}(A, x)$$
$$\mathcal{E} : (A{:}Type)(x{:}A)\ (C : (\mathbf{1}(A, x))Type)(c : C(*(A, x))(z : \mathbf{1}(A, x))C(z)$$

with the computation rule $\mathcal{E}(A, x, C, c, *(A, x)) = c$.

Remark 1. The type theories thus specified are intensional type theories as those implemented in the proof assistants Agda [3], Coq [12], Lego [29] and Matita [33]. They have nice meta-theoretic properties including Church-Rosser, Subject Reduction and Strong Normalisation. (See Goguen's thesis on the meta-theory of UTT [15].) In particular, the inductive types do not have the η-like equality rules. As an example, the above unit type is different from the singleton type [5] in that, for a variable $x : \mathbf{1}(A, a)$, x is not computationally equal to $*(A, a)$. □

[2] We use $A \rightarrow B$ and $A \times B$ for the non-dependent Π-type and Σ-type, respectively. Also, see Appendix A for a further explanation for the notation of untyped pairs.

Coercive Subtyping. Coercive subtyping for dependent type theories was first considered in [2] for overloading and has been developed and studied as a general approach to abbreviation and subtyping in type theories with inductive types [24,25]. Coercions have been implemented in the proof assistants Coq [12,39], Lego [29,6], Plastic [9] and Matita [33]. Here, we explain the main idea and introduce necessary terminologies. For a formal presentation with complete rules, see [25].

In coercive subtyping, A is a subtype of B if there is a coercion $c : (A)B$, expressed by $\Gamma \vdash A \leq_c B : Type$.[3] The main idea is reflected by the following *coercive definition rule*, expressing that an appropriate coercion can be inserted to fill up the gap in a term:

$$\frac{\Gamma \vdash f : (x{:}B)C \quad \Gamma \vdash a : A \quad \Gamma \vdash A \leq_c B : Type}{\Gamma \vdash f(a) = f(c(a)) : [c(a)/x]C}$$

In other words, if A is a subtype of B *via* coercion c, then any object a of type A can be regarded as (an abbreviation of) the object $c(a)$ of type B.

Coercions may be declared by the users. They must be *coherent* to be employed correctly. Essentially, coherence expresses that the coercions between any two types are unique. Formally, given a type theory T specified in LF, a set R of coercion rules is coherent if the following rule is admissible in $T[R]_0$:[4]

$$\frac{\Gamma \vdash A \leq_c B : Type \quad \Gamma \vdash A \leq_{c'} B : Type}{\Gamma \vdash c = c' : (A)B}$$

Coherence is a crucial property. Incoherence would imply that the extension with coercive subtyping is not conservative in the sense that more judgements of the original type theory T can be derived. In most cases, coherence does imply conservativity (e.g., the proof method in [40] can be used to show this). When the employed coercions are coherent, one can always insert coercions correctly into a derivation in the extension to obtain a derivation in the original type theory. For an intensional type theory, coercive subtyping is an *intensional extension*. In particular, for an intensional type theory with nice meta-theoretic properties, its extension with coercive subtyping has those nice properties, too.

Remark 2. Coercive subtyping corresponds to the view of types as consisting of canonical objects while 'subsumptive subtyping' (the more traditional approach with the subsumption rule) to the view of type assignment [28]. Coercive subtyping can be introduced for inductive types in a natural way [28,27], but this would be difficult, if not impossible, for subsumptive subtyping. Furthermore, coercive

[3] In this paper, we use $\leq_c$, rather than the strict relation $<_c$, for coercion judgements and assume that the identity is always a coercion: if $\Gamma \vdash A : Type$, then $\Gamma \vdash A \leq_{id_A} A : Type$, where $id_A \equiv [x{:}A]x$. This does not make an essential difference but simplifies the component-wise coercion rules in Sections 3.1 and 4.2.

[4] $T[R]_0$ is an extension of T with the subtyping rules in R together with the congruence, substitution and transitivity rules for the subtyping judgements, but *without* the coercive definition rule. See [25] for formal details.

subtyping is not only suitable for structural subtyping, but for non-structural subtyping. The use in this paper of the coercion concerning the unit type is such an example. □

2 Manifest Fields via Extensional Constructs

Σ-types or dependent record types can be used to represent types of modules. For instance, a type of some kind of abstract algebras may be represented as

$$M \equiv \sum[S : U,\ op : S \to S,\ ...],$$

where S stands for (the type of) the carrier set with U being a type universe. (For simplicity, we omit the details such as the equality over S etc.) Sometimes, one may use *manifest fields* [20] to specify that the data in a field is not only of a given type, but a *specific* object of that type. For instance, for $m : M$, we want to define a subtype of M the carrier set of whose objects must be the same as that of m (i.e., $m.1$ – see Section 1.1). This module type can be defined as

$$\sum[S = m.1 : U,\ op : S \to S,\ ...],$$

which is the same as M except that the first field is *manifest*, specifying that the data in that field must be the same as $m.1$.

Traditionally, when manifest fields are considered, they introduce an *extensional* notion of equality: in the above example, (the variable) S and $m.1$ are judgementally equal and, in particular, they are interchangeable in type-checking. Such *extensional manifest fields* can be introduced directly [16,37,13] and the associated notion of equality is a strong form of η-like equality which makes the meta-theoretic studies rather difficult.

Manifest fields can be coded by means of other extensional constructs, including the extensional equality *Eq*, which was first introduced in Martin-Löf's extensional type theory (ETT) [32] and adopted by NuPRL [10]. In ETT, the propositional equality $Eq(A, a, b)$ is equivalent to the judgemental equality: $\Gamma \vdash p : Eq(A, a, b)$ if and only if $\Gamma \vdash a = b : A$. With *Eq*, one may express a manifest field $v = a : A$ with two fields: '$v : A,\ x : Eq(A, v, a)$', where the second guarantees that v is judgementally equal to a [11]. As is known, because of the strength of *Eq*, the judgemental equality and type checking in ETT are undecidable.

Another extensional construct that can be used to express manifest fields is the singleton type [5,17]. For $a : A$, M is an object of the singleton type $\{a\}_A$ if and only if M and a are judgementally equal. With this, a manifest field $v = a : A$ can simply be represented as the field $v : \{a\}_A$. The singleton types also introduce a strong form of η-like equality (among other things such as subtyping) and are difficult in meta-theory. (See [14] for a sophisticated proof of strong normalisation of a simple type system with singleton types.)

It has been thought that it would be difficult, if not impossible, to have manifest fields in type theory without such extensional constructs. This is partly because that, in an intensional type theory, the propositional equality is not equivalent to the computational (judgemental) equality in a non-empty context and, therefore, to express $v = a : A$, it is not enough to just have a proof that v is propositionally equal to a; we would need a way to make them judgementally equal (for example, for type-checking).

However, we shall show that manifest fields can be expressed in an intensional type theory, with the help of coercive subtyping.

3 Intensional Manifest Fields in Σ-Types

An *intensional manifest field* (IMF) in a Σ-type is a field of the form

$$x : \mathbf{1}(A, a),$$

where $\mathbf{1}(A, a)$ is the unit type parameterised by $A : Type$ and $a : A$. It will be written by means of the following notation:

$$x \sim a : A.$$

In other words, we write $\sum[... \ x \sim a : A, \ ...]$ for $\sum[... \ x : \mathbf{1}(A, a), \ ...]$.

The IMFs (and the Σ-types involved) are well-defined and behave as intended with the help of the following two coercions:

- $\xi_{A,a}$, associated with $\mathbf{1}(A, a)$, maps the objects of $\mathbf{1}(A, a)$ to a. In a context where an object of type A is required (e.g., in the Σ-type but after the field $v \sim a : A$), v is coerced into a and behaves as an abbreviation of a.
- The component-wise coercion d_Σ propagates subtyping relations, including those specified by ξ, through Σ-types so that the IMFs can be used properly in larger contexts.

Example 1. Here is an example of how the coercion ξ is used to support IMFs. Consider the module type M in Section 2, repeated here:

$$M \equiv \sum[S : U, \ op : S \to S, \ ...].$$

For $m : M$, we can change its first field into an IMF by specifying that the carrier set must be 'the same' as (or, more precisely, abbreviate) the carrier set of m:

$$M_w \equiv \sum[S \sim m.1 : U, \ op : S \to S, \ ...].$$

Note that S is now of type $\mathbf{1}(U, m.1)$ and is not a type. The reason that $S \to S$ is well-typed is that S is now coerced into the type $\xi_{U,m.1}(S) = m.1$. □

3.1 Coercions ξ and d_Σ and Their Coherence

The coercion rule for ξ concerning the unit type is:

$$(\xi) \qquad \frac{\Gamma \vdash A : Type \quad \Gamma \vdash a : A}{\Gamma \vdash \mathbf{1}(A, a) \leq_{\xi_{A,a}} A : Type}$$

where $\xi_{A,a}(x) = a$ for any $x : \mathbf{1}(A, a)$.

The component-wise coercion expresses the idea that the subtyping relations propagate through the module types. For Σ-types, if A is a subtype of A' and B is a 'subtype' of B', then $\Sigma(A, B)$ is a subtype of $\Sigma(A', B')$. Formally, this is formulated by means of the following rule:

$$(d_\Sigma) \quad \frac{\Gamma \vdash B : (A)Type \quad \Gamma \vdash B' : (A')Type \qquad \Gamma \vdash A \leq_c A' : Type \quad \Gamma, x{:}A \vdash B(x) \leq_{c'[x]} B'(c(x)) : Type}{\Gamma \vdash \Sigma(A, B) \leq_{d_\Sigma} \Sigma(A', B') : Type}$$

where d_Σ maps (a, b) to $(c(a), c'[a](b))$ and is formally defined as $d_\Sigma(z) = (c(\pi_1(z)), c'[\pi_1(z)](\pi_2(z)))$, for any $z : \Sigma(A, B)$.

Remark 3. In the literature, the component-wise rules for Σ-types are usually formulated by means of $<_c$, rather than $\leq_c$. Similar rules can be recovered. For instance, when $B \equiv B' \circ c$ and $c'[x] \equiv id_{B'(c(x))}$ and if we omit the coercion judgement for the identity coercion $c'[x]$, the above rule becomes

$$\frac{\Gamma \vdash B' : (A')Type \quad \Gamma \vdash A \leq_c A' : Type}{\Gamma \vdash \Sigma(A, B' \circ c) \leq_{d_1} \Sigma(A', B') : Type}$$

where d_1 maps (a, b) to $(c(a), b)$. □

Proposition 1 (Coherence). *Let* $\mathcal{R} = \{(\xi), (d_\Sigma)\}$. *Then* $\mathcal{R}$ *is coherent.*

Proof. By induction on derivations, we prove the more general statement:

- if $\Gamma \vdash A \leq_c B : Type$ and $\Gamma \vdash A' \leq_{c'} B' : Type$, where $\Gamma \vdash A = A' : Type$ and $\Gamma \vdash B = B' : Type$, then $\Gamma \vdash c = c' : (A)B$.

For example, in the case that the last rules to derive $A \leq_c B$ and $A' \leq_{c'} B'$ are both (ξ) with $c \equiv \xi_{C,a}$ and $c' \equiv \xi_{C',b}$, we have that $\mathbf{1}(C, a) \equiv A = A' \equiv \mathbf{1}(C', b)$. Then, by Church-Rosser, $C = C'$, $a = b$, and $\xi_{C,a}(x) = a = b = \xi_{C',b}(x)$ for any $x : \mathbf{1}(C, a)$. Therefore, $\xi_{C,a} = \xi_{C',b}$ by the ξ-rule and η-rule in LF (see Chapter 9 of [23]). □

3.2 `with`-Clauses and Properties

Manifest fields can be introduced by means of the `with`-clauses (see, e.g., [37]). Usually, they introduce extensional manifest fields with new computation rules. We shall instead consider them with the intensional manifest fields.

Intuitively, given a Σ-type with a field $v : A$, a `with`-clause modifies it into the same Σ-type except that the corresponding field becomes manifest: $v \sim a : A$ (i.e., $v : \mathbf{1}(A, a)$). For instance, the module type M_w in Example 1 can be obtained from M as follows: $M_w = M$ <u>`with`</u> <u>`field`</u> 1 <u>`as`</u> $m.1$.

Definition 1 (`with`-clause for Σ-types). *Let* $M \equiv \sum[x_1 : A_1,\ ...,\ x_n : A_n]$, $i \in \{1, ..., n\}$ *and* $x_1 : A_1,\ ...,\ x_{i-1} : A_{i-1} \vdash a : A_i$. *Then,*

M <u>`with`</u> <u>`field`</u> i <u>`as`</u> $(x_1, ..., x_{i-1})a$
$=_{\text{df}} \sum[\ x_1 : A_1,\ ...,\ x_{i-1} : A_{i-1},\ \ x_i \sim a : A_i,\ \ x_{i+1} : A_{i+1},\ ...,\ x_n : A_n\ ].$

When $x_j \notin FV(a)$ $(j = 1, ..., i-1)$, we omit the variables x_j and simply write $(M\ \underline{\texttt{with}}\ \underline{\texttt{field}}\ i\ \underline{\texttt{as}}\ a)$ for $(M\ \underline{\texttt{with}}\ \underline{\texttt{field}}\ i\ \underline{\texttt{as}}\ (x_1, ..., x_{i-1})a)$. □

The fields in tuples (objects of Σ-types) can be modified similarly.

Definition 2 (|-operation for Σ-types). *Let $M \equiv \sum[x_1 : A_1,\ ...,\ x_n : A_n]$ and $m : M$. Then $m|_i =_{\mathrm{df}} (m.1, ..., m.(i-1),\ *(A'_i, m.i),\ m.(i+1), ..., m.n)$, where $A'_i \equiv [m.1, ..., m.(i-1)/x_1, ..., x_{i-1}]A_i$.* □

Remark 4. It is obvious that `with`-clauses can be nested. For instance, M $\underline{\texttt{with}}\ (\underline{\texttt{field}}\ i\ \underline{\texttt{as}}\ a\ \underline{\texttt{and}}\ \underline{\texttt{field}}\ j\ \underline{\texttt{as}}\ b)$ is $(M\ \underline{\texttt{with}}\ \underline{\texttt{field}}\ i\ \underline{\texttt{as}}\ a)\ \underline{\texttt{with}}\ \underline{\texttt{field}}\ j\ \underline{\texttt{as}}\ b$. This is similar for |-operations. E.g., $m|_{i,j}$ is $(m|_i)|_j$. □

The above definitions are adequate as the following proposition shows.

Proposition 2. *Let $M \equiv \sum[x_1 : A_1,\ ...,\ x_n : A_n]$ and $m : M$. Then,*

1. *For $i = 1, ..., n$, if $M_i \equiv (M\ \underline{\texttt{with}}\ \underline{\texttt{field}}\ i\ \underline{\texttt{as}}\ m.i)$ is well-typed, then $m|_i : M_i$, and*
2. *If $x_1, ..., x_{i-1} \notin FV(a)$, then $m.i = a$ if and only if $m|_i : (M\ \underline{\texttt{with}}\ \underline{\texttt{field}}\ i\ \underline{\texttt{as}}\ a)$.*

Proof. (1) is proved by induction on n, using the coercion ξ. (2) is a corollary of (1) and proved using the fact that, by type uniqueness, $m.i = a$ if and only if $M\ \underline{\texttt{with}}\ \underline{\texttt{field}}\ i\ \underline{\texttt{as}}\ m.i = M\ \underline{\texttt{with}}\ \underline{\texttt{field}}\ i\ \underline{\texttt{as}}\ a$. □

The following proposition shows that, if we modify a Σ-type by a `with`-clause appropriately, we obtain a subtype and, therefore, Σ-types with IMFs can be used adequately in any context.

Proposition 3. *Let M and $M_w \equiv (M\ \underline{\texttt{with}}\ \underline{\texttt{field}}\ i\ \underline{\texttt{as}}\ a)$ be Σ-types. Then, $M_w \leq M$ (i.e., $M_w \leq_c M$ for some c).*

Proof. The proof uses both coercions ξ and d_Σ. □

4 Dependent Record Types and Intensional Manifest Fields

Dependent record types are labelled Σ-types. For instance, $\langle n : Nat,\ v : Vect(n)\rangle$ is the dependent record type with objects (called *records*) such as $\langle n = 2,\ v = [5, 6]\rangle$, where the dependency has to be respected: $[5, 6]$ must be of type $Vect(2)$. It can be argued that record types are more natural than Σ-types to be considered as types of modules.

In this section, we shall give a new formulation of dependent record types, study intensional manifest fields in record types and illustrate their uses in expressing the module mechanism with ML-sharing and in modelling OO-programs.

Formation rules

$$\frac{\Gamma \ valid}{\Gamma \vdash \langle\rangle : RType[\emptyset]} \qquad \frac{\Gamma \vdash R : RType[L] \quad \Gamma \vdash A : (R)Type \quad l \notin L}{\Gamma \vdash \langle R,\ l : A\rangle : RType[L \cup \{l\}]}$$

Introduction rules

$$\frac{\Gamma \ valid}{\Gamma \vdash \langle\rangle : \langle\rangle} \qquad \frac{\Gamma \vdash \langle R,\ l : A\rangle : RType \quad \Gamma \vdash r : R \quad \Gamma \vdash a : A(r)}{\Gamma \vdash \langle r,\ l = a : A\rangle : \langle R,\ l : A\rangle}$$

Elimination rules

$$\frac{\Gamma \vdash r : \langle R,\ l : A\rangle}{\Gamma \vdash [r] : R} \qquad \frac{\Gamma \vdash r : \langle R,\ l : A\rangle}{\Gamma \vdash r.l : A([r])} \qquad \frac{\Gamma \vdash r : \langle R,\ l : A\rangle \quad \Gamma \vdash [r].l' : B \quad l \neq l'}{\Gamma \vdash r.l' : B}$$

Computation rules

$$\frac{\Gamma \vdash \langle r,\ l = a : A\rangle : \langle R,\ l : A\rangle}{\Gamma \vdash [\langle r,\ l = a : A\rangle] = r : R} \qquad \frac{\Gamma \vdash \langle r,\ l = a : A\rangle : \langle R,\ l : A\rangle}{\Gamma \vdash \langle r,\ l = a : A\rangle.l = a : A(r)}$$

$$\frac{\Gamma \vdash \langle r,\ l = a : A\rangle : R \quad \Gamma \vdash r.l' : B \quad l \neq l'}{\Gamma \vdash \langle r,\ l = a : A\rangle.l' = r.l' : B}$$

Fig. 1. The main inference rules for dependent record types

4.1 Dependent Record Types

Formally, we formulate dependent record types as an extension of the intensional type theory such as Martin-Löf's type theory or UTT, as specified in the logical framework LF. The syntax is extended with record types $\langle\rangle$ and $\langle R,\ l : A\rangle$ and records $\langle\rangle$ and $\langle r,\ l = a : A\rangle$, where we overload $\langle\rangle$ to stand for both the empty record type and the empty record. Records are associated with two operations: *restriction* (or *first projection*) $[r]$ that removes the last component of record r and *field selection* $r.l$ that selects the field labelled by l.

For every finite set of labels L, we introduce a kind $RType[L]$, the kind of the record types whose (top-level) labels are all in L. We shall also introduce the kind $RType$ of all record types. These kinds obey obvious subkinding relationships:

$$\frac{\Gamma \vdash R : RType[L] \quad L \subseteq L'}{\Gamma \vdash R : RType[L']} \qquad \frac{\Gamma \vdash R : RType[L]}{\Gamma \vdash R : RType} \qquad \frac{\Gamma \vdash R : RType}{\Gamma \vdash R : Type}$$

Equalities are also inherited by superkinds in the sense that, if $\Gamma \vdash k = k' : K$ and K is a subkind of K', then $\Gamma \vdash k = k' : K'$. The obvious rules are omitted.

The main inference rules for dependent record types are given in Figure 1. Note that, in record type $\langle R,\ l : A\rangle$, A is a family of types, indexed by the objects of R, and this is how dependency is embodied in the formulation.

Notation. For record types, we write $\langle l_1 : A_1,\ ...,\ l_n : A_n\rangle$ for $\langle\langle\langle\rangle,\ l_1 : A_1\rangle,\ ...,\ l_n : A_n\rangle$ and often use label occurrences and label non-occurrences to express dependency and non-dependency, respectively. For instance, we write

$\langle n : Nat,\ v : Vect(n)\rangle$ for $\langle n : [\_:\langle\rangle]Nat,\ v : [x:\langle n : [\_:\langle\rangle]Nat\rangle]Vect(x.n)\rangle$ and $\langle R,\ l : Vect(2)\rangle$ for $\langle R,\ l : [\_:R]Vect(2)\rangle$.

For records, we often omit the type information to write $\langle r,\ l = a\rangle$ for either $\langle r,\ l = a : [\_:R]A(r)\rangle$ or $\langle r,\ l = a : A\rangle$. Such a simplification is available thanks to coercive subtyping. A further explanation is given in Appendix A. □

The notion of equality between records is weakly extensional in the sense that two records are equal if their components are. This is reflected in the following two rules (similar rules are used in [7]):

$$\frac{\Gamma \vdash r : \langle\rangle}{\Gamma \vdash r = \langle\rangle : \langle\rangle} \qquad \frac{\Gamma \vdash r : \langle R,\ l : A\rangle \quad \Gamma \vdash r' : \langle R,\ l : A\rangle \quad \Gamma \vdash [r] = [r'] : R \quad \Gamma \vdash r.l = r'.l : A([r])}{\Gamma \vdash r = r' : \langle R,\ l : A\rangle}$$

For example, for any $r : \langle R,\ l : A\rangle$ (r can be a variable), we have, by the second rule above, that $r = \langle [r],\ l = r.l : A\rangle : \langle R,\ l : A\rangle$.

There are also congruence rules for record types and their objects, which we omit here. However, it is worth remarking that we pay special attention to the equality between record types. In particular, record types with different labels are not equal. For example, $\langle n : Nat\rangle \neq \langle n' : Nat\rangle$ if $n \neq n'$.

Remark 5. These remarks are mainly for people who are familiar with previous work on dependent records. First, it is worth pointing out that, as in [37], our formulation of record types is *independent* of structural subtyping; this is different from the other previous formulations [16,7,13], which have all made an essential use of structural subtyping. We consider this independence as a significant advantage, mainly because it allows one to adopt more flexible subtyping relations in formalisation and modelling. We also comment that, although it might have its own advantages (e.g., the economy in some rule formulations), mixing subtyping with dependent records is not an easy matter: it is meta-theoretically difficult and sometimes may lead to undecidability [16].

Our formulation is different from that in [37] which allows label repetitions ('label shadowing'). We have introduced kinds $RType[L]$ and this gives a satisfactory solution to the problem of how to ensure label distinctness (or to avoid label repetition) in record types. For example, this is used essentially in Appendix A when notational coercions are defined. Also, ensuring label distinctness makes it possible for us to employ structural coercions such as projections coherently in some applications such as OO-modelling discussed in Section 4.3.

Note that we have formulated record *types*, not record *kinds*. In the terminology used in this paper, both [7] and [13] study record *kinds* – their 'record types' are studied at the level of kinds in the logical framework. Since kinds have a much simpler structure than types, it is easier to add record kinds (e.g., to ensure label distinctness) than record types, while the latter is more powerful.

Finally, we should mention that, in the context of extensional type theory, people have studied encodings of record types by means of other constructs. For example, in NuPRL, 'very dependent function types' and intersection types have been studied to encode dependent record types [11,4]. However, it is difficult to see how this can be done in intensional type theories. □

4.2 Intensional Manifest Fields in Record Types

Intensional manifest fields can be defined for record types similarly as we did for Σ-types in Section 3. In record types, for $A : (R)Type$ and $a : (r{:}R)A(r)$,

$$l \sim a : A \;\text{ stands for }\; l : [r{:}R]\mathbf{1}(A(r), a(r)).$$

In the simpler situation, for $A : Type$ and $a : A$, $l \sim a : A$ stands for $l : \mathbf{1}(A, a)$. In records, for $b : B$,

$$l \sim_B b \;\text{ stands for }\; l = *(B, b).$$

Example 2. For $R = \langle S : U,\ op : S \to S\rangle$ and $r : R$, the S-field of the record type $\langle S \sim r.S : U,\ op : S \to S\rangle$ is manifest; intuitively, it insists that, for any record of this type, its S-field must be the same as the S-field of r. □

The component-wise coercion d_R for record types is given by the rule

$$\frac{\begin{array}{c}\Gamma \vdash R : RType[L] \quad \Gamma \vdash R' : RType[L] \quad \Gamma \vdash R \leq_c R' : RType \\ \Gamma \vdash A : (R)Type \quad \Gamma \vdash A' : (R')Type \quad \Gamma, x{:}R \vdash A(x) \leq_{c'[x]} A'(c(x)) : Type\end{array}}{\Gamma \vdash \langle R,\ l : A\rangle \leq_{d_R} \langle R',\ l : A'\rangle : RType} \quad (l \notin L)$$

where d_R maps $\langle r,\ l = a\rangle$ to $\langle c(r),\ l = c'[r](a)\rangle$ and is formally defined as, for any $r' : \langle R,\ l : A\rangle$, $d_R(r') =_{\mathrm{df}} \langle c(r_0),\ l = c'[r_0](r'.l)\rangle$, where $r_0 \equiv [r']$.

Remark 6. Note that a component-wise coercion only exists between the record types that have the same corresponding labels. For example, if $l \neq l'$, there is no component-wise coercion between $\langle l : A\rangle$ and $\langle l' : B\rangle$ even if $A \leq_c B$. □

Note that different applications employ different coercions and, thanks to the independence of the formulation of record types with subtyping, it is flexible to use different coercions. For example, in OO-modelling as illustrated in Section 4.3 below, we also employ the projections as coercions, with the following rules:

$$\frac{\Gamma \vdash \langle R,\ l : A\rangle : RType}{\Gamma \vdash \langle R,\ l : A\rangle \leq_{[\_]} R : RType} \qquad \frac{\Gamma \vdash A : Type \quad \Gamma \vdash \langle R,\ l : A\rangle : RType}{\Gamma \vdash \langle R,\ l : A\rangle \leq_{Snd} \langle l : A\rangle : RType}$$

where, in the second rule, A is a type, $\langle R,\ l : A\rangle$ stands for $\langle R,\ l : [\_{:}R]A\rangle$, and the kind of the second projection Snd is the non-dependent kind $(\langle R,\ l : A\rangle)\langle l : A\rangle$.[5]

Remark 7. Assuming that the extension with dependent record types has nice meta-theoretic properties such as Church-Rosser, we can show that the coercions ξ, d_R, $[\_]$ and Snd are coherent together. It is worth remarking that the

[5] In general, $Snd : (r{:}\langle R,\ l : A\rangle)\langle l : A([r])\rangle$ maps r to $\langle l = r.l\rangle$. First, note that the kind of Snd is different from that of field selection: the codomain of Snd is $\langle l : A([r])\rangle$, rather than simply $A([r])$. This makes an important difference: Snd is coherent with the first projection and the component-wise coercion, while field selection is not. Secondly, only non-dependent coercions (and, in this case, the non-dependent second projection) are studied in this paper. (*Dependent coercions*, where the codomain of a coercion may depend on its argument, are studied in [30].)

labels in record types play an important role for this coherence – the projections together are not coherent coercions for Σ-types [21]. Also, if one allowed label repetitions in record types, as in [37], the projection coercions $[\_]$ and Snd would be incoherent together. □

As for Σ-types, we can modify a record type by means of a `with`-clause. For $R \equiv \langle l_1 : A_1, ..., \ l_n : A_n\rangle$, $i \in \{1, ..., n\}$ and $a : (x : R_{i-1})A_i(x)$, where $R_{i-1} \equiv \langle l_1 : A_1, ..., \ l_{i-1} : A_{i-1}\rangle$,

$$\begin{aligned} & R \ \underline{\texttt{with}} \ l_i \ \underline{\texttt{as}} \ a \\ =_{\mathrm{df}} & \ \langle \ l_1 : A_1, \ \ ..., \ \ l_{i-1} : A_{i-1}, \ \ l_i \sim a : A_i, \ \ l_{i+1} : A_{i+1}, \ \ ..., \ \ l_n : A_n \ \rangle. \end{aligned}$$

And, for $r : R$,

$$r|_{l_i} =_{\mathrm{df}} \langle l_1 = r.l_1, ..., l_{i-1} = r.l_{i-1}, \ l_i \sim_{A_i(r_{i-1})} r.l_i, \ l_{i+1} = r.l_{i+1}, ..., l_n = r.l_n\rangle,$$

where $r_{i-1} \equiv \langle l_1 = r.l_1, \ ..., \ l_{i-1} = r.l_{i-1}\rangle$.

Remark 8. Similar propositions as Propositions 2 and 3 would show that the above definitions are adequate. It is also easy to see that the `with`-clauses and the $|$-operations can be nested. □

4.3 Modules and OO-Modelling in Intensional Type Theory

In this subsection, we show how to use the module types with intensional manifest fields to capture ML-style sharing [31,20] and to model classes in OO-style programs. We shall use record types in our examples.

Modules with ML-Style Sharing. In the language design for programming and formalisation, the topic of developing a suitable and powerful module mechanism has been attracting a lot of interests. A module mechanism that supports structure sharing has been of particular interest. For example, one may want to share a point of a circle and a point of a rectangle in developing a facility for bit-mapped graphics or to share the carrier set of a semigroup and that of an abelian group when constructing rings in a formal development of abstract mathematics.

For functional programming languages, two approaches to sharing have been studied: one is *sharing by parameterisation* or the Pebble-style sharing [8,19] and the other *sharing by equations* or the ML-style sharing [31,34]. Both have been studied in the context of formalisation of mathematics as well, especially in designing and using type theory based proof assistants.

It is known that ML-style sharing cannot be captured in an intensional type theory by the propositional equality, since it is not equivalent to the computational equality in a non-empty context [22]. Contrary to the common belief (cf., Section 2), however, ML-style sharing can be captured using the IMFs in intensional type theory, as the following example illustrates.

```
class inc_cell is
  var contents : Integer;
  method get(): Integer is
    return contents end;
  method set(n:Integer) is
    if get() < n
    then contents := n end;
end;
```

```
subclass re_inc_cell of inc_cell is
  var backup : Integer;
  method restore() is
    contents := backup end;
  override set(n:Integer) is
    if get() < n
    then { backup := contents;
           contents := n } end;
end;
```

Fig. 2. The class inc_cell and its subclass re_inc_cell

Example 3. A ring R is composed of an abelian group $(R, +)$ and a semigroup $(R, *)$, with extra distributive laws. One can construct a ring from an abelian group and a semigroup. When doing this, one must make sure that the abelian group and the semigroup share the same carrier set. One of the ways to specify such sharing is to use an 'equation' to indicate that the carrier sets are the same. This example shows that this can be done by means of the IMFs.

The signature types of abelian groups, semigroups and rings can be represented as the following record types, respectively, where U is a type universe:

$$\begin{aligned} AG &\equiv \langle A : U,\ + : A \to A \to A,\ 0 : A,\ inv : A \to A \rangle \\ SG &\equiv \langle B : U,\ * : B \to B \to B \rangle \\ Ring &\equiv \langle C : U,\ + : C \to C \to C,\ 0 : C,\ inv : C \to C,\ * : C \to C \to C \rangle \end{aligned}$$

Note that an abelian group and a semigroup do not have to share their carrier sets. In order to make this happen, we introduce the following record type, which is parameterised by an AG-signature and defined by means of a `with`-clause that specifies an IMF to ensure the sharing of the carrier sets:

$$\begin{aligned} SGw(ag) &= SG\ \underline{\texttt{with}}\ B\ \underline{\text{as}}\ ag.A \\ &= \langle B \sim ag.A : U,\ * : B \to B \to B \rangle, \end{aligned}$$

where $ag : AG$. Then, the function that generates the Ring-signature from those of abelian groups and semigroups can now be defined as:

$$\begin{aligned} & ringGen(ag, sg) \\ =_{\text{df}}\ & \langle C = ag.A,\ + = ag.+,\ 0 = ag.0,\ inv = ag.inv,\ * = sg.* \rangle, \end{aligned}$$

where the arguments $ag : AG$ and $sg : SGw(ag)$ share their carrier set. □

Modelling OO-Style Classes. Since definitional entries can be specified by means of IMFs, record types can be used to model the modular entities like classes in an object-oriented language, where methods are modelled as IMFs.

Example 4. Consider the class `inc_cell` in Figure 2, representing a memory cell whose content only increases. `inc_cell` can be interpreted as:

$$Cell_0 = \langle\ c : S_{cell},\ get \sim f_g : T_g,\ set \sim f_s : T_s\ \rangle$$

where

- $S_{cell} \equiv \langle\ contents : Int\ \rangle$ interprets the states of `inc_cell`;
- $f_g \equiv \lambda(s{:}S_{cell})s.contents$, of type $T_g \equiv S_{cell} \rightarrow Int$, interprets `get`; and
- $f_s \equiv \lambda(n{:}Int, s{:}S_{cell})$ `if` $get(s) < n$ `then` $\langle contents = n\rangle$ `else` s, of type $T_s \equiv Int \rightarrow S_{cell} \rightarrow S_{cell}$, interprets `set`.

Note that, in f_s, get can be applied to $s : S_{cell}$ because that it is coerced into $\xi_{T_g,f_g}(get) = f_g$ of type $T_g \equiv S_{cell} \rightarrow Int$. □

The interpretation in the above example follows the basic idea in functional interpretations of OO-languages [18,36]. In particular, when a method is interpreted, the type of states (S_{cell} in the above example) is used as argument and result types to model the effect of retrieving from and modifying the memory. There is a known problem with such a way of using state types in the interpretation when subclasses are considered [1]: the contravariance of subtyping does not lead to the natural subtyping relations between the interface types. For instance, if the subclass `re_inc_cell` in Figure 2 is interpreted similarly as in Example 4, the states of `re_inc_cell` would be interpreted as $S_{recell} \equiv \langle\ contents : Int,\ backup : Int\ \rangle$ and, for example, the method `set` as a function of type $T'_s \equiv Int \rightarrow S_{recell} \rightarrow S_{recell}$. Although $S_{recell} \leq S_{cell}$ (*via* projection coercions), T'_s is not a subtype of T_s ('the problem of contravariance'). As a consequence, the interface type of `re_inc_cell` is not a subtype of that of its superclass `inc_cell`. However, this would be very desirable and the problem can be solved in our setting by introducing a notion of universal state.

The universal state Ω. Given an object-oriented program, its classes form a DAG (directed acyclic graph), where an arrow from `C` to `C'` means that `C'` is a subclass of `C`. Let `C1`, ..., `Cn` be the leaves of the DAG and their states be interpreted as types S_1, ..., S_n, respectively. Then, the *universal state* (or, more precisely, the universal type of states) Ω is defined as the following (non-dependent) record type:

$$\Omega =_{\mathrm{df}} \langle\ s_1 : S_1,\ \ldots,\ s_n : S_n\ \rangle.$$

Now, with Ω, a method is interpreted as a function that takes a value from Ω and, if it modifies the state, returns a value to Ω. For instance, the method `set` in `inc_cell` is interpreted as a function of type $Int \rightarrow \Omega \rightarrow \Omega$ (rather than $Int \rightarrow S_{cell} \rightarrow S_{cell}$) and `set` in `re_inc_cell` as a (different) function of the same type. Therefore, the subtyping relationships between the interface types of `inc_cell` and `re_inc_cell` are as expected.

In general, the model enjoys desirable subtyping relationships between classes and their interface types. If a class `C` is interpreted as $C = \langle\ c : S_C,\ m_1 \sim a_1 : A_1,\ \ldots,\ m_n \sim a_n : A_n\ \rangle$, where S_C interprets the states of `C`, its interface type `I_C` is interpreted as $I_C = \langle\ c : S_C,\ m_1 : A_1,\ \ldots,\ m_n : A_n\ \rangle$. Therefore, we have $C \leq_c I_C$, where the coercion c is derived from ξ and d_R. Furthermore, if `C'` is a subclass of `C`, then $S_{C'}$ is a subtype of S_C (*via* projection coercions), and: $I_{C'} \leq_c I_C$, where the coercion c is derived from the structural coercions (the projection and component-wise coercions) for record types.

Remark 9. In the model construction of OO-classes as sketched above, subtype polymorphism is correctly captured and methods are invoked according to dynamic dispatch. We omit the detailed explanations here. □

Experiments in Proof Assistants. Experiments on the above applications have been done in the proof assistants Plastic [9] and Coq [12], both supporting the use of coercions. In Plastic, one can define parameterised coercions such as ξ and coercion rules for the structural coercions: we only have to declare ξ and the component-wise coercion (and the projection coercions for the application of OO-modelling), then Plastic obtains automatically all of the derivable coercions, as intended. However, Plastic does not support record types; so Σ-types were used for our experiments in Plastic, at the risk of incoherence of the coercions!

Coq supports a macro for dependent record types[6] and a limited form of coercions. In Coq, we have to use the identity $ID(A) = A$ on types to force Coq to accept the coercion ξ and to use type-casting as a trick to make it happen. Also, since Coq does not support user-defined coercion rules, we cannot implement the rule for the component-wise coercion; instead, we have to specify its effects on the record types individually. The Coq code for the Ring example in Example 3 can be found in Appendix B.

In the proof assistants such as Coq, verification of object-oriented programs can be done based on the formalisation of the model sketched above. For instance, one can show that, for the class `re_inc_cell`, it is an invariant that the backup value is always smaller than or equal to the contents value; formally, we prove in Coq, for every method m,

$$\begin{aligned} \forall s : \Omega.\ Pre(m) &\Rightarrow s.(backup) \leq s.(contents) \\ &\Rightarrow S(m,s).(backup) \leq S(m,s).(contents), \end{aligned}$$

where $Pre(m)$ stands for the precondition of m, $S(m,s)$ for the resulting state obtained from executing m with the initial state s, and $s.(\_)$ is the Coq-notation for field selection. Currently, we can only do small examples in formalisation and verification, partly because the manual encoding is rather tedious (and error-prone). We are working on the automated translations that will generate the Coq models of object-oriented programs and the Coq propositions of the specifications, and this will hopefully make the whole process much easier.

5 Conclusion

We have shown that manifest fields can be expressed in intensional type theory with the help of coercive subtyping. The intensional manifest fields strengthen

[6] It is a macro in the sense that dependent record types are actually implemented as inductive types with labels defined as global names (and, therefore, the labels of different 'record types' must be different). Coq [12] also supports a preliminary (but improper) form of 'manifest fields' by means of the let-construct, which we do not use in our experiments.

the module types such as record types and provide higher-order module mechanisms for modular development of proofs and dependently-typed programs and powerful representation mechanisms for, for example, formalisation and verification of OO-style programs.

Recently, it has come to our attention that, studying formalisation of mathematical structures, Sacerdoti-Coen and Tassi [38] have attempted to represent R with l as a by means of $\Sigma r : R.\ (r.l = a)$, where $=$ is the Leibniz equality, and to employ the so-called 'manifesting coercions' in order to approximate manifest fields. We remark that using an equality relation in this way is not completely satisfactory and seems unnecessarily complicated. Our notion of intensional manifest field is simple and desirable and, coupled with the record types as formulated in this paper, provides us a powerful tool in intensional type theory.

As to future work, we mention that our formulation of dependent record types forms a promising basis for investigations on the meta-theory of dependent record types. We also hope that the proof assistants will implement dependent record types properly so that they can be used effectively in practice.

Acknowledgement. I am grateful to Robin Adams who, among other things, has suggested the phrase 'intensional manifest field' to me.

References

1. Abadi, M., Cardelli, L.: A Theory of Objects. Springer, Heidelberg (1996)
2. Aczel, P.: Simple overloading for type theories (manuscript, 1994)
3. The Agda proof assistant (version 2) (2008), `http://appserv.cs.chalmers.se/users/ulfn/wiki/agda.php`
4. Allen, S., et al.: Innovations in computational type theory using Nuprl. Journal of Applied Logic 4(4) (2006)
5. Aspinall, D.: Subtyping with singleton types. In: Pacholski, L., Tiuryn, J. (eds.) CSL 1994. LNCS, vol. 933. Springer, Heidelberg (1995)
6. Bailey, A.: The Machine-checked Literate Formalisation of Algebra in Type Theory. PhD thesis, University of Manchester (1999)
7. Betarte, G., Tasistro, A.: Extension of Martin-Löf's type theory with record types and subtyping. In: Sambin, G., Smith, J. (eds.) Twenty-five Years of Constructive Type Theory, Oxford University Press, Oxford (1998)
8. Burstall, R., Lampson, B.: Pebble, a kernel language for modules and abstract data types. In: Plotkin, G., MacQueen, D.B., Kahn, G. (eds.) Semantics of Data Types 1984. LNCS, vol. 173. Springer, Heidelberg (1984)
9. Callaghan, P., Luo, Z.: An implementation of LF with coercive subtyping and universes. Journal of Automated Reasoning 27(1), 3–27 (2001)
10. Constable, R., et al.: Implementing Mathematics with the NuPRL Proof Development System. Prentice-Hall, Englewood Cliffs (1986)
11. Constable, R., Hickey, J.: Nuprl's class theory and its applications. In: Foundations of Secure Computation. IOS Press, Amsterdam (2000)
12. The Coq Development Team. The Coq Proof Assistant Reference Manual (Version 8.1), INRIA (2007)

13. Coquand, T., Pollack, R., Takeyama, M.: A logical framework with dependently typed records. Fundamenta Informaticae 65(1-2) (2005)
14. Courant, J.: Strong normalisation with singleton types. Electronic Notes in Theoretical Computer Science 70(1) (2002)
15. Goguen, H.: A Typed Operational Semantics for Type Theory. PhD thesis, University of Edinburgh (1994)
16. Harper, R., Lillibridge, M.: A type-theoretic approach to higher-order modules with sharing. In: POPL 1994 (1994)
17. Hayashi, S.: Singleton, union and intersection types for program extraction. Information and Computation 109(1/2), 174–210 (1994)
18. Kamin, S.: Inheritance in Smalltalk-80: a denotational definition. In: POPL 1988 (1988)
19. Lampson, B., Burstall, R.: Pebble, a kernel language for modules and abstract data types. Information and Computation 76(2/3) (1988)
20. Leroy, X.: Manifest types, modules and separate compilation. In: POPL 1994 (1994)
21. Luo, Y.: Coherence and Transitivity in Coercive Subtyping. PhD thesis, University of Durham (2005)
22. Luo, Z.: A higher-order calculus and theory abstraction. Information and Computation 90(1) (1991)
23. Luo, Z.: Computation and Reasoning: A Type Theory for Computer Science. Oxford University Press, Oxford (1994)
24. Luo, Z.: Coercive subtyping in type theory. In: van Dalen, D., Bezem, M. (eds.) CSL 1996. LNCS, vol. 1258. Springer, Heidelberg (1997)
25. Luo, Z.: Coercive subtyping. J. of Logic and Computation 9(1), 105–130 (1999)
26. Luo, Z.: Coercions in a polymorphic type system. Mathematical Structures in Computer Science 18(4) (2008)
27. Luo, Z., Adams, R.: Structural subtyping for inductive types with functorial equality rules. Mathematical Structures in Computer Science 18(5) (2008)
28. Luo, Z., Luo, Y.: Transitivity in coercive subtyping. Information and Computation 197(1-2), 122–144 (2005)
29. Luo, Z., Pollack, R.: LEGO Proof Development System: User's Manual. LFCS Report ECS-LFCS-92-211, Dept. of Computer Science, Univ. of Edinburgh (1992)
30. Luo, Z., Soloviev, S.: Dependent coercions. In: CTCS 1999, ENTCS 1929 (1999)
31. MacQueen, D.: Modules for standard ML. In: ACM Symp. on Lisp and Functional Programming (1984)
32. Martin-Löf, P.: Intuitionistic Type Theory. Bibliopolis (1984)
33. The Matita proof assistant (2008), `http://matita.cs.unibo.it/`
34. Milner, R., Harper, R., Tofts, M., MacQueen, D.: The Definition of Standard ML (revised). MIT, Cambridge (1997)
35. Nordström, B., Petersson, K., Smith, J.: Programming in Martin-Löf's Type Theory: An Introduction. Oxford University Press, Oxford (1990)
36. Pierce, B.C., Turner, D.N.: Simple type-theoretic foundations for object-oriented programming. J. of Functional Programming 4(2), 207–247 (1994)
37. Pollack, R.: Dependently typed records in type theory. Formal Aspects of Computing 13, 386–402 (2002)
38. Sacerdoti-Coen, C., Tassi, E.: Working with mathematical structures in type theory. In: Miculan, M., Scagnetto, I., Honsell, F. (eds.) TYPES 2007. LNCS, vol. 4941, pp. 157–172. Springer, Heidelberg (2008)
39. Saïbi, A.: Typing algorithm in type theory with inheritance. In: POPL 1997(1997)
40. Soloviev, S., Luo, Z.: Coercion completion and conservativity in coercive subtyping. Annals of Pure and Applied Logic 113(1-3), 297–322 (2002)

A Untyped Notations for Pairs and Records

Coercive subtyping can be used to explain and facilitate overloading [25,6,26]. The adoption of the untyped notations for pairs and records is a typical example.

Formally, the notations for pairs in Martin-Löf's type theory or UTT and for records in Section 4.1 are fully annotated with type information: they are of the 'typed' forms $pair(A, B, a, b)$ and $\langle r,\ l = a : A\rangle$, rather than the 'untyped' (a, b) and $\langle r,\ l = a\rangle$, respectively. A reason for this is that, in a dependent type theory, a pair or a record without type information may have two or more incompatible types. For example, the record $\langle n = 2,\ v = [5, 6]\rangle$ has both $\langle n : Nat,\ v : Vect(2)\rangle$ and $\langle n : Nat,\ v : Vect(n)\rangle$ as its types. The presence of the type information in the typed forms allows a straightforward algorithm for type-checking, but it is clumsy and impractical.

Can we use the simpler untyped notations instead? The answer is yes: this can be done with the help of coercive subtyping. We illustrate it for records (see Section 5.4 of [25] for a treatment of pairs). Let r' be the 'intended typed version' of r. We want to use $\langle r,\ l = a\rangle$ to stand for either of the following records:

$$r_1 \equiv \langle r',\ l = a : [\_{:}R]A(r')\rangle \ :\ \langle R,\ l : [\_{:}R]A(r')\rangle$$
$$r_2 \equiv \langle r',\ l = a : A\rangle \ :\ \langle R,\ l : A\rangle$$

and to be able to decide which it stands for in the context. This can be done as follows. Let L be any finite set of labels such that $l \notin L$. Consider the family

$$U_L : (R : RType[L])(A : (R)Type)(x : R)(a : A(x))Type$$

of inductive unit types $U_L(R, A, x, a)$ with the only object $u_L(R, A, x, a)$. We then declare coercions δ_1^L and δ_2^L:

$$U_L(R, A, x, a) \leq_{\delta_1^L} \langle R,\ l : [\_{:}R]A(x)\rangle$$
$$U_L(R, A, x, a) \leq_{\delta_2^L} \langle R,\ l : A\rangle$$

inductively defined as: $\delta_1^L(u_L(R, A, x, a)) = \langle x,\ l = a : [\_{:}R]A(x)\rangle$ and $\delta_2^L(u_L(R, A, x, a)) = \langle x,\ l = a : A\rangle$. Then the notation $\langle r,\ l = a\rangle$ can be used to denote the object $u_L(R, A, r', a)$ and, in a context, it will be coerced into the appropriate record r_1 or r_2 according to the contextual typing requirement.

B Coq Code for the Ring Example

The following is the Coq code for the Ring example – the construction of rings from abelian groups and semi-groups that share the domains. Note that we have only formalised the signatures of the algebras, omitting their axiomatic parts.

```
(* The parameterised unit type -- Unit/unit for 1/* *)
Inductive Unit (A:Type)(a:A) : Type := unit : Unit A a.
(* Coercion for the unit type; Use ID as trick to define it in Coq *)
```

```
Definition ID (A:Type) : Type := A.
Coercion unit_c (A:Type)(a:A)(_:Unit A a) := a : ID A.
(* Abelian Groups, Semi-groups and Rings -- signatures only *)
Record AG : Type := mkAG
  { A : Set; plus : A->A->A; zero : A; inv : A->A }.
Record SG : Type := mkSG
  { B : Set; times : B->B->B }.
Record Ring : Type := mkRing
  { C : Set; plus' : C->C->C; zero' : C; inv' : C->C; times' : C->C->C }.
(* Domain-sharing semi-groups; type-casting to make unit_c happen in Coq *)
Record SGw (ag : AG) : Type := mkSGw
  { B' : Unit Set ag.(A); times'' : let B' := (B' : ID Set) in B'->B'->B' }.
Implicit Arguments B'.  Implicit Arguments times''.
(* function to generate rings from abelian/semi-groups with shared domain *)
Definition ringGen (ag : AG)(sg : SGw ag) : Ring :=
  mkRing ag.(A) ag.(plus) ag.(zero) ag.(inv) sg.(times'').
```

A Machine-Checked Proof of the Average-Case Complexity of Quicksort in Coq

Eelis van der Weegen* and James McKinna

Institute for Computing and Information Sciences
Radboud University Nijmegen
Heijendaalseweg 135, 6525 AJ Nijmegen, The Netherlands
eelis@eelis.net, james.mckinna@cs.ru.nl

Abstract. As a case-study in machine-checked reasoning about the complexity of algorithms in type theory, we describe a proof of the average-case complexity of Quicksort in Coq. The proof attempts to follow a textbook development, at the heart of which lies a technical lemma about the behaviour of the algorithm for which the original proof only gives an intuitive justification.

We introduce a general framework for algorithmic complexity in type theory, combining some existing and novel techniques: algorithms are given a shallow embedding as monadically expressed functional programs; we introduce a variety of operation-counting monads to capture worst- and average-case complexity of deterministic and nondeterministic programs, including the generalization to count in an arbitrary monoid; and we give a small theory of expectation for such non-deterministic computations, featuring both general map-fusion like results, and specific counting arguments for computing bounds.

Our formalization of the average-case complexity of Quicksort includes a fully formal treatment of the 'tricky' textbook lemma, exploiting the generality of our monadic framework to support a key step in the proof, where the expected comparison count is translated into the expected length of a recorded list of all comparisons.

1 Introduction

Proofs of the $O(n \log n)$ average-case complexity of Quicksort [1] are included in many textbooks on computational complexity [9, for example]. This paper documents what the authors believe to be the first fully formal machine-checked version of such a proof, developed using the Coq proof assistant [2].

The formalisation is based on the "paper proof" in [9], which consists of three parts. The first part shows that the total number of comparisons performed by the algorithm (the usual complexity metric for sorting algorithms) can be written as a sum of expected comparison counts for individual pairs of input list elements. The second part derives from the algorithm a specific formula for

* Research carried out as part of the Radboud Master's programme in "Foundations".

S. Berardi, F. Damiani, and U. de'Liguoro (Eds.): TYPES 2008, LNCS 5497, pp. 256–271, 2009.

this expectation. The third and last part employs some analysis involving the harmonic series to derive the $O(n \log n)$ bound from the sum-of-expectations.

Of these three parts, only the first two involve the actual algorithm itself—the third part is strictly numerical. While the original proof provides a thorough treatment of the third part, its treatment of the first two parts is informal in two major ways.

First, it never actually justifies anything in terms of the algorithm's formal semantics. Indeed, it does not even formally define the algorithm in the first place, relying instead on assertions which are taken to be intuitively true. While this practice is common and perfectly reasonable for paper proofs intended for human consumption, it is a luxury we can not afford ourselves.

Second, the original proof (implicitly) assumes that the input list does not contain any duplicate elements, which significantly simplifies its derivation of the formula for the expected comparison count for pairs of individual input list elements. We take care to avoid appeals to such an assumption.

The key to giving a proper formal treatment of both these aspects lies in using an appropriate representation of the algorithm, capable of capturing its computational behaviour—specifically, its use of comparisons—in a way suitable for subsequent formal reasoning. The approach we take is to consider such operation-counting as a *side effect*, and to use the general framework of *monads* for representing side-effecting computation in pure functional languages. Accordingly we use a shallow embedding, in which the algorithm, here Quicksort, is written as a monadically expressed functional program in Coq. This definition is then instantiated with refinements of operation-counting monads to make the comparison count observable.

The embedding is introduced in section 2, where we demonstrate its use by first giving a simple deterministic monadic Quicksort definition, and then instantiating it with a simple operation counting monad that lets us prove its quadratic worst-case complexity.

For the purposes of the more complex average-case theorem, we then give (in section 3) a potentially-nondeterministic monadic Quicksort definition, and compose a monad that combines operation counting with nondeterminism, supporting a formal definition of the notion of the *expected* comparison count, with which we state the main theorem in section 4.

The next two sections detail the actual formalised proof. Section 5 corresponds to the first part in the original proof described above, showing how the main theorem can be split into a lemma (stated in terms of another specialized monad) giving a formula for the expected comparison count for individual pairs of input elements, and a strictly numerical part. Since we were able to fairly directly transcribe the latter from the paper proof, using the existing real number theory in the Coq standard library with few complications and additions, we omit discussion of it here and refer the interested reader to the paper proof.

Section 6 finishes the proof by proving the lemma about the expected comparison count for individual input list elements. Since this is the part where the original proof omits the most detail, and makes the assumption regarding duplicate elements, and where we really have to reason in detail about the behaviour of the algorithm, it is by far the most involved part of the formalisation.

Section 7 ends with conclusions and final remarks.

The Coq source files containing the entire formalisation can be downloaded from `http://www.eelis.net/research/quicksort/`. We used Coq version 8.2.

Related work. In his Ph.D thesis [12], Hurd presents an approach to formal analysis of probabilistic programs based on a comprehensive formalisation of measure-theoretic constructions of probability spaces, representing probabilistic programs using a state-transforming monad in which bits from an infinite supply of random bits may be consumed. He even mentions the problem of proving the average-case complexity of Quicksort, but leaves it for future work.

In [11], Audebaud and Paulin-Mohring describe a different monadic approach in which programs are interpreted directly as measures representing probability distributions. A set of axiomatic rules is defined for estimating the probability that programs interpreted this way satisfy certain properties.

Compared to these approaches, our infrastructure for reasoning about nondeterministic programs is rather less ambitious, in that we only consider finite expectation based on naïve counting probability, using a monad for nondeterminism which correctly supports weighted expectation. In particular, we do not need to reason explicitly with probability distributions.

A completely different approach to type-theoretic analysis of computational complexity is to devise a special-purpose type theory in which the types of terms include some form of complexity guarantees. Such an approach is taken in [4], for example.

2 A Shallow Monadic Embedding

As stated before, the key to giving a proper formal treatment of those parts of the proof for which the original contents itself with appeals to intuition, lies in the use of an appropriate representation of the algorithm. Indeed, we cannot even formally state the main theorem until we have both an algorithm definition and the means to denote its use of comparisons.

Since we are working in Coq, we already have at our disposal a full functional programming language, in the form of Coq's CIC [3]. However, just writing the algorithm as an ordinary Coq function would not let us observe its use of comparisons. We can however see comparison counting as a *side effect*. As is well known and standard practice in functional languages such as Haskell, side effects can be represented using *monads*: a side-effecting function f from A to B is represented as a function $A \to M\ B$ where M is a type constructor encapsulating the side effects. "Identity" and "composition" for such functions are given by *ret* (for "return") of type $A \to M\ A$ and *bind* (infix: $\gg=$) of type $M\ A \to (A \to M\ B) \to M\ B$ satisfying certain identities (the *monad laws*). For a general introduction to monadic programming and monad laws, see [5].

Furthermore, we use Haskell's "do-notation", declared in Coq as follows

Notation "x <- y ; z" := $(bind\ y\ (\lambda x : \_ \Rightarrow z))$

and freely use standard monadic functions such as:

$$liftM : \forall\ (M : Monad)\ (A\ B : Set), (A \to B) \to (M\ A \to M\ B)$$
$$filterM : \forall\ (M : Monad)\ (A : Set), (A \to M\ bool) \to list\ A \to M\ (list\ A)$$

Here, the Coq type *Monad* is a dependent record containing the (coercible) carrier of type $Set \to Set$, along with the *bind* and *ret* operations, and proofs of the three monad laws.

We now express Quicksort in this style, parameterizing it on both the monad itself and on the comparison operation. A deterministic Quicksort that simply selects the head of the input list as its pivot element, and uses two simple filter passes to partition the input list, looks as follows:

```
Variables (M : Monad) (T : Set) (le : T → T → M bool).
Definition gt (x y : T) : M bool := liftM negb (le x y).
Program Fixpoint qs (l : list T) {measure length l} : M (list T) :=
  match l with
  | nil ⇒ ret nil
  | pivot :: t ⇒
    lower ← filterM (gt pivot) t ≫= qs;
    upper ← filterM (le pivot) t ≫= qs;
    ret (lower ++ pivot :: upper)
  end.
```

We use Coq's **Program Fixpoint** facility [7] to cope with Quicksort's non-structural recursion, specifying list length as an input measure function that is separately shown to strongly decrease for each recursive call. For this definition of *qs*, these proof obligations are trivial enough for Coq to prove mostly by itself.

For recursive functions defined this way, Coq does not automatically define corresponding induction principles matching the recursive call structure. Hence, for this *qs* definition as well as the one we will introduce in section 3, we had to define these induction principles manually. To make their use as convenient as possible, we further customized and specialized them to take advantage of specific monad properties. We will omit further discussion of these issues in this paper, and will henceforth simply say: "by induction on *qs*, ...".

By instantiating the above definitions with the right monad, we can transparently insert comparison-counting instrumentation into the algorithm, which will prove to be sufficient to let us reason about its complexity. But before we do so, let us note that if the above definitions are instead instantiated with the identity monad and an ordinary elementwise comparison on T, then the monadic scaffolding melts away, and the result is equivalent to an ordinary non-instrumented, non-monadic version, suitable for extraction and correctness proofs (which are included in the formalisation for completeness). This means that while we will instantiate the definitions with less trivial monads to support our complexity proofs, we can take some comfort in knowing that the object of those proofs is, in a very concrete sense, the actual Quicksort algorithm (as one would write it in a functional programming language), rather than some idealized model thereof.

For reasons that will become clear in later sections, we construct the monad with which we will instantiate the above definitions using a monad transformer [8] *MMT* (for "monoid monad transformer"), which piggybacks a monoid onto an existing monad by pairing.

Variables $(monoid : Monoid)$ $(monad : Monad)$.
Let C_{MMT} $(T : Set) : Set := monad\ (monoid \times T)$.
Let ret_{MMT} $(T : Set) : T \rightarrow C_{MMT}\ T := ret \circ pair\ (monoid\_zero\ monoid)$.
Let $bind_{MMT}$ $(A\ B : Set)\ (a : C_{MMT}\ A)\ (ab : A \rightarrow C_{MMT}\ B) : C_{MMT}\ B :=$
$x \leftarrow a; y \leftarrow ab\ (snd\ x); ret\ (monoid\_mult\ monoid\ (fst\ x)\ (fst\ y), snd\ y)$.
Definition $MMT : Monad := Build\_Monad\ C_{MMT}\ bind_{MMT}\ ret_{MMT}$.

(In the interest of brevity, we omit proofs of the monad laws for *MMT* and all other monads defined in this paper. These proofs can all be found in the Coq code.)

We now use *MMT* to piggyback the additive monoid structure on $\mathbb{N}$ onto the identity monad, and lift elementwise comparison into the resulting monad, which we call *SP* (for "simply-profiled").

Definition $SP : Monad := MMT\ (\mathbb{N}, 0, +)\ IdMonad$.
Definition $le_{SP}\ (x\ y : \mathbb{N}) : SP\ bool := (1, le\ x\ y)$.

When instantiated with this monad and comparison operation, *qs* produces the comparison count as part of its result.

Definition $qs_{SP} := qs\ SP\ le_{SP}$.
Eval *compute* **in** $qs_{SP}\ (3 :: 1 :: 0 :: 4 :: 5 :: 2 :: nil)$.
$= (16, 0 :: 1 :: 2 :: 3 :: 4 :: 5 :: nil)$

Defining *cost* and *result* as the first and second projection, respectively, we trivially have identities such as $cost\ (ret_{SP}\ x) = 0$, $cost\ (le_{SP}\ x\ y) = 1$, and $cost\ (x \gg=_{SP} f) = cost\ x + cost\ (f\ (result\ x))$. This very modest amount of machinery is sufficient for a straightforward proof of Quicksort's quadratic worst-case complexity.

Proposition. $qs\_worst : \forall\ l, cost\ (qs_{SP}\ l) \leqslant (length\ l)^2$.[1]

Proof. The proof is by induction on *qs*. For $l = nil$, we have $cost\ (qs_{SP}\ nil) = cost\ (ret\ nil) = 0 \leqslant (length\ l)^2$. For $l = h :: t$, the cost decomposes into

[1] We do not use big-O notation for this simple statement, as it would only obfuscate. Big-O complexity is discussed in section 4.

$$\begin{aligned}&\mathit{cost}\ (\mathit{filter}\ (\mathit{le}\ h)\ t) + \mathit{cost}\ (\mathit{qs}_{SP}\ (\mathit{result}\ (\mathit{filter}\ (\mathit{le}\ h)\ t))) +\\ &\mathit{cost}\ (\mathit{filter}\ (\mathit{gt}\ h)\ t) + \mathit{cost}\ (\mathit{qs}_{SP}\ (\mathit{result}\ (\mathit{filter}\ (\mathit{gt}\ h)\ t))) +\\ &\mathit{cost}\ (\mathit{ret}\ (\mathit{result}\ (\mathit{qs}_{SP}\ (\mathit{result}\ (\mathit{filter}\ (\mathit{le}\ h)\ t))) +\!\!+\\ &\quad h :: \mathit{result}\ (\mathit{qs}_{SP}\ (\mathit{result}\ (\mathit{filter}\ (\mathit{gt}\ h)\ t))))).\end{aligned}$$

The *filter* costs are easily proved (by induction on t) to be *length* t each. The cost of the final *ret* is 0 by definition. The induction hypothesis applies to the recursive qs_{SP} calls. Furthermore, by induction on t, we can easily prove

$$\mathit{length}\ (\mathit{result}\ (\mathit{filter}\ (\mathit{le}\ h)\ t)) + \mathit{length}\ (\mathit{result}\ (\mathit{filter}\ (\mathit{gt}\ h)\ t)) \leqslant \mathit{length}\ t,$$

because the two predicates filtered on are mutually exclusive. Abstracting the filter terms as *flt* and *flt′*, this leaves

$$\begin{aligned}&\mathit{length}\ \mathit{flt} + \mathit{length}\ \mathit{flt}' \leqslant \mathit{length}\ t \rightarrow\\ &\quad \mathit{length}\ t + (\mathit{length}\ \mathit{flt})^2 + \mathit{length}\ t + (\mathit{length}\ \mathit{flt}')^2 + 0 \leqslant (S\ (\mathit{length}\ t))^2,\end{aligned}$$

which is true by elementary arithmetic. □

We now extend the technique to prepare for the average-case proof.

3 Nondeterminism and Expected Values

The version of Quicksort used in the average-case complexity proof in [9] differs from the one presented in the last section in two ways. This is also reflected in our formalisation.

First, the definition of *qs* is modified to use a single three-way partition pass, instead of two calls to *filter*, thus avoiding the pathological quadratic behaviour which can arise when the input list does not consist of distinct elements.

Second, and more significantly, we use *nondeterministic* pivot selection, thus avoiding the pathological quadratic behaviour from which any deterministic pivot selection strategy inevitably suffers. While this means that we have proved our result for a subtly different presentation of Quicksort, this nevertheless follows the textbook treatment, in line with common practice.

These two modifications together greatly simplify the formalisation, because they remove the need to carefully track input distributions in order to show that 'good' inputs (for which the original deterministic version of the algorithm performs well) sufficiently outnumber 'bad' inputs (for which the original version performs poorly). They further ensure that the $O(n \log n)$ average-case bound holds not just averaged over all possible input lists, but for each individual input list as well. In particular, it means that once we prove that the bound holds for an arbitrary input, the global bound immediately follows.

This also means that for a key lemma near the end of our proof, we can use straightforward induction over the algorithm's recursive call structure, without having to show that given appropriately distributed inputs, the partition step yields lists that are again appropriately distributed. Such issues are a major technical concern in more ambitious approaches to average-case complexity analysis [10, for example] and to the analysis of probabilistic algorithms.

The second modification is based on a new monad (again defined using *MMT*, but this time transforming a nondeterminism monad) with which the new definition can be instantiated, capturing the *expected* comparison count.

The first modification is relatively straightforward. Instead of calling *filterM*, which uses a two-way comparison operation producing a monadic *bool*, we define a function *partition*. It takes a three-way comparison operation producing a monadic *comparison*, which is an enumeration with values *Lt*, *Eq*, and *Gt*. We represent the resulting partitioning by a function of type *comparison* → *list T* rather than a record or tuple type containing three lists, because in the actual formalisation, this saves us from having to constantly map *comparison* values to corresponding record field accessors or tuple projections. This is only a matter of minor convenience; a record or tuple could have been used instead without problems.

Variables $(T : Set)$ $(M : Monad)$ $(cmp : T \to T \to M\ comparison)$.
Fixpoint $partition\ (t : T)\ (l : list\ T) : M\ (comparison \to list\ T) :=$
  match l **with**
  | $nil \Rightarrow ret\ (const\ nil)$
  | $h :: l' \Rightarrow$
    $c \leftarrow cmp\ h\ t; f \leftarrow partition\ t\ l';$
    $ret\ (\lambda c' \Rightarrow$ **if** $c = c'$ **then** $h :: f\ c'$ **else** $f\ c')$
  end.

Next, we redefine *qs* to use *partition*, and have it take as an additional parameter a *pick* operation, representing nondeterministic selection of an element of a non-empty list of choices. An *ne_list T* is a non-empty list of *T*'s, inductively defined in the obvious way.

Variable $pick : \forall\ A : Set, ne\_list\ A \to M\ A$.
Program Fixpoint $qs\ (l : list\ T)$ $\{$**measure** $length\ l\} : M\ (list\ T) :=$
  match l **with**
  | $nil \Rightarrow ret\ nil$
  | $\_ \Rightarrow$
    $i \leftarrow pick\ [0 \ldots length\ l - 1];$
    let $pivot := nth\ l\ i$ **in**
    $part \leftarrow partition\ pivot\ (remove\ l\ i);$
    $low \leftarrow qs\ (part\ Lt);$
    $upp \leftarrow qs\ (part\ Gt);$
    $ret\ (low +\!\!+ pivot :: part\ Eq +\!\!+ upp)$
  end.

The functions *nth* and *remove* select and remove the *n*th element of a list, respectively.

Note that the deterministic Quicksort definition in section 2 could also have been implemented with a *partition* pass instead, which might well have made the worst-case proof even simpler. We chose not to do this, in order to emphasise that the properties the average-case proof demands of the algorithm rule out the naïve but familiar implementation using *filter* passes.

Nondeterminism can now be emulated by instantiating these definitions with a suitable monad and *pick* operation. A deterministic, non-instrumented version can still be obtained, simply by using the identity monad and any deterministic *pick* operation, such as *head* or 'median-of-three' (not considered here).

Let us now consider what kind of nondeterminism monad would be suitable for reasoning about the expected value of a nondeterministic program like

$$x \leftarrow \mathit{pick}\ [0,1]; \mathbf{if}\ x = 0\ \mathbf{then}\ \mathit{ret}\ 0\ \mathbf{else}\ \mathit{pick}\ [1,2].$$

When executed in the list monad (commonly used to emulate nondeterministic computation), this program produces $[0,1,2]$ as its list of possible outcomes. Unfortunately, the information that 0 is a more likely outcome than 1 or 2 has been lost. Such relative probabilities are critical to the notion of an expected value: the expected value of the program above is $\mathit{avg}\ [0, \mathit{avg}\ [1,2]] = \frac{3}{4} \neq 1 = \mathit{avg}\ [0,1,2]$. This makes list nondeterminism unsuitable for our purposes.

Using tree nondeterminism instead solves the problem: we introduce the type *ne_tree* of non-empty trees, building on *ne_list*:

Inductive *ne_tree* $(T : \mathit{Set}) : \mathit{Set} :=$
  | *Leaf* $: T \rightarrow$ *ne_tree* T
  | *Node* : *ne_list* (*ne_tree* T) $\rightarrow$ *ne_tree* T.
Definition $\mathit{ret}_{\mathit{ne\_tree}}\ \{A : \mathit{Set}\} : A \rightarrow \mathit{ne\_tree}\ A := \mathit{Leaf}.$
Fixpoint $\mathit{bind}_{\mathit{ne\_tree}}\ (A\ B : \mathit{Set})$
  $(m : \mathit{ne\_tree}\ A)\ (k : A \rightarrow \mathit{ne\_tree}\ B) : \mathit{ne\_tree}\ B :=$
    match m **with**
    | *Leaf* $a \Rightarrow k\ a$
    | *Node* $ts \Rightarrow$ *Node* $(\mathit{ne\_list.map}\ (\lambda x \Rightarrow \mathit{bind}_{\mathit{ne\_tree}}\ x\ k)\ ts)$
    end.
Definition $M_{\mathit{ne\_tree}} : \mathit{Monad} := \mathit{Build\_Monad}\ \mathit{ne\_tree}\ \mathit{bind}_{\mathit{ne\_tree}}\ \mathit{ret}_{\mathit{ne\_tree}}.$
Definition $\mathit{pick}_{\mathit{ne\_tree}}\ (T : \mathit{Set}) : \mathit{ne\_list}\ T \rightarrow M_{\mathit{ne\_tree}}\ T$
  $:= \mathit{Node} \circ \mathit{ne\_list.map}\ \mathit{Leaf}.$

We use non-empty trees because we do not consider partial functions, and using potentially empty trees would complicate the definition of a tree's average value below. This is also why we used *ne_list* for *pick*.

With this monad and pick operation, the same program now produces the tree *Node* [*Leaf* 0, *Node* [*Leaf* 1, *Leaf* 2]], which preserves the relative probabilities. The expected value now coincides with the weighted average of these trees:

Definition $\mathit{ne\_tree.avg} : \mathit{ne\_tree}\ \mathbb{R} \rightarrow \mathbb{R} := \mathit{ne\_tree.fold}\ \mathit{id}\ \mathit{ne\_list.avg}.$

Relative probabilities are also the reason we use an n-ary choice primitive rather than a binary one, because correctly emulating (that is, without skewing the relative probabilities) an n-ary choice by a sequence of binary choices is only possible when n is a power of two.

To denote the expected value of a discrete measure f of the output of a program, we define

Definition $expec\ (T : Set)\ (f : T \to \mathbb{N}) : ne\_tree\ T \to \mathbb{R}$
$:= ne\_tree.avg \circ ne\_tree.map\ f.$

Thus, given a program P of type M_{ne_tree} (*list bool*), *expec length* P denotes the expected length of the result list, if we interpret values of type $M_{ne\_tree}\ T$ as nondeterministically computed values of type T.

The function *expec* gives rise to a host of identities, such as

$$\begin{aligned}
0 &\leq expec\ f\ t \\
expec\ (\lambda x \Rightarrow f\ x + g\ x)\ t &= expec\ f\ t + expec\ g\ t \\
expec\ ((*c) \circ f) &= (*c) \circ expec\ f \\
(\forall\ x \in t \to f\ x \leqslant g\ x) \to expec\ f\ t &\leq expec\ g\ t \\
(\forall\ x \in t \to f\ x = c) \to expec\ f\ t &= c \\
(\forall\ x \in t \to f\ x = 0) \leftrightarrow expec\ f\ t &= 0 \\
expec\ f\ (t \gg\!= (ret \circ g)) &= expec\ (f \circ g)\ t \\
expec\ (f \circ g)\ t &= expec\ f\ (ne_tree.map\ g\ t) \qquad (1)
\end{aligned}$$

To form the monad with which we will instantiate *qs* for the main theorem, we now piggyback the additive monoid on $\mathbb{N}$ onto M_{ne_tree} using *MMT*, and call the result *NDP* (for "nondeterministically profiled"):

Definition $M_{NDP} : Monad := MMT\ (\mathbb{N}, 0, +)\ M_{ne\_tree}.$

Definition $cmp_{NDP}\ (x\ y : T) : M_{NDP}\ bool := ret_{ne\_tree}\ (1, cmp\ x\ y).$

Definition $qs_{NDP} := qs\ M_{NDP}\ cmp_{NDP}\ (lift\ pick_{ne\_tree}).$

We can now denote the expected comparison count for a qs_{NDP} application by $expec\ cost\ (qs_{NDP}\ l)$, and will use this in our statement of the main theorem in the next section.

But before we do so, we define a slight refinement of *expec* that specifically observes the monoid component of computations in monads formed by transforming M_{ne_tree} using *MMT* (like *NDP*).

Definition $monoid\_expec\ (m : Monoid)\ (f : m \to \mathbb{N})\ \{A : Set\}$
$: (MMT\ m\ M_{ne\_tree}\ A) \to \mathbb{R} := expec\ (f \circ fst).$

Since $cost = fst$, we have $expec\ cost\ t = monoid\_expec\ id\ t$.

In addition to all the identities *monoid_expec* inherits from *expec*, it has some of its own. One identity states that if one transforms M_{ne_tree} using a monoid m, then for a monoid homomorphism h from m to the additive monoid on $\mathbb{N}$, *monoid_expec h* distributes over *bind*, provided that the expected monoid value of the right hand side does not depend on the computed value of the left hand side:

$monoid\_expec\_plus : \forall\ (m : Monoid)\ (h : m \to (\mathbb{N}, 0, +)),$
$monoid\_homo\ h \to \forall\ (A\ B : Set)$
$(f : MMT\ m\ M_{ne\_tree}\ A)\ (g : A \to MMT\ m\ M_{ne\_tree}\ B) :$
$(\forall\ x\ y \in f \to monoid\_expec\ h\ (g\ (snd\ x)) = monoid\_expec\ h\ (g\ (snd\ y))),$
$monoid\_expec\ h\ (f \gg\!= g) =$
$monoid\_expec\ h\ f + monoid\_expec\ h\ (g\ (snd\ (ne\_tree.head\ f))).$

Since *id* is a monoid homomorphism, *monoid_expec_plus* applies to *NDP* and *expec cost*. In section 5, we will use *monoid_expec_plus* with another monoid and homomorphism.

4 The Statement

The last thing needed before the main theorem can be stated, is the notion of big-O complexity. We use the standard textbook definition, except that we make explicit how we measure inputs to f, namely with respect to a measure function m:

Definition $bigO\ (X : Set)\ (m : X \to \mathbb{N})\ (f : X \to \mathbb{R})\ (g : \mathbb{N} \to \mathbb{R}) : Prop$
$:= \exists\ c\ n, \forall\ x, n \leqslant m\ x \to f\ x \leqslant c * g\ (m\ x).$
Notation "$wrt\ m,\ f = O\ (g)$" $:= bigO\ m\ f\ g.$

We now state the main theorem.

Theorem $qs\_avg : wrt\ length, expec\ cost\ \circ\ qs_{NDP} = O\ (\lambda n \Rightarrow n * \log_2 n).$

Thanks to the property discussed at the start of the previous section, *qs_avg* follows as a corollary from the stronger statement

$$qs\_expec\_cost : \forall\ l,\ expec\ cost\ (qs_{NDP}\ l) \leqslant 2 * length\ l * (1 + \log_2 (length\ l)),$$

the proof of which is described in the next two sections.

5 Reduction to Pairwise Comparison Counts

As described in the introduction, the key ingredient in the proof is a lemma giving a formula for the expected comparison count for individual pairs of input list elements, indexed a certain way. More specifically, if $X \equiv X_{I_0} \ldots X_{I_{n-1}}$ is the input list, with I a permutation of $[0 \ldots n-1]$ such that $X_0 \ldots X_n$ is sorted, then the expected comparison count for any X_i and X_j with $i < j$ is at most $2/(1+j-i)$. In other words, the expected comparison count for two input list elements is bounded by a simple function of the number of list elements that separate the two in the sort order. We prove this fact in the next section, but first show how *qs_expec_cost* follows from it.

Combined with the observation that the total expected comparison count ought to equal the sum of the expected comparison count for each individual pair of input elements, the property described above suggests breaking up the inequality into

$$expec\ cost\ (qs_{NDP}\ l) \leqslant \sum_{(i,j) \in IJ} ecc\ i\ j \leqslant 2 * length\ l * (1 + \log_2 (length\ l)),$$

where $IJ := \{(i,j) \in [0, length\ l) \mid i < j\}$, and $ecc\ i\ j := 2\ /\ (1 + j - i)$.

The right-hand inequality is a strictly numerical affair, requiring a bit of analysis involving the harmonic series. As stated before, this part of the proof was fairly directly transcribed from the paper proof, with few complications and additions, and so we will not discuss it.

The left inequality is the challenging one. To bring it closer to the index summation, we first write l on the left-hand side as $map\ (nth\ (sort\ l))\ li$, where $sort$ may be any sorting function (including qs itself), and where li is a permutation of $[0 \ldots n-1]$ such that $map\ (nth\ (sort\ l))\ li = l$ (such an li can easily be proven to exist).

Next, we introduce a specialized monad and comparison operation that go one step further in focusing specifically on these indices.

Definition $Monoid_U : Monoid := (list\ (\mathbb{N} \times \mathbb{N}), nil, +\!\!+)$.

Definition $U : Monad := MMT\ Monoid_U\ M_{ne\_tree}$.

Definition $lookup\_cmp\ (x\ y : \mathbb{N}) : comparison :=$
$cmp\ (nth\ (sort\ l)\ x)\ (nth\ (sort\ l)\ y)$.

Definition $unordered\_nat\_pair\ (x\ y : \mathbb{N}) : \mathbb{N} \times \mathbb{N} :=$
if $x \leqslant y$ **then** (x, y) **else** (y, x).

Definition $cmp_U\ (x\ y : \mathbb{N}) : U\ comparison :=$
$ret\ (unordered\_nat\_pair\ x\ y :: nil, lookup\_cmp\ x\ y)$.

Definition $qs_U : list\ \mathbb{N} \to list\ \mathbb{N} := qs\ U\ cmp_U\ pick_U$.

The function qs_U operates directly on lists of indices into $sort\ l$. Comparison of indices is defined by comparison of the values they denote in $sort\ l$. Furthermore, rather than producing a grand total comparison count the way NDP does, U records every pair of indices compared, by using MMT with $Monoid_U$, the free monoid over $\mathbb{N} \times \mathbb{N}$ pairs, instead of the additive monoid on $\mathbb{N}$ we used until now.

We now rewrite

$$\begin{aligned} &expec\ cost\ (qs_{NDP}\ (map\ (nth\ (sort\ l))\ li)) \\ &\quad = monoid\_expec\ length\ (qs_U\ li) = expec\ (length \circ fst)\ (qs_U\ li). \end{aligned}$$

The first equality expresses that the expected number of comparisons counted by NDP is equal to the expected length of the list of comparisons recorded by U. In the formalisation, this is a separate lemma proved by induction on qs. The second equality merely unfolds the definition of $monoid\_expec$.

After rewriting with identity 1 in section 3 on page 264, the goal becomes

$$expec\ length\ (ne\_tree.map\ fst\ (qs_U\ li)) \leqslant \sum_{(i,j) \in IJ} ecc\ i\ j.$$

We now invoke another lemma which bounds a nondeterministically computed list's expected length by the expected number of occurrences of specific values in that list. More specifically, it states that

$$\begin{aligned} &\forall\ (X : Set)\ (fr : X \to \mathbb{R})\ (q : list\ X)\ (t : ne\_tree\ (list\ X)), \\ &\quad (\forall\ x \in q, expec\ (count\ x)\ t \leqslant fr\ x) \to \\ &\quad (\forall\ x \notin q, expec\ (count\ x)\ t = 0) \to expec\ length\ t \leqslant \sum_{x \in q} fr\ x. \end{aligned}$$

We end up with two subgoals, the first of which is

$$\forall\,(i,j) \notin IJ,\ expec\ (count\ (i,j))\ (ne\_tree.map\ fst\ (qs_U\ li)) = 0.$$

Rewriting this using identity 1 from section 3 in reverse, then rewriting the *expec* as a *monoid_expec*, and then generalizing the premise, results in

$$\forall\ i\ j\ li, (i \notin li \vee j \notin li) \rightarrow monoid\_expec\ (count\ (i,j))\ (qs_U\ li) = 0 \qquad (2)$$

which can be shown by induction on *qs*, although we will not do so in this paper. We *will* use this property again in the next section.

The second subgoal, expressed with *monoid_expec*, becomes

$$\forall\,(i,j) \in IJ, monoid\_expec\ (count\ (i,j))\ (qs_U\ li) \leqslant ecc\ i\ j \qquad (3)$$

which corresponds exactly to the property described at the beginning of this section. We prove it in the next section.

6 Finishing the Proof

Again, the proof of (3) is by induction on *qs*. But to get a better induction hypothesis, we drop the $(i,j) \in IJ$ premise (because as was shown in the last section, the statement is also true if $(i,j) \notin IJ$), and add a premise saying *li* is a permutation of a contiguous sequence of indices.

$$\forall\ i\ j, i < j \rightarrow \forall\ (li : list\ \mathbb{N})\ (b : \mathbb{N}), Permutation\ [b \ldots b + length\ li - 1]\ li \rightarrow$$
$$monoid\_expec\ (count\ (i,j))\ (qs_U\ li) \leqslant ecc\ i\ j.$$

In the base case, *li* is *nil*, and the left-hand side of the inequality reduces to 0. In the recursive case, *qs* unfolds:

$$\begin{array}{l} monoid\_expec\ (count\ (i,j))\ (\\ \quad pi \leftarrow pick\ [0 \ldots n-1]; \\ \quad \mathbf{let}\ pivot := nth\ li\ pi\ \mathbf{in} \\ \quad part \leftarrow partition_U\ pivot\ (remove\ li\ pi); \\ \quad lower \leftarrow qs_U\ (part\ Lt); \\ \quad upper \leftarrow qs_U\ (part\ Gt); \\ \quad ret\ (lower +\!\!+\ pivot :: part\ Eq +\!\!+\ upper) \\ \quad) \leqslant ecc\ i\ j. \end{array}$$

Since cmp_U is deterministic, $partition_U$ is as well. Furthermore, since we know exactly what monadic effects $partition_U$ has, we can split those effects off and revert to simple effect-free *filter* passes. Finally, we rewrite using the following *monoid_expec* identity:

$$monoid\_expec\ f\ (pick\ l \gg\!= m) = avg\ (map\ (monoid\_expec\ f \circ m)\ l).$$

This way, the goal ends up in a form using less monadic indirection:

$$
\begin{array}{l}
avg\ (map\ (monoid\_expec\ (count\ (i,j))\ \circ\ (\lambda pi \Rightarrow \\
\quad \mathbf{let}\ pivot := nth\ li\ pi\ \mathbf{in} \\
\quad \mathbf{let}\ rest := remove\ li\ pi\ \mathbf{in} \\
\quad \mathbf{let}\ flt := \lambda c \Rightarrow filter\ ((= c)\ \circ\ lookup\_cmp\ pivot)\ rest\ \mathbf{in} \\
\quad ne\_tree.map\ (map\_fst\ (+\!\!+ map\ (unordered\_nat\_pair\ pivot)\ rest))\ (\\
\qquad lower \leftarrow qs_U\ (flt\ Lt); \\
\qquad upper \leftarrow qs_U\ (flt\ Gt); \\
\qquad ret\ (lower +\!\!+ pivot :: flt\ Eq +\!\!+ upper) \\
\quad)))\ [0 \ldots n-1]) \leqslant ecc\ i\ j.
\end{array}
$$

Here, *map_fst* applies a function to a pair's first component.

We now distinguish between five different cases that can occur for the nondeterministically picked *pivot* (which, because we are in the U monad, is an index). It can either be less than i, equal to i, between i and j, equal to j, or greater than j. Each case occurs a certain number of times, and has an associated expected number of (i, j) comparisons (coming either from the *map_fst* term representing the *partition* pass, or from the two recursive qs_U calls). To represent this split, we first rewrite the right-hand side of the inequality to

$$
\frac{ecc\ i\ j * (i - b) + 1 + 0 + 1 + ecc\ i\ j * (b + n - j)}{n}.
$$

This form reflects the facts that

- the case where *pivot* is less than i occurs $i - b$ times, and in each instance, the expected number of (i, j) comparisons is no more than $ecc\ i\ j$;
- the case where the *pivot* is equal to i occurs once, and in this case no more than a single (i, j) comparison is expected;
- in the case where *pivot* lies between i and j, the number of expected (i, j) comparisons is 0, and hence it does not matter how often this case occurs;
- the case where the *pivot* is equal to j occurs once, and in this case no more than a single (i, j) comparison is expected;
- the case where the *pivot* is greater than j occurs $b + n - j$ times, and in each instance, the expected number of (i, j) comparisons is no more than $ecc\ i\ j$.

With the right-hand side of the inequality in this form, we unfold the *avg* application on the left into *sum* (...) / n, and then cancel the division by n on both sides. Next, to actually realize the split, we apply a specialized lemma stating that

$$
\begin{array}{l}
\forall\ b\ i\ j\ X\ f\ n\ (li : list\ \mathbb{N}) \\
\quad (g : [0 \ldots n-1] \rightarrow U\ X), Permutation\ [b \ldots b + length\ li - 1]\ li \rightarrow \\
\quad b \leqslant i < j < b + S\ n \rightarrow \forall\ ca\ cb, 0 \leqslant ca \rightarrow 0 \leqslant cb \rightarrow \\
\quad (\forall\ pi, nth\ li\ pi < i \rightarrow expec\ f\ (g\ pi) \leqslant ca) \rightarrow \\
\quad (\forall\ pi, nth\ li\ pi = i \rightarrow expec\ f\ (g\ pi) \leqslant cb) \rightarrow \\
\quad (\forall\ pi, i < nth\ li\ pi < j \rightarrow expec\ f\ (g\ pi) = 0) \rightarrow \\
\quad (\forall\ pi, nth\ li\ pi = j \rightarrow expec\ f\ (g\ pi) \leqslant cb) \rightarrow \\
\quad (\forall\ pi, j < nth\ li\ pi \rightarrow expec\ f\ (g\ pi) \leqslant ca) \rightarrow \\
\qquad sum\ (map\ (expec\ f\ \circ\ g)\ [0\,.\,.\,n]) \leqslant \\
\qquad\quad ca * (i - b) + cb + 0 + cb + ca * (b + n - j).
\end{array}
$$

Five subgoals remain after applying this lemma—one for each listed case. The first one reads

$$\begin{array}{l}
\forall\ \mathit{pi},\\
\quad \mathbf{let}\ \mathit{pivot} := \mathit{nth}\ \mathit{li}\ \mathit{pi}\ \mathbf{in}\\
\quad \mathbf{let}\ \mathit{rest} := \mathit{remove}\ \mathit{li}\ \mathit{pi}\ \mathbf{in}\\
\qquad \mathit{pivot} < i \rightarrow\\
\qquad \mathit{monoid_expec}\ (\mathit{count}\ (i,j))\\
\qquad\quad (\mathit{ne_tree.map}\ (\mathit{map_fst}\ (+\!\!+\mathit{map}\ (\mathit{unordered_nat_pair}\ \mathit{pivot})\ \mathit{rest}))\ (\\
\qquad\qquad \mathit{foo} \leftarrow \mathit{qs}_U\ (\mathit{filter}\ ((=\mathit{Lt}) \circ \mathit{lookup_cmp}\ \mathit{pivot})\ \mathit{rest});\\
\qquad\qquad \mathit{bar} \leftarrow \mathit{qs}_U\ (\mathit{filter}\ ((=\mathit{Gt}) \circ \mathit{lookup_cmp}\ \mathit{pivot})\ \mathit{rest});\\
\qquad\qquad \mathit{ret}\ (\mathit{foo} +\!\!+ (\mathit{pivot} :: \mathit{filter}\ ((=\mathit{Gt}) \circ \mathit{lookup_cmp}\ \mathit{pivot})\ \mathit{rest}) +\!\!+ \mathit{bar})))\\
\qquad\quad \leqslant \mathit{ecc}\ i\ j.
\end{array}$$

Since *count* (i,j) is a monoid homomorphism, we may rewrite using another lemma saying that

$$\begin{array}{l}
\forall\ (m : \mathit{Monoid})\ (h : m \rightarrow (\mathbb{N}, 0, +)), \mathit{monoid_homo}\ h \rightarrow\\
\forall\ (g : m)\ (A : \mathit{Set})\ (t : \mathit{MMT}\ m\ M_{\mathit{ne_tree}}\ A),\\
\quad \mathit{monoid_expec}\ h\ (\mathit{ne_tree.map}\ (\mathit{map_fst}\ (\mathit{monoid_mult}\ m\ g))\ t) =\\
\quad h\ g + \mathit{monoid_expec}\ h\ t.
\end{array}$$

This leaves

$$\begin{array}{l}
\mathit{count}\ (i,j)\ (\mathit{map}\ (\mathit{unordered_nat_pair}\ \mathit{pivot})\ \mathit{rest}) +\\
\mathit{monoid_expec}\ (\mathit{count}\ (i,j))\\
\quad (\mathit{foo} \leftarrow \mathit{qs}_U\ (\mathit{filter}\ ((=\mathit{Lt}) \circ \mathit{lookup_cmp}\ \mathit{pivot})\ \mathit{rest});\\
\quad \mathit{bar} \leftarrow \mathit{qs}_U\ (\mathit{filter}\ ((=\mathit{Gt}) \circ \mathit{lookup_cmp}\ \mathit{pivot})\ \mathit{rest});\\
\quad \mathit{ret}\ (\mathit{foo} +\!\!+ (\mathit{nth}\ v\ \mathit{pi} :: \mathit{filter}\ ((=\mathit{Eq}) \circ \mathit{lookup_cmp}\ \mathit{pivot})\ \mathit{rest}) +\!\!+ \mathit{bar}))\\
\leqslant \mathit{ecc}\ i\ j.
\end{array}$$

From $\mathit{pivot} < i$ and $i < j$, we have $\mathit{pivot} < j$. Since each of the comparisons in *map (unordered_nat_pair pivot) rest* involves the pivot element, it follows that none of them can represent comparisons between i and j. Hence, the first term vanishes. Furthermore, *monoid_expec_plus* lets us distribute *monoid_expec* over the *bind* applications. Since the *ret* term does not produce any comparisons either (by definition), its *monoid_expec* term vanishes, too. What remains are the two recursive calls:

$$\begin{array}{l}
\mathit{monoid_expec}\ (\mathit{count}\ (i,j))\ (\mathit{qs}_U\ (\mathit{filter}\ ((=\mathit{Lt}) \circ \mathit{lookup_cmp}\ \mathit{pivot})\ \mathit{rest})) +\\
\mathit{monoid_expec}\ (\mathit{count}\ (i,j))\ (\mathit{qs}_U\ (\mathit{filter}\ ((=\mathit{Gt}) \circ \mathit{lookup_cmp}\ \mathit{pivot})\ \mathit{rest}))\\
\quad \leqslant \mathit{ecc}\ i\ j.
\end{array}$$

All indices in the first filtered list denote elements less than the element denoted by the pivot. Since the former precede the latter in *sort l*, it must be the case that these indices are all less than *pivot*. And since $\mathit{pivot} < i$, it follows that the first qs_U term will produce no (i,j) comparisons (using property (2) at the end of section 5 on page 267). Hence, the first *monoid_expec* term vanishes, leaving

$$\begin{array}{l}
\mathit{monoid_expec}\ (\mathit{count}\ (i,j))\\
\quad (\mathit{qs}_U\ (\mathit{filter}\ ((=\mathit{Gt}) \circ \mathit{lookup_cmp}\ \mathit{pivot})\ \mathit{rest})) \leqslant \mathit{ecc}\ i\ j.
\end{array}$$

We now compare *nth* (*sort* l) i with *nth* (*sort* l) *pivot*.

- If the two are equal, then i will not occur in the *filter* term, and so (again) no (i, j) comparisons are performed.
- If *nth* (*sort* l) $i <$ *nth* (*sort* l) *pivot*, then we must have $i <$ *pivot*, contradicting the assumption that *pivot* $< i$.
- If *nth* (*sort* l) $i >$ *nth* (*sort* l) *pivot*, then we apply the induction hypothesis. For this, it must be shown that filtering the list of indices preserves contiguity, which follows from the fact that the indices share the order of the elements they denote in *sort* l.

This concludes the case where *pivot* $< i$. The case where $j <$ *pivot* is symmetric. The other three cases use similar arguments. The proof is now complete.

7 Final Remarks

In the interest of brevity, we have omitted lots of detail and various lemmas in the description of the proof. Still, the parts shown are reasonably faithful to the actual formalisation, with two notable exceptions.

First, we have pretended to have used ordinary natural numbers as indices into ordinary lists, completely ignoring issues of index validity that could not be ignored in the actual formalisation. There, we use vectors (lists whose size is part of their type) and bounded natural numbers in many places instead. Using these substantially reduces the amount of $i <$ *length* l proofs that need to be produced, converted, and passed around, but this solution is still far from painless.

Second, using the **Program** facility to deal with Quicksort's non-structural recursion is not completely as trivial as we made it out to be. Since the recursive calls are nested in lambda abstractions passed to the *bind* operation of an unspecified monad, the relation between their arguments and the function's parameters is not locally known, resulting in unprovable proof obligations. To make these provable, we Σ-decorated the types of *filter* and *partition* in the actual formalisation with modest length guarantees.

The formalised development successfully adopted from the original proof the idea of using a nondeterministic version of the algorithm to make the $O(n \log n)$ bound hold for any input list, the idea of taking an order-indexed perspective to reduce the problem to a sum-of-expected-comparison-counts, and the use of the standard bound for harmonic series for the strictly numerical part. However, for the actual reduction and the derivation of the formula for the expected comparison count, the intuitive arguments essentially had to be reworked from scratch, building on the monadic representation of the algorithm and the various comparison counting/nondeterminism monads.

The shallow monadic embedding provides a simple but effective representation of the algorithm. Being parameterized on the monad used, it allows a single definition to be instantiated either with basic monads (like the identity monad or bare nondeterminism monads) to get a non-instrumented version suitable for extraction and correctness proofs, or with MMT-transformed monads to support complexity proofs. Furthermore, since this approach lets us re-use all standard Coq data types and facilities, including the powerful **Program Fixpoint** command, the actual algorithm definition itself is reasonably clean.

We have shown that it is straightforward to give a fully formal treatment in type theory of a classical result in complexity theory. This clearly shows the utility and applicability of the general monadic approach we have developed.

References

1. Hoare, C.: Quicksort. The Computer Journal 5, 10–15 (1962)
2. The Coq Development Team: The Coq Proof Assistant Reference Manual – Version V8.2 (February 2009), `http://coq.inria.fr`
3. Bertot, Y., Castéran, P.: Coq'Art: Interactive Theorem Proving and Program Development. Texts in Theoretical Computer Science. Springer, Heidelberg (2004)
4. Constable, R.L.: Expressing computational complexity in constructive type theory. In: Leivant, D. (ed.) LCC 1994. LNCS, vol. 960, pp. 131–144. Springer, Heidelberg (1995)
5. Wadler, P.: Monads for functional programming. In: Jeuring, J., Meijer, E. (eds.) AFP 1995. LNCS, vol. 925, pp. 24–52. Springer, Heidelberg (1995)
6. Sedgewick, R.: The analysis of quicksort programs. Acta Inf. 7, 327–355 (1977)
7. Sozeau, M.: Subset coercions in Coq. In: Altenkirch, T., McBride, C. (eds.) TYPES 2006. LNCS, vol. 4502, pp. 237–252. Springer, Heidelberg (2007)
8. Liang, S., Hudak, P., Jones, M.P.: Monad transformers and modular interpreters. In: POPL 1995, pp. 333–343. ACM, New York (1995)
9. Cormen, T., Leiserson, C., Rivest, R., Stein, C.: Introduction to Algorithms, 2nd edn. MIT Press, Cambridge (2001)
10. Schellekens, M.: A Modular Calculus for the Average Cost of Data Structuring. Springer, Heidelberg (2008)
11. Audebaud, P., Paulin-Mohring, C.: Proofs of Randomized Algorithms in Coq. In: Uustalu, T. (ed.) MPC 2006. LNCS, vol. 4014, pp. 49–68. Springer, Heidelberg (2006)
12. Hurd, J.: Formal Verification of Probabilistic Algorithms. PhD thesis, University of Cambridge (2002)

Coalgebraic Reasoning in Coq: Bisimulation and the λ-Coiteration Scheme

Milad Niqui*

Department of Software Engineering
Centrum Wiskunde & Informatica, The Netherlands
M.Niqui@cwi.nl

Abstract. In this work we present a modular theory of the coalgebras and bisimulation in the intensional type theory implemented in *Coq*. On top of that we build the theory of weakly final coalgebras and develop the λ-coiteration scheme, thereby extending the class of specifications definable in *Coq*. We provide an instantiation of the theory for the coalgebra of streams and show how some of the productive specifications violating the guardedness condition of *Coq* can be formalised using our library.

Keywords: Coinduction Coalgebra Bisimulation Coiteration Coq.

1 Introduction

Coinduction is a method for proving properties of infinite objects such as streams and infinite trees. It is dual to the usual approach of using induction both for computation and reasoning and can be studied from a category theoretical [20] or type theoretical point of view [12]. Coinduction provides a verification paradigm for programs written in a lazy functional programming e.g. *Haskell*, and hence it is implemented in many theorem proving tools. Among the many tools used for coinductive reasoning are the ones based on constructive type theory such as *Coq* and *Agda* where *coinductive types* serve this purpose.

The invocation of coinductive types in these tools is quite similar (at least on surface) to the functional programming syntax, these types are defined using their constructors akin to the general recipe for defining algebraic data types. However, type theories where termination is crucial for finite objects enforce similar restrictions for ensuring *productivity*. These restrictions are syntactic tests and will inevitably exclude some legitimate productive definitions. Thus not all *Haskell* programs, even those describing total functions, are accepted in coinductive type theories. Several workarounds exist such as [10] where topological properties of fixed points are used, [5] where advanced type theoretic techniques are used or [1] where type theory is extended and refined with type-based termination to facilitate dealing with the productivity issues.

* Supported by a VENI grant from the Netherlands Organisation for Scientific Research (NWO).

S. Berardi, F. Damiani, and U. de'Liguoro (Eds.): TYPES 2008, LNCS 5497, pp. 272–288, 2009.

One approach that deserves more attention is the direct formalisation of various categorical schemes from the theory of coalgebras.[1] The present work indicates that this is a relatively low-cost and generic approach and by formalising a single scheme a large class of total functions can be programmed in coinductive type theory. The scheme we choose is the λ-coiteration scheme of Bartels [3] which is one in a family of schemes intended to expand the basic iteration scheme of final coalgebras. This is a *scheme* i.e., it allows the formalisation of a class of specifications (or *Haskell*-like programs) satisfying a specific syntactic form.

Coalgebraic semantics is so close to coinductive types that in many situations the proof techniques are identical, making coinductive reasoning merely a translation of coalgebras. However in the case of *intensional* type theories this is not the case. In intensional type theory the two objects being provably equal does not entail that they are *convertible*. This restriction is necessary for the decidability of type checking and although it is not a theoretical obstacle for programming, it can be practically inconvenient. In particular, formalising category theory is susceptible to this inconvenience, as proving the *uniqueness* of arrows in universal properties of limits adheres to extensional properties of functions. The main workarounds for working extensionally in intensional type theories, is to use *setoids* and work modulo an extensionally defined equality. We will partly follow this through our use of bisimulation equivalence though we will not directly use setoids because we still prefer to benefit from convenient computational properties of intensional equality. The present work shows how one may exploit the intensional equality to the maximum and meanwhile tackle the difficulties through two important tools: (1) working (if necessary) modulo bisimulation equivalence, (2) requiring the functors to satisfy some sort of extensionality. Note that (1) is specific to coalgebras while (2) is applicable to general categorical constructions.

We use the machinery of the proof assistant *Coq* to present the formalisation, but the article is applicable to other intensional incarnations of coinductive types (such as *Agda*). Coinductive types themselves will only be used in Section 7 when we instantiate the theory. Throughout the article we use a syntax loosely based on *Coq*'s syntax, adapted for presenting in an article. In particular we use the uncurried version of the functions when they are presented in mathematical formulae. A complete *Coq* formalisation of the material in this paper can be found in [19].

Related Work. McBride [16] and Matthes [15] use extensional functors for formalising category theoretical notions in the intensional type theory. Hancock and Setzer develop the weakly final coalgebra and bisimulation for a very powerful functor capable of representing interactive IO programs in intensional type theory [12]. Their work is formalised in *Agda* by Michelbrink [18] but the latter formalisation uses inductive–recursive universe which is beyond the **CIC**. Neither of [12, 18] study various definition schemes but their work is so expressive

[1] In this respect Anton Setzer's proposals advocating coalgebraic alternative to coinductive types are very promising and should be mentioned.

that a development of schemes for their functor would considerably extend the class of specifications definable in *Coq*. Our work can be seen as a first step in the formalisation of [12, 18] in *Coq* while along the way extending the class of definition for basic polynomial functors such as streams.

The work by Bertot and Komendantskaya [5] and more recently in [6] uses advanced type theoretic techniques to bypass the guardedness condition of *Coq* for a larger class of functions than those covered by syntactic schemes, including partial functions. Especially in [6] the stream functions that we define in Examples 1–4 (Section 7) are formalised in *Coq* by viewing them as functions on natural numbers and using structural recursion.

2 Coinductive Types

The *Coq* proof assistant [8] is an implementation of the Calculus of Inductive Constructions (**CIC**) extended with coinductive types. Coinductive types were added to *Coq* by Giménez [11]. Their implementation follows the same philosophy as that of inductive types in **CIC**, namely there is a general scheme that allows for formation of coinductive types if their constructors are given, and if these constructors satisfy a strict positivity condition. For example, the type of streams of elements of a set A can be defined using[2] its constructor `Cons` as

```
CoInductive Streams (A : Set) : Set :=
| Cons : A → Streams A → Streams A.
```

From now on we shall use A^ω to denote the type `Streams A`.

After a coinductive type is defined one can introduce its inhabitants and functions. Such definitions are given by a *cofixed point* operator cofix. This operator, when given a well-typed definition that satisfies a *guardedness condition*, will introduce an inhabitant of the coinductive type. Assuming that I is a coinductive type, when defining a function $f\colon T \longrightarrow I$ this condition requires each recursive occurrence of f in the body of f to be the immediate argument of a constructor of some inductive or coinductive type [9, 11]. Finally there is a reduction (in fact expansion) rule corresponding to the cofix operator that allows the expansion of a cofixed point only when a case analysis of the cofixed point is done.

Like other syntactic criteria, the guardedness condition of *Coq* excludes some productive functions. An example is the *shuffle product* of streams of numbers defined as:

$$(x :: xs) \otimes (y :: ys) := \quad x \cdot y \quad :: \quad (xs \otimes (y :: ys)) \;\oplus\; ((x :: xs) \otimes ys) \tag{2.1}$$

with $\oplus$ being the pointwise addition:

$$(x :: xs) \oplus (y :: ys) := \quad x + y \quad :: \quad xs \oplus ys$$

[2] Note that, as it is the case with algebraic and inductive data types, the type `Stream` and its constructor `Cons` are defined simultaneously.

This is a type of convolution product that corresponds to the lazy computation of the product of two power series in Maclaurin form [17]. Symbolically it corresponds to the derivation of power series and plays an important role in the stream calculus of [21]. We give a *Coq* formalisation of this and another type of convolution product in Section 7. We do this by developing the λ-coiteration scheme of [4] that when applied to streams is completely definable in terms of cofix. This is due to the fact that the guardedness condition captures the coiteration scheme of weakly final coalgebras [12]. However, we do not restrict ourselves to streams; we take a more generic approach that is reusable for other similar functors.

3 Extensional Functors and Coalgebras

We are interested in endofunctors on Set.[3] For this we need a type of Set-functors inhabiting a dependent pair[4] of operations $F\colon \mathsf{Set} \longrightarrow \mathsf{Set}$ (on objects) and $l_{F_{X,Y}} : (X{\to}Y){\to}F(X){\to}F(Y)$ (on arrows) satisfying the following properties (ignoring the subscripts X, Y in l_F when there is no risk of confusion).

---

l_F_id: $\forall X\ (x\colon\ F(X)),\ x\ =\ l_F\ (\lambda z.z)\ x.$

l_F_compose: $\forall XYZ\ (g{:}X{\to}Y)\ (f{:}Y{\to}Z)\ x,$
$$(\lambda z.l_F\ f\ (l_F\ g\ z))\ x\ =\ l_F\ (\lambda z.f\ (g\ z))\ x.$$

l_F_ext: $\forall XY\ (f\ g{:}X{\to}Y)\ x,\ (\forall z, f\ z = g\ z) \to l_F\ f\ x\ =\ l_F\ g\ x.$

The first two conditions are standard functorial properties; while l_F_ext will be very helpful in dealing with extensional properties of functor compositions. A functor satisfying l_F_ext is called an *extensional functor* [15]. Obviously if one is working in a setting where *functional extensionality* holds i.e.,

$$\forall XY(fg : X{\to}Y), (\forall z, fz = gz) \to f = g\ , \tag{Ext}$$

then all functors, in fact all operations on sets, are extensional. So in **CIC** +Ext trivially all functors are extensional. But in **CIC** this is not the case. It is unclear whether the assumption that all functors are extensional is a weaker axiom than Ext, but we can prove that assuming extensionality of some specific functors is tantamount to Ext in the presence of η which is the rule:

$$\forall XY(f : X{\to}Y), \lambda z.(fz) = f\ . \tag{η}$$

---

[3] This is the type Set of *Coq* which corresponds to constructive sets and computations. It can be identified with any categorical model of type theory.

[4] In fact in our work these are formalised as *module types* in *Coq* (see [19]). Matthes gives two different formalisations using record types and type classes [15].

Proposition 1. *Assume η. For a given set A the functor $F(X) := X^A$ is extensional if and only if for all X all functions $A \to X$ satisfy* Ext.

Proof. ($\Leftarrow$) is trivial, for ($\Rightarrow$) assume l_F_ext and let $f, g \colon A \longrightarrow A$ be given s.t.

$$\forall z, fz = gz \ . \tag{3.1}$$

Then by applying $l_{F_{A,X}}$_ext to $f, g, x := \lambda z.z \colon F(A)$ and (3.1) we have $\lambda z.fz = \lambda z.gz$. Now by ($\eta$) we obtain $f = g$, so f and g satisfy Ext. □

It is well-known that **CIC**+η is weaker than **CIC**+Ext [13] and hence the above shows that $F(X) := X^A$ cannot be proven to be an extensional functor inside **CIC**. On the other hand each functor composed of finite sums and products seems to be extensional in **CIC**. In fact we can prove the following lemma in **CIC**.

Lemma 1

(i) The constant functor, sending every set to a fixed set U and each arrow to the identity on U is extensional.
(ii) The identity functor, identity on objects and arrows, is extensional.
(iii) Disjoint sum of two extensional functors obtained by case analysis on arrows is extensional.
(iv) Product of two extensional functors obtained by pairing on arrows is extensional.
(v) Composition of two extensional functors obtained by composition on arrows is extensional.
(vi) For each $n \in \mathbb{N}$ the n-th iteration of an extensional functor is extensional.

This lemma also appears in [15]. A proof in the form of parametric modules can be found in [19]. This lemma ensures the extensionality of most polynomial functions used in practice bar those based on exponential. In particular it holds for functors used in Examples 1–4.

The main advantage of l_F_ext is that it eliminates the need for Ext without having to resort to setoids functors and hence leaving us with a lightweight formalisation. This is in contrast with the formalisation of category theory in [14] where setoids are used.

After this we define the notion of F-coalgebra for extensional functors as a set together with a transition structure.

```
Record F_coalg : Type := { st : Set ;
                           tr : st → F st }
```

Then we need to define the lifting of a relation R on the image of functor F; this will later be used for expressing the commutativity of diagrams involving the weakly final coalgebra.

$$\texttt{Definition}\ l_{Rel(F)}\ (S_1\ S_2 : F\_\texttt{coalg})\ (R : S_1.st \to S_2.st \to \mathsf{Prop})$$
$$(z_x : F\ S_1.st)\ (z_y : F\ S_2.st)\ :\ \mathsf{Prop}\ :=$$
$$\exists xy,\ R\ x\ y\ \wedge\ z_x = S_1.tr\ x\ \wedge\ z_y = S_2.tr\ y.$$

Here Prop is the **CIC**'s universe of propositions which is a subtype of Set. The Set-Prop distinction is *not* essential in our work; Prop could be replaced by Set everywhere. However, exploiting the distinction we can consider bisimilarity as a computationally irrelevant object akin to the other forms of equality.

4 Bisimulation

Bisimulation is the basic tool for studying the elements in a coalgebra. First we recall the usual categorical definition of F-*bisimulation* [20]: given two sets X, Y, a relation $R \subseteq X \times Y$ is a bisimulation between X and Y if there is a map $\gamma : R \longrightarrow F(R)$ s.t. both squares in this diagram commute (by π_i we denote the i-th projection of a tuple):

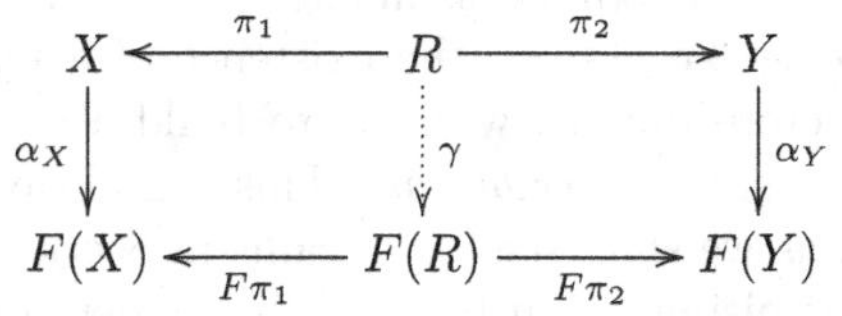

In **CIC** though, where we use dependent types for subsets, there is a distinction between a Prop-valued relation $R\colon X \to Y \to \mathsf{Prop}$ and the set of pairs in $\{(x, y) \in X \times Y | Rxy\}$. The latter is a set of dependent pairs also called a Σ-type. Because we will be composing Σ-types built on a relation in Prop with Σ-types built on other Σ-types we need to fix the notation. By $\{\overline{\exists} x\colon X, \phi(x)\}$ we denote the set of elements of X satisfying $\phi : X \to \mathsf{Prop}$, and by $\{\Sigma x\colon X, f(x)\}$ we denote the set of elements X for which $f(x)$ is inhabited (here $f\colon X \to \mathsf{Set}$ is an X-indexing of sets). We shall use variable R for Prop-valued and variable ρ for Set-valued ones, i.e., $\rho\colon X \to Y \to \mathsf{Set}$. Given a relation R (resp. ρ) we write $\{\overline{\exists}(R)\}$ (resp. $\{\Sigma(\rho)\}$ as a shorthand for $\{\overline{\exists} u\colon X \times Y,\ R\ \ \pi_1(u)\ \ \pi_2(u)\}$ (resp. $\{\Sigma u\colon X \times Y,\ \rho\ \ \pi_1(u)\ \ \pi_2(u)\}$). Note that an element of $\{\overline{\exists}(R)\}$ is a 3-tuple consisting a pair from $X \times Y$ and a proof that they satisfy R.

With the above notation an F-bisimulation for an extensional functor F will be determined by the existence of $\gamma\colon \{\overline{\exists}(R)\} \to F\{\overline{\exists}(R)\}$. Set theoretically this is equivalent to $\gamma\colon R \to F(R)$ but in **CIC** the distinction is necessary. But this is not the only discrepancy: the above diagram for bisimulation is an existential statement. In order to formalise the existence of the γ in a way that can be later used as a witness, in **CIC** we have to define the *set* of all F-bisimulations between X and Y.

Following the above we define two predicates; first when a Prop-valued relation is bisimulation and second for a Set-valued relation:

`Definition` $F\_bisim?\ (S_1\ S_2\colon F\_\mathtt{coalg})\ (R\colon S_1.st{\to}S_2.st{\to}\mathsf{Prop})\ :=$
$$\Big\{\Sigma\gamma\colon \{\overline{\exists} R\}{\to}F\{\overline{\exists} R\},\ \forall y\colon \{\overline{\exists} R\},\quad l_F\ \pi_1\ \gamma(y) = S_1.tr\ (\pi_1(y))\ \bigwedge\ l_F\ \pi_2\ \gamma(y) = S_2.tr\ (\pi_2(y))\Big\}.$$

`Definition` $F\_\sigma bisim?\ (S_1\ S_2\colon F\_\mathtt{coalg})\ (\rho\colon S_1.st{\to}S_2.st{\to}\mathsf{Set})\ :=$
$$\Big\{\Sigma\gamma\colon \{\Sigma\rho\}{\to}F\{\Sigma\rho\},\ \forall y\colon \{\Sigma\rho\},\quad l_F\ \pi_1\ \gamma(y) = S_1.tr\ (\pi_1(y))\ \bigwedge\ l_F\ \pi_2\ \gamma(y) = S_2.tr\ (\pi_2(y))\Big\}.$$

We usually ignore the first two arguments of $F\_bisim?$ and $F\_\sigma bisim?$ and simply write $F\_bisim?(R)$. Now we can define when a bisimulation is maximal.

`Definition` $F\_max\_bisim?\ (S_1\ S_2 : F\_\mathtt{coalg})\ (R : S_1.st{\to}S_2.st{\to}\mathsf{Prop})\ :=$
$$F\_bisim?(R)\ \wedge\ \forall\rho,\ F\_\sigma bisim?(\rho){\to}\forall s_1 s_2,\ \rho\ s_1\ s_2{\to}R\ s_1\ s_2.$$

As we will see later the subtle occurrence of a Set-valued relation ρ is crucial in the proof of the fact that bisimulation is closed under composition.

It is well-known that the maximal bisimulation between any two F-coalgebras exists [20]. In our theory we assume the existence of a maximal bisimulation. Later on for each concrete functor we have to build a concrete relation which should be proven to satisfy $F\_max\_bisim?$. This can always be built using as a coinductive type [12], as we shall see for streams in Section 7.

It is known that for bisimulation to be closed under composition functor F should satisfy some additional property, e.g. in [20] F is required to preserve weak pullbacks. We require a similar albeit weaker restriction. First we define the carrier set of the weak pullback of $f\colon X \longrightarrow Z$ and $g\colon Y \longrightarrow Z$ to be the set

$$\mathsf{WP}(f,g) := \{\overline{\exists} u\colon X\times Y, f(\pi_1(u)) = g(\pi_2(u))\}\ .$$

Subsequently we require that a function

$$i_{wpF}\colon \mathsf{WP}(l_F(f), l_F(g)) \longrightarrow F(\mathsf{WP}(f,g))$$

satisfying the following property exist.

$$\mathsf{WP}_F^{\rightarrow}\ :\ \forall XYZ\ \ (f : X{\to}Z)\ \ (g : Y{\to}Z)\ \ (u : \mathsf{WP}(l_F(f), l_F(g))),\quad l_F\ \pi_1\ i_{wpF}(u) = \pi_1(u)\ \bigwedge\ l_F\ \pi_2\ i_{wpF}(u) = \pi_2(u).$$

Evidently this is weaker than the assumption that F preserves weak pullbacks because we only require the preservation of the pullback arrows, and even then up to the existence of a one-way map i_{wpF} which is not required to be an isomorphism.

Given the above requirements, i.e., a maximal F-bisimulation between coalgebras S_1 and S_2 and a map i_{wpF} satisfying $\mathsf{WP}_F^{\rightarrow}$ we can develop a theory of bisimulation. First we need the following properties.[5]

[5] This theorem and all the following ones are all formalised in *Coq* and available in [19].

Lemma 2

i) $F\_bisim?(S_1, S_2, R) \implies F\_\sigma bisim?(S_1, S_2, R)$.
ii) $F\_bisim?(S, S, =)$, *i.e., propositional equality is a bisimulation relation.*
iii) $F\_bisim?(S_1, S_2, R) \implies F\_bisim?(S_2, S_1, \lambda xy.Ryx)$.
iv) $F\_bisim?(S_1, S_2, R_{12}) \wedge F\_bisim?(S_2, S_3, R_{23}) \implies$
$F\_\sigma bisim?(S_1, S_3, \lambda xz.\{\overline{\exists} y, R_{12}xy \wedge R_{23}yz\})$, *i.e., bisimulation preserves composition.*

Note that in (i) the subtyping relation between Prop and Set is used. The only technical part of the proof Lemma 2 is part (iv). The relation $R_{12}\bar{\circ}R_{23} := \lambda xz.\{\overline{\exists} y, R_{12}xy \wedge R_{23}yz\}$ is the counterpart of the set-theoretic composition of two relations $\lambda xz.\exists y, R_{12}xy \wedge R_{23}yz$. For the rest we follow the proof in [20], defining the maps in the following diagram. Here $X := \mathsf{WP}(\pi_2^{R_{12}}, \pi_1^{R_{23}})$ i.e., the weak pullback for $\pi_2^{R_{12}}\colon \{\overline{\exists}(R_{12})\} \longrightarrow S_2$ and $\pi_1^{R_{23}}\colon \{\overline{\exists}(R_{23})\} \longrightarrow S_2$.

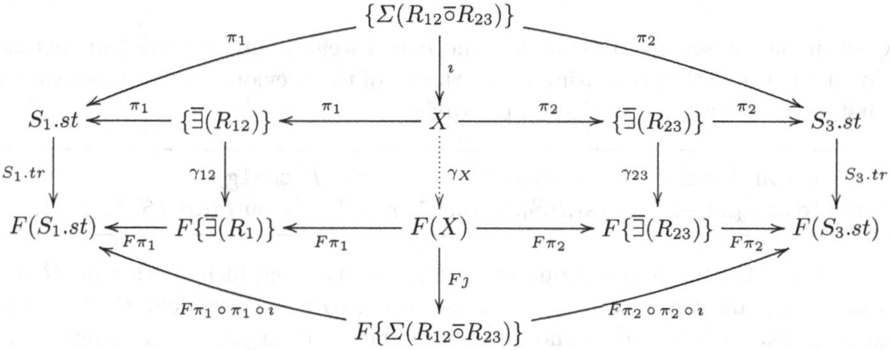

In this diagram $\jmath$ is the map sending an element $\langle s_1, s_2, \phi_{12}, s_2', s_3, \phi_{23}, \phi_= \rangle$ of X to $\langle s_1, s_3, s_2, \phi_{123}\rangle$, where ϕ's are proof obligations and in particular $\phi_=$ is a proof that $s_2 = s_2'$. Likewise $\imath$ is the 'inverse' of $\jmath$ and

$$\imath\langle s_1, s_3, s_2, \phi_{123}\rangle := \langle s_1, s_2, \phi_{12}, s_2, s_3, \phi_{23}, \phi_{\text{refl}}\rangle \ .$$

The main part is defining a coalgebraic structure on X to obtain the transition map γ_X. For this we use the map $p\colon X \longrightarrow \mathsf{WP}(l_F(\pi_2^{R_{12}}), l_F(\pi_1^{R_{23}}))$ defined as

$$p\langle s_1, s_2, \phi_{12}, s_2', s_3, \phi_{23}, \phi_= \rangle := \langle \gamma_{12}\langle s_1, s_2, \phi_{12}\rangle, \gamma_{23}\langle s_2', s_3, \phi_{23}\rangle, \phi_{l_F}\rangle$$

where ϕ_{l_F} is the proof that

$$l_F(\pi_2^{R_{12}})(\gamma_{12}\langle s_1, s_2, \phi_{12}\rangle) = l_F(\pi_1^{R_{23}})(\gamma_{23}\langle s_2', s_3, \phi_{23}\rangle) \ ,$$

and is obtained by $\phi_=$ and the commutativity of the bisimulation diagrams for $\{\overline{\exists}(R_{12})\}$ and $\{\overline{\exists}(R_{23})\}$. Now taking $\gamma_X := i_{wpF} \circ p$ we can prove that γ_X is indeed a homomorphism of coalgebras i.e., the small squares in the above diagram commute. Subsequently the entire diagram above commutes. Which means that $F\jmath \circ \gamma_X \circ \imath$ is the map making $\{\Sigma(R_{12}\bar{\circ}R_{23})\}$ an F-bisimulation and thus completing the proof.

Using Lemma 2 we can easily derive the following theorem.

Theorem 1. *For any coalgebra S the maximal bisimulation on S is an equivalence relation.*

Theorem above is the main tool for a generic definition of bisimulation as an extensional equality on coalgebras: due to our modular formalisation in **CIC**, each time we instantiate the theory of this section with an extensional Set-functor satisfying $\mathsf{WP}_F^{\rightarrow}$ and a maximal bisimulation relation we get this theorem for free.

As a final remark we note that all the machinery based on weak pullbacks and Σ-types is necessitated by the proof of transitivity which in turn is based on Lemma 2.iv. In other words, the reflexivity and the symmetry of maximal bisimulation holds for any extensional functor and for the weaker notion of maximality obtained by replacing $F\_\sigma bisim?$ by $F\_bisim?$ in the definition of $F\_max\_bisim?$.

5 Weakly Final Coalgebras

Continuing the set-up so far we assume F is a weak pullback preserving extensional functor so that the bisimulation theory of the previous section is derivable. First we define when a coalgebra is *weakly final*:

$$\texttt{Definition}\ F\_\mathrm{wfin?}\ (S_0\colon F\_\texttt{coalg})\ :=\ \forall\ (S_1\colon F\_\texttt{coalg}),$$
$$\{\overline{\exists}\texttt{unfld}_F\colon S_1.st\to S_0.st, \forall s_1,\ \ S_0.tr\ \ (\texttt{unfld}_F\ \ s_1) = l_F\ \ \texttt{unfld}_F\ \ (S_1.tr\ \ s_1)\}.$$

If Ω satisfies the above property we call the maximal bisimulation on Ω the *bisimilarity* and we denote it by $\cong$. According to the above definition the existence of a coalgebra homomorphism originating from any other coalgebra is enough. For concrete functors $S_0.st$ can be taken to be a suitably chosen coinductive type with $S_0.tr$ being the inverse of the constructors. In each concrete case we cannot prove the uniqueness of $\texttt{unfld}_F(S_1)$ up to intensional equality without assuming EXT as an axiom. However, the following form of uniqueness can be proven for concrete functors that we deal with (for streams see Lemma 4).

$$\begin{aligned}\Omega_{\text{unique}}\colon \forall\ (S\colon F\_\texttt{coalg})\ (f\ \ g\colon S.st\to\Omega.st),&\\ (\forall s_0,\ \ \Omega.tr\ \ (f\ \ s_0)\ \ =\ \ l_F(f)\ \ (S.tr\ \ s_0)\ \ )\ \to&\\ (\forall s_0,\ \ \Omega.tr\ \ (g\ \ s_0)\ \ =\ \ l_F(g)\ \ (S.tr\ \ s_0)\ \ )\ \to\ \forall s,\ \ f\ \ s\ \cong\ g\ \ s.&\end{aligned}$$

Finally, we need another requirement that is needed when proving commutativity of diagrams up to bisimilarity (cf. Section 6).

$$\begin{aligned}l_F^{\cong}\_\mathrm{ext}\colon \forall X\ \ (f\ \ g: X\to\Omega.st)\ \ (y\colon F\ \ X),\ \ (\forall x,\ \ f(x)\ \cong\ g(x))\ \to&\\ l_{Rel(F)}\big(\Omega,\Omega,\cong,\ \ l_F(f)(y),\ \ l_F(g)(y)\big).&\end{aligned}$$

Note that here we take as argument an arbitrary set X which does not need to have a coalgebraic structure. It allows us to use this property in more general situations, e.g. in next section we use this on a bi-algebraic structure.

So far we have always used the (intensional)[6] propositional equality to use the commutativity of diagrams. However it is well-known that for weakly final coalgebras the natural equality is the bisimilarity [12] which can be used for proofs based on *coinduction principle*. The coinduction principle states that maximal bisimulation is the equality. In **CIC** this may be stated as 'the maximal bisimulation on weakly final coalgebra is propositional equality', but it is not provable. I.e., for concrete functors we cannot prove that $\cong$ and $=$ coincide. But given Theorem 1 we know that any weakly final coalgebra can be turned into a setoid with a corresponding coinduction proof principle. And thus, finding bisimulation will result in equality in that setoid. This enables us to translate and verify in **CIC** the proofs by coinduction principle.

6 λ-Coiteration Scheme

Our theory so far has the *coiteration scheme* which is the existence of the arrow in $F\_\text{wfin?}$. The scheme of λ-coiteration was developed in order to extend the class of *Haskell*-like specifications beyond coiteration [4]. Our purpose is to develop the λ-coiteration scheme inside **CIC** in the theory of previous sections.

First we sketch the scheme given in [4]. Let B, T be two extensional functors and Ω be a weakly final B-coalgebra. Let $\Lambda\colon TB \Longrightarrow BT$ be a natural transformation. Given a map $g\colon X \longrightarrow BTX$ if the diagram below commutes then f is called λ-*coiterative arrow induced by* g.

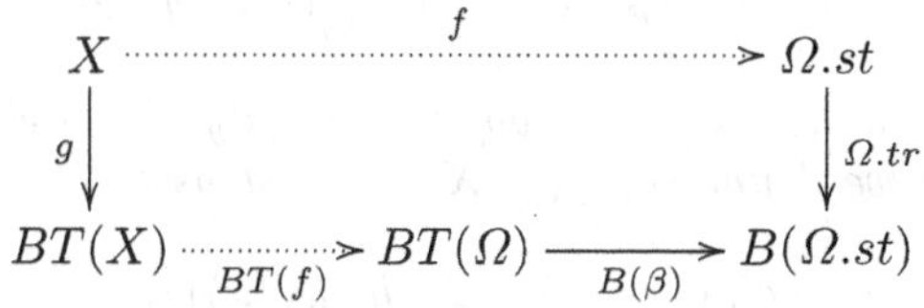

$$
\begin{array}{ccc}
X & \xrightarrow{\quad\quad f \quad\quad} & \Omega.st \\[2pt]
{\scriptstyle g}\big\downarrow & & \big\downarrow{\scriptstyle \Omega.tr} \\[2pt]
BT(X) & \underset{BT(f)}{\dashrightarrow} BT(\Omega) \xrightarrow{\; B(\beta) \;} & B(\Omega.st)
\end{array}
$$

Here $\beta := \pi_1(\phi \;\; S_0)$ where ϕ is a proof of $F\_\text{wfin?}(\Omega)$ and S_0 the coalgebra with carrier $T(\Omega)$ and transition function $\Lambda_\Omega \circ T(\Omega.tr)\colon T(\Omega) \longrightarrow BT(\Omega)$.

In [4] it is proven that if the ambient category possesses countable coproducts then given g, Λ a unique λ-coiterative arrow exists. In **CIC** a countable coproduct is an $\mathbb{N}$-indexed family of sets and always exists (see T^* below), and thus we can prove the existence of λ-coiterative arrow for B and T without further assumptions.

Our proof follows the one in [4] with some simple modifications with respect to equality. For presenting the λ-coiterative arrow we need to formalise several structures of [4] in **CIC**. The translation of these structures is straightforward. Let $T^* := \lambda X.\{\Sigma j\colon \mathbb{N}, T^j(X)\}$ where T^j is the recursively defined j-th iteration of T. We can prove that T^j (by induction) and T^* are extensional Set-functors.[7]

---

[6] Propositional equality of **CIC** in the empty context is indistinguishable from the intensional equality of the conversion rules of the type theory [2].

[7] Extensionality of T^* is *not* needed. One of the referees suggested simplifying the definition of T^* in the original manuscript to this extensional functor.

For each j and any set Y with $y \in T^j(Y)$ let

$$\imath_{jY} : T^j(Y) \longrightarrow T^*(Y)$$
$$\imath_{jY}(y) := \langle j, y \rangle \ .$$

Furthermore, for a sets Y, Z and $\mathbb{N}$-indexed family of functions $f_j : T^j Y \longrightarrow Z$ let $[f_j]_0^\infty : T^*(Y) \longrightarrow Z$ be

$$[fj]_0^\infty := \lambda x. f_{\pi_1(x)}(\pi_2(x)) \ .$$

Next let $\chi_X := [\imath_{(j+1)X}]_0^\infty$. We define the iteration of Λ recursively as:

$$\Lambda_X^0 = \lambda x.x$$
$$\Lambda_X^{j+1} = \lambda x \colon T^j(TB(X)).\Lambda_{T(X)}(l_{T^j} \quad \Lambda_X \quad x) \ .$$

Finally let $\Lambda_X^* : T^*B(X) \longrightarrow BT^*(X)$ be

$$\Lambda_X^* := [\lambda x \colon T^j(X).\, l_B \quad \imath_{jX} \quad (\Lambda_X^j(x))]_0^\infty \ .$$

Now we can define the function making the above diagram commute.

Definition 1. *Given Λ and g as above let S_1 be the coalgebra with carrier $T^*(X)$ and the transition function*

$$\lambda x \colon T^*(X).\, l_B \quad \chi_X \quad \left(\Lambda_{T(X)}^*(l_{T^*} \quad g \quad x)\right) \ .$$

Let $h := \pi_1(\phi \quad S_1)$ be the map given by weak finality of Ω (where ϕ is a proof of $F\_wfin?(\Omega)$). Then we define $coit_{\Lambda X g} \colon X \longrightarrow \Omega.st$ as

$$coit_{\Lambda X g} := \lambda x \colon X.\, h\big(\imath_{0X}(x)\big) \ .$$

In our setting we should state the commutativity using bisimilarity.

Theorem 2. *The map $coit_{\Lambda X g}$ is the* λ-coiterative arrow induced by *g up to bisimilarity, i.e., for all $x \colon X$*

$$l_{Rel(F)}\Big(\Omega, \Omega, \cong, \Omega.tr\big(coit_{\Lambda X g}(x)\big), l_B \quad \beta \quad (l_B \quad (l_T(coit_{\Lambda X g})) \quad (g(x)))\Big) \ .$$

The proof of Theorem 2 is quite technical and can be found in the *Coq* formalisation [19]. It follows to a great extent the paper proof in [4]. However working in **CIC** results in some minor differences. As we mentioned above in **CIC** we have T^* for free, on the other hand we must explicitly assume that the functor B is extensional and satisfies $l_B^{\cong}$_ext. Another technical difference is that for each j we need a map $!_{jX} \colon T^j(T(X)) \longrightarrow T(T^j(X))$ recursively defined as

$$!_{0X} = \lambda x.x$$
$$!_{(j+1)X} = !_{jT(X)} \ .$$

The role of this map is to replace the reasoning steps that rely on the intensional equality $T^j(T(X)) = T(T^j(X))$. This is because although this equality is provable in **CIC** as a propositional equality, the two sides when considered as types are not convertible.[8] Such non-convertibility would be an obstacle in proving the commutativity of diagrams by naturality laws, which are otherwise automatically proven by the conversion mechanism of *Coq*. Our use of $!_{jX}$ is a workaround that, although making proofs more tedious, works suitably.

7 Streams

In this section we show that the theory of Sections 3–6 can be instantiated by the important case of streams, and hence the requirements that we put on functors are reasonable. Note that in those sections we did not use coinductive types, while in this section we will use the coinductive types of *Coq*.

Fix a set B. Already from Lemma 1 we know that the stream functor defined as $F(X) := B\times X$ with $l_{B\times}(f)\langle b, x\rangle = \langle b, f(x)\rangle$ is an extensional functor. This allows us to build the coalgebra $B\times$`_coalg` of the functor above with the obvious components of the transition map:

$$\mathrm{hd}_S\colon S.st \to B := \lambda s.\pi_1(S.tr(s)) \ , \qquad \mathrm{tl}_S\colon S.st \to S.st := \lambda s.\pi_2(S.tr(s)).$$

Now we need a maximal bisimulation between any two $B\times$-coalgebras. This will be a coinductive type defined as:

`CoInductive` $max^{S_1S_2}_{B\times\_bisim}$ $(s_1 : S_1.st)$ $(s_2 : S_2.st)$: Prop :=

| $max_{B\times\_bisim0}$: $\mathrm{hd}_{S_1}(s_1) = \mathrm{hd}_{S_2}(s_2) \rightarrow max^{S_1S_2}_{B\times\_bisim}\ \mathrm{tl}(s_1)\ \mathrm{tl}(s_2) \rightarrow$

$max^{S_1S_2}_{B\times\_bisim}\ s_1\ s_2.$

Subsequently we can prove this lemma:

Lemma 3. *Let S_1, S_2 be two $B\times$-coalgebras. Then*

$$B\times\_max\_bisim?(S_1, S_2, max^{S_1S_2}_{B\times\_bisim})$$

Proof. The proof has two parts. First to prove that $max^{S_1S_2}_{B\times\_bisim}$ is a bisimulation take

$$\gamma := \lambda x.\langle \mathrm{hd}_{S_1}(\pi_1(x)), \langle \mathrm{tl}_{S_1}(\pi_1(x)), \mathrm{tl}_{S_2}(\pi_2(x)), \phi\rangle\rangle$$

where ϕ is a proof that

$$\langle \mathrm{hd}_{S_1}(\pi_1(x)), \mathrm{tl}_{S_1}(\pi_1(x))\rangle = S_1.st(\pi_1(x)) \ ,$$
$$\langle \mathrm{hd}_{S_1}(\pi_1(x)), \mathrm{tl}_{S_2}(\pi_2(x))\rangle = S_2.st(\pi_2(x)) \ .$$

In the second part for each ρ satisfying $F\_\sigma bisim?(\rho)$ and each s_1, s_2 for which the set $\rho s_1 s_2$ is inhabited, we ought to build an element of the coinductive type

[8] Obviously there are two possible ways to define T^j. No matter which of the two ways we take we will always need $!_{jX}$ or its inverse.

$max^{S_1 S_2}_{B\times\_bisim}(s_1, s_2)$. That means we employ the constructor $max_{B\times\_bisim0}$ and use the facts that $\mathrm{hd}_{S_1}(s_1) = \mathrm{hd}_{S_2}(s_2)$ and $max^{S_1 S_2}_{B\times\_bisim}(\mathrm{tl}(s_1), \mathrm{tl}(s_2))$. Both of these are provable using the commutativity of bisimulation for ρ. The latter also uses the fact that

$$\rho \ \ \mathrm{tl}_1(s_1) \ \ \mathrm{tl}_2(s_2) \ \neq \ \emptyset \ . \qquad \square$$

Next we define the map $i_{wpB\times}$ as follows (again ϕ's are proof obligations).

$$i_{wpB\times}\langle\langle b_0, s_0\rangle, \langle b_1, s_1\rangle, \phi_{01}\rangle := \langle b_0, \langle s_0, s_1, \phi_{\mathrm{refl}}\rangle\rangle \ .$$

With this definition we can prove $\mathsf{WP}_{\overrightarrow{B\times}}$ and hence the ingredients of the bisimulation theory are all supplied. This means that we get Theorem 1 for free.

At this point we focus on weakly final coalgebra of streams. Consider the coinductively defined set B^ω of streams over B introduced in Section 2. Taking $\nu_{B\times} := \langle B^\omega, \langle \mathrm{hd}_{B^\omega}, \mathrm{tl}_{B^\omega}\rangle\rangle$ to be the coalgebra of streams, it is easy to prove that $B\times\_\mathrm{wfin?}(\nu_{B\times})$ holds: the witness is the *unfold* map for streams which is easily defined using the cofix operator of *Coq*:

$\texttt{CoFixpoint unfld}_{B\times}$ $(S_1\colon B\times \texttt{\_coalg})$ $(s_1\colon S_1.st)\colon$ $B^\omega :=$
$\qquad\qquad\qquad\qquad \texttt{Cons}\ \mathrm{hd}_{S_1}(s_1)\ \big(\texttt{unfld}_{B\times}\ \ S_1\ \ \mathrm{tl}_{S_1}(s_1)\big).$

Proving the uniqueness $\nu_{B\times_{\mathrm{unique}}}$ needs the following lemma.

Lemma 4. *Let be a $B\times$-coalgebra. Then*

i) $\texttt{unfld}_{B\times}\ \ S\ \ s = \texttt{Cons}\ \ \mathrm{hd}_S(s)\ \ \big(\texttt{unfld}_{B\times}\ \ S\ \ \mathrm{tl}_S(s)\big)$.
ii) Let $f\colon S.st \longrightarrow B^\omega$ *be such that*

$$\forall s\colon S.st, f(s) = \texttt{Cons}\ \ \mathrm{hd}_S(s)\ \ f\big(\mathrm{tl}_S(s)\big) \ .$$

Then for all s in S we have

$$\texttt{unfld}_{B\times}\ \ S\ \ s\ \cong\ f(s) \ .$$

Part (i) is trivial (see definition of $\texttt{unfld}_{B\times}$), while part (ii) uses constructor of the coinductive type $max^{\nu_{B\times}\nu_{B\times}}_{B\times\_bisim}$ and the cofix operator of *Coq* to build the bisimilarity [19].

Finally the proof of $l^{\cong}_{B\times}\_\mathrm{ext}$ is a routine use of properties of $\cong$ as an equivalence relation.

Then we can apply the scheme developed in the previous section to define streams and functions on streams. We illustrate this by some examples. For each example we mention which parameters for the λ-coiteration scheme should be taken. All the choices for functor T in these examples are extensional by Lemma 1. Some of the examples require B to be a semi-ring. Each example contains a *Haskell*-like specification; applying Theorem 2 and replacing the definition of $l_{Rel(F)}$ enables us to derive the specifications as a bisimilarity.

Example 1. For *shuffle product* defined in Section 2 choose:

$$\begin{aligned}
T &:= \lambda X.X \times X \\
\Lambda_X &:= \lambda x.\langle \pi_1(x) + \pi_3(x), \langle \pi_2(x), \pi_4(x)\rangle\rangle \\
g &:= \lambda x\colon B^\omega \times B^\omega.\langle \mathrm{hd}_{B^\omega}(\pi_1(x)) \cdot \mathrm{hd}_{B^\omega}(\pi_2(x)), \\
&\qquad \langle \mathrm{tl}_{B^\omega}(\pi_1(x)), \pi_2(x), \pi_1(x), \mathrm{tl}_{B^\omega}(\pi_2(x))\rangle\rangle \ .
\end{aligned}$$

Then given two streams xs, ys we can define $xs \otimes ys$ as $\mathrm{coit}_{\Lambda X g}\langle xs, ys\rangle$. In this case Theorem 2 leads to the following bisimilarity for $\otimes$ which is the counterpart of (2.1) in the intensional setting of *Coq*.

$$xs \otimes ys \cong \mathtt{Cons}\ \ (\mathrm{hd}_{B^\omega}(xs) \cdot \mathrm{hd}_{B^\omega}(ys))\ \ (\ \mathrm{tl}_{B^\omega}(xs) \otimes ys\ \ \oplus\ \ xs \otimes \mathrm{tl}_{B^\omega}(ys)\ )$$

Here $xs \oplus ys := \beta\langle xs, ys\rangle$ can also be proven, by Lemma 4.(i), to satisfy

$$xs \oplus ys\ =\ \mathtt{Cons}\ \ (\mathrm{hd}_{B^\omega}(xs) + \mathrm{hd}_{B^\omega}(ys))\ \ (\ \mathrm{tl}_{B^\omega}(xs)\ \ \oplus\ \ \mathrm{tl}_{B^\omega}(ys)\ )$$

Note that for $\oplus$ we get an equality, which by Lemma 2.(ii) leads to a bisimilarity. □

Example 2. The *ordinary convolution product* that is used for computing the product of formal power series is definable as

$$(x :: xs) \times (y :: ys) :=\ \ x \cdot y\ \ ::\ \ (xs \times (y :: ys))\ \oplus\ ((x :: \bar{0}) \times ys)$$

with $\bar{0}$ the constant zero stream [21]. For the λ-coiteration scheme this a trivial variant of the shuffle product; with T, Λ as in Example 1 we redefine g:

$$\begin{aligned}
g := \lambda x\colon B^\omega \times B^\omega.&\langle \mathrm{hd}_{B^\omega}(\pi_1(x)) \cdot \mathrm{hd}_{B^\omega}(\pi_2(x)), \\
&\langle \mathrm{tl}_{B^\omega}(\pi_1(x)), \pi_2(x), \mathtt{Cons}\ \ \mathrm{hd}_{B^\omega}(\pi_1(x))\ \ \bar{0}, \mathrm{tl}_{B^\omega}(\pi_2(x))\rangle\rangle \ .
\end{aligned}$$ □

Example 3. The stream of natural numbers with the specification

$$\mathtt{nats} := 0 :: \mathit{map}\ \ \lambda n.n{+}1\ \ \mathtt{nats}$$

is a well-known example of a stream definition not accepted by the guardedness condition of *Coq* [11]. This is definable in *Coq* by taking $B := \mathbb{N}$ and

$$\begin{aligned}
T &:= \lambda X.X \times \mathbb{N}^{\mathbb{N}} \times X \\
\Lambda_X &:= \lambda x.\langle \pi_1(x)(\pi_2(x)), \langle \pi_1(x), \pi_3(x)\rangle\rangle \\
g &:= \lambda x{:}\mathbf{1}.\langle 0, \langle \lambda n.n{+}1, *\rangle\rangle \\
\mathtt{nats} &:= \mathrm{coit}_{\Lambda X g}(*) \ .
\end{aligned}$$ □

Example 4. The Fibonacci specification studied in [6] can also be defined using the λ-coiteration scheme but the specification should be slightly unwound as

$$\mathtt{fibs} := 0 :: \ \oplus_3 (1, \mathtt{fibs}, \mathtt{fibs}) \tag{7.1}$$

where $\oplus_3$ is a ternary unwinding of $\oplus$:

$$\oplus_3(x_0,\ x :: xs,\ y :: ys) := x_0 + y ::\ \oplus_3\ (x, xs, ys)$$

We define this by putting a coalgebraic structure on the unit set $\mathbf{1} = \{*\}$:

$$\begin{aligned}
&T := \lambda X.B \times X \times X \\
&\Lambda_X := \lambda x.\langle \pi_1(x) + \pi_4(x), \langle \pi_2(x), \pi_3(x), \pi_5(x)\rangle\rangle \\
&g := \lambda x{:}\mathbf{1}.\langle 0, \langle 1, *, *\rangle\rangle \\
&\texttt{fibs} := \mathrm{coit}_{\Lambda X g}(*)
\end{aligned}$$

Again Theorem 2 gives us (7.1) up to bisimilarity. Furthermore we can prove the following bisimilarity which corresponds to the specification used in [6] as a definition of stream of Fibonacci numbers.

$$\texttt{fibs} \cong \texttt{Cons}\ \ 0\ \ \Big(\texttt{Cons}\ \ 1\ \ (\ \mathrm{tl}_{B^\omega}(\texttt{fibs}) \oplus \texttt{fibs}\ )\Big)\ .$$

Note that we are using $\oplus$ from Example 1. The proof of this bisimilarity is based on the following properties of $\oplus$ and $\oplus_3$.

$$\begin{aligned}
\oplus_3\ (x, xs, ys) &\cong (\texttt{Cons}\ \ x\ \ xs)\ \oplus\ ys\ ; \\
xs \oplus ys &\cong ys \oplus xs\ .
\end{aligned}$$

We can prove both bisimilarities in two different ways [19], either by using cofix to build an inhabitant of the coinductive type $max^{\nu_{B\times}\nu_{B\times}}_{B\times\_bisim}$, or by explicitly providing the bisimulation relations and using Lemma 3. In the latter case the two bisimulation relations are given respectively by:

$$\begin{aligned}
R_1\sigma\tau &:= \exists x\ xs\ ys,\ \ \sigma = \oplus_3(x, xs, ys)\ \ \wedge\ \ \tau = (\texttt{Cons}\ \ x\ \ xs) \oplus ys\ ; \\
R_2\sigma\tau &:= \exists xs\ ys,\ \ \sigma = xs \oplus ys\ \ \wedge\ \ \tau = ys \oplus xs\ . \qquad \square
\end{aligned}$$

As seen in these examples Theorem 2 provides the bisimilarity equation to recover the specification that was used to forge the parameters T, Λ and g. In general if we want to prove a bisimilarity in *Coq* we have several additional tools:

(i) using the properties of bisimilarities as an equivalence relations and perform equational reasoning;
(ii) using 'type theoretic coinduction', i.e., using cofix and the constructors of coinductive type of $max^{\nu_{B\times}\nu_{B\times}}_{B\times\_bisim}$;
(iii) using the conventional coinduction principle and explicitly providing a bisimulation relation between the two sides of the bisimilarity.

We usually apply a combination of the above techniques, but each has characteristics that make it suitable in specific contexts. For example (i) is especially useful when dealing with bisimilarity as a setoid equality, and in combination with other reasoning tools for setoids. Technique (ii) seems to be more suitable for mechanisation as it follows the shape of specifications and leads to smaller

Coq proof scripts while (iii) is usually more verbose. On the other hand applying (ii) entails that one has to be wary of the guardedness condition as one is using cofix operator of *Coq*, while in (iii) no guardedness check is performed.

As we see the λ-coiteration scheme considerably extends the class of functions definable in *Coq*, giving their behavioural equations for free. However, like all syntactic schemes, there is limitation to this scheme, e.g. in [3] it is shown that a specification for the stream of *Hamming* numbers is not accepted by this scheme.

8 Conclusions and Further Work

We have provided a modular theory of coalgebras in the intensional setting of **CIC** which can be instantiated for specific functors built out of finite sums and products. Each instantiation will give us a theory of bisimilarity which can then be used to build a setoid and work extensionally. Furthermore we showed the usefulness of our coalgebraic setting by developing the λ-coiteration scheme in it and thus extending the class of productive specifications definable in *Coq*. We demonstrated this by an instantiation of our theory for streams and showed some concrete specifications refused by the guardedness condition but accepted using the λ-coiteration scheme in *Coq*.

Our work eases future coalgebraic developments in *Coq*. It is a good evidence that once some technicalities with respect to dependent types are handled most categorical schemes can be translated into intensional type theory. On the other hand it shows that the schemes from category theory can provide suitable workarounds for the restrictions of the guardedness condition without changing the underlying type theory.

The future work would be to build a larger library of results on weakly final coalgebras and developing more powerful definition schemes. Immediate would be the schemes obtained by adding monadic, pointed or cofree structure on the bi-algebraic nature of λ-coiteration [4, 7]. The long-term challenge would be to extend the formalisation to the powerful functor of Hancock–Setzer [12, 18] and investigating the various schemes there.

Acknowledgements. The author wishes to thank Ralph Matthes, Jan Rutten, Christian Koehler and the anonymous referees for their helpful comments, and pointers to the literature.

References

[1] Abel, A.: Semi-continuous sized types and termination. Logic. Methods in Comput. Sci. 4(2:3), 1–33 (2008)
[2] Altenkirch, T., McBride, C., Swierstra, W.: Observational equality, now! In: Stump, A., Xi, H. (eds.) Proc. of PLPV 2007, pp. 57–68. ACM Press, New York (2007)
[3] Bartels, F.: Generalised coinduction. In: Corradini, A., Lenisa, M., Montanari, U. (eds.) Proc. of CMCS 2001. ENTCS, vol. 44(1), pp. 67–87. Elsevier, Amsterdam (2001)

[4] Bartels, F.: On Generalised Coinduction and Probabilistic specification Formats: Distributive laws in coalgebraic modelling. PhD thesis, Vrije Universiteit Amsterdam (2004)
[5] Bertot, Y., Komendantskaya, E.: Inductive and coinductive components of corecursive functions in coq. In: Adámek, J., Kupke, C. (eds.) Proc. of CMCS 2008. ENTCS, vol. 203(5), pp. 25–47. Elsevier, Amsterdam (2008)
[6] Bertot, Y., Komendantskaya, E.: Using structural recursion for corecursion. Technical report, INRIA (September 2008), `http://hal.inria.fr/inria-00322331/` (cited 18 Febrary 2009)
[7] Cancila, D., Honsell, F., Lenisa, M.: Generalized coiteration schemata. In: Gumm, P. (ed.) Proc. of CMCS 2003. ENTCS, vol. 82(1), pp. 76–93. Elsevier, Amsterdam (2003)
[8] The Coq Development Team. *Reference Manual, Version 8.2.* LogiCal Project (July 2006), `http://coq.inria.fr/V8.2/doc/html/refman/` (cited 18 February 2009)
[9] Coquand, T.: Infinite objects in type theory. In: Barendregt, H., Nipkow, T. (eds.) TYPES 1993. LNCS, vol. 806, pp. 62–78. Springer, Heidelberg (1994)
[10] Di Gianantonio, P., Miculan, M.: A unifying approach to recursive and corecursive definitions. In: Geuvers, H., Wiedijk, F. (eds.) TYPES 2002. LNCS, vol. 2646, pp. 148–161. Springer, Heidelberg (2003)
[11] Giménez, E.: Un Calcul de Constructions Infinies et son Application a la Verification des Systemes Communicants. PhD thesis PhD 96-11, Laboratoire de l'Informatique du Parallélisme, Ecole Normale Supérieure de Lyon (December 1996)
[12] Hancock, P., Setzer, A.: Interactive programs and weakly final coalgebras in dependent type theory. In: Crosilla, L., Schuster, P. (eds.) From Sets and Types to Topology and Analysis. Towards Practicable Foundations for Constructive Mathematics. Oxford Logic Guides, vol. 48, pp. 115–134. Oxford University Press, Oxford (2005)
[13] Hofmann, M.: Conservativity of equality reflection over intensional type theory. In: Berardi, S., Coppo, M. (eds.) TYPES 1995. LNCS, vol. 1158, pp. 153–164. Springer, Heidelberg (1996)
[14] Huet, G., Saïbi, A.: Constructive category theory. In: Plotkin, G.D., Stirling, C., Tofte, M. (eds.) Proof, Language, and Interaction, Essays in Honour of Robin Milner, pp. 239–276. MIT Press, Cambridge (2000)
[15] Matthes, R.: An induction principle for nested datatypes in intensional type theory. J. Funct. Programming, 29 page (to appear, 2009), `http://www.irit.fr/~Ralph.Matthes/papers/VNestITTfinal.pdf` (cited 18 February 2009)
[16] McBride, C.: Dependently Typed Programs and their Proofs. PhD thesis, University of Edinburgh (1999)
[17] McIlroy, M.D.: The music of streams. Inform. Process. Lett. 77(2–4), 189–195 (2001)
[18] Michelbrink, M.: Interfaces as functors, programs as coalgebras - a final coalgebra theorem in intensional type theory. Theoret. Comput. Sci. 360(1–3), 415–439 (2006)
[19] Niqui, M.: Files for Coq v. 8.2 (October 2008), `http://coq.inria.fr/contribs/Coalgebras` (cited 18 February 2009)
[20] Rutten, J.J.M.M.: Universal coalgebra: a theory of systems. Theoret. Comput. Sci. 249(1), 3–80 (2000)
[21] Rutten, J.J.M.M.: A coinductive calculus of streams. Math. Structures Comput. Sci. 15(1), 93–147 (2005)

A Process-Model for Linear Programs*

Luca Paolini[1] and Mauro Piccolo[1,2]

[1] Dipartimento di Informatica, Università di Torino, Italia
[2] Preuves, Programmes et Systèmes, Paris VII, France

Abstract. We use ℓinProc (i.e. a typed process calculus based on the calculus of solos) in order to express computational processes generated by $\mathcal{S}\ell$PCF$^-$, namely a simple programming language conceived in order to program only linear functions. We define a faithful translation of $\mathcal{S}\ell$PCF$^-$ on ℓinProc which enables us to process redexes of $\mathcal{S}\ell$PCF$^-$ in a parallel way. Afterward, we prove that a suitable observational equivalence between processes is correct w.r.t the operational semantics of $\mathcal{S}\ell$PCF$^-$, via our interpretation.

1 Introduction

Harold Abelson, Gerald Jay and Julie Sussman, in their famous book "Structure and Interpretation of Computer Programs" [1, Ch.1] state:

> "We are about to study the idea of a computational process. Computational processes are abstract beings that inhabit computers. As they evolve, processes manipulate other abstract things called data. The evolution of a process is directed by a pattern of rules called a program. People create programs to direct processes. In effect, we conjure the spirits of the computer with our spells."

In the first chapter, they introduce a programming language and a simple way to describe the *dynamical becoming* of the evaluation of applications of programs to inputs. The dynamic of process evolution is represented by sequences of programs related by means of rewriting rules. Fingerprints of λ-calculus pervade the book and, indeed, we are interested in a model of processes conjured by a typed λ-calculus. Unfortunately, λ-calculus lacks a satisfactory description of *interaction* between processes cooperating, competing and synchronizing between themselves. To overcome such limitations, many calculi have been proposed which focus on dynamical aspects of computation with particular regards for interaction, see for instance [8, 15, 27]. Such calculi are more intensional than λ-calculus, which instead focuses on functional aspects of computation. Although a main motivation for the comparison between lambda and process calculi has been to study the expressiveness of process calculi, another worthy motivation is to study theories induced by equivalences in the world of processes conjured

* Paper partially supported by MIUR-Cofin'07 CONCERTO Project.

S. Berardi, F. Damiani, and U. de'Liguoro (Eds.): TYPES 2008, LNCS 5497, pp. 289–305, 2009.

by programs, see [16, 17, 25]. The main results presented in this paper are set in the latter research-line.

The interaction is the key idea behind the introduction of game semantics for programming language [2, 11], more precisely interaction between a program and the environment (where the program itself is intended to be executed). A seminal work exploring correspondences between process-calculi and game semantics has been presented in [10], where Hyland-Ong strategies are represented by processes of an appropriately sorted polyadic π-calculus. More recently, such result has been improved by introducing an elaborate type discipline for the π-calculus in [4], where the use of linear modalities [12, 13] is crucial. A game-independent process-based language for game strategies has been formalized, where strategies are normalized processes. Programs of PCF can be directly interpreted on such processes in a fully abstract way.

Our purpose is to deepen and advance such explorations by proposing a process-model, namely a syntactical model, built on a suitable process-calculus, inducing a corresponding semantics for a typed λ-calculus..

We are convinced that a key aspect of such explorations is linearity in many respects, moreover linearity make analysis simpler and clearer. Accordingly, we tackle the construction of a process-model for $\mathbb{S}\ell\mathrm{PCF}^-$, namely a simple programming language conceived in order to program "linear" functions between coherence spaces. The least full sub-category of coherence spaces, including the infinite flat domain (representing natural numbers) and the coherence spaces representing linear functions between domains in the model itself (by avoiding the use of exponential domain constructors) forms a fully abstract model for such programming language [20]. In order to build a process model for $\mathbb{S}\ell\mathrm{PCF}^-$, we introduce ℓinProc, namely a process calculus based on a typed calculus of Solos [14]. The calculus of Solos is a modification of the asynchronous π-calculus where explicit causal dependency is forbidden by avoiding prefixes and binding guards. We give an encoding of $\mathbb{S}\ell\mathrm{PCF}^-$ in ℓinProc which respects the operational equivalence between programs, i.e. it does not equate operationally different programs.

We note that the semantic nature of strategies in [4, 10] is explicitly reflected by the use of an infinitary syntax for the π-processes, on which no parallel reductions can be performed. Since we want to study processes induced by programs, our interpretation is actually given on finite processes. Moreover, we introduce some simplifications in the linear typing discipline. Hence, we break with the classical game semantic approaches, for another classical computer science approach: translation of programs (finite terms) of a language in finite terms of another language.

A translation of a calculus into another calculus is *faithful* whenever a reduction in the source calculus can be mimicked by some reductions in the target calculus. The encodings given in [4, 10] impose deterministic reduction strategies, so they are not faithful. This means that, there are programs $\mathtt{M}, \mathtt{N}$ translated, respectively, by processes $\mathtt{P}, \mathtt{Q}$ such that $\mathtt{M} \rightarrow_\beta \mathtt{N}$, but $\mathtt{P}$ cannot be process-reduced to $\mathtt{Q}$ (although $\mathtt{P}$ and $\mathtt{Q}$ are observationally equivalent). ℓinProc enables us to simulate the reduction of all redexes of $\mathcal{S}\ell$PCF$^-$, giving us a faithful encoding. Thus, no evaluation strategy is determined in advance and reduction can be actually done in an asynchronous way[1]. Questions on which λ-calculus reduction strategies can be encoded in process calculi has been posed in [17, Sect.8]. Moreover, a minimal requirement of λ-models is to induce a congruence equivalence which contains the β-equivalence, see [24]. This latter requirement is trivially induced by faithful embedding of λ-calculus. Also, we note that in [27, (p.467)] and [28] in order to faithful mimic the β-reduction, the usual reduction-rules for replication and prefixes of process calculi are extended.

There are several motivations behind this paper. From a programming language point of view we provide a tool for study both parallel evaluation strategies and equivalence of programs. From a process calculus point of view we give a fine representation of a sequential language where the redexes can actually be reduced in parallel. From a game-semantics point of view, we suggest a parallel description of strategies by avoiding useless causality.

The results of this paper are the starting points for many further developments. We are characterizing the relevant contexts of ℓinProc (i.e. contexts that are able to separate processes corresponding to different programs) in order to tackle the full abstraction of our syntactical model. We are working on a characterization of processes corresponding to the interpretation of programs, by adapting the proof-nets correctness criterion. We are already able to extend our results on the pair $\mathcal{S}\ell$PCF and ℓinProc at the price of some additional technicalities due to pairing-projections codifications. However, we plan to explore process-languages inducing similar results for language more complex languages such as PCF [23] or StPCF [19]. We plan to interpret processes directly on linear coherence spaces, following techniques developed for proof-nets and proof-structures. We plan to define a new kind of game semantics with a more flexible structure, where useless sequentialization is relaxed. Moreover, we want to explore the application of Levy's optimality theory to the evaluation of programs inside ℓinProc. In particular, to tackle the relations between that theory with the notion of operational linearity.

Outline of the Paper. In Section 2, we recall $\mathcal{S}\ell$PCF$^-$ i.e. a linear programming language. In Section 3, we present ℓinProc, i.e. a process language based on the calculus of Solos. In Section 4, we formalize a faithful translation of $\mathcal{S}\ell$PCF$^-$ into ℓinProc. In Section 5, we prove the correctness of the obtained process-model.

2 A Semantically Linear λ-Calculus

$\mathcal{S}\ell$PCF$^-$ is the fragment of the language $\mathcal{S}\ell$PCF presented in [20] avoiding the use of `which?`. We remark that $\mathcal{S}\ell$PCF$^-$ is a Turing-complete syntactical restriction

[1] Gordon Plotkin in [22] remarked that the call-by-value parameter passing is hardly in accord with a strategy on (call-by-name) λ-calculus, thus he introduced $\lambda\beta_v$-calculus.

Table 1. Type assignment system for $\mathcal{S}\ell$PCF$^-$

$$\frac{}{\_ \vdash \underline{0} : \iota} \qquad \frac{}{\_ \vdash \mathtt{succ} : \iota \multimap \iota} \qquad \frac{}{\_ \vdash \mathtt{pred} : \iota \multimap \iota} \qquad \frac{\mathtt{x} \in \mathrm{Var}^\iota}{\_ \vdash \mathtt{x} : \iota} \qquad \frac{}{\mathtt{f}^{\sigma \multimap \tau} \vdash \mathtt{f} : \sigma \multimap \tau}$$

$$\frac{\mathcal{F} \in \mathrm{SVar}^{\sigma \multimap \tau}}{\_ \vdash \mathcal{F} : \sigma \multimap \tau} \qquad \frac{\_ \vdash \mathtt{M} : \sigma \quad \mathcal{F} \in \mathrm{SVar}^\sigma}{\_ \vdash \mu\mathcal{F}.\mathtt{M} : \sigma} \qquad \frac{\Gamma, \mathtt{f}^\sigma \vdash \mathtt{M} : \tau}{\Gamma \vdash \lambda \mathtt{f}^\sigma.\mathtt{M} : \sigma \multimap \tau} \qquad \frac{\Gamma \vdash \mathtt{M} : \tau}{\Gamma \vdash \lambda \mathtt{x}^\iota.\mathtt{M} : \iota \multimap \tau}$$

$$\frac{\Gamma_{\mathtt{M}} \cap \Gamma_{\mathtt{N}} = \emptyset \quad \Gamma_{\mathtt{M}} \vdash \mathtt{M} : \sigma \multimap \tau \quad \Gamma_{\mathtt{N}} \vdash \mathtt{N} : \sigma}{\Gamma_{\mathtt{M}} \cup \Gamma_{\mathtt{N}} \vdash \mathtt{MN} : \tau} \qquad \frac{\Gamma_{\mathtt{M}} \cap \Gamma = \emptyset \quad \Gamma_{\mathtt{M}} \vdash \mathtt{M} : \iota \quad \Gamma \vdash \mathtt{L} : \iota \quad \Gamma \vdash \mathtt{R} : \iota}{\Gamma_{\mathtt{M}} \cup \Gamma \vdash \ell\mathtt{if}\ \mathtt{M}\ \mathtt{L}\ \mathtt{R} : \iota}$$

of PCF [23] and it is fully abstract with respect to the model of linear function between coherence spaces considered in [20] (as noted just in page 104 of [20], just before Theorem 4). Hence $\mathcal{S}\ell$PCF$^-$ is denotationally linear.

A (paradigmatic) programming language rests on a syntax together with an evaluation strategy and a notion of observables on which the evaluation eventually stops. We abuse the name of a programming language even by meaning its syntax and its related calculus.

Truth-values of $\mathcal{S}\ell$PCF$^-$ are encoded as integers (zero encodes "true" while any other numeral stands for "false"). The set $\mathbb{T}$ of *types* is defined as follows, $\sigma, \tau ::= \iota \mid (\sigma \multimap \tau)$ where ι is the only *ground* type (i.e. natural numbers) and $\sigma, \tau, ...$ are meta-variables ranging over types. As customary $\multimap$ associates to right. Hence $\sigma_1 \multimap \sigma_2 \multimap \sigma_3$ is an abbreviation for $\sigma_1 \multimap (\sigma_2 \multimap \sigma_3)$. It is easy to see that all types τ have the shape $\tau_1 \multimap ... \multimap \tau_n \multimap \iota$, for some type $\tau_1, ..., \tau_n$ where $n \geq 0$.

Let $\mathrm{Var}^\sigma, \mathrm{SVar}^\sigma$ be denumerable sets of variables of type σ. The set of *ground* variables is Var^ι, the set of *higher-order* variables is $\mathrm{HVar} = \bigcup_{\sigma,\tau \in \mathbb{T}} \mathrm{Var}^{\sigma \multimap \tau}$, and the whole set of variables is $\mathrm{Var} = \mathrm{Var}^\iota \cup \mathrm{HVar} \cup \mathrm{SVar}$. Letters $\mathtt{x}^\sigma$ range over variables in Var^σ, letters $\mathtt{y}^\iota, \mathtt{z}^\iota, \ldots$ range over variables in Var^ι, letters $\mathtt{f}^{\sigma \multimap \tau}, \mathtt{g}^{\sigma \multimap \tau}, \ldots$ range over variables in HVar, while $\mathcal{F}_0^\sigma, \mathcal{F}_1^\sigma, \mathcal{F}_2^\sigma, \ldots$ range over stable variables, namely variables in SVar^σ. Latin letters $\mathtt{M}, \mathtt{N}, \mathtt{L}, \ldots$ range over terms.

Definition 1. *Let* $\Gamma \subseteq \mathrm{HVar}$. *Typed* terms *of* $\mathcal{S}\ell$PCF *are defined by using a type assignment proving judgment of the shape* $\Gamma \vdash \mathtt{M} : \sigma$, *in Table 1.*

Note that only higher-order variables are subject to syntactical constraints. Except for the $\ell\mathtt{if}$ construction typed by an additive rule doing an implicit contraction, higher-order variables are treated linearly. Ground and stable variables belong to distinct kinds only for sake of simplicity. Their free use implies that $\mathcal{S}\ell$PCF$^-$ is not syntactically linear (in the sense of [20]).

Sometimes types will be omitted when they are clear from the context or uninteresting. Note that given types of all variables of a term $\mathtt{M}$, there is a unique σ such that $\mathtt{M}$ has type σ (sometimes denoted with $\mathtt{M}^\sigma$). Sometimes, parentheses are omitted, always by respecting the following conventions: application associates to the left and application binds more tightly than abstraction, i.e. $\lambda\mathtt{x}.\mathtt{MNL} = (\lambda\mathtt{x}.((\mathtt{MN})\mathtt{L}))$. Free variables of terms are defined as expected. A term

$\mathtt{M}$ is *closed* if and only if $\mathrm{FV}(\mathtt{M}) = \emptyset$, otherwise $\mathtt{M}$ is *open*. Terms are considered up to α-equivalence, namely a bound variable can be renamed provided no free variable is captured. Moreover, $\mathtt{M}[\underline{\mathtt{n}}/\mathtt{y}]$, $\mathtt{M}[\mathtt{N}/\mathtt{f}]$ and $\mathtt{M}[\mathtt{N}/F]$ denote the expected capture-free substitutions. We define $\underline{\mathtt{n}}$ to be $\mathtt{succ}(\cdots(\mathtt{succ}(0))\cdots)$ where $\mathtt{succ}$ is applied n-times to 0. Let $\mathcal{P} = \{\mathtt{M}^{\iota} \in \mathcal{S}\ell\mathrm{PCF} \mid \mathrm{FV}(\mathtt{M}^{\iota}) = \emptyset\}$ be the set of *programs* and let $\mathcal{N} = \{0, \ldots, \underline{\mathtt{n}}, \ldots\}$ be the set of *numerals*.

Definition 2. *We denote $\rightsquigarrow$ the firing (without any context-closure) of one of the following rules:*

$(\lambda \mathtt{f}^{\sigma \multimap \tau}.\mathtt{M})\mathtt{N} \rightsquigarrow_{\beta} \mathtt{M}[\mathtt{N}/\mathtt{f}]$	$(\lambda \mathtt{z}^{\iota}.\mathtt{M})\underline{\mathtt{n}} \rightsquigarrow_{\iota} \mathtt{M}[\underline{\mathtt{n}}/\mathtt{z}]$	$\mu F.\mathtt{M} \rightsquigarrow_{\mathrm{Y}} \mathtt{M}[\mu F.\mathtt{M}/F]$
$\mathtt{pred}\,(\mathtt{succ}\,\underline{\mathtt{n}}) \rightsquigarrow_{\delta} \underline{\mathtt{n}}$	$\ell\mathtt{if}\ \underline{0}\ \mathtt{L}\ \mathtt{R} \rightsquigarrow_{\delta} \mathtt{L}$	$\ell\mathtt{if}\ \underline{\mathtt{n+1}}\ \mathtt{L}\ \mathtt{R} \rightsquigarrow_{\delta} \mathtt{R}$

We call redex each term or sub-term having the shape of a left-hand side of rules defined above. We denote $\rightarrow_{\mathcal{S}\ell}$ the contextual closure of $\rightsquigarrow$. Moreover, we denote $\rightarrow^{*}_{\mathcal{S}\ell}$ the reflexive and transitive closure of $\rightarrow_{\mathcal{S}\ell}$. Let $\mathtt{M} \in \mathcal{P}$; we write $\mathtt{M} \Downarrow \underline{\mathtt{n}}$ when $\mathtt{M} \rightarrow^{*}_{\mathcal{S}\ell} \underline{\mathtt{n}}$ according to lazy leftmost strategy[2].

We remark that $\rightsquigarrow_{\beta}$ formalizes a call-by-name parameter passing in case of an higher-order argument. On the other hand, $\rightsquigarrow_{\iota}$ formalize a call-by-value parameter passing, namely the reduction can fire only when the argument is a numeral. As done in [5], it is easy to prove properties as subject-reduction, post-position of δ-rules in a sequence of reductions, the confluence and a standardization theorem.

Let $[\sigma]$ be a special constant of type σ. The set of σ-*contexts* Ctx_{σ} is generated by the following grammar: $\mathtt{C}[\sigma] ::= [\sigma] \mid \mathtt{x}^{\tau} \mid F^{\tau} \mid 0 \mid \mathtt{succ} \mid \mathtt{pred} \mid \ell\mathtt{if}\ \mathtt{C}[\sigma]\ \mathtt{C}[\sigma]\ \mathtt{C}[\sigma] \mid (\lambda \mathtt{x}.\mathtt{C}[\sigma]) \mid (\mathtt{C}[\sigma]\mathtt{C}[\sigma]) \mid \mu F.\mathtt{C}[\sigma]$. So, $C[\mathtt{N}^{\sigma}]$ denotes the result obtained by replacing all the occurrences of $[\sigma]$ in the context $\mathtt{C}[\sigma]$ by the term $\mathtt{N}^{\sigma}$ and by allowing the capture of free variables in $\mathtt{N}^{\sigma}$. Clearly $\mathtt{N}^{\sigma} \in \mathcal{S}\ell\mathrm{PCF}$ and $\mathtt{C}[\sigma] \in \mathrm{Ctx}_{\sigma}$ doesn't imply that $C[\mathtt{N}^{\sigma}] \in \mathcal{S}\ell\mathrm{PCF}$, because of our linear constraints.

Let $\mathtt{M}^{\sigma}, \mathtt{N}^{\sigma}$ be terms of $\mathcal{S}\ell\mathrm{PCF}^{-}$ such that $\mathtt{M} \rightsquigarrow \mathtt{N}$ and $\mathtt{C}[\mathtt{M}], \mathtt{C}[\mathtt{N}] \in \mathcal{P}$, for a context $\mathtt{C}[\sigma]$. It is easy to check that, if $\mathtt{C}[\mathtt{N}] \Downarrow \underline{\mathtt{n}}$ then $\mathtt{C}[\mathtt{M}] \Downarrow \underline{\mathtt{n}}$.

Theorem 3. $\mathtt{M} \Downarrow \underline{\mathtt{n}}$ *if and only if* $\mathtt{M} \rightarrow^{*}_{\mathcal{S}\ell} \underline{\mathtt{n}}$*, for all term* $\mathtt{M}$.

Proof. $\mathtt{M} \Downarrow \underline{\mathtt{n}}$ implies $\mathtt{M} \rightarrow^{*}_{\mathcal{S}\ell} \underline{\mathtt{n}}$ straightforwardly. The other direction follows by induction on the number of reduction steps of $\mathtt{M} \rightarrow^{*}_{\mathcal{S}\ell} \underline{\mathtt{n}}$. The case of zero steps is immediate. The inductive case follows by remark just before this Theorem. □

Now, let us define the theory induced by the operational semantics. If $\mathtt{M}^{\sigma}, \mathtt{N}^{\sigma} \in \mathcal{S}\ell\mathrm{PCF}$ then $\mathtt{M} \lessapprox_{\sigma} \mathtt{N}$ whenever $\forall \mathtt{C}[\sigma]$ s.t. $\mathtt{C}[\mathtt{M}], \mathtt{C}[\mathtt{N}] \in \mathcal{P}$, if $\mathtt{C}[\mathtt{M}] \Downarrow \underline{\mathtt{n}}$ then $\mathtt{C}[\mathtt{N}] \Downarrow \underline{\mathtt{n}}$. Moreover, $\mathtt{M} \approx_{\sigma} \mathtt{N}$ if and only if $\mathtt{M} \lessapprox_{\sigma} \mathtt{N}$ and $\mathtt{N} \lessapprox_{\sigma} \mathtt{M}$.

We put $\Omega^{\iota} = \mu F^{\iota}.F^{\iota}$ and if $\sigma_0 = \mu_1 \multimap \ldots \multimap \mu_{\mathtt{m}} \multimap \iota$, for some $m \in \mathbb{N}$, then $\Omega^{\sigma_0 \multimap \ldots \multimap \sigma_{\mathtt{n}} \multimap \iota} = \lambda \mathtt{x}_0^{\sigma_0} \ldots \mathtt{x}_{\mathtt{n}}^{\sigma_{\mathtt{n}}}.\ell\mathtt{if}(\Omega^{\sigma_1 \multimap \ldots \multimap \sigma_{\mathtt{n}} \multimap \iota}\mathtt{x}_1^{\sigma_1} \ldots \mathtt{x}_{\mathtt{n}}^{\sigma_{\mathtt{n}}})(\mathtt{x}_0 \Omega^{\mu_1} \ldots \Omega^{\mu_{\mathtt{m}}})(\mathtt{x}_0 \Omega^{\mu_1} \ldots \Omega^{\mu_{\mathtt{m}}})$.

[2] The lazy reduction strategy reduces first the leftmost redex which is not under a λ-abstraction. For instance the term $(\lambda \mathtt{x}^{\iota}.((\lambda \mathtt{z}^{\iota}.\mathtt{z})\,\underline{5}))(\mathtt{pred}\,\underline{3})$ reduces to $(\lambda \mathtt{x}^{\iota}.((\lambda \mathtt{z}^{\iota}.\mathtt{z})\,\underline{5}))\ \underline{2}$ according to such a strategy.

By using Ω^σ it is possible to define approximants of a fix-point $\mu F.\mathtt{M}^\sigma$ as follows,

$$\mu^0 F.\mathtt{M}^\sigma = \Omega^\sigma, \qquad \mu^{n+1} F.\mathtt{M}^\sigma = \mathtt{M}[\mu^n F.\mathtt{M}/F].$$

Lemma 4. *Let $\mathtt{M}_0^{\sigma_0}, ..., \mathtt{M}_m^{\sigma_m}$ be a sequence of closed terms ($m \geq 0$).*

1. $\Omega^{\sigma_0 \multimap \dots \multimap \sigma_m \multimap \iota}\mathtt{M}_0 \dots \mathtt{M}_m$ *is a program and there is no* $\underline{\mathtt{n}}$ *such that* $\Omega^{\sigma_0 \multimap \dots \multimap \sigma_m \multimap \iota}\mathtt{M}_0 \dots \mathtt{M}_m \Downarrow \underline{\mathtt{n}}$.
2. *Let* $(\mu F.\mathtt{P}^\sigma)\mathtt{M}_0 \dots \mathtt{M}_m$ *be a program.* $(\mu F.\mathtt{P}^\sigma)\mathtt{M}_0 \dots \mathtt{M}_m \Downarrow \underline{\mathtt{n}}$ *if and only if* $(\mu^{k+1} F.\mathtt{P}^\sigma)\mathtt{M}_0 \dots \mathtt{M}_m \Downarrow \underline{\mathtt{n}}$, *for some* $k \in \mathbb{N}$.

3 A Linear Process Calculus

ℓinProc is a typed process calculus, based on Solos calculus [14] and extended to treat explicit constants. In this calculus we give the possibility to communicate names as well as ground values (i.e. integers). Since, ℓinProc is conceived in order to model SℓPCF, some notations of the previous Section are overloaded.

We use two sets of names, $\mathcal{Q}$ for *question-names* and $\mathcal{A}$ for *answer-names*. Letters $\mathtt{p}, \mathtt{q}, \mathtt{u}, \mathtt{v}, \mathtt{w}, \mathtt{f}, \mathtt{g}, \dots$ range over question-names and $\mathtt{a}, \mathtt{b}, \mathtt{c}, \dots$ range over answer-names. For sake of simplicity, when useful we use $\varkappa$ to denote all kinds of names. We write $\vec{\varkappa}$ for a (possible empty) finite sequence $\varkappa_1, \dots, \varkappa_n$ of names and $|\vec{\varkappa}|$ for its length. In order to treat ground information we use a set of variables Var ranged over by $\mathtt{x}, \mathtt{y}, \mathtt{z}, \dots$. Last, we use a set $\mathcal{H}$ of *process variables* ranged over by F. ℓinProc is based on three syntactical categories.

Expressions:	$\mathtt{E} ::= \mathtt{n} \mid \mathtt{x} \mid \mathtt{succ(E)} \mid \mathtt{pred(E)}$
Solos	$\mathfrak{s} ::= \mathtt{q}\langle\vec{\mathtt{p}}, \mathtt{a}\rangle \mid \overline{\mathtt{q}}\langle\vec{\mathtt{p}}, \mathtt{a}\rangle$
Processes	$\mathtt{P} ::= \mathfrak{s} \mid \mathtt{a(x)}.(\overline{\mathtt{q}}\langle\vec{\mathtt{p}}, \mathtt{b}\rangle; \mathtt{P}) \mid \overline{\mathtt{a}}\langle\mathtt{E}\rangle \mid \mathbf{0} \mid \mathtt{P}\Vert\mathtt{Q} \mid (\nu\varkappa^{\mathtt{S}})\mathtt{P} \mid \lfloor\mathtt{P} \overset{\mathtt{E}}{\oplus} \mathtt{Q}\rfloor \mid F^{\mathtt{q}} \mid \mathtt{rec}F.\mathtt{P}^{\ddagger}$

‡ We assume that P contains exactly one free question-name, as a constraint on $\mathtt{rec}F.\mathtt{P}$ construction.

The Solos calculus has been introduced in [14] as a simplification of the asynchronous π-calculus replacing prefixes by solos (not binding actions) and maintaining the restriction as unique binder[3]. The name $\mathtt{q}$ is the *subject* and $\vec{\mathtt{p}}, \mathtt{a}$ are the *objects* of both solos $\mathtt{q}\langle\vec{\mathtt{p}}, \mathtt{a}\rangle$ and $\overline{\mathtt{q}}\langle\vec{\mathtt{p}}, \mathtt{a}\rangle$. Parallel composition of processes $\mathtt{P}\Vert\mathtt{Q}$ is as usual, commutative, associative and it has the termination $\mathbf{0}$ as neutral element. The restriction $(\nu\varkappa^{\mathtt{S}})\mathtt{P}$ limits the scope of the name $\varkappa$ to $\mathtt{P}$. Here S is a syntactical type annotation, that will be formalized ahead in this section. We avoid the type annotations when they are clear from the context or uninteresting.

[3] In process calculi, the purpose of a prefixing $\alpha.\mathtt{P}$ is to freeze the continuation agent P until the action α has been consumed in a reduction. In other words, reductions involving P causally depends on the reduction involving the prefix α. Apart from this explicit causal dependency, mobile calculi possess another, implicit form of causal dependency, which relies on the scope (or restriction) operator (causal dependencies in the π-calculus have been studied in several works, see [26, 6, 7]). Consider, for instance, the following agent: $(\nu\mathtt{v})(\overline{\mathtt{u}}\langle\mathtt{v}\rangle\Vert\mathtt{v}\langle\mathtt{y}\rangle)$. In this agent, the subprocess $\mathtt{v}\langle\mathtt{y}\rangle$ can in no way react before the solo $\overline{\mathtt{u}}\langle\mathtt{v}\rangle$ because the subject name v is bound. When $\overline{\mathtt{u}}\langle\mathtt{v}\rangle$ reacts, the scope of v is extended, possibly enabling a reaction with $\mathtt{v}\langle\mathtt{y}\rangle$.

We denote with FN($\mathtt{P}$) the set of free names of $\mathtt{P}$, defined in the standard way. ℓinProc extends the Solos Calculus, in order to manipulate also ground entities and recursion. *Expressions* are build on numerals, ground variables, successor and predecessor. The *ground-output* process $\overline{\mathtt{a}}\langle\mathtt{E}\rangle$ will be used to send numerals. The *ground-input* process $\mathtt{a(x).(\mathfrak{s};P)}$ consists of a prefix $\mathtt{a(x)}$ that both binds the occurrences of the ground variable $\mathtt{x}$ in its body $\mathtt{P}$, and freezes the solo $\mathfrak{s}$ (by forbidding its interaction). The solo will be unfrozen only after a reduction involving the corresponding prefix, while reductions in the body $\mathtt{P}$ are allowed, even if a value on $\mathtt{a}$ has not been received yet. The set of ground free variables GFV($\mathtt{P}$) of a process $\mathtt{P}$ is defined as expected. We remark that the ground-prefix is the only binder acting on variables. Such a prefix allows the control of causal dependencies needed for modelling a call-by-value parameter passing policy on ground values. We emphasize that, ground-input makes us able to express an explicit constant-driven causality: a ground value reception enables a (potential) solo communication. The ground-driven *sum* $\lfloor \mathtt{P} \overset{\mathtt{E}}{\oplus} \mathtt{Q} \rfloor$ is used to model the conditional. It acts like $\mathtt{P}$, if $\mathtt{E}$ evaluates to $\mathtt{0}$, while it acts like $\mathtt{Q}$, if $\mathtt{E}$ is evaluated to a numeral different from $\mathtt{0}$. We denote $\mathcal{F}^{\mathtt{q}}$ a hold-place for a process, namely a process variable bringing the free question name $\mathtt{q}$ with it. Last, the *recursion* $\mathtt{rec}\mathcal{F}.\mathtt{P}$ binds all the free occurrences of the process variable $\mathcal{F}$ in $\mathtt{P}$. We remark that as a syntactical constraint, it is assumed that $\mathtt{P}$ contains exactly one free question-name. The set of free process variables FPV($\mathtt{P}$) is defined as expected. The substitution of a process $\mathtt{P}$ to a process hold-place $\mathcal{F}^{\mathtt{q}}$ will be done by an higher-order process-substitution.

Linearity for process calculi has already been studied in [4, 12, 13, 28], mainly to ensure properties like determinism and strong normalization. A bounded name occurs exactly once in "input" and exactly once in "output". The main idea is to use a very elementary form of typing for names in processes, namely **action modalities**, denoted with ϵ: $+$ is the *output modality*, $-$ is the *input modality* and $\updownarrow$ is the *neutral modality* (meaning the use of a name in both input and output mode). Each name can possess a unique modality in a process. A *duality* operation (denoted by an over-line) is defined on modalities as follows $\overline{+} = -$ and $\overline{-} = +$. Remark that $\overline{\updownarrow}$ is undefined. A partial match operator $\odot$ on modalities is defined as

$$+ \odot - = - \odot + = \updownarrow .$$

Sorting [18] together with an action modality are ingredients to indicate possible usages of names in ℓinProc. We denote $\mathring{\iota}$ the *atomic sorting* for name delivering ground data, i.e. the sort related to answer-names. Since we want model a programming language, following [2, 11], our sortings will ask questions and will receive an unique answer. Thus, *composite sortings* are defined by $[\vec{\phi}, \mathring{\iota}]$, where $\vec{\phi}$ is a list (possibly empty) of composite sortings. Composite sortings are associated to question-names. The sorting shape respects straightforwardly $\mathcal{S}\ell$PCF type. **Sorting** (denoted with $\mathtt{S}$) and **channel types** (denoted with α) are generated by the following grammars:

Sorting. $\mathtt{S} ::= \mathring{\iota} \mid [\vec{\phi}, \mathring{\iota}]$	*Channel Type.* $\alpha ::= \mathtt{S}^{\epsilon}$

Table 2. Typing rules for ℓinProc

$$\frac{}{\Gamma \vdash \mathbf{0} \rhd \_}\ (\mathbf{z}) \qquad \frac{\Gamma \vdash \mathtt{P} \rhd \mathtt{A} \quad \Gamma \vdash \mathtt{Q} \rhd \mathtt{B} \quad \text{☢}(\mathtt{A},\mathtt{B})}{\Gamma \vdash \mathtt{P} \| \mathtt{Q} \rhd \mathtt{A} \odot \mathtt{B}}\ (\mathbf{par}) \qquad \frac{\Gamma \vdash \mathtt{P} \rhd \mathtt{A} \quad \Gamma \vdash \mathtt{Q} \rhd \mathtt{A}}{\Gamma \vdash \lfloor \mathtt{P} \overset{\mathtt{E}}{\oplus} \mathtt{Q} \rfloor \rhd \mathtt{A}}\ (\mathbf{sum})$$

$$\frac{\vec{\mathtt{p}}, \mathtt{a} \text{ distinct}}{\Gamma \vdash \mathtt{q}\langle \vec{\mathtt{p}}, \mathtt{a}\rangle \rhd \mathtt{q} : [\vec{\phi}, \iota]^-, \vec{\mathtt{p}} : \vec{\phi}^-, \mathtt{a} : \iota^-}\ (\mathbf{in}) \qquad \frac{\vec{\mathtt{p}}, \mathtt{a} \text{ distinct}}{\Gamma \vdash \overline{\mathtt{q}}\langle \vec{\mathtt{p}}, \mathtt{a}\rangle \rhd \mathtt{q} : [\vec{\phi}, \iota]^+, \vec{\mathtt{p}} : \vec{\phi}^+, \mathtt{a} : \iota^+}\ (\mathbf{out})$$

$$\frac{\Gamma \vdash \overline{\mathtt{q}}\langle \vec{\mathtt{p}}, \mathtt{b}\rangle \rhd \mathtt{B} \quad \Gamma \vdash \mathtt{P} \rhd \mathtt{A} \quad \text{☢}(\mathtt{a} : \iota^-, \mathtt{A}, \mathtt{B})}{\Gamma \vdash \mathtt{a}(\mathtt{x}).(\overline{\mathtt{q}}\langle \vec{\mathtt{p}}, \mathtt{b}\rangle; \mathtt{P}) \rhd (\mathtt{a} : \iota^-) \odot \mathtt{A} \odot \mathtt{B}}\ (\mathbf{gi}) \qquad \frac{}{\Gamma \vdash \overline{\mathtt{a}}\langle \mathtt{E}\rangle \rhd \mathtt{a} : \iota^+}\ (\mathbf{go}) \qquad \frac{\Gamma \vdash \mathtt{P} \rhd \mathtt{A}}{\Gamma \vdash \mathtt{P} \rhd \mathtt{A}, \varkappa : \mathtt{S}^{\updownarrow}}\ (\mathbf{w})$$

$$\frac{}{\Gamma, \mathcal{F} : \phi \vdash \mathcal{F}^{\mathtt{q}} \rhd \mathtt{q} : \phi^-}\ (\mathbf{pv}) \qquad \frac{\Gamma, \mathcal{F} : \phi \vdash \mathtt{P} \rhd \mathtt{q} : \phi^-}{\Gamma \vdash \mathtt{rec}\mathcal{F}.\mathtt{P} \rhd \mathtt{q} : \phi^-}\ (\mathbf{rec}) \qquad \frac{\Gamma \vdash \mathtt{P} \rhd \mathtt{A}, \varkappa : \mathtt{S}^{\updownarrow}}{\Gamma \vdash (\nu \varkappa^{\mathtt{S}})\mathtt{P} \rhd \mathtt{A}}\ (\mathbf{res})$$

Sorting and (non-neutral) modalities are straightforwardly related to the game-semantic notions of arenas and player/opponent of game semantics as formalized in [4, 10]. If ϕ is a composite sorting then ϕ^ϵ is a *question type* while ι^ϵ is an *answer type*. A *dual* of a type $\alpha = \mathtt{S}^\epsilon$, is defined by $\overline{\alpha} = \mathtt{S}^{\overline{\epsilon}}$. Differently to systems presented in [4, 12, 13, 28] our modalities do not occur inside sorting, for sake of simplicity.

In order to define our typing system we need two kinds of environments, respectively bringing type-information of free names and process variables in a process. An **action type** is a finite set of pairs name plus channel type, in which each name appears at most once. We remark that question names are paired with question types while answer names are paired with answer types. $\mathtt{A}, \mathtt{B}, \ldots$ range over action types and, as usual, $\mathtt{A}, \mathtt{v} : \alpha$ denotes the action type $\mathtt{A} \cup \{(\mathtt{v} : \alpha)\}$, with $(\mathtt{v} : \alpha) \notin \mathtt{A}$. It can be seen as partial function from names to types. We define $\mathrm{FN}(\mathtt{A})$ as the set of names appearing in $\mathtt{A}$ and if $\mathtt{A} = \mathtt{v} : \alpha, \mathtt{A}'$ we define $\mathtt{A}(\mathtt{v}) = \alpha$. We define an extension of the match operator $\odot$ to action types. Intuitively, the operation performs the union of action types and it manages the match of the names that appear in both action types. More formally, let $\mathtt{A}, \mathtt{B}$ such that for all $\mathtt{v} \in \mathrm{FN}(\mathtt{A}) \cap \mathrm{FN}(\mathtt{B})$, $\mathtt{A}(\mathtt{v}) = \overline{\mathtt{B}(\mathtt{v})}$. Under this hypothesis, we define $\mathtt{A} \odot \mathtt{B} = \mathtt{C}$ where $\mathrm{FN}(\mathtt{C}) = \mathrm{FN}(\mathtt{A}) \cup \mathrm{FN}(\mathtt{B})$ and given $\mathtt{v} \in \mathrm{FN}(\mathtt{C})$

$$\mathtt{C}(\mathtt{v}) = \begin{cases} \mathtt{A}(\mathtt{v}) \text{ if } \mathtt{v} \in \mathrm{FN}(\mathtt{A}) \setminus \mathrm{FN}(\mathtt{B}) \\ \mathtt{B}(\mathtt{v}) \text{ if } \mathtt{v} \in \mathrm{FN}(\mathtt{B}) \setminus \mathrm{FN}(\mathtt{A}) \\ \mathtt{S}^{\updownarrow} \quad \text{if } \mathtt{v} \in \mathrm{FN}(\mathtt{A}) \cap \mathrm{FN}(\mathtt{B}) \wedge \mathtt{A}(\mathtt{v}) = \mathtt{S}^\epsilon \end{cases}$$

It is easy to show that $\odot$ is a partial commutative associative operator. We write $\text{☢}(\mathtt{A}, \mathtt{B})$ when $\mathtt{A} \odot \mathtt{B}$ is defined. A **process environments** is a set of pairs process variables plus composite sorting, in which each variable appears at most once. It is denoted by Γ and we apply to it similar conventions as those introduced for action types. Valid typing judgments have the shape $\Gamma \vdash \mathtt{P} \rhd \mathtt{A}$ where Γ is a process environment, $\mathtt{P}$ is a process and $\mathtt{A}$ is an action type.

Definition 5. *A typing judgment is valid when it is conclusion of a derivation respecting the typing rules in Table 2.*

We allow weakening and contraction on process environments, while we treat linearly action types. The rules **(z)**, **(in)** , **(out)** and **(go)** impose correct modalities on our typing. The key rules are **(par)** and **(gi)** that check composability of subprocesses, ensuring the linearity policy. In **(gi)** we choose that the solo frozen by the ground prefix has to be an output. The only further rule composing different subprocesses is **(sum)** which is managed in an additive way. Rules **(res)** and **(w)** manage neutral names. Process environments bring the type information of process variables, then **(pv)** and **(rec)** impose the use of $\mathcal{F}$ by respecting the chosen type. In **(rec)** we allow rec-abstraction of a process variable $\mathcal{F}$, on a process having $\mathtt{q}$ as unique free name.

Given a process P, a *slice* of P is the process obtained from P by replacing to each summation construct one of its branches. Intuitively a slice should be thought as a possible evolution of the process. Linearity guarantees that for each slices, names are used exactly once in input or in output way.

Proposition 6 (Linearity). *If* $\Gamma \vdash \mathtt{P} \rhd \mathtt{A}, \varkappa : \mathtt{S}^{\epsilon}$ *then*

1. $\epsilon \in \{+,-\}$ *implies that* $\varkappa$ *occurs exactly once in all slices of* P,
2. $\epsilon = \updownarrow$ *implies that* $\varkappa$ *occurs either zero times or twice in all slices of* P,
3. $\epsilon = \updownarrow$ *if and only if* $\Gamma \vdash (\nu\varkappa^{\mathtt{S}})\mathtt{P} \rhd \mathtt{A}$,
4. $\mathcal{F} \notin \Gamma$ *implies that* $\Gamma, \mathcal{F} : \phi \vdash \mathtt{P} \rhd \mathtt{A}$,
5. $\Gamma = \Gamma', \mathcal{F} : \phi$ *and* $\mathcal{F} \notin \mathrm{FPV}(\mathtt{P})$ *imply that* $\Gamma' \vdash \mathtt{P} \rhd \mathtt{A}$.

3.1 *ℓin*Proc Reductions and Congruences

We need three different substitutions: ground substitution, name substitution and process substitution. We denote $\mathtt{P}[\vec{\mathtt{n}}/\vec{\mathtt{x}}]$ the expected substitution of integers to variables in a process P. Recall that the ground-prefix is the only variable binder of *ℓin*Proc and note that no free variable can be captured in such substitution. We denote $\mathtt{P}[\vec{\varkappa}/\vec{\varkappa}']$ the expected substitution of names to names in a process, in a capture-free way. Last, process substitution is a straightforward adaptation of higher-order π-calculus substitution [27]. The process substitution of a process P in another process to all occurrences of $\mathcal{F}$ is defined as follows, $\mathfrak{s}\{\mathtt{P}/\mathcal{F}\} = \mathfrak{s}$, $(\mathtt{Q}_0 \| \mathtt{Q}_1)\{\mathtt{P}/\mathcal{F}\} = \mathtt{Q}_0\{\mathtt{P}/\mathcal{F}\} \| \mathtt{Q}_1\{\mathtt{P}/\mathcal{F}\}$, $((\nu\varkappa)\mathtt{Q})\{\mathtt{P}/\mathcal{F}\} = (\nu\varkappa)\mathtt{Q}\{\mathtt{P}/\mathcal{F}\}$, $(\lfloor \mathtt{Q}_0 \overset{\mathtt{E}}{\oplus} \mathtt{Q}_1 \rfloor)$ $\{\mathtt{P}/\mathcal{F}\} = \lfloor (\mathtt{Q}_0)\{\mathtt{P}/\mathcal{F}\} \overset{\mathtt{E}}{\oplus} (\mathtt{Q}_1)\{\mathtt{P}/\mathcal{F}\} \rfloor$, $(\mathtt{a(x)\,Q})\{\mathtt{P}/\mathcal{F}\} = \mathtt{a(x)\,Q}\{\mathtt{P}/\mathcal{F}\}$, $(\overline{\mathtt{a}}\langle \mathtt{E} \rangle)\{\mathtt{P}/\mathcal{F}\}$ $= \overline{\mathtt{a}}\langle \mathtt{E} \rangle$, if $\mathcal{F}_0 \neq \mathcal{F}$ then $\mathcal{F}_0{}^{\mathtt{p}}\!\!\smile\{\mathtt{P}/\mathcal{F}\} = \mathcal{F}_0{}^{\mathtt{p}}\!\!\smile$, if $\mathrm{FN}(\mathtt{P}) = \{\mathtt{q}'\}$ then $\mathcal{F}^{\mathtt{q}}\!\!\smile\{\mathtt{P}/\mathcal{F}\} =$ $\mathtt{P}\{\mathtt{q}/\mathtt{q}'\}$ and $(\mathtt{rec}\mathcal{F}'.\mathtt{P}')\{\mathtt{P}/\mathcal{F}\} = \begin{cases} \mathtt{rec}\mathcal{F}'.\mathtt{P}' & \text{if } \mathcal{F}' = \mathcal{F}, \\ \mathtt{rec}\mathcal{F}'.(\mathtt{P}'\{\mathtt{P}/\mathcal{F}\}) & \text{otherwise.} \end{cases}$

Definition 7. *The* ***structural congruence*** $\equiv$ *between expressions and processes is the least congruence containing* α*-equivalences on variables, names and process-variable, and satisfying the following laws*

1. $\mathtt{pred(succ\,n)} \equiv \mathtt{n}$
2. $\mathtt{P}\|\mathbf{0} \equiv \mathtt{P}$, $\mathtt{P}\|\mathtt{Q} \equiv \mathtt{Q}\|\mathtt{P}$, $(\mathtt{P}\|\mathtt{Q})\|\mathtt{R} \equiv \mathtt{P}\|(\mathtt{Q}\|\mathtt{R})$, $(\nu\varkappa)\mathbf{0} \equiv \mathbf{0}$, $\mathtt{rec}\mathcal{F}.\mathtt{P} \equiv \mathtt{P}\{\mathtt{rec}\mathcal{F}.\mathtt{P}/\mathcal{F}\}$, $(\nu\varkappa_1)(\nu\varkappa_2)\mathtt{P} \equiv (\nu\varkappa_2)(\nu\varkappa_1)\mathtt{P}$, $(\nu\varkappa)(\mathtt{P}\|\mathtt{Q}) \equiv \mathtt{P}\|(\nu\varkappa)\mathtt{Q}$ *if* $\varkappa \notin \mathrm{FN}(\mathtt{P})$.
3. $\lfloor \mathbf{0} \overset{\mathtt{E}}{\oplus} \mathbf{0} \rfloor \equiv \mathbf{0}$, $\mathtt{P}\|\lfloor \mathtt{Q} \overset{\mathtt{E}}{\oplus} \mathtt{R} \rfloor \equiv \lfloor \mathtt{P}\|\mathtt{Q} \overset{\mathtt{E}}{\oplus} \mathtt{P}\|\mathtt{R} \rfloor$, $(\nu\varkappa^{\mathtt{S}})\lfloor \mathtt{P} \overset{\mathtt{E}}{\oplus} \mathtt{Q} \rfloor \equiv \lfloor (\nu\varkappa^{\mathtt{S}})\mathtt{P} \overset{\mathtt{E}}{\oplus} (\nu\varkappa^{\mathtt{S}})\mathtt{Q} \rfloor$, $\mathtt{a(x)}.(\mathfrak{s}; \lfloor \mathtt{P} \overset{\mathtt{E}}{\oplus} \mathtt{Q} \rfloor) \equiv \lfloor \mathtt{a(x)}.(\mathfrak{s};\mathtt{P}) \overset{\mathtt{E}}{\oplus} \mathtt{a(x)}.(\mathfrak{s};\mathtt{Q}) \rfloor$ *if* $\mathtt{x} \notin \mathrm{GFV}(\mathtt{E})$.

Table 3. Operational Semantics of ℓinProc

$$\frac{\forall k \le n \quad \{u_k, u'_k\} = \{p_k, p'_k\} \text{ and } \{b, b'\} = \{a, a'\}}{(\nu u_1, \dots, u_n, b)(q\langle p_1, \dots, p_n, a\rangle \| \overline{q}\langle p'_1, \dots, p'_n, a'\rangle \| R) \rightarrow_{\circledS} R[u'_1/u_1, \dots, u'_n/u_n, b'/b]}$$

$$\frac{P \rightarrow_{\circledS} P'}{a(x).(\mathfrak{s}; P) \rightarrow_{\circledS} a(x).(\mathfrak{s}; P')} \qquad \frac{}{a(x).(\mathfrak{s}; P) \| \overline{a}\langle n\rangle \longrightarrow \mathfrak{s} \| P[n/x]}$$

$$\frac{P \rightarrow_{\circledS} P'}{(\nu \varkappa)P \rightarrow_{\circledS} (\nu \varkappa)P'} \qquad \frac{P \equiv P' \rightarrow_{\circledS} Q' \equiv Q}{P \rightarrow_{\circledS} Q} \qquad \frac{P \rightarrow_{\circledS} P'}{P \| Q \rightarrow_{\circledS} P' \| Q} \qquad \frac{P \rightarrow_{\circledS} P'}{\mathtt{rec}\mathcal{F}.P \rightarrow_{\circledS} \mathtt{rec}\mathcal{F}.P'}$$

$$\frac{}{\lfloor P \overset{0}{\oplus} Q \rfloor \rightarrow_{\circledS} P} \qquad \frac{n \neq 0}{\lfloor P \overset{n}{\oplus} Q \rfloor \rightarrow_{\circledS} Q} \qquad \frac{Q \rightarrow_{\circledS} Q'}{\lfloor P \overset{E}{\oplus} Q \rfloor \rightarrow_{\circledS} \lfloor P \overset{E}{\oplus} Q' \rfloor} \qquad \frac{P \rightarrow_{\circledS} P'}{\lfloor P \overset{E}{\oplus} Q \rfloor \rightarrow_{\circledS} \lfloor P' \overset{E}{\oplus} Q \rfloor}$$

4. $(\nu \varkappa)\big(a(x).(\mathfrak{s}; P)\big) \equiv a(x).(\mathfrak{s}; (\nu \varkappa)P)$ *if* $\varkappa \notin \{a\} \cup \mathrm{FN}(\mathfrak{s})$
$Q \| a(x).(\mathfrak{s}; P) \equiv a(x).(\mathfrak{s}; Q \| P)$ *if* $x \notin \mathrm{GFV}(Q)$
$a(x).\big(\mathfrak{s}; b(y).(\mathfrak{s}'; P)\big) \equiv b(y).\big(\mathfrak{s}'; a(x).(\mathfrak{s}; P)\big)$ *if* $x \neq y$

The structural congruence axioms presented above deserve some explanation. Rules in (1) are the usual axioms dealing with the evaluation of ground expressions. Rules in (2) are the usual structural rules of π-calculus, plus the rule dealing with recursion. Rules in (3) are the structural rules dealing with the sum. They impose the distributive property of the sum with respect to parallel composition, restriction and ground input prefix. These laws allow us to derive congruences like $P \equiv \lfloor P \overset{E}{\oplus} P \rfloor$. Similar laws were introduced in [3]. Rules in (4) are the structural rules dealing with our ground prefix, which is not as the usual input-prefix. Such rules leave untouched the solo $\mathfrak{s}$ of the prefix $a(x).(\mathfrak{s}; P)$ and they formalize that the body P is in parallel with the prefix itself, taking care only of potential occurrences of x. The implicit causality coming from the underlying calculus of solos, together with rules in (4) will give us the possibility to mimic all reduction strategies of $\mathcal{S}\ell$PCF inside our ℓinProc.

Definition 8. *We endow* ℓinProc *of a reduction relation* $\rightarrow_{\circledS}$*, namely the least relation* $\rightarrow_{\circledS}$ *satisfying the rules of Table 3. As usual, we define* $\rightarrow^*_{\circledS}$ *to be the reflexive transitive closure of* $\rightarrow_{\circledS}$*.*

The rules describing the interaction between solos is a simplification (valid only under our linear constraints) of a more general rule[4] presented in [14] which exploit the unification in a very clever way. Some example can help the reader,

[4] We write θ for a total endo-function on $\mathcal{Q} \cup \mathcal{A}$ such that $\varkappa \neq \theta(\varkappa)$ for finitely many names $\varkappa$. We use $\mathrm{Dom}(\theta) = \{\varkappa | \theta(\varkappa) \neq \varkappa\}$ and $\mathrm{Ran}(\theta) = \{\theta(\varkappa) | \varkappa \neq \theta(\varkappa)\}$. Let $\{\vec{v} = \vec{w}\}$ be the smallest equivalence relation on $\mathcal{Q} \cup \mathcal{A}$ relating each v_i with w_i and let us assume a name substitution θ *agrees with* the equivalence φ if for every v, w, $v \varphi w$ iff $\theta(v) = \theta(w)$. The general rule presented in [14] is

$$\frac{|\vec{\varkappa}_1| = |\vec{\varkappa}_2| \quad \theta \text{ agrees with } \{\vec{\varkappa}_1 = \vec{\varkappa}_2\} \quad \mathrm{Ran}(\theta) \cap \vec{\varkappa}^* = \emptyset \quad \mathrm{Dom}(\theta) = \vec{\varkappa}^*}{(\nu \vec{\varkappa}^*)(u\langle \vec{\varkappa}_1\rangle \| \overline{u}\langle \vec{\varkappa}_2\rangle \| R) \rightarrow_{\circledS} R\theta}.$$

$$(\nu \mathtt{p}_0, \mathtt{p}_1, \mathtt{a})(\mathtt{q}\langle \mathtt{q}_0, \mathtt{q}_1, \mathtt{a}\rangle \| \overline{\mathtt{q}}\langle \mathtt{p}_0, \mathtt{p}_1, \mathtt{b}\rangle \| \mathtt{R}) \rightarrow_{\circledS} \mathtt{R}[\mathtt{q}_0/\mathtt{p}_0, \mathtt{q}_1/\mathtt{p}_1, \mathtt{b}/\mathtt{a}]\ ,$$
$$(\nu \mathtt{p}_0, \mathtt{q}_1, \mathtt{b})(\mathtt{q}\langle \mathtt{q}_0, \mathtt{q}_1, \mathtt{a}\rangle \| \overline{\mathtt{q}}\langle \mathtt{p}_0, \mathtt{p}_1, \mathtt{b}\rangle \| \mathtt{R}) \rightarrow_{\circledS} \mathtt{R}[\mathtt{q}_0/\mathtt{p}_0, \mathtt{p}_1/\mathtt{q}_1, \mathtt{a}/\mathtt{b}]\ ,$$

but $(\nu \mathtt{p})(\mathtt{q}\langle \mathtt{p}, \mathtt{p}, \mathtt{a}\rangle \| \overline{\mathtt{q}}\langle \mathtt{p}_0, \mathtt{p}_1, \mathtt{b}\rangle \| \mathtt{R})$ cannot be reduced. We observe a difference between the managing of name passing and the managing of ground-values passing. The communication of names uses the unification mechanism, which is usual both in solos calculus and in the calculus of fusion [21]; there is a perfect symmetry between name-emission and name-reception. The communication of ground values instead is asymmetric; when two answer names synchronize, all the occurrences of the ground variable bounded by the prefix-construct, are substituted with the corresponding value.

Theorem 9 (Subject Reduction). *Let* $\Gamma \vdash \mathtt{P} \rhd \mathtt{A}$. *If* $\mathtt{P} \rightarrow_{\circledS} \mathtt{Q}$ *then* $\Gamma \vdash \mathtt{Q} \rhd \mathtt{A}$.

Proof. The proof is by induction on the derivation proving $\mathtt{P} \rightarrow_{\circledS} \mathtt{Q}$. □

Linearity implies that the reduction in our calculus is confluent.

Lemma 10 (Confluence). *Let* $\Gamma \vdash \mathtt{P} \rhd \mathtt{A}$.
If $\mathtt{P} \rightarrow_{\circledS} \mathtt{Q}_\mathtt{i}$ *for all* $\mathtt{i} \in \{0, 1\}$, *then either* $\mathtt{Q}_0 \equiv \mathtt{Q}_1$ *or there is* $\mathtt{Q}$ *such that* $\mathtt{Q}_\mathtt{i} \rightarrow_{\circledS} \mathtt{Q}$.

Proof. The proof is by induction on the derivation proving $\mathtt{P} \rightarrow_{\circledS} \mathtt{Q}_0$. The base case essentially follow by Proposition 6 (1). All the inductive steps are easy. □

Since we are interested in the extensional behavior of terms, we define a Morris like contextual equivalence as basic equality over processes.

Definition 11 (Observability). *Let* $\Gamma \vdash \mathtt{P} \rhd \mathtt{a} : \iota^+$. *We use* $\Gamma \vdash \mathtt{P} \Downarrow_{\overline{\mathtt{a}}\langle \mathtt{n}\rangle}$ *to denote that* $\mathtt{P} \rightarrow^*_{\circledS} \mathtt{P}' \equiv (\nu \vec{\varkappa})(\overline{\mathtt{a}}\langle \mathtt{E}\rangle \| \mathtt{P}'')$ *where* $\mathtt{E} \equiv \mathtt{n}$ *and* $\mathtt{a} \notin \vec{\varkappa}$.

A relation $\cong$ is a *typed congruence* when $\equiv \subseteq \cong$ and it is a typed equality closed under typed contexts. Moreover, $\Gamma \vdash \mathtt{P} \cong \mathtt{Q} \rhd \mathtt{A}$ is an abbreviation for $\Gamma \vdash \mathtt{P} \rhd \mathtt{A}$, $\Gamma \vdash \mathtt{Q} \rhd \mathtt{A}$ and $\mathtt{P} \cong \mathtt{Q}$. We use ρ to denote a total function from the set of ground variables to the set of numerals. Given a process $\mathtt{P}$, we denote $\mathtt{P}\rho$ to be the process obtained applying the substitution ρ to ground free variables of $\mathtt{P}$.

Definition 12. $\cong_\mathtt{E}$ *is the greatest typed congruence on processes such that,* $\Gamma \vdash \mathtt{P} \cong_\mathtt{E} \mathtt{Q} \rhd \mathtt{A}$ *and* $\Gamma \vdash \mathtt{P}\rho \Downarrow_{\overline{\mathtt{a}}\langle \mathtt{n}\rangle}$ *imply* $\Gamma \vdash \mathtt{Q}\rho \Downarrow_{\overline{\mathtt{a}}\langle \mathtt{n}\rangle}$, *for all* ρ.

Following [9, 28], we can prove that $\cong_\mathtt{E}$ is consistent (i.e. does not equate all process), it is reduction closed, it is maximally consistent (i.e. the only typed congruence which strictly includes $\cong_\mathtt{E}$ is not consistent) and it equates all insensitive processes (i.e. processes that does not produce any observation) with the same type.

4 Processing Programs

The encoding of $\mathcal{S}\ell$PCF into ℓinProc is an adaptation of the encoding presented by Hyland and Ong in [10]. First of all, types can be translated in sortings as follows, $[\![\tau_1 \multimap \ldots \multimap \tau_k \multimap \iota]\!] = [[\![\tau_1]\!], \ldots [\![\tau_k]\!], \hat{\iota}]$ $(k \geq 0)$. Before formalizing the translation of programs, we give some hints.

Table 4. Translation of $\mathcal{S}\ell$PCF$^-$ on ℓinProc

To lighten the notation, from now on, we do not annotate type explicitly on processes and we denote $\mathtt{q}(\mathtt{q}_1,\ldots,\mathtt{q}_n,\mathtt{a})\,\mathtt{P} = (\nu\mathtt{q}_1,\ldots,\mathtt{q}_n,\mathtt{a})(\mathtt{q}\langle\mathtt{q}_1,\ldots,\mathtt{q}_n,\mathtt{a}\rangle\|\mathtt{P})$.

$$\left.\begin{array}{l}[\![\mathtt{pred}^{\iota\multimap\iota}]\!]^{\mathtt{q}_\epsilon} = \mathtt{q}_\epsilon(\mathtt{q}_1,\mathtt{a}_\epsilon)\Big(\overline{\mathtt{q}}_1(\mathtt{a}_1)(\nu\mathtt{q})\,\mathtt{a}_1(\mathtt{x}).(\overline{\mathtt{q}}\langle\mathtt{a}_\epsilon\rangle;\mathtt{q}(\mathtt{a})\,\overline{\mathtt{a}}\langle\mathtt{pred}(\mathtt{x})\rangle)\Big)\\ [\![\mathtt{succ}^{\iota\multimap\iota}]\!]^{\mathtt{q}_\epsilon} = \mathtt{q}_\epsilon(\mathtt{q}_1,\mathtt{a}_\epsilon)\Big(\overline{\mathtt{q}}_1(\mathtt{a}_1)(\nu\mathtt{q})\,\mathtt{a}_1(\mathtt{x}).(\overline{\mathtt{q}}\langle\mathtt{a}_\epsilon\rangle;\mathtt{q}(\mathtt{a})\,\overline{\mathtt{a}}\langle\mathtt{succ}(\mathtt{x})\rangle)\Big)\end{array}\right\} \begin{array}{l}[\![\underline{\mathtt{n}}^\iota]\!]^{\mathtt{q}_\epsilon} = \mathtt{q}_\epsilon(\mathtt{a}_\epsilon)\overline{\mathtt{a}}_\epsilon\langle\mathtt{n}\rangle\\ [\![\mathtt{x}^\iota]\!]^{\mathtt{q}_\epsilon} = \mathtt{q}_\epsilon(\mathtt{a}_\epsilon)\,\overline{\mathtt{a}}_\epsilon\langle\mathtt{x}\rangle\end{array}$$

$$[\![\lambda\mathtt{x}^\iota.\mathtt{M}^{\sigma_1\multimap\ldots\multimap\sigma_n\multimap\iota}]\!]^{\mathtt{q}_\epsilon} = \mathtt{q}_\epsilon(\mathtt{q}_0,\mathtt{q}_1,\ldots,\mathtt{q}_n,\mathtt{a}_\epsilon)\,\overline{\mathtt{q}}_0(\mathtt{a}_0)\,(\nu\mathtt{q})\,\mathtt{a}_0(\mathtt{x}).\Big(\overline{\mathtt{q}}\langle\mathtt{q}_1,\ldots,\mathtt{q}_n,\mathtt{a}_\epsilon\rangle;[\![\mathtt{M}]\!]^{\mathtt{q}}\Big)$$

$$[\![\mathtt{f}^{\sigma_1\multimap\ldots\multimap\sigma_n\multimap\iota}]\!]^{\mathtt{q}_\epsilon} = \mathtt{q}_\epsilon(\mathtt{q}_1,\ldots,\mathtt{q}_n,\mathtt{a}_\epsilon)\,\overline{\mathtt{f}}\langle\mathtt{q}_1,\ldots,\mathtt{q}_n,\mathtt{a}_\epsilon\rangle$$

$$[\![\mathtt{M}^{\sigma_1\multimap\ldots\multimap\sigma_n\multimap\iota}\mathtt{N}^{\sigma_1}]\!]^{\mathtt{q}_\epsilon} = \mathtt{q}_\epsilon(\mathtt{q}_2,\ldots,\mathtt{q}_n,\mathtt{a})\,(\nu\mathtt{p},\mathtt{q}_1)(\overline{\mathtt{p}}\langle\mathtt{q}_1,\ldots,\mathtt{q}_n,\mathtt{a}\rangle\|[\![\mathtt{M}]\!]^{\mathtt{p}}\|[\![\mathtt{N}]\!]^{\mathtt{q}_1})$$

$$[\![\ell\mathtt{if}\ \mathtt{M}_1^\iota\ \mathtt{M}_2^\iota\ \mathtt{M}_3^\iota]\!]^{\mathtt{q}_\epsilon} = \mathtt{q}_\epsilon(\mathtt{a}_\epsilon)(\nu\mathtt{q}_1)\Big([\![\mathtt{M}_1^\iota]\!]^{\mathtt{q}_1}\|\overline{\mathtt{q}_1}(\mathtt{a}_1)\,(\nu\mathtt{q}_2)\,\mathtt{a}_1(\mathtt{x}).(\overline{\mathtt{q}}_2\langle\mathtt{a}_\epsilon\rangle;\lfloor[\![\mathtt{M}_2]\!]^{\mathtt{q}_2}\overset{\mathtt{x}}{\oplus}[\![\mathtt{M}_3]\!]^{\mathtt{q}_2}\rfloor)\Big)$$

$$[\![\lambda\mathtt{f}^\tau.\mathtt{M}^{\sigma_1\multimap\ldots\multimap\sigma_n\multimap\iota}]\!]^{\mathtt{q}_\epsilon} = \mathtt{q}_\epsilon(\mathtt{f},\mathtt{q}_1,\ldots,\mathtt{q}_n,\mathtt{a})\,(\nu\mathtt{p})(\overline{\mathtt{p}}\langle\mathtt{q}_1,\ldots,\mathtt{q}_n,\mathtt{a}\rangle\|[\![\mathtt{M}]\!]^{\mathtt{p}})$$

$$[\![\mathcal{F}^{\sigma_1\multimap\ldots\multimap\sigma_n\multimap\iota}]\!]^{\mathtt{q}_\epsilon} = \mathtt{q}_\epsilon(\mathtt{q}_1,\ldots,\mathtt{q}_n,\mathtt{a})\,(\nu\mathtt{p})(\overline{\mathtt{p}}\langle\mathtt{q}_1,\ldots,\mathtt{q}_n,\mathtt{a}\rangle\|\mathcal{F}^{\mathtt{p}})\quad [\![\mu\mathcal{F}^\sigma.\mathtt{M}^\sigma]\!]^{\mathtt{q}_\epsilon} = \mathtt{rec}\mathcal{F}.[\![\mathtt{M}]\!]^{\mathtt{q}_\epsilon}$$

Let $\mathtt{M}$ be a $\mathcal{S}\ell$PCF term such that $\Gamma\vdash\mathtt{M}:\sigma_1\multimap\ldots\multimap\sigma_n\multimap\iota$ where $n\geq 0$. Encoding exploits overloads of symbols for variables of $\mathcal{S}\ell$PCF and symbols for names and variables of ℓinProc. The interpretation of $\mathtt{M}$ is given on a process $\mathtt{P}$ such that $\mathrm{GFV}(\mathtt{P}) = \mathrm{FV}(\mathtt{M})\cap\mathrm{Var}^\iota$, $\mathrm{SFV}(\mathtt{P}) = \mathrm{FV}(\mathtt{M})\cap\mathrm{SVar}$ and $\mathrm{FN}(\mathtt{P}) = (\mathrm{FV}(\mathtt{M})\cap\mathrm{HVar})\cup\{\mathtt{q}\}$ where $\mathtt{q}$ is a fresh name, called *access-name*. Channel type of free names of $\mathtt{P}$, but the access-name, is obtained by translating the type of corresponding variable in $\mathtt{M}$ in a sorting together with positive modality. Channel type of the access-name is the translation of the type of $\mathtt{M}$ together with a negative modality. The sorting associated to process variables is obtained by translating the type of corresponding stable variable. More formal details are in Theorem 14. If $\mathtt{M}$ is closed then $\mathtt{P}$ contains a unique free name, namely the access-name $\mathtt{q}$ typed by negative modality and the sorting $[[\![\sigma_1]\!],\ldots[\![\sigma_n]\!],\mathring{\iota}]$ corresponding to the type of $\mathtt{M}$. The translation of $\mathtt{M}$ can be questioned by the solo $\overline{\mathtt{q}}\langle\mathtt{p}_1,...,\mathtt{p}_n,\mathtt{a}\rangle$ communicating a list of n question-names $\mathtt{p}_1,...,\mathtt{p}_n$ where processes mimicking "actual arguments" can be questioned in its turn by $\mathtt{P}$, and an answer-name $\mathtt{a}$ where $\mathtt{P}$ can communicate the computation result.

Definition 13. *We denote by* $[\![\mathtt{M}]\!]^{\mathtt{q}}$ *the process encoding a program* $\Gamma\vdash\mathtt{M}:\sigma$ *on the access-name* $\mathtt{q}$. *The recursive definition of the translation from* $\mathcal{S}\ell$PCF$^-$ *to* ℓinProc *is given in Table 4.*

We use the ground-prefix in order to model the causal dependency needed in order to respect call-by-value computations. As instances, we remark that $\mathtt{q}_\epsilon(\mathtt{a}_\epsilon)\overline{\mathtt{a}}_\epsilon\langle\underline{\mathtt{n}}\rangle$ is an abbreviation for $(\nu\mathtt{a}_\epsilon^{\mathring{\iota}})(\mathtt{q}_\epsilon\langle\mathtt{a}_\epsilon\rangle\|\overline{\mathtt{a}}_\epsilon\langle\underline{\mathtt{n}}\rangle)$ and the translation of $\mathtt{pred}^{\iota\multimap\iota}$ is an abbreviation for

$$(\nu\mathtt{q}_1^{[\mathring{\iota}]})(\nu\mathtt{a}_\epsilon^{\mathring{\iota}})\Big(\mathtt{q}_\epsilon\langle\mathtt{q}_1,\mathtt{a}_\epsilon\rangle\|(\nu\mathtt{a}_1^{\mathring{\iota}})(\overline{\mathtt{q}}_1\langle\mathtt{a}_1\rangle\|(\nu\mathtt{q}^{[\mathring{\iota}]})\,\mathtt{a}_1(\mathtt{x}).(\overline{\mathtt{q}}\langle\mathtt{a}_\epsilon\rangle\|(\nu\mathtt{a}^{\mathring{\iota}})(\mathtt{q}\langle\mathtt{a}\rangle\,\overline{\mathtt{a}}\langle\mathtt{pred}(\mathtt{x})\rangle))))\Big).$$

Theorem 14 (Typing soundness). *Let* $\mathtt{f}_1:\sigma_1,\ldots\mathtt{f}_n:\sigma_n\vdash\mathtt{M}:\sigma$ *be a term of* $\mathcal{S}\ell$PCF$^-$ *such that* $\mathrm{FV}(\mathtt{M})\cap\mathrm{SVar} = \{\mathcal{F}_1^{\tau_1},\ldots,\mathcal{F}_m^{\tau_m}\}$ *and* $\mathrm{FV}(\mathtt{M})\cap\mathrm{Var}^\iota = \{\mathtt{x}_1,\ldots,\mathtt{x}_h\}$. *Then* $\mathcal{F}_1:[\![\tau_1]\!],\ldots,\mathcal{F}_m:[\![\tau_m]\!]\vdash[\![\mathtt{M}]\!]^{\mathtt{q}}\triangleright\mathtt{q}:[\![\sigma]\!]^-,\mathtt{f}_1:[\![\sigma_1]\!]^+,\ldots,\mathtt{f}_n:[\![\sigma_n]\!]^+$ *where* $\mathrm{GFV}([\![\mathtt{M}]\!]^{\mathtt{q}}) = \{\mathtt{x}_1,\ldots,\mathtt{x}_h\}$.

Proof. By induction on the derivation of $\Gamma\vdash\mathtt{M}:\sigma$. □

We say that a translation from a calculus to another is *faithful* whenever each reduction on the first calculus can be mimicked by some reductions in the second one. We can prove that our translation is faithful, more precisely $\mathtt{M} \rightarrow_{\mathcal{S}\ell} \mathtt{N}$ implies $[\![\mathtt{M}]\!]^{\mathtt{q}} \rightarrow^{*}_{\circledS} [\![\mathtt{N}]\!]^{\mathtt{q}}$.

Lemma 15

1. $[\![\mathtt{pred}\,(\mathtt{succ}\;\underline{\mathtt{n}})]\!]^{\mathtt{q}} \rightarrow^{*}_{\circledS} [\![\underline{\mathtt{n}}]\!]^{\mathtt{q}}$.
2. *Let* $\Gamma \vdash [\![\mathtt{L}]\!]^{\mathtt{q}_2}, [\![\mathtt{R}]\!]^{\mathtt{q}_2} \rhd \mathtt{q}_2 : [\![\iota]\!]^{-}, \mathtt{A}$.
   Both $[\![\ell\mathtt{if}\;\underline{\mathtt{0}}\;\mathtt{L}\;\mathtt{R}]\!]^{\mathtt{q}} \rightarrow^{*}_{\circledS} [\![\mathtt{L}]\!]^{\mathtt{q}}$ *and* $[\![\ell\mathtt{if}\;\underline{\mathtt{n+1}}\;\mathtt{L}\;\mathtt{R}]\!]^{\mathtt{q}} \rightarrow^{*}_{\circledS} [\![\mathtt{R}]\!]^{\mathtt{q}}$.

Proof

1. $[\![\mathtt{succ}\,\underline{\mathtt{n}}]\!]^{\mathtt{q_s}} = \mathtt{q_s}(\mathtt{a_s})(\nu \mathtt{q}'_{\mathtt{s}}, \mathtt{q_n}) \begin{pmatrix} \mathtt{q}'_{\mathtt{s}}(\mathtt{q}'_{\mathtt{n}}, \mathtt{a}'_{\mathtt{s}})\, \overline{\mathtt{q}}'_{\mathtt{n}}(\mathtt{a}'_{\mathtt{n}})\, (\nu \mathtt{q})\, \mathtt{a}'_{\mathtt{n}}(\mathtt{x}).(\overline{\mathtt{q}}\langle \mathtt{a}'_{\mathtt{s}}\rangle; \mathtt{q}(\mathtt{a})\, \overline{\mathtt{a}}\langle \mathtt{succ}(\mathtt{x})\rangle) \,\| \\ \overline{\mathtt{q}}'_{\mathtt{s}}\langle \mathtt{q_n}, \mathtt{a_s}\rangle \| \mathtt{q_n}(\mathtt{a_n}) \overline{\mathtt{a}}_{\mathtt{n}}\langle \mathtt{n}\rangle \end{pmatrix}$,
   $[\![\mathtt{pred}(\mathtt{succ}\,\underline{\mathtt{n}})]\!]^{\mathtt{q}_\epsilon} = \mathtt{q}_\epsilon(\mathtt{a}_\epsilon)\, (\nu \mathtt{q}^{*}_{\mathtt{p}}, \mathtt{q_s}) \begin{pmatrix} \mathtt{q}^{*}_{\mathtt{p}}(\mathtt{q}^{*}_{\mathtt{s}}, \mathtt{a}^{*}_{\mathtt{p}})\, \overline{\mathtt{q}}^{*}_{\mathtt{s}}(\mathtt{a}^{*}_{\mathtt{s}})\, (\nu \mathtt{q})\, \mathtt{a}^{*}_{\mathtt{s}}(\mathtt{y}).(\overline{\mathtt{q}}\langle \mathtt{a}^{*}_{\mathtt{p}}\rangle; \mathtt{q}(\mathtt{a})\, \overline{\mathtt{a}}\langle \mathtt{pred}(\mathtt{y})\rangle) \,\| \\ \overline{\mathtt{q}}^{*}_{\mathtt{p}}\langle \mathtt{q_s}, \mathtt{a}_\epsilon\rangle \| [\![\mathtt{succ}\,\underline{\mathtt{n}}]\!]^{\mathtt{q_s}} \end{pmatrix}$
   $\rightarrow^{*}_{\circledS} \mathtt{q}_\epsilon(\mathtt{a}_\epsilon)\, \overline{\mathtt{a}}_\epsilon\langle \mathtt{n}\rangle = [\![\underline{\mathtt{n}}]\!]^{\mathtt{q}_\epsilon}$
2. $[\![\ell\mathtt{if}\;\underline{\mathtt{0}}\;\mathtt{L}\;\mathtt{R}]\!]^{\mathtt{q}_\epsilon} = \mathtt{q}_\epsilon(\mathtt{a}_\epsilon)\, (\nu \mathtt{q}_1) \left(\mathtt{q}_1(\mathtt{a}_1)\, \overline{\mathtt{a}}_1\langle 0\rangle \middle\| \overline{\mathtt{q}}_1(\mathtt{a}_1)\, (\nu \mathtt{q}_2)\, \mathtt{a}_1(\mathtt{x}) (\overline{\mathtt{q}}_2\langle \mathtt{a}_\epsilon\rangle; \lfloor [\![\mathtt{L}]\!]^{\mathtt{q}_2} \overset{\mathtt{x}}{\oplus} [\![\mathtt{R}]\!]^{\mathtt{q}_2} \rfloor)\right) \rightarrow^{*}_{\circledS} [\![\mathtt{L}]\!]^{\mathtt{q}_\epsilon}$.
   The other case is similar to the previous one. □

Ground substitution and stable substitution does not present particular problems with respect to faithful translation.

Lemma 16. *If* $\Gamma \vdash \mathtt{M} : \sigma_1 \multimap \ldots \sigma_{\mathtt{n}} \multimap \iota$ *(*$n \geq 0$*) then*
$\mathtt{q}_\epsilon(\mathtt{q}_1, \ldots, \mathtt{q}_{\mathtt{n}}, \mathtt{a})\, (\nu \mathtt{p})(\overline{\mathtt{p}}\langle \mathtt{q}_1, \ldots, \mathtt{q}_{\mathtt{n}}, \mathtt{a}\rangle \| [\![\mathtt{M}]\!]^{\mathtt{p}}) \rightarrow^{*}_{\circledS} [\![\mathtt{M}]\!]^{\mathtt{q}_\epsilon}$

Proof. The proof is by cases on the definition of interpretation. All cases are straightforward, except the one dealing with recursion. In fact, for all $\mathtt{M}$, the process $[\![\mathtt{M}]\!]^{\mathtt{p}}$ always start with a question on $\mathtt{p}$, except for the case $\mathtt{M} = \mu F.\mathtt{M}'$. Thus let us consider the case $\mathtt{M} = \mu F_1 \ldots \mu F_{\mathtt{m}}.\mathtt{M}'$ $(m \geq 1)$ where $\mathtt{M}' \neq \mu F.\mathtt{M}''$. We can check, by induction on m, that for an opportune process $\mathtt{P}$ we have $[\![\mu F_1 \ldots \mu F_{\mathtt{m}}.\mathtt{M}']\!]^{\mathtt{p}} \equiv \mathtt{p}(\mathtt{q}_1, \ldots, \mathtt{q}_{\mathtt{n}}, \mathtt{a})\, \mathtt{P}$. So, we get

$$\begin{aligned} &\mathtt{q}_\epsilon(\mathtt{q}_1, \ldots, \mathtt{q}_{\mathtt{n}}, \mathtt{a})\, (\nu \mathtt{p})(\overline{\mathtt{p}}\langle \mathtt{q}_1, \ldots, \mathtt{q}_{\mathtt{n}}, \mathtt{a}\rangle \| [\![\mu F_1 \ldots \mu F_{\mathtt{m}}.\mathtt{M}']\!]^{\mathtt{p}}) \equiv \\ &\qquad \mathtt{q}_\epsilon(\mathtt{q}_1, \ldots, \mathtt{q}_{\mathtt{n}}, \mathtt{a})\, (\nu \mathtt{p})(\overline{\mathtt{p}}\langle \mathtt{q}_1, \ldots, \mathtt{q}_{\mathtt{n}}, \mathtt{a}\rangle \| \mathtt{p}(\mathtt{q}_1, \ldots, \mathtt{q}_{\mathtt{n}}, \mathtt{a})\, \mathtt{P}) \\ &\rightarrow_{\circledS} \mathtt{q}_\epsilon(\mathtt{q}_1, \ldots, \mathtt{q}_{\mathtt{n}}, \mathtt{a})\, \mathtt{P} \equiv [\![\mu F_1 \ldots \mu F_{\mathtt{m}}.\mathtt{M}']\!]^{\mathtt{q}_\epsilon}. \end{aligned}$$ □

Lemma 17. *If* $\Gamma \vdash (\lambda \mathtt{x}^{\iota}.\mathtt{M})\underline{\mathtt{n}} : \sigma$ *then* $[\![(\lambda \mathtt{x}^{\iota}.\mathtt{M})\underline{\mathtt{n}}]\!]^{\mathtt{q}} \rightarrow^{*}_{\circledS} [\![\mathtt{M}[\mathtt{n}/\mathtt{x}]]\!]^{\mathtt{q}}$.

Proof. $[\![\mathtt{M}]\!]^{\mathtt{q}}[\underline{\mathtt{n}}/\mathtt{x}] \equiv [\![\mathtt{M}[\underline{\mathtt{n}}/\mathtt{x}]]\!]^{\mathtt{q}}$ follows easily by interpretation, hence

$$\begin{aligned} &[\![(\lambda \mathtt{x}^{\iota}.\mathtt{M})\underline{\mathtt{n}}]\!]^{\mathtt{q}_\epsilon} = \mathtt{q}_\epsilon(\vec{\varkappa}, \mathtt{a}_\epsilon)\, (\nu \mathtt{p}_{\mathtt{M}}, \mathtt{q}) \begin{pmatrix} \overline{\mathtt{p}_{\mathtt{M}}}\langle \mathtt{q}, \vec{\varkappa}, \mathtt{a}_\epsilon\rangle \| \mathtt{q}(\mathtt{a})\, \overline{\mathtt{a}}\langle \mathtt{n}\rangle \\ \mathtt{p}_{\mathtt{M}}(\mathtt{q}, \vec{\varkappa}, \mathtt{a}_\epsilon)\, \overline{\mathtt{q}}(\mathtt{a})\, (\nu \mathtt{p}'_{\mathtt{M}})\, \mathtt{a}(\mathtt{x}).(\overline{\mathtt{p}}'_{\mathtt{M}}\langle \vec{\varkappa}, \mathtt{a}_\epsilon\rangle; [\![\mathtt{M}]\!]^{\mathtt{p}'_{\mathtt{M}}}) \end{pmatrix} \\ &\rightarrow^{*}_{\circledS} \mathtt{q}_\epsilon(\vec{\varkappa}, \mathtt{a}_\epsilon)(\nu \mathtt{p}'_{\mathtt{M}})(\overline{\mathtt{p}}'_{\mathtt{M}}\langle \vec{\varkappa}, \mathtt{a}_\epsilon\rangle \| [\![\mathtt{M}]\!]^{\mathtt{p}'_{\mathtt{M}}}[\mathtt{n}/\mathtt{x}]) \rightarrow^{*}_{\circledS} [\![\mathtt{M}]\!]^{\mathtt{q}_\epsilon}[\mathtt{n}/\mathtt{x}] \rightarrow^{*}_{\circledS} [\![\mathtt{M}[\underline{\mathtt{n}}/\mathtt{x}]]\!]^{\mathtt{q}_\epsilon} \end{aligned}$$

where the second reduction follows by Lemma 16. □

Lemma 18. *Let* $\Gamma \vdash \mathtt{M} : \sigma$.

1. *If* $F \in \mathrm{SFV}^{\tau}(\mathtt{M})$ *and* $\_ \vdash \mathtt{N} : \tau$ *then* $[\![\mathtt{M}]\!]^{\mathtt{q}}\{[\![\mathtt{N}]\!]^{\mathtt{q}}/F\} \rightarrow^{*}_{\circledS} [\![\mathtt{M}[\mathtt{N}/F]]\!]^{\mathtt{q}}$.
2. *If* $\mathtt{M} \rightsquigarrow_{\mathtt{Y}} \mathtt{N}$ *then* $[\![\mathtt{M}]\!]^{\mathtt{q}} \rightarrow^{*}_{\circledS} [\![\mathtt{N}]\!]^{\mathtt{q}}$.

Proof

1. By induction on the derivation of $\Gamma \vdash \mathtt{M} : \sigma$ and by Lemma 16.
2. Let $\mathtt{M} = \mu F.\mathtt{M}'$, thus $[\![\mu F.\mathtt{M}']\!]^{\mathtt{q}} \equiv [\![\mathtt{M}']\!]^{\mathtt{q}}\{[\![\mu F.\mathtt{M}']\!]^{\mathtt{q}}/F\} \rightarrow^*_{Ⓢ} [\![\mathtt{M}'[\mu F.\mathtt{M}'/F]]\!]^{\mathtt{q}}$. □

To mimic substitutions of programs to higher-order variables, we need to use the ground prefix together with its structural rules.

Lemma 19. *Let* $\Gamma, \mathtt{f} : \sigma \multimap \tau \vdash \mathtt{M} : \sigma'$ *and* $\Delta \vdash \mathtt{N} : \sigma \multimap \tau$ *two* $\mathcal{S}\ell$PCF$^-$*-terms.*

1. $(\nu\mathtt{f})([\![\mathtt{M}]\!]^{\mathtt{q}} \| [\![\mathtt{N}]\!]^{\mathtt{f}}) \rightarrow^*_{Ⓢ} [\![\mathtt{M}[\mathtt{N}/\mathtt{f}]]\!]^{\mathtt{q}}$.
2. *If* $(\lambda\mathtt{f}^{\sigma\multimap\tau}.\mathtt{M})\mathtt{N} \leadsto_\beta \mathtt{M}[\mathtt{N}/\mathtt{f}]$ *then* $[\![(\lambda\mathtt{f}^{\sigma\multimap\tau}.\mathtt{M})\mathtt{N}]\!]^{\mathtt{q}} \rightarrow^*_{Ⓢ} [\![\mathtt{M}[\mathtt{N}/\mathtt{f}]]\!]^{\mathtt{q}}$.

Proof

1. The proof is by induction on the derivation of $\Gamma \vdash \mathtt{M} : \tau$. For the base case $\mathtt{M} = \mathtt{f}$, we make use of Lemma 16. For the inductive step, there are two non-trivial cases, that are the ground λ-abstraction case and the ℓ**if**-case. In case of ground λ-abstraction, we need only to use the structural rule $\mathtt{P}\|\mathtt{a}(\mathtt{x}).(\overline{\mathtt{q}}\langle\vec{\varkappa}\rangle;\mathtt{Q}) \equiv \mathtt{a}(\mathtt{x}).(\overline{\mathtt{q}}\langle\vec{\varkappa}\rangle;\mathtt{P}\|\mathtt{Q})$ providing that $\mathtt{x} \notin \mathrm{GFV}(\mathtt{P})$. In case of $\ell\mathtt{if}$, it is necessary to use also the distributive laws for the sum. All further cases are straightforward.
2. $[\![(\lambda\mathtt{f}^{\sigma\multimap\tau}.\mathtt{M})\mathtt{N}]\!]^{\mathtt{q}_\epsilon} =$
   $\mathtt{q}_\epsilon(\vec{\varkappa},\mathtt{a}_\epsilon)\,(\nu\mathtt{q},\mathtt{f})\big(\overline{\mathtt{q}}\langle\mathtt{f},\vec{\varkappa},\mathtt{a}_\epsilon\rangle \;\|\; \mathtt{q}(\mathtt{f},\vec{\varkappa},\mathtt{a})(\nu\mathtt{p})(\overline{\mathtt{p}}\langle\vec{\varkappa},\mathtt{a}\rangle\|[\![\mathtt{M}]\!]^{\mathtt{p}}) \;\|\; [\![\mathtt{N}]\!]^{\mathtt{f}}\big) \rightarrow^*_{Ⓢ}$
   $(\nu\mathtt{f})([\![\mathtt{M}]\!]^{\mathtt{q}_\epsilon}\|[\![\mathtt{N}]\!]^{\mathtt{f}}) \rightarrow^*_{Ⓢ} [\![\mathtt{M}[\mathtt{N}/\mathtt{f}]]\!]^{\mathtt{q}_\epsilon}$. □

Note that our translation actually maps a calculus (i.e. $\mathcal{S}\ell$PCF$^-$ with $\rightarrow_{\mathcal{S}\ell}$-reduction) into another calculus (i.e ℓinProc with $\rightarrow_{Ⓢ}$-reduction).

Theorem 20. *Our translation is faithful, i.e. if* $\mathtt{M} \rightarrow_{\mathcal{S}\ell} \mathtt{N}$ *then* $[\![\mathtt{M}]\!]^{\mathtt{q}} \rightarrow^*_{Ⓢ} [\![\mathtt{N}]\!]^{\mathtt{q}}$.

Proof. Note that $\rightarrow_{Ⓢ}$ is closed under all context, by Table 3. Thus, the proof follows by previous lemmas. □

The result established by Theorem 20 is very strong and it overcomes the traditional encodings from programs to processes [27]: in our encoding *no reduction strategy is determined in advance.* Gordon Plotkin in [22] remarked that the call-by-value parameter passing is hardly in accord with a strategy on (call-by-name) λ-calculus.

Corollary 21. *If* $\mathtt{M} \Downarrow \underline{\mathtt{n}}$ *then* $[\![\mathtt{M}]\!]^{\mathtt{P}} \cong_{\mathtt{E}} [\![\underline{\mathtt{n}}]\!]^{\mathtt{P}}$.

Proof. Since $\rightarrow_{Ⓢ} \subseteq \cong_{\mathtt{E}}$ by Lemma 10, the proof follows by Definition 12. □

5 Soundness and Correctness

An interpretation is said to be *adequate* when $[\![\mathtt{M}]\!] \cong_{\mathtt{E}} [\![\underline{\mathtt{n}}]\!]$ and $\mathtt{M} \Downarrow \underline{\mathtt{n}}$ are logically equivalent for any program $\mathtt{M}$, numeral $\underline{\mathtt{n}}$. Actually, we prove a stronger form of adequacy result, namely that $\mathtt{M} \rightarrow^*_{\mathcal{SL}} \underline{\mathtt{n}}$, $[\![\mathtt{M}]\!] \rightarrow^*_{\circledS} [\![\underline{\mathtt{n}}]\!]$ and $[\![\mathtt{M}]\!] \cong_{\mathtt{E}} [\![\underline{\mathtt{n}}]\!]$ are logically equivalent, for any program M and numeral $\underline{\mathtt{n}}$.

In order to complete the proof of adequacy for our interpretation, we straightforward adapt the Tait's computability argument likewise to that done in [19, 23] for denotational semantics. In the following, we use $\mathcal{X}$ to denote a generic variables of $\mathcal{SL}$PCF$^-$ (ground, high-order or stable variable of $\mathcal{SL}$PCF$^-$).

Definition 22. *The "computability predicate" is defined by the following cases.*

- *Case* $\mathrm{FV}(\mathtt{M}) = \emptyset$.
  - *Subcase* $\sigma = \iota$. $\mathrm{Comp}(\mathtt{M}^\iota)$ *if and only if* $[\![\mathtt{M}]\!]^{\mathtt{q}} \cong_{\mathtt{E}} [\![\underline{\mathtt{n}}]\!]^{\mathtt{q}}$ *implies* $\mathtt{M} \rightarrow^*_{\mathcal{SL}} \underline{\mathtt{n}}$.
  - *Subcase* $\sigma = \mu \multimap \tau$. $\mathrm{Comp}(\mathtt{M}^{\mu\multimap\tau})$ *if and only if* $\mathrm{Comp}(\mathtt{M}^{\mu\multimap\tau}\mathtt{N}^\mu)$ *for each closed* $\mathtt{N}^\mu$ *such that* $\mathrm{Comp}(\mathtt{N}^\mu)$.
- *Case* $\mathrm{FV}(\mathtt{M}^\sigma) = \{\mathcal{X}_1^{\tau_1}, \ldots, \mathcal{X}_n^{\tau_n}\}$, *for some* $n \geq 1$.
  $\mathrm{Comp}(\mathtt{M}^\sigma)$ *if and only if* $\mathrm{Comp}(\mathtt{M}[\mathtt{N}_1/\mathcal{X}_1, \ldots, \mathtt{N}_n/\mathcal{X}_n])$ *for each closed* $\mathtt{N}_i^{\tau_i}$ *such that* $\mathrm{Comp}(\mathtt{N}_i^{\tau_i})$.

Lemma 23 states a standard equivalent formulation of computability predicate.

Lemma 23. *Let* $\mathtt{M}^{\tau_1\multimap\ldots\multimap\tau_m\multimap\iota} \in \mathcal{SL}\mathrm{PCF}$ *and* $\mathrm{FV}(\mathtt{M}) = \{\mathcal{X}_1^{\mu_1}, ..., \mathcal{X}_n^{\mu_n}\}$ $(n, m \in \mathbb{N})$. $\mathrm{Comp}(\mathtt{M})$ *if and only if* $[\![\mathtt{M}[\mathtt{N}_1/\mathcal{X}_1, \ldots, \mathtt{N}_n/\mathcal{X}_n]\mathtt{P}_1 \ldots \mathtt{P}_m]\!]^{\mathtt{q}} \cong_{\mathtt{E}} [\![\underline{\mathtt{n}}]\!]^{\mathtt{q}}$ *implies* $\mathtt{M}[\mathtt{N}_1/\mathcal{X}_1, \ldots, \mathtt{N}_n/\mathcal{X}_n]\mathtt{P}_1 \ldots \mathtt{P}_m \rightarrow^*_{\mathcal{SL}} \underline{\mathtt{n}}$ *for each closed terms* $\mathtt{N}_i^{\mu_i}$ *and* $\mathtt{P}_j^{\tau_j}$ *such that* $\mathrm{Comp}(\mathtt{N}_i)$ *and* $\mathrm{Comp}(\mathtt{P}_j)$ *where* $i \leq n, j \leq m$.

The proof is an adaptation of the proof given by Plotkin in [23].

Lemma 24. *If* $\mathtt{M}^\sigma \in \mathcal{SL}\mathrm{PCF}$ *then* $\mathrm{Comp}(\mathtt{M}^\sigma)$.

Proof. The proof is by induction on the "untyped syntax shape" of M. □

Corollary 25 (Strong adequacy). *Let* M *be a program and* $\underline{\mathtt{n}}$ *be a numeral.* $[\![\mathtt{M}]\!]^{\mathtt{q}} \cong_{\mathtt{E}} [\![\mathtt{n}]\!]^{\mathtt{q}}$, $\mathtt{M} \rightarrow^*_{\mathcal{SL}} \mathtt{n}$ *and* $[\![\mathtt{M}]\!]^{\mathtt{q}} \rightarrow^*_{\circledS} [\![\underline{\mathtt{n}}]\!]^{\mathtt{q}}$ *are logically equivalent.*

Proof. $[\![\mathtt{M}]\!]^{\mathtt{q}} \cong_{\mathtt{E}} [\![\mathtt{n}]\!]^{\mathtt{q}}$ implies $\mathtt{M} \rightarrow^*_{\mathcal{SL}} \mathtt{n}$ by Lemma 24. Moreover, $\mathtt{M} \rightarrow^*_{\mathcal{SL}} \mathtt{n}$ implies $[\![\mathtt{M}]\!]^{\mathtt{q}} \rightarrow^*_{\circledS} [\![\underline{\mathtt{n}}]\!]^{\mathtt{q}}$ by Theorem 20. Thus, since $[\![\mathtt{M}]\!]^{\mathtt{q}} \rightarrow^*_{\circledS} [\![\underline{\mathtt{n}}]\!]^{\mathtt{q}}$ implies $[\![\mathtt{M}]\!]^{\mathtt{q}} \cong_{\mathtt{E}} [\![\mathtt{n}]\!]^{\mathtt{q}}$ the proof is done. □

Consequently, ℓinProc give us a syntactical model where we can study the operational equivalence between $\mathcal{SL}$PCF-programs. Motivations are the game-semantics goals. To provide tools for proving properties of our language and programs. But also, to provide rigorous definitions of implementation instance with good parallel and optimal evaluation features.

Theorem 26 (Correctness). *If* $[\![\mathtt{M}^\sigma]\!]^{\mathtt{q}} \cong_{\mathtt{E}} [\![\mathtt{N}^\sigma]\!]$ *then* $\mathtt{M} \approx_\sigma \mathtt{N}$.

Proof. Let $\mathtt{B} \vdash \mathtt{M} : \sigma$ and $\mathtt{B} \vdash \mathtt{N} : \sigma$ such that $[\![\mathtt{M}]\!]^{\mathtt{q}} \cong_{\mathtt{E}} [\![\mathtt{N}]\!]^{\mathtt{q}}$. If $\mathtt{C}[\sigma]$ is a closing context such that both $\mathtt{C}[\mathtt{M}]$ and $\mathtt{C}[\mathtt{N}]$ are programs and $\mathtt{C}[\mathtt{M}] \rightarrow^*_{\mathcal{SL}} \underline{\mathtt{n}}$ for some value $\underline{\mathtt{n}}$, then $[\![\mathtt{C}[\mathtt{M}]]\!] \rightarrow_{\circledS} [\![\underline{\mathtt{n}}]\!]$ by Corollary 25. So $[\![\mathtt{C}[\mathtt{N}]]\!]^{\mathtt{q}} \cong_{\mathtt{E}} [\![\mathtt{C}[\mathtt{M}]]\!] \cong_{\mathtt{E}} [\![\underline{\mathtt{n}}]\!]^{\mathtt{q}}$, implies $[\![\mathtt{C}[\mathtt{N}]]\!]^{\mathtt{q}} \rightarrow^*_{\circledS} [\![\underline{\mathtt{n}}]\!]^{\mathtt{q}}$ by Corollary 25, which implies $\mathtt{C}[\mathtt{N}] \rightarrow^*_{\mathcal{SL}} \underline{\mathtt{n}}$ by strong adequacy. By definition of operational equivalence the proof is done. □

As a final remark, we proposed a process model of $\mathcal{S}\ell$PCF showing that, as a consequence of faithfulness, all evaluation strategies of $\mathcal{S}\ell$PCF-programs can be represented by our ℓinProc-processes. Furthermore, we show that our process model is adequate w.r.t. the operational equivalence of $\mathcal{S}\ell$PCF.

One of the key point to obtain faithfulness is the introduction of ground-input prefix in ℓinProc, in order to model the call-by value computation. In this way, we got a translation from a calculus ($\mathcal{S}\ell$PCF with $\to_{\mathcal{S}\ell}$-reduction) into an other calculus (ℓinProc with $\to_{\circledS}$-reduction). A reviewer asked whether it is possible to take a modification of $\mathcal{S}\ell$PCF with absolutely no constraint on β-reduction, i.e. a language where both ground and high-order arguments are treated using a call-by name policy, and get a faithfulness result w.r.t. an opportune process language. We can answer positively to such a question; it is not difficult to see that a suitable process calculus could be a modified version of ℓinProc in which ground-input prefix is replaced with a completely asynchronous construct. However we should observe that the so obtained source language does not enjoy denotational linearity in the sense of [20]. Our purpose includes in fact to use processes to relate the classical denotational models focussing on functional aspects of computation and the new game models focussing on the dynamical (operational) aspects of computation. In particular, the proposed source language does not have a clear denotational status; however it could be an interesting example of functional calculus whose reduction can be mimicked by a fully asynchronous process calculus.

Acknowledgements. We would like to thank the anonymous reviewers for the useful suggestions they pointed out.

References

1. Abelson, H., Sussman, G.J., Sussman, J.: Structure and Interpretation of Computer Programs. MIT Press, Cambridge (1985), `http://mitpress.mit.edu/sicp/full-text/book/book.html`
2. Abramsky, S., Malacaria, P., Jagadeesan, R.: Full abstraction for PCF. Information and Computation 163(2), 409–470 (2000)
3. Beffara, E.: An algebraic process calculus. In: Proceedings of LICS 2008 (to appear, 2008)
4. Berger, M., Honda, K., Yoshida, N.: Sequentiality and the pi-calculus. In: Abramsky, S. (ed.) TLCA 2001. LNCS, vol. 2044, pp. 29–45. Springer, Heidelberg (2001)
5. Berry, G., Curien, P.-L., Lévy, J.-J.: Full abstraction for sequential languages: the state of the art. In: Nivat, M., Reynolds, J. (eds.) Algebraic Semantics, pp. 89–132. Cambridge University Press, Cambridge (1985)
6. Boreale, M., Sangiorgi, D.: A fully abstract semantics for causality in the π-calculus. Acta Informatica 35(5), 353–400 (1998)
7. Degano, P., Priami, C.: Causality for mobile processes. In: Fülöp, Z., Gécseg, F. (eds.) ICALP 1995. LNCS, vol. 944, pp. 660–671. Springer, Heidelberg (1995)
8. Hoare, C.: Communicating Sequential Processes. Prentice Hall, Englewood Cliffs (1985)

[9] Honda, K., Yoshida, N.: On reduction-based process semantics. Theoretical Computer Science 151(2), 437–486 (1995)
[10] Hyland, J.M.E., Ong, L.C.-H.: Pi-calculus, dialogue games and PCF. In: Proceedings of FPCA 1995, pp. 96–107. ACM Press, New York (1995)
[11] Hyland, J.M.E., Ong, L.C.-H.: On full abstraction for PCF: I, II, and III. Information and Computation 163(2), 285–408 (2000)
[12] Kobayashi, N., Pierce, B.C., Turner, D.N.: Linear types and the Pi-Calculus. In: 23rd ACM SIGPLAN-SIGACT Symposium on Principles of Programming Languages - POPL, pp. 358–371. ACM Press, New York (1996)
[13] Kobayashi, N., Pierce, B.C., Turner, D.N.: Linearity and the Pi-Calculus. ACM Transactions on Computational Logic 21(5), 914–947 (1999)
[14] Laneve, C., Victor, B.: Solos in concert. Mathematical Structures in Computer Science 13(5), 657–683 (2003)
[15] Milner, R.: Communication and Concurrency. Prentice-Hall, Englewood Cliffs (1889)
[16] Milner, R.: Functions as processes. In: Paterson, M. (ed.) ICALP 1990. LNCS, vol. 443, pp. 167–180. Springer, Heidelberg (1990)
[17] Milner, R.: Functions as processes. Mathematical Structures in Computer Science 2(2), 119–141 (1992)
[18] Milner, R.: The polyadic π-calculus: A tutorial. In: Bauer, F.L., Brauer, W., Schwichtenberg, H. (eds.) Logic and Algebra of Specification, Proceedings of International NATO Summer School, Marktoberdorf, Germany, 1991, vol. 94. Springer, Heidelberg (1993)
[19] Paolini, L.: A stable programming language. Information and Computation 204(3), 339–375 (2006)
[20] Paolini, L., Piccolo, M.: Semantically linear programming languages. In: 10th International ACM SIGPLAN Symposium on Principles and Practice of Declarative Programming, Valencia, Spain, pp. 97–107. ACM, New York (2008)
[21] Parrow, J., Victor, B.: The fusion calculus: Expressiveness and symmetry in mobile processes. In: Thirteenth Annual Symposium on Logic in Computer Science (LICS), Indiana, pp. 176–185. IEEE Computer Society Press, Computer Society Press, Los Alamitos (1998)
[22] Plotkin, G.D.: Call-by-name, call-by-value and the λ-calculus. Theoretical Computer Science 1, 125–159 (1975)
[23] Plotkin, G.D.: LCF considerd as a programming language. Theoretical Computer Science 5, 225–255 (1977)
[24] Ronchi Della Rocca, S., Paolini, L.: The Parametric λ-Calculus: a Metamodel for Computation. In: Texts in Theoretical Computer Science. An EATCS Series, Springer, Berlin (2004)
[25] Sangiorgi, D.: An investigation into functions as processes. In: Brookes, S.D., Main, M.G., Melton, A., Mislove, M.W., Schmidt, D.A. (eds.) MFPS 1993. LNCS, vol. 802, pp. 143–159. Springer, Heidelberg (1994)
[26] Sangiorgi, D.: Internal mobility and agent-passing calculi. In: Fülöp, Z., Gecseg, F. (eds.) ICALP 1995. LNCS, vol. 944, pp. 672–683. Springer, Heidelberg (1995)
[27] Sangiorgi, D., Walker, D.: The π-calculus: a theory of mobile processes. Cambridge University Press, Cambridge (2001)
[28] Yoshida, N., Berger, M., Honda, K.: Strong normalisation in the PI-calculus. Information and Computation 191(2), 145–202 (2004)

Some Complexity and Expressiveness Results on Multimodal and Stratified Proof Nets

Luca Roversi[1] and Luca Vercelli[2,*]

[1] Dip. di Informatica, Univ. di Torino
http://www.di.unito.it/~rover/
[2] Dip. di Matematica, Univ. di Torino
http://www.di.unito.it/~vercelli/

Abstract. We introduce a multimodal stratified framework MS that generalizes an idea hidden in the definitions of Light Linear/Affine logical systems: "More modalities means more expressiveness". MS is a set of building-rule schemes that depend on parameters. We interpret the values of the parameters as modalities. Fixing the parameters yields deductive systems as instances of MS, that we call *subsystems*. Every subsystem generates stratified proof nets whose normalization preserves *stratification*, a structural property of nodes and edges, like in Light Linear/Affine logical systems. A first result is a sufficient condition for determining when a subsystem is strongly polynomial time sound. A second one shows that the ability to choose which modalities are used and how can be rewarding. We give a family of subsystems as complex as Multiplicative Linear Logic — they are linear time and space sound — that can represent Church numerals and some common combinators on them.

Keywords: Implicit Computational Complexity, Structural Proof-theory, Linear Logic, Polynomial Time Computations.

1 Introduction

This work relates to Implicit Computational Complexity (ICC), an area of Theoretical Computer Science that explores machine-independent characterizations of complexity classes.

Motivations. We are interested in polynomial time computations and in their characterizations by means of restrictions of Linear Logic (LL) [1]. Specifically, we focus on Light Affine Logic (LAL) [2], a simplification of Light Linear Logic (LLL) [3]. LAL is: (i) strongly polynomial time sound, and (ii) polynomial time complete, under the Curry-Howard (CH) correspondence. (i) means that every derivation Π of LAL normalizes in a time bounded by a polynomial in the size $|\Pi|$ of Π, under any normalization strategy. (ii) says that every polynomial time Turing machine can be represented as a derivation of LAL.

Stratification is the key feature to obtain the bound in (i). Stratification is a property of nodes and edges, invariant under cut elimination, and follows from a careful interplay

* Both the authors have been supported by MIUR PRIN CONCERTO — protocol number 2007BHXCFH.

S. Berardi, F. Damiani, and U. de'Liguoro (Eds.): TYPES 2008, LNCS 5497, pp. 306–322, 2009.

of bang (!) and paragraph (§) boxes. !-boxes are those we know since the introduction of LL, while §-boxes are specific to LAL. Operationally, both kinds of boxes can merge. In particular, !-boxes can merge into §-boxes. As usual, !-boxes are the sub-proof nets that can be duplicated. The key role of merging ! and §-boxes is twofold. On one side, merging realizes the stratification: the number of boxes enclosing a node/edge cannot change under cut elimination. On the other side, it gives expressiveness to the logical system. Without §-boxes Church numerals, used as iterators, could not exist in LAL.

Contributions. This work should be viewed as a first step towards making the slogan: "More modalities, more expressiveness" effective. We introduce the framework MS, a set of *proof nets building-rule schemes* that compose *nodes* to obtain proof nets. The edges labelling these formulæ, besides the usual LL connectives $\multimap, \otimes, \forall$, may contain logical operators to be interpreted as modalities. The choice of which modalities using is, in principle, arbitrary. So, we can have many more than the two ones of LAL inside our formulæ. This is why we dub MS as being "multimodal".

Part of the nodes of MS are *standard*, corresponding to axiom, cut, linear implication, tensor and universal quantification of LL proof nets. Instead, the *modal* nodes serve to contract or weaken modal formulæ and to build boxes relatively to all the modalities we might be interested to deal with, in some fixed set of multimodal proof nets. Indeed, modal nodes and building-rule schemes of MS depend on parameters to be instantiated with modalities. For example, the contraction node of MS is $\mathbf{Y}_q(n, m)$, with q, n, m modalities. If we let $q, n, m = 1$, its instance $\mathbf{Y}_1(1, 1)$ represents the contraction node of LAL, under the assumption that $!^1$ is ! inside MS. Analogously, we can represent the rule that builds "bang" boxes of LAL inside MS. Indeed, MS has a *Promotion building-rule scheme* $P_q(m_1, \ldots, m_k)$, with $q, m_1, \ldots, m_k$ arbitrary modalities. $P_q(m_1, \ldots, m_k)$ can be applied to every proof net $\Pi : A_1, \ldots, A_k \vdash B$ of MS, with assumptions of type $A_1, \ldots, A_k$ and conclusion of type B. Its application yields a new proof net $\Pi' : !^{m_1} A_1, \ldots, !^{m_k} A_k \vdash !^q B$. If we set $k \leq 1$, and $q, m_k = 1$, then both $P_1(1), P_1()$ represent the !-box building-rules of LAL in MS. In particular, since we have $P_q(m_1, \ldots, m_k)$, which puts a modality in front of all $A_1, \ldots, A_k, B$ of any Π, and since we do not have nodes corresponding to dereliction and digging, the sets of proof nets, generated by instantiating the schemes of MS, is stratified. This is why, besides "multimodal", we dub MS also as "stratified".

Our goal is to propose MS *as a generator of strong polynomial time sound systems, obtained as subsystems of* MS *by instances of node and building-rule schemes. The subsystems would generate stratified proof nets, typed with formulæ whose modalities control the normalization complexity.*

It should not be surprising that not every subsystem $\mathcal{P}$ of MS can be polynomial time sound. We shall prove that both the use of a finite number of modalities and a bounded number of *spindles* in the proof nets of $\mathcal{P}$, assures $\mathcal{P}$ is strongly polynomial time sound. "Spindle" is a technical notion pinpointing the proof net structure that, if iteratively composed an unbounded number of times, yields unsound polynomial time normalizations. The proof of strong polynomial time soundness exploits the Context Semantics in [4].

We conclude by showing how MS may be potentially useful to discover systems with interesting complexity bounds, just "playing" with modalities. We show how to instantiate the schemes of MS to obtain a family $\{\mathcal{P}^{M}_{\mathbb{LTS}} | M \in \mathbb{N}\}$ of systems such that every $\mathcal{P}^{M}_{\mathbb{LTS}}$ is *linear time-space sound* subsystem: every proof net $\Pi \in \mathcal{P}^{M}_{\mathbb{LTS}}$ both normalizes in linear time and in linear space. So $\mathcal{P}^{M}_{\mathbb{LTS}}$ belongs to the same complexity class as Multiplicative Linear Logic (MLL), but it extends MLL, since some Church numerals and some common combinators on them exist as proof nets of $\mathcal{P}^{M}_{\mathbb{LTS}}$, for every M.

Related works. Our multimodal framework for polynomial time computations may recall ramified systems [5,6]. The formal relation is still officially unclear, even though the results claimed in [7] support the idea that modalities will result in a formal, logically founded, refinement of ramification.

[8] introduces a *by-levels* analysis of elementary and polynomial time computations with a motivation orthogonal to ours: levels, in the spirit of 2-Sequents [9,10], rule out the stratified structure of boxes. The effect seems analogous to the one we obtain with MS, since the resulting logical systems generalize the original ones. The relation among levels in [8] and our multimodalities has to be clarified.

Acknowledgments. We thank Ugo Dal Lago and Marco Gaboardi for the useful discussions leading to this work and the three referees for their high quality comments and suggestions.

Paper outline. We introduce MS in Section 2, the sufficient geometrical condition to determine which are the polynomial time subsystems of MS in 3, and the linear time-space subsystem of MS, strictly extending MLL, in 4. Conclusions, and further work, are in Section 5.

2 The Framework MS

We proceed by: (i) giving the logical formulæ that label the edges of the proof nets, (ii) introducing the *nodes* and the inductive process to build the proof nets, (iii) defining some static measures on the proof nets, (iv) fixing the main normalization steps on the proof nets, and (v) setting the dynamic measures to assess the normalization cost.

***Formulae of* MS.** Let $M \in \mathbb{N}$ and $\mathcal{V}$ be a countable set of propositional variables, ranged over by $x, y, w, \ldots$. For every fixed M, $\mathcal{F}^M$ is the set of the *formulæ* generated by

$$F ::= L \mid E \quad L ::= x \mid F \otimes F \mid F \multimap F \mid \forall x.F \quad E ::= !^n F \quad (n \in \{1,2,\ldots,M\})$$

using F as start symbol. $\mathcal{F}$ is $\bigcup_{M \in \mathbb{N}} \mathcal{F}^M$. E is the start symbol of *modal* formulæ; L the one of *linear* or *non-modal* formulæ. $A, B, C, \ldots$ will range over formulæ belonging to $\mathcal{F}$, and $\Gamma, \Delta, \Phi, \Psi$ over multisets of formulæ. $A\left[{}^{B}/_{y}\right]$ will denote substitution of B for y in A.

***The nodes for the proof nets of* MS.** They are in Figure 1 together with their *abbreviated* and *long names*, respectively. $u, v, w, \ldots$ will range over (occurrences) of nodes.

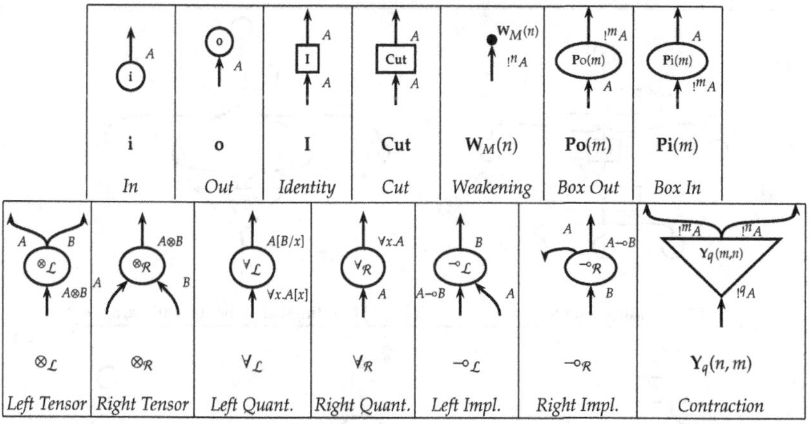

Fig. 1. The nodes, with $M, m, n, q \in \mathbb{N}$

The proof nets of MS. The nodes $\mathbf{i}, \mathbf{I}$, and $\mathbf{o}$, connected as in the leftmost graph of Figure 2 form a proof net. Moreover, if the two rightmost graphs in Figure 2 are proof nets, then all the graphs in Figure 3 are proof nets as well, built using the associated *building-rule scheme*. Namely, we build the proof nets as we were using a sequent calculus rule schemes. For example, "$\mathbf{Y}_q(n,m)$ *building-rule scheme with* $m, n, q \in \mathbb{N}$" says that we build a new proof net starting from the leftmost generic proof net in Figure 2, eliminating two of its $\mathbf{i}$ nodes, plugging the two outgoing edges of a $\mathbf{Y}_q(n,m)$ node into the two dangling edges of Π, labeled $!^n A$ and $!^m A$.

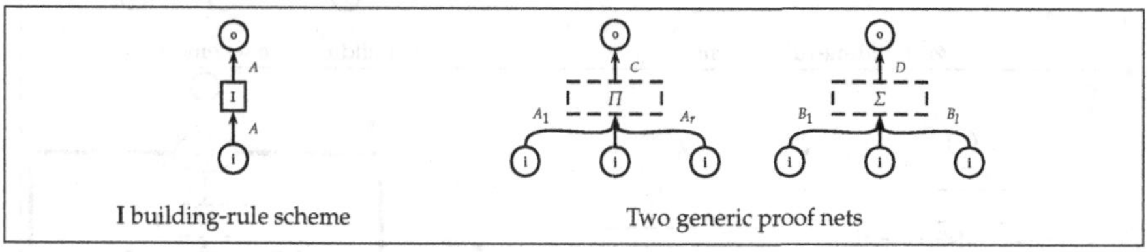

Fig. 2. The basic proof net and two generic proof nets

Another example is "$P_q(m_1, \ldots, m_r)$ *building-rule scheme with* $q, m_1, \ldots, m_r, r \in \mathbb{N}$". It *simultaneously* introduces exactly a single $\mathbf{Po}(m)$ node and a, possibly empty, sequence of $\mathbf{Pi}(m)$ nodes, to form a *(modal) box*, starting from the leftmost generic proof net in Figure 2. The notation $P_q(m_1, \ldots, m_r)$ summarizes the parameters of the box it introduces: q is the (index of the) modality of the edge outgoing $\mathbf{Po}(m)$, and every m_i, with $0 \leq i \leq r$, is the (index of the) modality associated to the edge incoming the ith instance of $\mathbf{Pi}(m)$, counting from the left. We shall tend to identify the unique instance of $\mathbf{Po}(m)$, used by $P_q(m_1, \ldots, m_r)$, with the box it introduces. Namely, we shall call *box* that instance of $\mathbf{Po}(m)$. Finally, we want to remark that the $\mathbf{W}_M(n)$ building-rule scheme only operates on a modal formula $!^n A$. This choice simplifies the

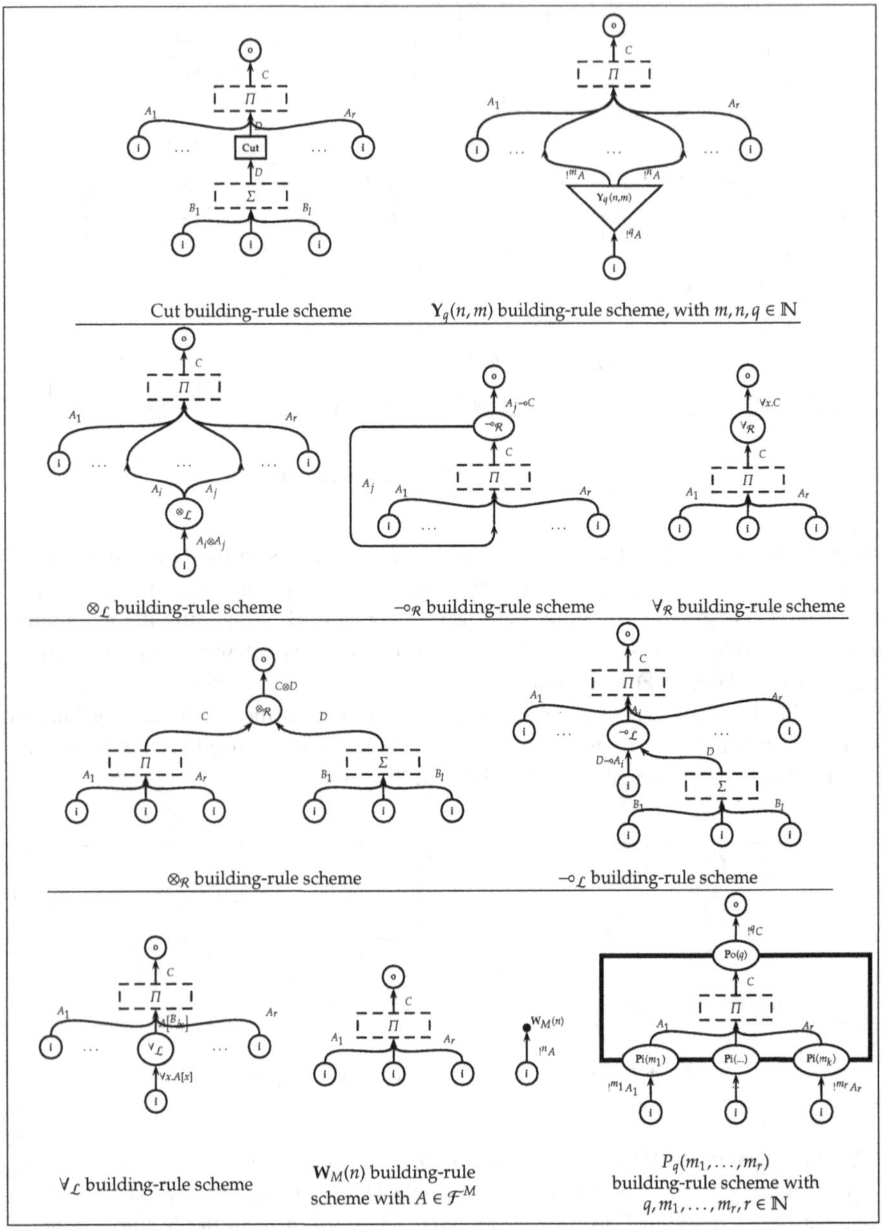

Fig. 3. The building-rule schemes that inductively define the proof nets

presentation of MS. However, we could relax the building scheme-rule of $\mathbf{W}_M(n)$ to any A, without affecting our results, concerning the complexity. Now, let Π be a proof net. V_Π is the set of its nodes. E_Π is the set of its edges. B_Π is the set of its $\mathbf{Po}(m)$ nodes. Moreover, $P_\Pi : B_\Pi \to \mathbb{N}$ counts the premises of a box.

Static measures. Let Π be a proof net. $\partial(u)$ denotes the *depth*, or *level*, for every $u \in V_\Pi \cup E_\Pi$. Specifically, $\partial(u)$ is the greatest number of (nested) boxes that contain u. $\partial(\Pi)$ denotes the *depth of* Π, the greatest $\partial(u)$, for every $u \in V_\Pi \cup E_\Pi$.

V_Π, E_Π, and B_Π can be made relative to a given level $0 \leq d \leq \partial(\Pi)$. Namely, V^d_Π is the set of nodes at level d. Analogous definitions hold for E^d_Π, B^d_Π.

$\mathrm{s}(\Pi)$ denotes the *size* of Π. It counts the number of the nodes in Π. $\mathrm{s}_d(\Pi)$ denotes the *size of* Π *at level* d. It counts the number of nodes that occur at level d. Finally, $b_d(\Pi)$ counts the number of boxes, or, equivalently, of occurrences of $\mathbf{Po}(m)$ nodes, at level d.

For example, for the net, say Π, in Figure 9(b) $\partial(\Pi) = 1$, $s_0(\Pi) = 15$ since the occurrences of $\mathbf{Pi}(m)$ and $\mathbf{Po}(m)$ are not inside the box they delimit. $s(\Pi) = 21$ and $b_0(\Pi) = 1$.

***Subsystems of* MS.** We call *subsystems of* MS all the sets of *instances* of the building rules in Figure 2 and 3. $\mathcal{P} \subseteq$ MS *denotes that* $\mathcal{P}$ *is a subsystem of* MS. Two examples are $\mathcal{P} = \{\otimes_\mathcal{L}, \mathbf{I}, \mathbf{Y}_3(1,2), P_1(1,1)\}$ and $\mathcal{P}' = \{\otimes_\mathcal{L}, \mathbf{I}, \mathbf{Y}_1(1,2), P_1(1), P_1(2,3)\}$. We also use regular expressions to identify sets of scheme instances. $P_q(a^?, m_1, \ldots, m_r)$ denotes $\{P_q(m_1, \ldots, m_r), P_q(a, m_1, \ldots, m_r)\}$, while $P_q(a^*, m_1, \ldots, m_r)$ denotes the infinite set $\{P_q(m_1, \ldots, m_r), P_q(a, m_1, \ldots, m_r), P_q(a, a, m_1, \ldots, m_r), \ldots\}$. Finally, the *number of modalities of any* $\mathcal{P} \subseteq$ MS is the number of modalities occurring in the instances of building rule-schemes that form $\mathcal{P}$.

Normalization. The *normalization steps (norm. step)* on the proof nets of MS are identified by: $[\otimes_\mathcal{R}/\otimes_\mathcal{L}]$, $[\multimap_\mathcal{R}\ /\ \multimap_\mathcal{L}]$, $[\forall_\mathcal{R}/\forall_\mathcal{L}]$, $[P_q(m_1, \ldots, m_r)/\mathbf{W}_M(q)]$, $[P_n(n_1, \ldots, n_l)/P_q(n, m_1, \ldots, m_r)]$, $[P_q(m_1, \ldots, m_r)/\mathbf{Y}_q(n, m)]$, $[\mathbf{I}/\_]$,$[\_/\mathbf{I}]$. The first three norm. steps are the standard ones, relative to the logical operators that occur in the name. The fourth one corresponds to a standard box erasure by a weakening node. The fifth norm. step merges two boxes in the obvious way. Figure 4 shows the details of the remaining two norm. steps. Also, the *linear norm. steps* are $[\otimes_\mathcal{R}/\otimes_\mathcal{L}]$, $[\multimap_\mathcal{R}\ /\ \multimap_\mathcal{L}]$, $[\forall_\mathcal{R}/\forall_\mathcal{L}]$, $[\mathbf{I}/\_]$, $[\_/\mathbf{I}]$. The *polynomial* ones are $[P_n(n_1, \ldots, n_l)/P_q(n, m_1, \ldots, m_r)]$, $[P_q(m_1, \ldots, m_r)/\mathbf{Y}_q(n, m)]$, $[\mathbf{I}/\_]$, $[\_/\mathbf{I}]$. The *garbage* steps are $[P_q(m_1, \ldots, m_r)/\mathbf{W}_M(q)]$, $[\mathbf{I}/\_]$, $[\_/\mathbf{I}]$. In particular, the underscore of $[\mathbf{I}/\_]$ and $[\_/\mathbf{I}]$ stands for any logical operator or building scheme-rule. $\rightarrow$ will be the contextual closure of $\triangleright$, so that $\Pi \rightarrow \Sigma$ whenever Π rewrites to Σ. $\rightarrow^*$ is the reflexive and transitive closure of $\rightarrow$, $\Pi \rightarrow^n \Sigma$ denotes *exactly* n steps of $\Pi \rightarrow^* \Sigma$, $\Pi \rightarrow_d \Sigma$ is a norm. step at depth d, and, finally, $\Pi \rightarrow^n_\mathrm{d} \Sigma$ denotes *exactly* n steps of $\Pi \rightarrow^* \Sigma$ at depth d. A proof net Π is *normal* when none of the above norm. steps rewrites it.

Notice that $[P_q(m_1, \ldots, m_r)/\mathbf{Y}_q(n, m)]$ implies that, in general, $\rightarrow$ is not deterministic. Nevertheless, every reduction always terminates. It is enough to consider the *forgetful map* from proof nets of MS to proof nets of ELL that transforms every modality $!^n$ into !. Under that mapping, every norm. step in MS becomes a cut elimination step in ELL.

We observe that the instances of the norm. steps $[P_n(n_1, \ldots, n_l)/P_q(n, m_1, \ldots, m_r)]$, and $[P_q(m_1, \ldots, m_r)/\mathbf{Y}_q(n, m)]$, induced by fixing some $\mathcal{P}$ in MS, may not lead to a proof net of $\mathcal{P}$, starting from a proof net of $\mathcal{P}$. This is why the set of norm. steps we can consider as valid for any given $\mathcal{P}$ in MS, are those ones that map a proof net of $\mathcal{P}$ to a proof net of $\mathcal{P}$.

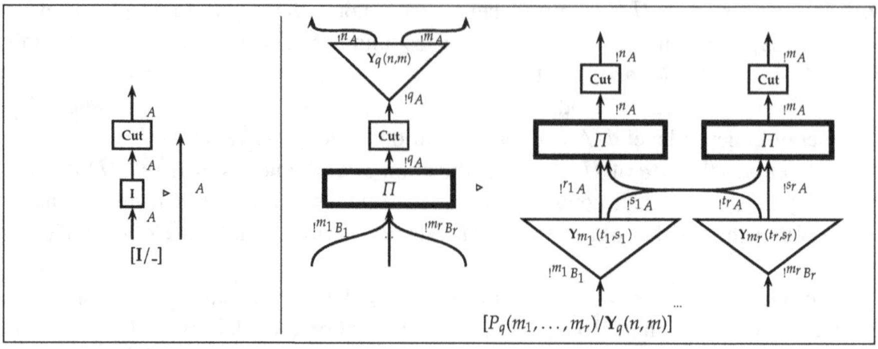

Fig. 4. The norm. steps [**I**/_] and $[P_q(m_1,\ldots,m_r)/\mathbf{Y}_q(n,m)]$

Focusing on any given $\Pi \to \Sigma$ we can see that every node x of Σ different from a cut can be mapped back to its (unique) *source* node y: x is the *residual* of y. Conversely, not every node has a residual. Finally, we remark that both the structure of proof nets and the behavior of the norm. steps make MS *stratified*: the residual x of any node y at level d different from a cut in a given proof net Π keeps being at level d.

Dynamic measures. Let Π be a proof net. $[\Pi]$ is the *reduction time* of Π. It is defined as $\max_{\Pi \to^n \Sigma} n$. $\|\Pi\|$ is the *used space* of Π. It is defined as $\max_{\Pi\to^*\Sigma} \mathrm{s}(\Sigma)$. If $0 \leq i, d \leq \partial(\Pi)$, then $[\Pi]^d = \max_{\Pi\to^n_d\Sigma} n$, $\|\Pi\|^d = \max_{\Pi\to^*_d\Sigma} \mathrm{s}(\Sigma)$ and $\|\Pi\|^d_i = \max_{\Pi\to^*_d\Sigma} \mathrm{s}_i(\Sigma)$.

3 Polynomial Time and MS

We proceed by: (i) defining what a *polynomial time sound* subsystem of MS is, (ii) recalling *Context Semantics* [4], (iii) defining the notion of *spindle*, and (iv) using (ii) and (iii) to assess the normalization cost.

Strong polynomial time sound subsystems. Let $\mathcal{P} \subseteq$ MS. $\mathcal{P}$ is *strongly polynomial step* whenever for every $d \in \mathbb{N}$ there is a polynomial $p_d(n)$ such that, for every $\Pi \in \mathbf{PN}(\mathcal{P})$, with $\partial(\Pi) \leq d$, $[\Pi] \leq p_d(\mathrm{s}(\Pi))$.

$\mathcal{P}$ is *strongly polynomial size* whenever for every $d \in \mathbb{N}$ there is a polynomial $p_d(n)$ such that, for every $\Pi \in \mathbf{PN}(\mathcal{P})$, with $\partial(\Pi) \leq d$, $\|\Pi\| \leq p_d(\mathrm{s}(\Pi))$.

$\mathcal{P}$ is *strongly polynomial time sound* (or ptime) if it is both polynomial step and size. PMS denotes the class of the ptime subsystems of MS. This distinction is meaningful since, for example, the multiplicative and exponential fragment of Elementary Affine Logic is a subsystem of MS. Our goal is to give a sufficient condition to say when any $\mathcal{P} \subseteq$ MS belongs to PMS.

Context Semantics. We recall and simplify Context Semantics in [4], developed for quantitative analysis of Linear Logic. The simplification works because we rule out both *digging* and *dereliction*, and we perform the reductions *level-by-level*. This is the reason why we need just one stack instead of the traditional two stacks. *Multimodality* is harmless because we manage all the modalities as they were equal.

Exponential signatures are generated by $t ::= \mathtt{e} \mid \mathtt{r}(t) \mid \mathtt{l}(t)$. $\mathcal{E}$ is their set. A *stack element* belongs to $S = \{\mathtt{a}, \mathtt{o}, \mathtt{f}, \mathtt{s}, \mathtt{x}\} \cup \mathcal{E}$. A *stack* s is a finite non empty sequence of stack elements, i.e. $s \in S^+$. A *polarity* is an element of $\mathcal{B} = \{+, -\}$. If c is a polarity, we denote $c \downarrow$ the other one. If Π is a proof net, a *context* of Π is an element of $C_\Pi = E_\Pi \times S^+ \times \mathcal{B}$. If $C, C' \in C_\Pi$ we write $C \mapsto_\Pi C'$ if C' can be obtained from C through the *rewriting relation* in Figure 5.

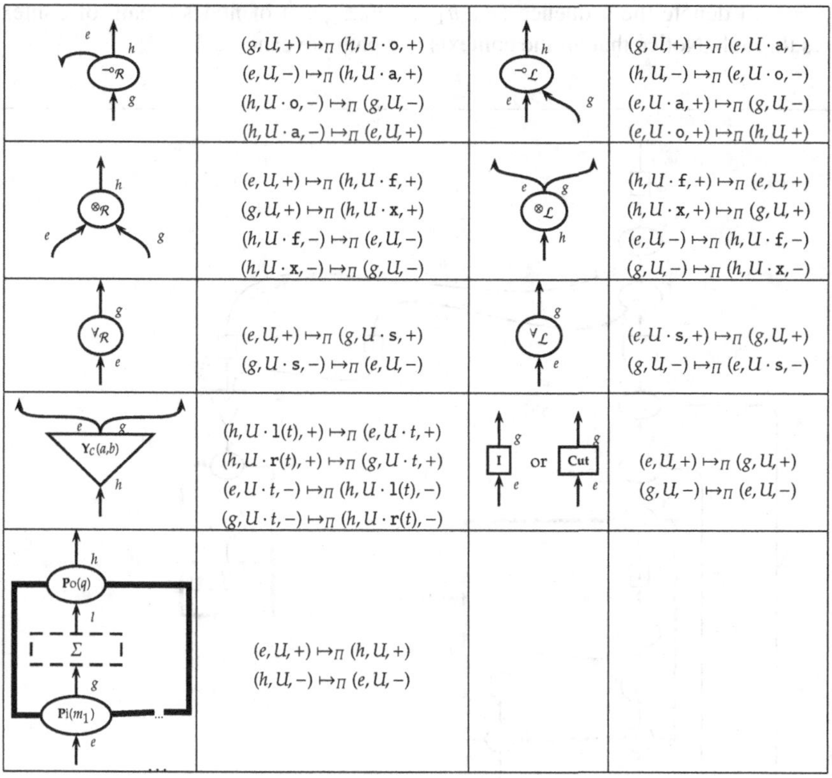

Fig. 5. Rewriting Relation among contexts. Notice that if $(e, U, b) \mapsto_\Pi (e', U', b')$ then also $(e', U', b' \downarrow) \mapsto_\Pi (e, U, b \downarrow)$.

We say that: (i) $(e, t \cdot U, b)$ is *canonical* if t is an exponential signature, U does not contain exponential signatures, and whenever U contains an even number of a's, then b is $+$, otherwise, whenever U contains an odd number of a's, then b is $-$, (ii) an *initial* context is $(e, t, +)$ where t is an exponential signature, and (iii) a context C is *final* if $\nexists D\,(C \mapsto_\Pi D)$.

Canonical, initial and final contexts compose paths to travel along the edges of a proof net. A path simulates the annihilation of pairs of nodes by a normalization step . The goal is to use paths to walk through a net from any box root to either a weakening node, erasing it, or to the terminal node of the whole net or of a proof net inside the box. Many canonical paths from the same box root means many contraction nodes that, possibly, will duplicate the box. An example of maximal path is in Figure 6.

Definition 1 (Paths and Maximal Paths). *Let $\Pi \in \mathbf{PN}(\mathsf{MS})$, $u_0 \in B_\Pi$, $u_n \in V_\Pi$. A* path *from u_0 to u_n is a finite sequence of contexts $\tau = (C_1, \ldots, C_n) \in {C_\Pi}^*$ such that (i) $C_1 = ((u_0, u_1), U_1, b_1)$ is initial, (ii) $C_n = ((u_{n-1}, u_n), U_n, b_n)$, and (iii) $C_i \mapsto_\Pi C_{i+1}$, for every $1 \leq i \leq n-1$. Moreover, we say that τ is* maximal *whenever C_n is final.*

Also, τ will denote the sequence $(u_0, u_1 \ldots, u_{n-1}, u_n)$ of nodes a path of contexts passes through. Notice that all the contexts in a path are canonical.

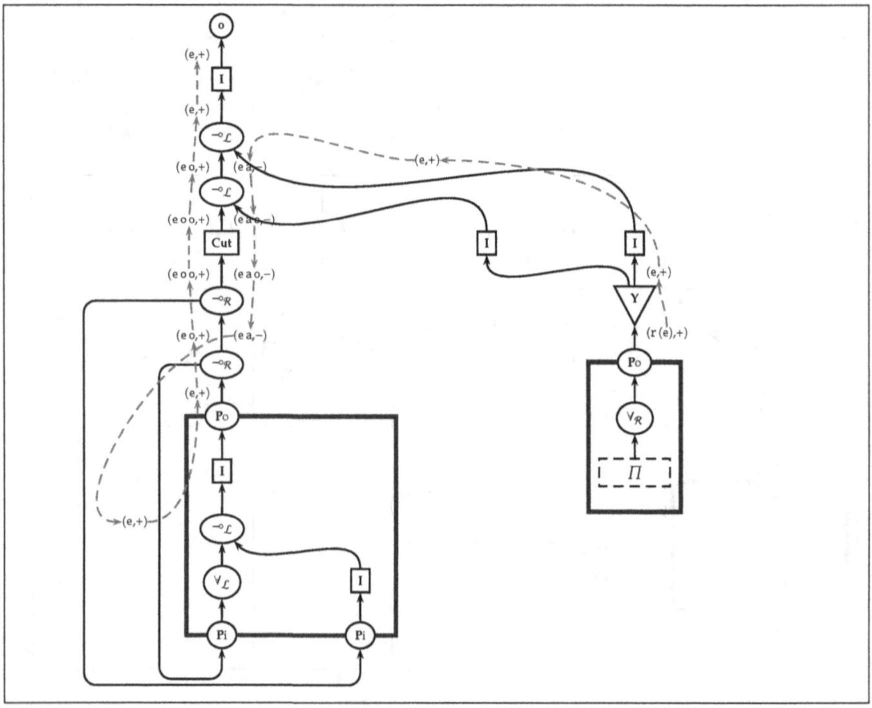

Fig. 6. An example of maximal path

Definition 2 (Number of Paths). *Let $\Pi \in \mathbf{PN}(\mathsf{MS})$, and $b \in B_\Pi$. $R_\Pi(b)$ is the number of maximal paths from b to some $v \in V_\Pi$. It is called* number of paths of b.

Lemma 1 (Number of Copies). *Let $\Pi \rightarrow^*_{\mathrm{d}} \Sigma$, $b \in B^d_\Pi$ and Θ the proof net inside b at depth $d+1$. Then, in Σ, there are at most $R_\Pi(b)$ equal residuals ("copies") of Θ.*

So, sometimes we will call *number of copies of Π* the number $R_\Pi(b)$. The proof generalizes the work of [4].

Canonical Reductions. A reduction sequence $\Pi = \Pi_0 \to^* \Pi'_0 \to^* \Pi_1 \to^* \Pi'_1 \to^* \ldots \to^* \Pi'_\partial \to^* \Pi_{\partial+1} \to^* \Sigma$ is *canonical* if $\partial = \partial(\Pi)$, $\Pi_i \to^*_i \Pi'_i$, reducing all the linear norm. steps at level i, $\Pi'_i \to^*_i \Pi_{i+1}$, reducing all the polynomial norm. steps at level i, and $\Pi_{\partial+1} \to^* \Sigma$, reducing all the garbage steps. The coming lemmas will imply Proposition 1. Firstly, a generalization of the polynomial soundness in [2], provable by induction on l, is:

Lemma 2 (Canonical Reductions Are The Worst Ones). *Let Π be a proof net. For every reduction $\sigma : \Pi = \Pi_0 \to \Pi_1 \to \ldots \to \Pi_l = \Sigma$ there exists another reduction $\tau : \Pi = \Pi'_0 \to \Pi'_1 \to \ldots \to \Pi'_{l'} = \Sigma$ such that (i) τ is canonical, (ii) $l' \geq l$, and (iii) $\max\{\mathrm{s}(\Pi'_i) \mid i \leq l'\} \geq \max\{\mathrm{s}(\Pi_j) \mid j \leq l\}$. Moreover, if σ is a reduction performed in some subsystem $\mathcal{P}$, then also τ is performed in $\mathcal{P}$.*

Lemma 3 (A Condition for Polynomiality). *Let $\mathcal{P} \subseteq$ MS. Let us assume that for every $\partial \in \mathbb{N}$ there exist two polynomials $p_\partial(x), q_\partial(x)$ such that, for every $\Pi \in \mathbf{PN}(\mathcal{P})$ with $\partial(\Pi) \leq \partial$ and for every $d \leq \partial(\Pi)$, $[\Pi]^d \leq p_\partial(\mathrm{s}(\Pi))$ and $\|\Pi\|^d \leq q_\partial(\mathrm{s}(\Pi))$. Then $\mathcal{P} \in$ PMS.*

Proof. The result holds for canonical reductions because iterating ∂ times a polynomial still gives a polynomial. Then, Lemma 2 implies the thesis. □

Weight for the proof nets. $\mathrm{T}_d(\Pi)$ is $\sum_{u \in V^d_\Pi} \mathrm{T}_d(\Pi, u)$, where[1]:

$$\mathrm{T}_d(\Pi, u) \stackrel{\text{def}}{=} \begin{cases} 1 & u \in \{\mathbf{I}, \mathbf{Cut}, \mathbf{W}_M(n), \mathbf{Pi}(m), \mathbf{i}, \mathbf{o}\} \\ 3 & u \in \{\otimes_\mathcal{L}, \otimes_\mathcal{R}, \forall_\mathcal{L}, \forall_\mathcal{R}, \multimap_\mathcal{L}, \multimap_\mathcal{R}, \mathbf{Y}_q(m,n)\} \\ 2 \cdot (P_\Pi(u)+1) \cdot R_\Pi(u)^2 & u \in \{\mathbf{Po}(m)\}. \end{cases}$$

Lemma 4 (The Weight bounds Time and Space). *Let Π be a proof net. (i) $[\Pi]^d$, $\|\Pi\|^d_d \leq T_d(\Pi)$, and (ii) $\|\Pi\|^d_i \leq \mathrm{s}_i(\Pi) \cdot T_d(\Pi)$. So, we also have $\|\Pi\|^d \leq \mathrm{s}(\Pi) \cdot T_d(\Pi)$.*

Proof

(i) Let us consider a reduction $\Pi \to^n_d \Sigma$. A case analysis on cuts shows that $T_d(\cdot)$ strictly decreases during reduction. So $n \leq T_d(\Pi)$. Since every node has weight at least 1, $\mathrm{s}_d(\Pi) \leq T_d(\Pi)$. So $\|\Pi\|^d_d \leq T_d(\Pi)$.

(ii) Let us consider a reduction $\Pi \to^*_d \Sigma$ and fix a level $i > d$. For every u node at level i, inside a box $b \in B^d_\Pi$, at most $R_\Pi(b)$ copies of u will appear in Σ (Lemma 1). Let $b_1, \ldots, b_k$ be all the boxes at level d, and $S_1, \ldots, S_k$ their sizes at level i. Then, $\|\Pi\|^d_i \leq S_1 \cdot R_\Pi(b_1) + \ldots + S_k \cdot R_\Pi(b_k) \leq \mathrm{s}_i(\Pi) \cdot T_d(\Pi)$, since $R_\Pi(b_i) \leq \mathrm{T}_d(\Pi, b_i)$. □

Proposition 1 (Polynomially bounded R_Π implies Ptime). *Let $\mathcal{P} \subseteq$ MS. Let us assume there exists a polynomial $r(x)$ such that, for every proof net Π of $\mathcal{P}$ and for every box $b \in B_\Pi$, $R_\Pi(b) \leq r(\mathrm{s}(\Pi))$. Then, $\mathcal{P} \in$ PMS.*

[1] We adapt the *modified weight* in [4]. We notice that we might have chosen a simpler definition of weight, for example with $\mathrm{T}_d(\Pi, u) = 1$, for every node, but $\mathbf{Po}(m)$. However, we plan to generalize MS with *unconstrained wakening*. So, we adopt a more *flexible* definition than the one strictly required.

Proof. Algebraic manipulations show that $T_d(\Pi) \leq 3\,\mathrm{s}_d(\Pi)+2\,\mathrm{s}_d(\Pi)\cdot\sum_{b\in B^d_\Pi} R_\Pi(b)$ $\leq 3\,\mathrm{s}(\Pi) + 2\,\mathrm{s}^2(\Pi)\cdot r(\mathrm{s}(\Pi))$. Lemma 4 and Lemma 3 imply the statement. □

The sufficient condition for ptime we prove relies on the technical notion of *spindle*, a particular configuration of nodes. Too many spindles composed in a proof net of some $\mathcal{P} \subseteq \mathsf{MS}$ may lead to exponential blow up.

Definition 3 (Spindles). *Let $\Pi \in$* **PN**$(\mathcal{P})$*, $u \in V_\Pi$ be a contraction, b be a Box Out node, and d another node (possibly $b = d$), with $\partial(u) = \partial(b) = \partial(d)$. Let τ, ρ, χ be three paths such that: (i) both τ and ρ connect u to b, (ii) τ starts from one of the conclusions of u, and ρ from the other one, (iii) no other paths exist connecting u to b, (iv) χ (possibly empty) is between b and d. The subgraph Σ that contains all and only the nodes of τ, ρ, χ is a* **spindle** *between u and d.*[2]

Definition 4 (Chains of Spindles). *Let $u_1, \ldots, u_r$ be contractions and $d_1, \ldots, d_r$ other nodes of $\Pi \in$* **PN**$(\mathcal{P})$*, all at the same depth. Let us assume that, for every $i \leq r$, both there is a spindle Σ_i between u_i and d_i, and d_i is connected through an edge to u_{i+1}. The subgraph Σ that contains all the nodes and edges of $\Sigma_1, \ldots, \Sigma_r$ is a* **chain of spindles between** u_1 **and** d_r *of length $|\Sigma| = r$. Σ is* **dangerous** *if its initial and final edges are labeled by two formulae $!^i A$ and $!^i B$.*

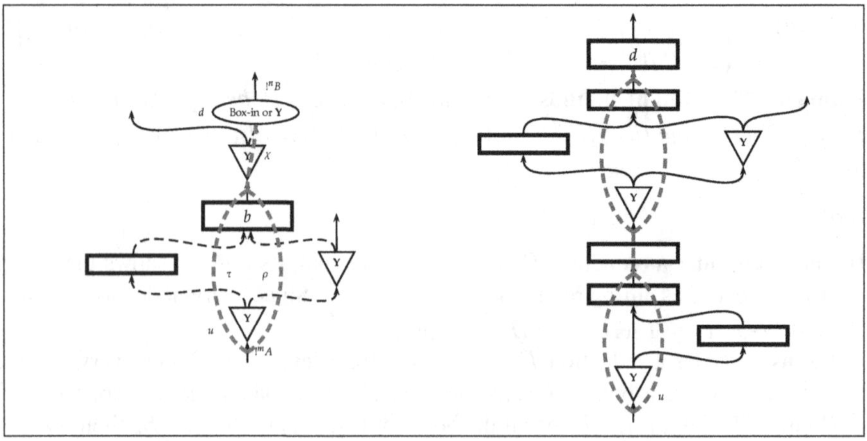

Fig. 7. An example: a spindle and a chain of spindles, both between u and d

If Σ is a dangerous chain of r spindles with both initial and final edges labeled by $!^i A$, using the **Cut** building-rule scheme, Σ can compose with itself an arbitrary number L of times, yielding a chain Θ_L of spindles with length $r \cdot L$.

[2] Point *(iii)* is equivalent to "*τ and ρ share only the node u and the box b, up to* ***linear reductions.***"

Lemma 5 (Bounding the Chains limits the Number of Paths). *Let $\mathcal{P} \subseteq$ MS. Let us assume that every chain of spindles inside a proof net $\Pi \in \mathbf{PN}(\mathcal{P})$ cannot be longer than L, for some fixed $L \in \mathbb{N}$. Then, there is a polynomial $p(x)$ such that $R_\Pi(b) \leq p(\mathrm{s}(\Pi))$, for every $\Pi \in \mathbf{PN}(\mathcal{P})$, and $b \in B_\Pi$.*

Proof. Let b be a fixed box at depth d in Π.

1. We want to classify the links $u \in V_\Pi^d$ in classes that we call class-0, class-1, class-2, ... according to *how many consecutive spindles there are between b and u*. Let us consider all the possible paths τ between b and u. Let us call Σ_b^u the sub-graph that contains all the nodes of all the τ's. Let us consider all the possible chains of spindles $\Phi_b^u, \Psi_b^u, \ldots, \Omega_b^u$ whose nodes are among the nodes of Σ_b^u. u is *class-i* if $i = \max\{|\Phi_b^u|, |\Psi_b^u|, \ldots, |\Omega_b^u|\}$. Notice that there are at most $L+1$ classes $(0, \ldots, L)$ by hypothesis. Also, notice that some links may remain unclassified (e.g. the links that cannot be reached from u).
2. Let m_i be the *number of Contractions* of class-i (they are all at depth d), and $m = m_0 + \ldots + m_L$. We observe that $m \leq \mathrm{s}_d(\Pi)$.
3. For $0 \leq i \leq L$, let Π_i be the *sub-graph containing all and only the class-j nodes*, for every $j \leq i$.
4. By definition, a path τ starting from b is *maximal relatively to Π_i* if there exists a maximal path τ' starting from b whose intersection with Π_i is exactly τ.
5. We call R_i, for $0 \leq i \leq L$, the number of paths starting from b and maximal relatively to Π_i. The definitions imply $R_\Pi(b) = R_L$.

We prove that $R_i \leq \prod_{j=0}^{i}(m_j + 1)$ by induction on i. Two paths separate when we traverse a contraction node, but *never* crossing $\otimes_\mathcal{L}$ or $\otimes_\mathcal{R}$ nodes, which, instead, force to go in one specific direction. In Π_0 there are no spindles, meaning $R_0 \leq m_0 + 1$. By induction, let $R_{i-1} \leq \prod_{j=0}^{i-1}(m_j + 1)$. Then, the paths maximal relatively to Π_i are at most $R_i \leq R_{i-1} \cdot (m_i + 1) \leq \prod_{j=0}^{i}(m_j + 1)$. So, $R_\Pi(b) = R_L \leq \prod_{j=0}^{L}(m_i + 1) \leq \mathrm{s}(\Pi)^{L+1}$. □

Some comments on the strategy we use to prove Lemma 5 are worth doing. Let $b \in B_\Pi^d$ be the box and Σ be the graph we deal with in the proof just concluded. The goal of the proof is to find a bound on the number of copies of b, produced by $\Pi \rightarrow_d^* \Sigma$. The intuition driving the proof strategy lies on the observation that the number of copies of b depends, essentially, on both a *vertical* and a *horizontal* component. The *vertical* component is the number of consecutive spindles that we can have moving upward in Σ, starting from b. Instead, if a spindle exists between b and some node u, then the *horizontal* component counts all the paths from b to u external to any spindle from b to u. Specifically, the proof of Lemma 5 shows that the reduction time is more affected by the vertical component than by the horizontal one. The reason is that the horizontal component can be bounded by $\mathrm{s}_d(\Pi)$, at least restricting to some right u's, as we do.

Proposition 1 and Lemma 5 imply:

Proposition 2 (Bounded Chains imply Ptime). *Let $\mathcal{P} \subseteq$ MS. Let us assume that every chain of spindles inside a proof net $\Pi \in \mathbf{PN}(\mathcal{P})$ cannot be longer than L, for some fixed $L \in \mathbb{N}$. Then $\mathcal{P} \in$ PMS.*

Proposition 3 (The absence of Dangerous Chains imply Ptime). *If the number of modalities that occur in the instances of the building-rule schemes that define $\mathcal{P} \subseteq \mathsf{MS}$ is finite, and no dangerous chain of spindles can occur in any $\Pi \in \mathbf{PN}(\mathcal{P})$, then $\mathcal{P} \in \mathsf{PMS}$.*

Proof (of Proposition 3). Let M be the number of modalities of $\mathcal{P}$. Proposition 2 implies the thesis, if we prove $\mathcal{P}$ only builds chains of spindles at most M long. By contradiction, let $\mathcal{P}$ build a chain of $r > M$ spindles, whose spindles begin with the contractions $u_1, \ldots, u_r$. By the pigeons-hole principle, at least two contractions u_i and u_j share the same input label. If Σ is the chain of spindles between u_i and u_j, Σ is dangerous, yielding a contradiction. □

4 Linear Time and Space Subsystems in MS

Here we want to usefully exploit MS. We define a partial order $(\mathbb{LTS}, \leq_{\mathbb{LTS}})$, and a family $\{\mathcal{P}^M_{\mathbb{LTS}} \mid M \in \mathbb{N}\}$ of subsystems of MS, such that, for every fixed M: (i) $\mathcal{P}^M_{\mathbb{LTS}}$ has at most M modalities, (ii) $\mathcal{P}^M_{\mathbb{LTS}}$ contains proof nets that normalize in linear time and space, and (iii) the first M Church numerals together with a successor, a sum and a function which, essentially, is a predecessor exist as proof nets of $\mathcal{P}^M_{\mathbb{LTS}}$. So, every $\mathcal{P}^M_{\mathbb{LTS}}$ strictly extends the expressiveness of Multiplicative Linear Logic (MLL), whose proof nets, recall, have linear time and space complexity as well.

The partial order $(\mathbb{LTS}, \leq_{\mathbb{LTS}})$. $(\mathbb{LTS}, \leq_{\mathbb{LTS}})$ is a partial order with $\mathbb{LTS}$ as carrier and $\leq_{\mathbb{LTS}}$ as order relation. The antireflexive restriction of $\leq_{\mathbb{LTS}}$ is $<_{\mathbb{LTS}}$. We write $x \gtrless_{\mathbb{LTS}} y$ whenever x, y cannot be compared under $\leq_{\mathbb{LTS}}$. $x \uparrow^z y$ means both that $x \gtrless_{\mathbb{LTS}} y$ and that $x, y <_{\mathbb{LTS}} z$. $x \sqcap y$ denotes the *greatest lower bound* (glb) of $x, y \in \mathbb{LTS}$ under $\leq_{\mathbb{LTS}}$. $(\mathbb{LTS}, \leq_{\mathbb{LTS}})$ is such that, for every $x, y \in \mathbb{LTS}$: (i) there is $\bot \in \mathbb{LTS}$ such that $\bot \leq_{\mathbb{LTS}} x$, and (ii) if $x \uparrow^z y$, for some z, then $x \sqcap y = \bot$.

The family $\{\mathcal{P}^M_{\mathbb{LTS}} \mid M \in \mathbb{N}\}$. Let $x_0 = \bot, x_1, \ldots, x_M$ be distinct elements of $\mathbb{LTS}$. $\mathcal{P}^M_{\mathbb{LTS}}$ is a subsystem of MS whose modal rules, in Figure 8, only use the modalities $!^0 = \S, !^1, \ldots, !^M$.

The structure of $(\mathbb{LTS}, \leq_{\mathbb{LTS}})$ might look somewhat "overdimensioned" w.r.t. the system we finally obtain. However, intuitively, a system which is linear time and space sound, and which is not the multiplicative fragment of Linear Logic, necessarily requires a careful control over the ways we can compose structure, the goal being forbidding any situation where the size grows too much. Specifically, $(\mathbb{LTS}, \leq_{\mathbb{LTS}})$ rules out chains of spindles of length > 1. So, every $\mathcal{P}^M_{\mathbb{LTS}}$ is ptime thanks to Proposition 2. The more, we can prove:

$P_0(0^*)$	$P_{m_0}(m_1, \ldots, m_k)$ whenever $\bot \neq x_{m_0} \leq_{\mathbb{LTS}} x_{m_1}, \ldots, x_{m_k}$
$\mathbf{Y}_k(0, j)$ whenever $x_j <_{\mathbb{LTS}} x_k$	$\mathbf{Y}_k(i, j)$ whenever $x_i \uparrow^{x_k} x_j$
$\mathbf{W}(i)$ for every $x_i \in \mathbb{LTS}$	

Fig. 8. The modal rules of $\mathcal{P}^M_{\mathbb{LTS}}$, with $m_0, \ldots, m_k \in \{1, \ldots, M\}$

Proposition 4 (Linear Time and Space Soundness of $\mathcal{P}^M_{\mathrm{LTS}}$.). *Let $\mathcal{P}^M_{\mathrm{LTS}}$ be given. There exist k_1, k_2 such that, for every $\Pi \in \mathbf{PN}\left(\mathcal{P}^M_{\mathrm{LTS}}\right)$, $\|\Pi\| \leq k_1 \cdot \mathrm{s}(\Pi)$ and $[\Pi] \leq k_2 \cdot \mathrm{s}(\Pi)$.*

Proof. Let b be a box at depth d in Π. We recall that the paths outgoing b split only in Contraction nodes. Also, we observe that they can build only spindles entering a §-box, and that paths, entering a §-box, cannot split anymore, since § never labels the premise of a contraction. So, the paths form a binary tree rooted at b. Finally, the modalities labelling edges along a path decrease. So, no more than M contractions may lie on a single path, and we get $R_\Pi(b) \leq 2^M$ $(\star)$. Now, we recall that $\mathrm{T}_d(\Pi)$ strictly decreases during the normalizations at level d. So:

$$\mathrm{T}_d(\Pi) \leq 3\,\mathrm{s}(\Pi) + 2\sum_{b \in B^d_\Pi} (P_\Pi(b)+1)\cdot R_\Pi(b)^2 \leq 3\,\mathrm{s}(\Pi) + 2\cdot 2^{2M} \sum_{b \in B^d_\Pi} (P_\Pi(b)+1) \leq \mathrm{s}(\Pi)\left(3 + 2^{2M+1}\right)$$

the first and third inequalities holding thanks to algebraic manipulations, and the second one thanks to $(\star)$. So, $[\Pi]^d$, $\|\Pi\|^d_d$ are linear in $\mathrm{s}(\Pi)$ because of Lemma 4.(i). However, for $i > d$, the bound $\|\Pi\|^d_i \leq \mathrm{s}_i(\Pi) \cdot T_d(\Pi)$ implied by Lemma 4.(ii) is too high. To lower it, recall that if u is a link at level $i > d$, inside a box b at level d, u can be copied at most $R_\Pi(b) \leq 2^M$ times: $\|\Pi\|^d_i \leq 2^M \cdot \mathrm{s}_i(\Pi)$ is linear, as well as $\|\Pi\|^d$. The final statement follows from composing, level by level, the obtained bounds. We obtain the constants k_1, k_2 that depend on M and $\partial(\Pi)$. □

Concerning Proposition 4, M limits the height of the chains we build, level by level, in the course of the normalization. An arbitrary M would yield a polynomial size bound, because the (level by level) size of the proof net would replace M as bound on the length of consecutive contractions.

Church numerals. The type of a Church numeral $\overline{n}$ can be $i\mathsf{CN} \equiv \forall\alpha.!^i(\alpha \multimap \alpha) \multimap §(\alpha \multimap \alpha)$, for every $i \in \{0,\ldots,M\}$. The *spine* in Figure 9(a) is the usual key structure to represent a Church numeral $\overline{m}$ as a proof net, like, for example, in LAL.

The sums. A type for the sum can be $j\mathsf{CN} \multimap i\mathsf{CN} \multimap k\mathsf{CN}$, whenever $x_j \uparrow^{x_k} x_i$. Figure 9(b) shows its proof net. It takes $\overline{m}$ and $\overline{n}$ to yield $\overline{m+n}$. A further type for the proof net summing two Church numerals can be $j\mathsf{CN} \multimap i\mathsf{CN} \multimap k\mathsf{CN}$, for every j, i such that $x_j \leq_{\mathrm{LTS}} x_i$, or $x_i \leq_{\mathrm{LTS}} x_j$, and $x_j, x_i <_{\mathrm{LTS}} x_k$. The final proof net would be analogous to the one in Figure 9(b).

A "moral" Predecessor. Every $i \in \mathcal{P}^M_{\mathrm{LTS}}$ contains a proof net with type $i\mathsf{CN} \multimap (\forall\alpha.!^i(\alpha \multimap \alpha) \multimap §(\alpha \multimap ((\alpha \multimap \alpha) \otimes \alpha)))$, taking $\overline{n}$ to yield a proof net of type $(\forall\alpha.!^i(\alpha \multimap \alpha) \multimap §(\alpha \multimap ((\alpha \multimap \alpha) \otimes \alpha)))$. The second element of the pair in the result would essentially correspond to $\overline{n-1}$. We cannot get exactly $\overline{n-1}$ because we lack the unconstrained weakening. Like in LAL, weakening would allow to erase the first element. Recall that we have ruled out unconstrained weakening from PMS, in order to keep this introductory work as simple as possible.

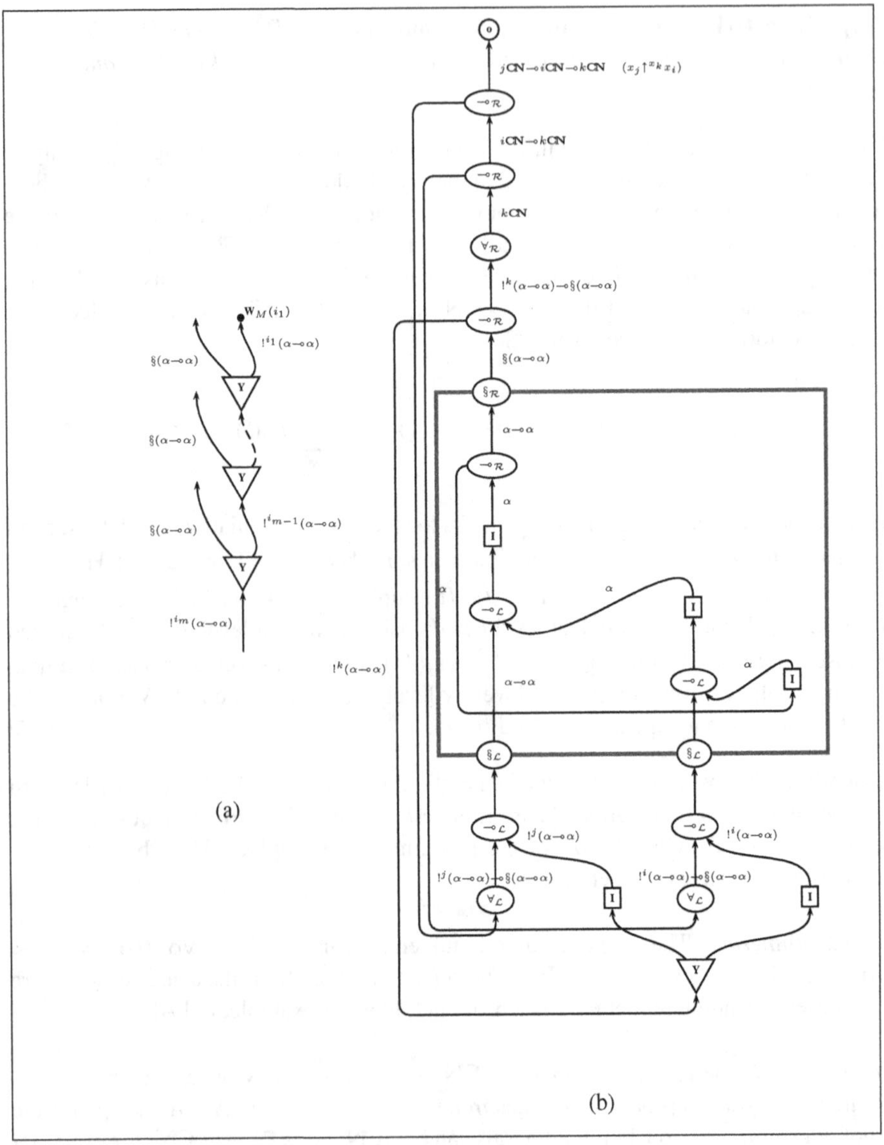

Fig. 9. (a) A spine of contractions inside a Church numeral. (b) The Sum between two Church numerals.

5 Conclusions and Further Work

We introduce MS, that we want to use as a framework to find deductive systems that, under the proof-as-programs analogy, only develop strongly polynomial time computations, and that, thanks to their multimodal nature, can be reasonably expressive.

Concerning strong polynomial time soundness, currently, we can supply a sufficient condition on subsystems of MS. Concerning expressiveness, we can show how to exploit multimodality to set a linear time and space sound subsystem of MS, which strictly extends MLL.

Active and future work can develop in many directions.

Certainly, it will be necessary to investigate the conditions under which the subsystems we can obtain from MS enjoy good computational properties, like confluence, for example.

However, we think that, currently, the most stimulating directions are the following ones.

We look for a ptime soundness criterion. In fact, we expect that the inverse of Proposition 3 holds as well.

Also, we want to exploit the expressiveness that seems implicitly supplied by the multimodality. "Playing" with the modalities, we are currently defining subsystems whose iteration schemes are more expressive that those we can represent in LAL. The least goal is to reformulate, at least in part, the results claimed in [7], or, symmetrically, to extend the results of [11].

We want to characterize *maximal* subsystems of MS. The intuition can be given by a simple example. The relevant fragment of LAL corresponds to a subsystem $\mathcal{P}_{\mathsf{LAL}}$ with $K = \{\mathbf{Y}_1(1,1), P_1(1^?), P_2(1^*,2^*)\}$ as modal rules. We can extend K as $K \cup \{\mathbf{Y}_1(1,2), \mathbf{Y}_1(2,2)\}$, which, interestingly enough, is still strongly ptime thanks to Proposition 3. Namely, $\mathcal{P}_{\mathsf{LAL}}$ is not maximal w.r.t. ptime.

Finally, we want to relate the class $\{\mathcal{P}^M_{\mathbb{LTS}} \mid M \in \mathbb{N}\}$ with the polynomial time and non-size-increasing system in [12]. The conjecture is that we can embed (at least) an additive-pair free fragment of it, with only boolean lists, simultaneously giving a proof-theoretical meaning to the type $\diamond$.

References

1. Girard, J.Y.: Linear Logic. Theo. Comp. Sci. 50, 1–102 (1987)
2. Asperti, A., Roversi, L.: Intuitionistic light affine logic. ACM Trans. Comput. Log. 3(1), 137–175 (2002)
3. Girard, J.Y.: Light linear logic. Inf. Comput. 143(2), 175–204 (1998)
4. Dal Lago, U.: Context semantics, linear logic and computational complexity. In: LICS 2006, pp. 169–178. IEEE, Los Alamitos (2006)
5. Leivant, D., Marion, J.Y.: Lambda-calculus characterisations of polytime. Fund. Inf. 19, 167–184 (1993)
6. Leivant, D.: Predicative recurrence and computational complexity i: word recurrence and poly-time. Feasible Mathematics II, 320–343 (1994)
7. Roversi, L.: Weak Affine Light Typing is complete with respect to Safe Recursion on Notation. Technical Report 104/08, Dipartimento di Informatica, Torino, C.so Svizzera, n.185 — 10149 Torino — Italy (April 2008)
8. Baillot, P., Mazza, D.: Linear logic by levels and bounded time complexity. Technical report (January 2008), `http://arxiv.org/abs/0801.1253v1`
9. Masini, A.: 2-Sequent Calculus: Intuitionism and Natural Deduction. J. Log. Comput. 3(5), 533–562 (1993)

10. Martini, S., Masini, A.: On the fine structure of the exponential rule. In: Girard, J.Y., Lafont, Y., Regnier, L. (eds.) Advances in Linear Logic, pp. 197–210. Cambridge University Press, Cambridge (1995)
11. Murawski, A.S., Ong, C.H.L.: On an interpretation of safe recursion in light affine logic. Theor. Comput. Sci. 318(1-2), 197–223 (2004)
12. Hofmann, M.: Linear types and non-size-increasing polynomial time computation. Inf. and Comp. 183, 57–85 (2003)

Author Index